INTERNATIONAL UNION OF CRYSTALLOGRAPHY
BOOK SERIES

IUCr Monographs on Crystallography

1 *Accurate molecular structures*
A. Domenicano, I. Hargittai, editors

2 *P.P. Ewald and his dynamical theory of Xray diffraction*
D.W.J. Cruickshank, H.J. Juretschke, N. Kato, editors

3 *Electron diffraction techniques, Vol. 1*
J.M. Cowley, editor

4 *Electron diffraction techniques, Vol. 2*
J.M. Cowley, editor

5 *The Rietveld method*
R.A. Young, editor

6 *Introduction to crystallographic statistics*
U. Shmueli, G.H. Weiss

7 Crystallographic instrumentation
L.A. Aslanov, G.V. Fetisov, G.A.K. Howard

8 *Direct phasing in crystallography*
C. Giacovazzo

9 *The weak hydrogen bond*
G.R. Desiraju, T. Steiner

10 *Defect and microstructure analysis by diffraction*
R.L. Snyder, J. Fiala and H.J. Bunge

11 *Dynamical theory of X-ray diffraction*
A. Authier

12 *The chemical bond in inorganic chemistry*
I.D. Brown

13 *Structure determination from powder diffraction data*
W.I.F. David, K. Shankland, L.B. McCusker, Ch. Baerlocher, editors

14 *Polymorphism in molecular crystals*
J. Bernstein
15 *Crystallography of Modular Materials*
G. Ferraris, E. Makovicky, S. Merlino

IUCr Texts on Crystallography

1 *The solid state*
A. Guinier, R. Julien
4 *X-ray charge densities and chemical bonding*
P. Coppens
5 *The basics of crystallography and diffraction, second edition*
C. Hammond
6 *Crystal structure analysis: principles and practice*
W. Clegg, editor
7 *Fundamentals of crystallography, second edition*
C. Giacovazzo, editor

Crystallography Of Modular Materials

GIOVANNI FERRARIS

Department of Mineralogical and Petrologic Sciences
University of Turin

EMIL MAKOVICKY

Mineralogical Department of the Geological Institute
University of Copenhagen

STEFANO MERLINO

Department of Earth Science
University of Pisa

OXFORD
UNIVERSITY PRESS

Great Clarendon Street, Oxford OX2 6DP

Oxford University Press is a department of the University of Oxford
It furthers the University's objective of excellence in research, scholarship,
and education by publishing worldwide in

Oxford New York

Auckland Bangkok Buenos Aires Cape Town Chennai
Dar es Salaam Delhi Hong Kong Istanbul Karachi Kolkata
Kuala Lumpur Madrid Melbourne Mexico City Mumbai Nairobi
São Paulo Shanghai Taipei Tokyo Toronto

Published in the United States
by Oxford University Press Inc., New York

First published 2004

A catalogue record for this title is available from the British Library

Library of Congress Cataloging in Publication Data
(Data available)

ISBN 0 19 852664 4

10 9 8 7 6 5 4 3 2 1

Typeset by Newgen Imaging Systems (P) Ltd., Chennai, India
Printed and bound in Great Britain
on acid-free paper by
Biddles Ltd. www.biddles.co.uk

To our families

Acknowledgements

A nominative list of colleagues who kindly let us have preprints and unpublished results would be at high risk of incompleteness: to all of them our warm gratitude. G.F. is indebted to Massimo Nespolo, for stimulating discussions, Angela Gula, Augusta Alberico, Cristiano Ferraris, and Gabriella Ivaldi for their precious collaboration in preparing the figures. E.M. thanks Camilla Sarantaris, Britta Munch, Claes C. Christiansen, and Ole Bang Berthelsen for their valuable assistance in preparation of the manuscript. S.M. thanks Slavo Ďurovič, for his constructive reading of parts related to the OD theory, Elena Bonaccorsi, Natale Perchiazzi ed Marco Bellezza for their generous help in preparation of the manuscript and, in particular, of the figures. We would like to thank Sonke Adlung, Marsha Filion, Tamsin Langrishe, Anita Petrie, Daniel Stewart, and Anja Tschoertner (who first contacted G.F. to initiate this book) of the OUP staff who, at various stages, assisted us in the technical production of this book.

Preface

La natura non opera con l'intervento di molte cose
quel che si può fare col mezo di poche.
Galileo Galilei
"Dialogo dei massimi sistemi – Seconda giornata" (Firenze 1632)

Nature does not act by means of many things
when it can do so by means of few.
Galileo Galilei
"Dialogue Concerning the Two Chief World Systems – The second day"

In recent years, attention has been drawn on a variety of complex modules which are shared by different structures spanning from the long known tetrahedral and octahedral layers in silicates to, more recently, perovskite-type slabs in superconductors and in structurally engineered organic–inorganic hybrids (chapters 1 and 4). The latter materials show that structural modularity is not confined to inorganic materials. The universal presence of modularity in any kind of crystalline structures becomes more evident when polytypism (chapters 2 and 3) and twinning (chapter 5) are considered. In general, the concepts of structural modularity prove to be particularly useful for at least three purposes: description and classification of known structures; modelling of unknown structures that are based on known modules; tailoring of new materials and tuning of their properties.

Fundamental for a fruitful application of the modular methods is a systematization of the relevant basic concepts, which are spread in hundreds of original papers and, a part from a pair of partial attempts, never have been organically analysed. Thus, the basis of homologous, polysomatic, and higher rank series are developed in the first chapter. Crucial to attain a systematic approach to polytypism and model related structures, is the OD (order/disorder) theory, which deals with those inorganic and molecular structures based on both ordered and disordered stacking of one or more layers (chapter 2). Related to the OD structures and, in general, to the building of modular structures, is twinning at unit-cell scale (chapter 1). At any scale, a twin can be considered a modular structure built up by the same differently oriented domain (module). Owing to the problems caused in structure determination, routine recognition of twinning has been recently implemented in some crystallographic software. In this book (chapter 5), the theory of twinning is expressly developed, with worked examples, in view of solving structural problems.

Encouraged by the positive comments that five anonymous referees expressed on the proposed layout of this book, we faced the not easy task of providing a unified treatment of results so far spread in different journals and original papers published mainly in the last 20 years. Our task has been supported by the conviction that the lack of a digested review of the matter, has so far limited to a minority of crystallographers, even less materials scientists, the appreciation and the application of the concepts of structural modularity. Such a limitation is also related to the barrier represented by use of different definitions and nomenclature to express the same concept, a situation that is typical of original papers published by different groups in fast developing fields.

The book is mainly intended for researchers in the field of crystalline materials at any level, from post-graduate students, to post-doc and senior people willing to explore new paths related to their scientific interests. Parts of the book can be integrated in general courses on solid state devoted to undergraduate students. The importance and implications of modular character exceeds by far the narrow field of structural chemistry or mineralogical systematics, to which most of the examples illustrated in this book belong (just because others are still rare). It has direct influence on nucleation of phases, crystal growth, physical properties and thermodynamic characteristics of the compounds involved. Little variations in, and frequent similarity of, thermodynamic properties of the members of a modular series lead to serious implications for their assemblages, be it in geology, chemistry or chemical engineering.

G. F.
E. M.
S. M.

May 2003

Contents

1

Modular series—principles and types

Some families of inorganic compounds form an array of more or less closely related, very complex structures with a number of independent atomic positions and unit cells of large dimensions. These structures can be broken up into a number of fragments (modules) that have relatively simple substructures but are joined into a more complex whole. Some modules occur in many members of these families, often in variably expanded/contracted forms as well as combined with elements of other kinds. From the point of view of the bond strength distribution, these modules/elements may form independent units, weakly bonded on the outside, or they are only geometrical (configurational) entities that otherwise merge into the bonding pattern of the structure.

The same category of complex structures can also be approached from the opposite direction. Increasingly dense arrays of iso- or allochemical, extended structural defects lead to a formation of progressively smaller modules of undisturbed, relatively simple structure. These modules decrease in size (Wadsley 1963) until they match in defect density and module (fragment) size those from the above mentioned, structurally and chemically complex families.

These opposite structural trends produced different schools of thought and modes of description. Applying the latter approach to oxide structures, a concept of **homologous series** was coined for such compounds by Magnéli in 1953; the former approach led to a concept of **polysomatic series**, devised for silicate examples by Thompson in 1970. In the realm of intermetallic structures a very similar concept was introduced by Krypiakevich and Gladyshevskii (1972). In the two latter publications, the starting point was a similarity (formal analogy) of the observed phenomena to polytypism, stressed also by Angel (1986). Experience in polytypes was also the point of departure for Zvyagin and Romanov (1990) and Vainshtein *et al.* (1994) in their treatment of layered superconductors, which is very similar to Krypiakevich and Gladyshevskii's 'structural series' approach.

Chemical twinning approach to the formation of homologous series was promoted by Takéuchi *et al.* (1974), Andersson and Hyde (1974), Makovicky and Karup-Møller (1977) and others. Makovicky (1981, 1985*a*, *b*) described a number of homologous series formed from archetypal sulfide structures by various structure building principles. **Chemical twinning** is synonymous with **unit-cell twinning** as defined by Andersson and Hyde (1974).

Modular description was applied successfully to many families of complex oxides (e.g. Magnéli 1953; Bursill *et al.* 1969; Ijima 1975) and complex sulphides (e.g. Otto and Strunz 1968; Takéuchi and Takagi 1974; Takéuchi 1978; Makovicky and Karup-Møller 1977; Makovicky 1981, 1985, 1989) as well as to selected groups of halides, borates, silicates, etc. (Thompson 1978; Ferraris *et al.* 1986; Ferraris 1997; Merlino and Pasero 1997; a detailed review by Veblen 1991). Recently, it found rich application in high-pressure silicates (Finger and Hazen 1991*a*,*b*), due to close packing of oxygen at high pressures and among selected high-temperature superconductors related to perovskite (as summarized by Mitchell 2002).

The homologous/polysomatic series are series of compounds with incrementally varying chemical composition. In the homologous approach, these changes result from incremental growth of basic modules; in the polysomatic description by varying the proportions of the two layer modules present in each intermediate member. However, as already briefly mentioned, a broad spectrum of modular structures exist—polytypes—in which variation in modular arrangement produces structural variations without—in principle—altering their chemistry. Besides ample publications on polytypes in general (e.g. Thompson 1978; Angel 1986; Zvyagin 1987, 1993), detailed studies have been produced of their principal subdivision, the OD polytypes that obey the symmetry theory of OD-groupoids (Ito 1950; Dornberger-Schiff 1956; Ďurovič 1992; Merlino 1990*a*,*b*). Recent research has revealed the importance and omnipresence of polytypy or of its subdued manifestations as twinning of desymmetrized structures in almost all categories of inorganic compounds and in an increasing number of organic compounds as well. This ubiquitous character of polytypy brings about a number of subtle questions in the application and delimitation of the concept; they will be addressed here.

1.1 Brief outline

The present section deals with modular structures and phenomena in which modularity is connected with incremental changes in chemical composition, yielding potentially series of structurally related but chemically distinct compounds. At first, modularity and non-modular crystal chemistry are interrelated by so-called configuration levels. A short account of non-modular categories of structural similarity facilitates the use of these categories in subsequent text and allows one to draw certain parallels with such modular categories. Enumeration of structure-building principles and elements relevant to modular structures is followed by the definition of various types of homologous series and by elucidation of the homologous *versus* polysomatic description principles. Typical examples from various realms of inorganic chemistry and mineralogy will illustrate these concepts.

More general categories of structural similarity on modular level are defined as so-called merotype and plesiotype series; again with a number of examples given. Finally, structures of selected series of ordered derivatives of solid solutions are described as modular structures.

1.2 Hierarchical classification of structures

1.2.1 *Configuration levels*

Classification of crystal structures is as old as the structure determination itself. The approaches to it vary according to the structural families classified, the authors classifying them and the purpose of the classification (teaching texts, mineral system, cement literature, etc.).

One of the approaches divides each structure into a clear **structural unit** that contributes most to the internal energy of the compound and the remaining **interstitial atoms** (Lima-de-Faria *et al.* 1990). Such structural configurations based on aggregates of selected coordination polyhedra often attain considerable complexity. The best example is the classification of silicates that developed from a simple, straightforward system by Bragg (1930) into an extensive, complex system by Liebau (1985). Similarly complex is the polyhedral classification of borates (Burns *et al.* 1995). Frequently, several distinct configurations occur in one structure, whereas the same configuration occurs in several structures that belong to quite distinct structure types. Distinct configurations can be found in different members of one and the same homologous series (a series of structures built on the same structural principle with certain module expanding in volume by regular increments—for example by addition of a row/layer of coordination polyhedra) or even in the ordered derivatives of one homologue that at high temperatures is a complete solid solution. Good examples of this occur in complex sulphides of Pb–Ag–(Sb,Bi) (Makovicky 1997*b*).

In the realm of complex sulphides and oxides, we often deal with structures that have only indistinctly expressed 'backbones' of strongly bonded units, that is, they have an approximately homogeneous bond strength distribution. Therefore, the classifications based on structural units (i.e. polyhedral groups) are less justified for these compounds than for the cases where structural units are well defined and play prominent role in the energy balance of the structure.

At the other end of the structural classification spectrum stands the modular classification that concentrates on large-scale structural features, that is, configurations that involve large assemblages of atoms/coordination polyhedra/molecules. It involves certain level of abstraction from the occupancy/bonding details.

1.2.1.1 *Hierarchical description*

There are several, up to five levels of configurational complexity between the structural-unit and modular approaches to structural classification. These form the basis of the principle of hierarchical description of structural configurations. Each level of configurational complexity may reveal structural relationships to a different group of compounds. **Primary** (or first-order) **configurations** are polyhedra BX_n common to the entire 'phyllum' of compounds. These may be coordination tetrahedra SiO_4, triangles and tetrahedra BO_3/BO_4 or BiS_5 coordination pyramids in complex

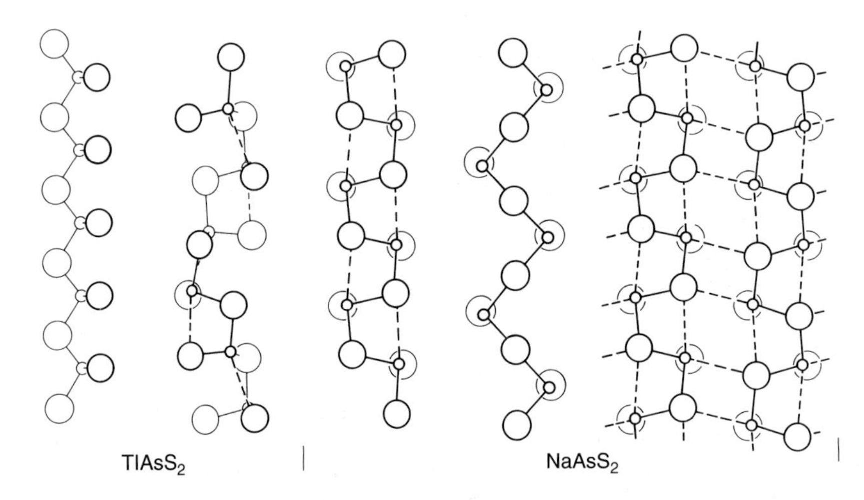

Fig. 1.1. The AsS_2 chains in the crystal structures of $NaAsS_2$ and $TlAsS_2$ in the idealized form (secondary configurations) and true, tertiary configuration involving weak, long interactions (crank-shaft chains and layers).

sulfides. **Secondary** (second-order) **configurations** are the groups (clusters, chains, layers, etc.) built from these polyhedra (or their strongest-bonded portions) using the strongest B–X bonds. These can be, for example, chains $(SiO_3)^{2-}$ of corner-sharing coordination polyhedra, similar AsS_2 chains of AsS_3 coordination pyramids, or rings $(Si_6O_{18})^{12+}$ in cyclosilicates.

Tertiary (third-order) **configurations** are dimensionally higher units obtained by composition/interplaiting of secondary configurations via secondary, weaker B–X bonds, by combination of several groups around a lone electron pair micelle, etc. The AsS_2 chains of AsS_3 pyramids in **$NaAsS_2$** and **$TlAsS_2$** (Fig. 1.1) are a typical example. The secondary configurations, the 'straight' AsS_2 single chains are modified into tertiary configurations—zig-zag or spiral chains (or entire layers for $NaAsS_2$)—when satisfying the longer As-S contacts. The same holds for $(SiO_3)^{2-}$ chains that, by virtue of satisfying bonding to other polyhedra in the structure, assume configurations of pyroxene chain, wollastonite chain, etc. The secondary and tertiary configurations may comprise a combination of different polyhedra with relatively strong bonds, such as octahedra and tetrahedra in the module construction in the modular classification by Hawthorne (1994) and Fleck *et al.* (2002).

The network of weak B–X contacts may be essential for the stability of the entire structure. This is the case for $TlSb_3S_5$ (Gostojič *et al.* 1982) and $TlSb_5S_8$ (Engel 1980) in which the tertiary configuration of a diagonal slab, that is, the interplay of short, intermediate, and long B–S contacts is essential both for the Tl : Sb accommodation and for the overall PbS-like structural scheme. The same arguments may be extended to B–X and X–X contacts in, for example, the sulphates/selenates/chromates, etc. of the kröhnkite family classified by Fleck *et al.* (2002) in which the chain distortion correlates with the chain interactions with surrounding cations, or in the silicates of nepheline–kalsilite family (Palmer 1994) with distorted tridymite-like framework.

1.2.1.2 *Modular classification*

Quaternary configurations (groups, formations) are obtained when the tertiary groups are enveloped by, or combined with coordination polyhedra of other elements. Normally, these are the modules (or, sometimes, submodules) of the modular description. **Quinary configuration** is the packing of quaternary (or ternary) configurations (modules) in a large-scale structural pattern. In this way, the hierarchical classification systematizes and interrelates various approaches that were used for the description and classification of sulphides, sulphosalts, high-T superconductors, etc.

In the crystal structure of **jamesonite $FePb_4Sb_6S_{14}$** (Niizeki and Buerger 1957), the primary configurations are three crystallographically non-equivalent SbS_3 pyramids (Fig. 1.2(a)). If strictly defined, secondary configurations comprise two parallel, infinite SbS_2 chains and independent SbS_3 groups stacked between them (Fig. 1.2(b)). Ternary configuration is the most characteristic one—ribbons (triple chains) Sb_3S_7 of edge-sharing square pyramids (Figs 1.2(c)). Two such chains, back to back, comprise lone electron pair micelles of Sb in **jamesonite**. Fourth-order configurations are lozenge-shaped rods of SnS-like structure, infinite along $[001]_{SnS}$, in which Sb, Pb, and Fe participate. At the quinary level, these rods interconnect via coordination octahedra of Fe into **rod-based layers**, which are stacked parallel to each other by means of non-commensurate interspaces between strips of surfaces of opposite kind (Fig. 1.3).

In the case of **biopyriboles** (chain silicates that comprise pyroxenes, amphiboles, with mica structure as the end member), primary configurations are the SiO_4 tetrahedra. Secondary configurations are the SiO_3, or the Si_4O_{11}, Si_6O_{16}, etc., ribbons in their idealized, simple form whereas tertiary configurations are the real pyroxene chains or amphibole, jimthompsonite, etc. ribbons fully adjusted to the structure. Fourth-order configurations are the T–O–T (tetrahedral ribbon–octahedral ribbon–tetrahedral ribbon) rods, so-called **'I-beams'** (Thompson 1978) that represent the most typical modular element of these structures. On quinary level, these modules interconnect via selective ligands and large coordination polyhedra of 'marginal' cations into a chess-board structure scheme.

1.3 Short recapitulation of non-modular categories of similarity

The IUCr report (Lima-de-Faria *et al.* 1990) clarified and defined concisely different degrees of similarity between inorganic crystal structures. The following hierarchy of terms was codified for the types and degrees of structural similarity: **isopointal**, **isoconfigurational**, **crystal-chemically isotypic**, and **homeotypic structures**. Their short recapitulation before embarking upon more complicated modular relationships is considered useful.

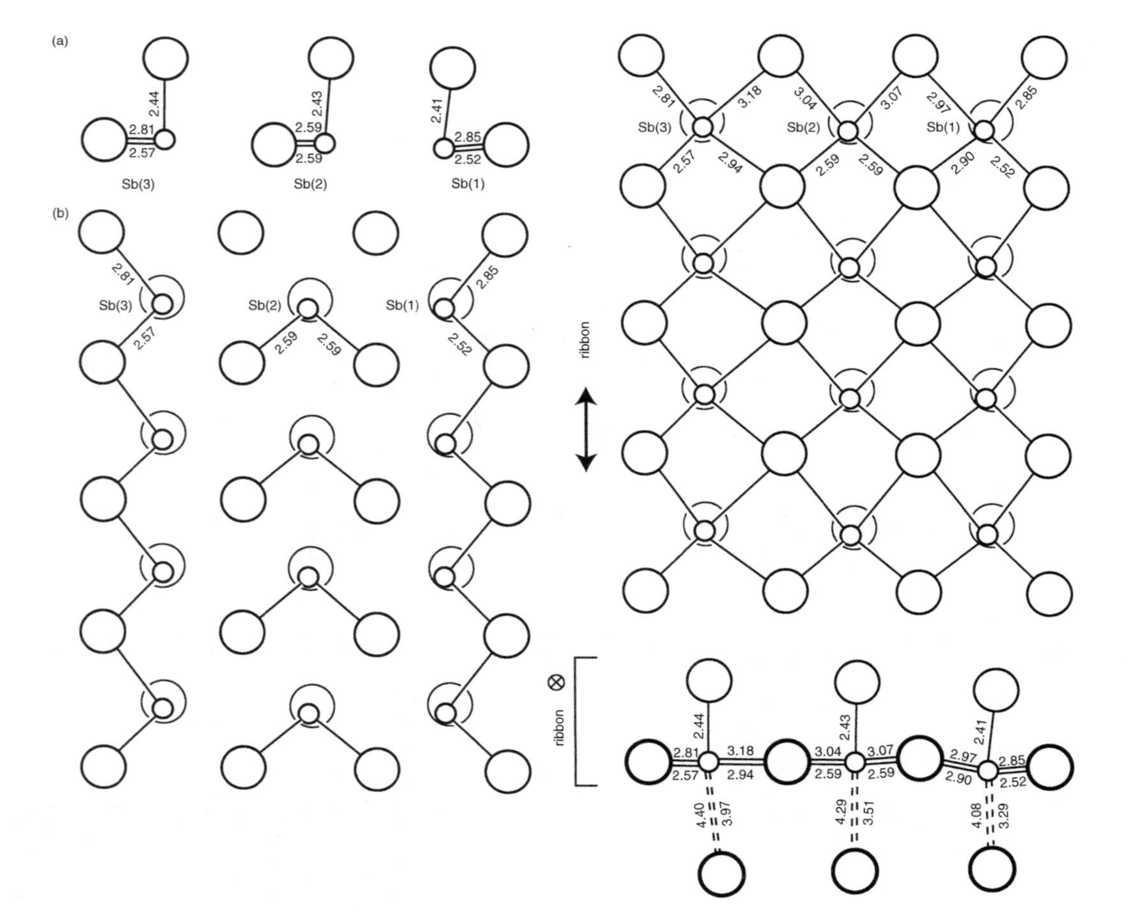

Fig. 1.2. Hierarchical description of the jamesonite ($FePb_4Sb_6S_{14}$) structure (Niizeki and Buerger 1957). (a) Primary configurations: coordination pyramids SbS_3; (b) secondary configurations: SbS_2 chains and SbS_3 groups; (c) tertiary configurations: Sb_3S_7 ribbons. Quaternary configurations: rods of archetypal structure and quinary configurations (structure with rod-based layers) are illustrated in Fig. 1.3.

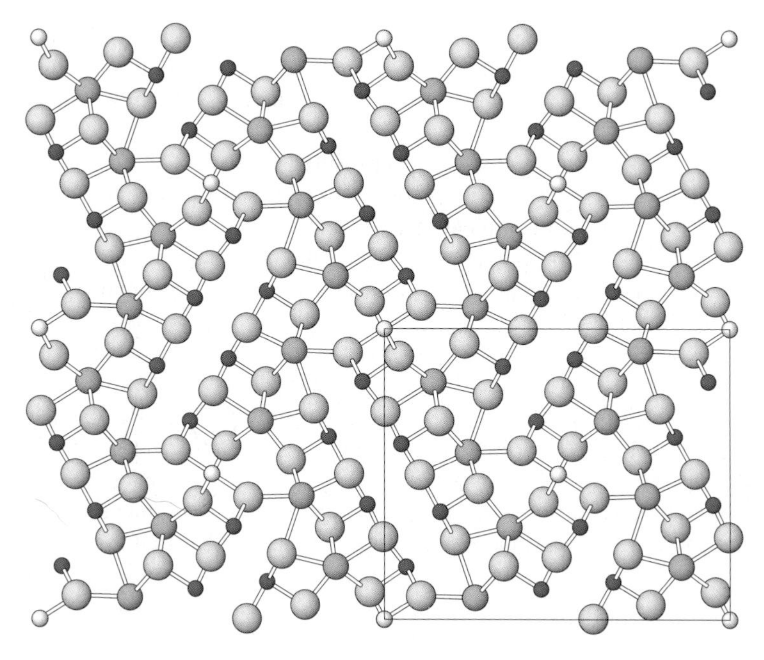

Fig. 1.3. The crystal structure of jamesonite $FePb_4Sb_6S_{14}$—the quaternary (rods) and quinary (the rod packing) configurations. Sb, black; Pb, grey; Fe, white; S, large, light atoms. Rows of Sb atoms define lone electron pair micelles.

Two structures are isopointal when they have the same space-group type or belong to a pair of enantiomorphic space-group types and they share a complete sequence of (sometimes only partly, at random) occupied Wyckoff positions. This definition puts no restrictions on unit cell parameters or on the adjustable x, y, z parameters of the Wyckoff positions. Therefore, **isopointal structures** may have different geometric arrangements and atomic coordinations.

It obviously is a 'computer category', a very convenient search principle in databases where it limits considerably, by computer means, the search for similar structures. An example is the pair **FeS_2 (pyrite)—CO_2**, both cubic $Pa\bar{3}$, with occupied Wyckoff positions 4(a) and 8(c)xxx (Lima-de-Faria *et al.* 1990).

Two structures are **isoconfigurational** (i.e. configurationally isotypic) if they (a) are isopointal, and (b) for all corresponding Wyckoff positions, both the crystallographic point configurations and their geometrical interrelationships are similar.

Translated into a structural language, this definition means that the entire configurations of the two structures are similar: axial ratios, unit-cell angles, values of adjustable x, y, z parameters for corresponding atoms and their coordination polyhedra. It should be noted that this definition speaks about configurations, not about the physical–chemical characteristics of the atoms. Thus, the ccp pairs noble metal

(Cu, Ag, Au)—noble gas (Ar, Ne) or the pair NaCl—PbS, respectively, are configurational isotypes. This definition is most useful when the differences in bonding characteristics and atom properties exist albeit less sharply and less well defined than in the above examples.

We define two structures as **crystal-chemically isotypic** if they (a) are **isoconfigurational** and (b) the corresponding atoms and corresponding bonds (interactions) have similar physical and chemical characteristics (e.g. electronegativities, radius ratios, electronic states, or bond-strength distribution).

There is no *a priori* definition of, and limits upon, geometric, chemical or physical similarity in the latter two definitions. These limits will change from investigation to investigation according to the family (type) of compounds studied, the physical or chemical properties we concentrate upon, and the purpose of study. Two basic approaches will always be (a) *a priori* model considerations, and (b) statistical studies of entire known groups of compounds, looking for clustering and gaps in various property diagrams.

If the important physical/chemical characteristics in the last category are reversed, we speak about a **type–antitype pair**, for example, **CaF_2–Li_2O** (Lima-de-Faria *et al.* 1990).

In a number of cases, the kinship among a series of structures is intuitively obvious although they do not correspond to the strict conditions of the above definitions. Among them, we can discern homeotypic structures, interstitial (i.e. 'stuffed') derivatives, polytypic structures and recombination structures; the latter two categories are among the principal topics of this book.

The most frequent and important is the first category. Two structures are **homeotypic** if one or more of the following conditions required for isotypism are relaxed:

(1) instead of identical (or enantiomorphic) space group (type), they can have a subgroup or a supergroup;
(2) limited variation in axial ratios, interaxial angles, x, y, z values and coordination properties is allowed;
(3) site occupancy limits are relaxed allowing given sites to be occupied by different atomic species (splitting of the original Wyckoff position due to (1).

Bergerhoff *et al.* (1999) include interstitial derivatives among homeotypes.

Examples are numerous: among site-ordering variants, we can quote C (**diamond**), ZnS (**sphalerite**), $CuFeS_2$ (**chalcopyrite**) or Cu_3SbS_4 (**famatinite**); among distortion derivatives, for example, the numerous distortion derivatives of ideal **perovskite** ($CaTiO_3$) (Mitchell 2002). The ideal, undistorted structure is called an **aristotype** (Megaw 1973), the derived distorted structures **hettotypes** (*ibid*).

Quite naturally, homeotypes with their variability of deviations from the 'type' compound are a complicated and difficult problem for a computer search as well as for systematic classifications.

1.4 Elements of modular description. Types of homologous/polysomatic series

1.4.1 *Structure building principles and operators*

Modular, or fragment-recombination (Lima-de-Faria *et al.* 1990) structures are composed of modules/fragments (blocks, rods or layers) of archetypal structures that are recombined in various orientations by the action of structure-building operators such as **reflection twinning**, **glide-reflection twinning** (**swinging twinning**), **axial twinning**, **cyclic twinning**, **non-commensurability**, **antiphase boundaries**, **crystallographic shear** or **unit-cell intergrowths**.

The **structure-building operators** are essentially allochemical (i.e. they change chemical composition of the bulk structure), mostly planar, defects which produce a definite geometric relationship between the structural portions they join; they result from localized chemical (crystal-chemical) changes in the structure.

For the large part, the **structure-building principles** (operations) according to which the rods, blocks or layers of archetypal structure are recombined into a complex 'recombination' structure are various types of unit-cell twinning (Andersson and Hyde 1974, Ito 1950). The most frequent type is the reflection (i.e. **mirror-reflection**) **twinning** that acts either on a full set of atoms or only on a 'contracted' (i.e. partial) set of atoms out of those present in the larger of the two mirror-related portions (Takéuchi 1978). Mirror-reflection twinning can be connected with the creation of a new type of coordination polyhedra straddling the mirror plane. Examples are the capped trigonal coordination prisms of two distinct kinds on composition planes of reflection twinning of ccp and hcp arrays, respectively. The next type is the **glide-reflection twinning** (Andersson and Hyde 1974) in which the two layers are reflected into each other by means of a glide-plane connected with a more or less profound change of coordinations along this plane. It can take place consecutively on different (hkl) planes of the archetypal structure (this being called swinging twinning by Bovin and Andersson (1977)). Unit-cell twinning on a set of symmetry axes instead of a symmetry plane was described by Takéuchi (1997) for the plagionite homologues (Pb-Sb sulphides). The last type of unit-cell twinning is the **cyclic twinning** (Hyde *et al.* 1974), a one-dimensional type of twinning that has to occur as a periodic array of twinning centres. For complex sulphides, **non-commensurability** between adjacent building blocks, etc. (i.e. the substructures in two adjacent blocks meet according to a vernier principle across the block interface) comes next in the order of importance (Makovicky and Hyde 1981, 1992), followed by **antiphase** and out-of-phase **boundaries**, **crystallographic shear** (Hyde *et al.* 1974), the **t/2-shear** (or slip) in the structures with a short pronounced *t* period, and the **intergrowth** of two different structure types on a unit cell scale (*ibid.*).

It should be stressed again that the definition of unit-cell twinning adopted here implies appreciable configurational and chemical changes on the twin composition planes, unlike that of classical twinning. This is true in general—the coordination states and, in the case of cations, often also the chemical species of atoms on the block surfaces or in the interfaces differ from those inside the building blocks. The structure-building principles, especially the reflection twinning, can apply either to all the details of the structure, including the chemical species, or only to its general topology (or even only to its anion array), something akin to the difference between isotypes and homeotypes.

Takéuchi (1997) distinguishes between (a) **chemical twinning** in the sense of Hyde *et al.* (1974) (i.e. **unit-cell twinning** of Andersson and Hyde (1974)), that is a structure-building operation in which blocks (slabs) of archetypal structure are reunited by polysynthetic twinning on unit cell level and (b) **tropochemical twinning** (Takéuchi 1978) in which the frequency of cell-twin boundaries (the frequency of cell twinning) changes from a phase to a phase, while the structure scheme is kept unchanged. The frequency of cell-twinning causes differences in chemical composition. Thus, this term covers primarily the 'homologous series' of other authors. As examples, especially the **Phase V series** and lillianite homologues as well as, other, less strictly defined examples in the system PbS–Bi_2S_3(–Ag_2S) are given, as well as, the **enstatite-IV homologues**, **pinakiolite polytypes**, and the **plagionite series of homologues $Pb_{3+2x}Sb_8S_{15+2x}$**, that is, a broad palette of diverse structure-building mechanisms.

In his treatment of 'polysynthetic structures', Ito (1950) enumerates all the symmetry operators that can act as operators of unit-cell twinning. However, practically all his examples come from the realm of polytypes.

1.4.2 *Archetypes*

Archetypes used in the modular description are those simpler structures that display/encompass all fundamental bonding and geometric properties of the structure portions in the interior of building blocks/moduli. For sulphides, they are, for example, PbS, SnS, TlI, ZnS, for oxides, ReO_2, rutile, tetragonal bronze, **spinel** or a deficient CaF_2 structure; for silicates, most often a stack of **T–O–T silicate layers**, or the pyroxene and wollastonite structures.

Modular structures (including polytypes) can be **monoarchetypal**, with all building blocks based on the same archetype or **diarchetypal** (triarchetypal, ...) with distinct types of blocks based on two (three, ...) different archetypes. The vast category of monoarchetypal structures comprises many oxides and silicates (the **humite series**, **spinelloids**, **ReO_3-type block structures**) or sulphides (**lillianite homologous series**, the **jordanite–kirkiite** pair, **rod-based complex sulphides**). Diarchetypal are, for example, complex sulphides of the **kobellite homologous series** in which blocks of PbS and SnS archetype combine, or saphirine-type compounds in which layers of **pyroxene archetype** combine with those from **spinel archetype**.

1.4.3 *Homologous series*

Many modular structures are members of homologous series. There are two principal types of homologous series which were named **accretional series** and **variable-fit series** by Makovicky (1988). These have only the traditional name in common; it implies incremental growth of some parameters while maintaining the same structural principles. Thus, there exists no abstract homologous series and we advocate use of the above adjective specifiers at least once, at the decisive point of any structure description.

1.4.3.1 *Accretional series/polysomatic series*

The accretional series is a series of structures in which the type(s) and general shapes of building blocks (rods, layers) as well as the principles that define their mutual relationships (the recombination operators) remain preserved but the size of these blocks varies incrementally by varying the number of fundamental coordination polyhedra in them in an exactly defined way. The order N of a homologue in this type of series can be defined by the number of coordination polyhedra (polyhedral layers) across a suitably defined diameter of the building block (rod, layer). The ratio of atoms on the block surfaces (interfaces) to those inside the blocks varies with N as also does the overall cation/anion ratio.

Every member of the accretional series has its own chemical formula, unit cell parameters and symmetry; a general chemical formula can be devised for the entire series. A given homologue can represent a single compound or a (dis)continuous solid solution, which can exsolve into a series of structurally ordered phases with well defined compositions (or compositional ranges) but with the same N. The ideal space group can then be reduced to subgroups in the process of (cation) ordering. It should be stressed that all these ordered compounds are members of the homologous series. There exist cases (e.g. **biopyriboles**) where accretional homology is complicated by the appearance of **polytypy** with the stacking direction different from that of the direction of block accretion. Homologues belonging to distinct polytypes will then form subseries (e.g. distinct clino- and ortho-series among biopyriboles).

Thompson (1978) described accretional homologous series using a different name, that of a **polysomatic series** and a different approach: the coordination polyhedra on the surfaces and in the interfaces of the blocks (in his case, layers) are treated as one type of (layer) module whereas the incrementally accreting polyhedral layers in the layer (block) interior as another type of layer modules. Thus, all accretional homologues are treated as **ordered intergrowths** of (usually) two structure types which occur in different proportions in different homologues (polysomes). The two descriptions are equally valid; the homologous approach stresses the accretional growth of the fundamental motif whereas the polysomatic approach leads to a simple slicing of structures and formulation of additive chemical formulae, a point appreciated by many authors. However, is should be clearly stated that the polysomatic approach obscures the important differences between the subtypes of accretional series defined in the next paragraph. It also presents difficulties for either rudimentary or very elaborate interfaces (Makovicky 1989) as well as for rod structures with two intersecting systems of interfaces (Veblen 1991). Depending on the details of the structure, certain series can better be described by the homologous (accretional) approach; others by a polysomatic approach. This point will be further advanced below.

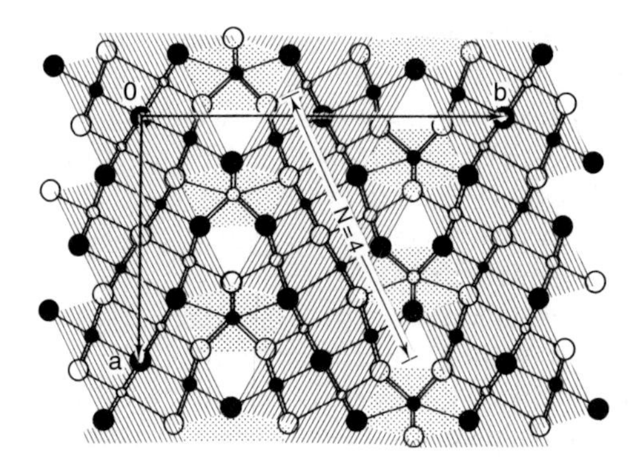

Fig. 1.4. The crystal structure of lillianite, $Pb_3Bi_2S_6$ (Takagi and Takeuchi 1972). Circles in order of decreasing size indicate S, Pb and the mixed (Pb,Bi) positions. Their colouring indicates atoms at $z = 0$ and $z = 1/2$, respectively. Simple connections are drawn between the atoms on the same level, double connections between atoms $\Delta z = 1/2$ apart. PbS-like layers $(311)_{PbS}$ are ruled, bicapped Pb coordination prisms stippled. Number of octahedra in a unit slab (i.e. homologue order N) is indicated.

The accretional approach allows further division of homologous series. Some accretional series are **extensive series** with N varying over a range of values. Besides the members with equal widths ($N_1 = N_2$) across the unit-cell twinning plane or an interface of another kind, those with unequal widths ($N_1 \neq N_2$) can occur. A number of accretional series are limited to only pairs of homologues (N_1, and $N_2 = N_1 + 1$) because of various local or global crystal chemical reasons. These pairs can be extended into **combinatorial series**, the members of which represent regular intergrowths of the above two accretional homologues: $N_1N_2N_1N_2\ldots$, $N_1N_1N_2N_1N_1N_2\ldots$, $N_1N_2N_2N_2N_1N_2N_2N_2\ldots$, etc. Several prominent examples of such series will be quoted below; in some instances only one intermediate member, $N_1N_2N_1N_2\ldots$, is known.

The accretional homologous (polysomatic) series are the most frequent type of homologous series. In a part of them, the intimate contact of two adjacent accreting blocks is most succinctly described as reflection- or glide-plane twinning on a unit cell scale; crystallographic shear or antiphase boundaries abound as well (Fig. 1.4). In other cases, there is a fairly thick interlayer between two adjacent accreting slabs of one kind, with configurations so different from those in the accreting slabs that the structure is a regular, coherent intergrowth of two structure types. When the two component structures represent real or reasonable compounds, these are the cases suitable for polysomatic description.

1.4.4 *Examples of accretional series*

1.4.4.1 *Lillianite homologous series*

A simple series produced by reflection twinning is the accretional **extensive series of lillianite homologues** (Makovicky and Karup-Møller 1977*a*,*b*). These are sulphides and oxides with ccp anion arrays with filled coordination octahedra, twinned on $(311)_{PbS}$ with corresponding coordination changes on twin planes. Due to the extensive investigations by a number of authors, this series is eminently suitable for the illustration of problems connected with a crystal-chemical rather than ideal, geometric homologous series. Its comparison with the immediately following pavonite homologous series illustrates how seemingly minor differences in a set of coordination polyhedra assume a decisive crystal-chemical role and bring about a well defined, distinct series of compounds with specific structural properties.

The most prominent members of this series are **Pb–Bi–Ag sulphosalts** with the structures consisting of alternating layers of **PbS archetype**, cut parallel to $(311)_{PbS}$ (Fig. 1.4). These planes also represent the reflection- and contact planes of unit-cell twinning. The overlapping octahedra of the adjacent, mirror-related layers are replaced by bicapped trigonal coordination prisms PbS_{6+2} with the Pb atoms positioned on the mirror planes (Otto and Strunz 1968; Takéuchi *et al.* 1974 etc.).

Distinct homologues differ in the thickness of the PbS-like layers. This is conveniently expressed as the number N of octahedra in the chain of octahedra that runs diagonally across an individual archetypal layer and is parallel to $[011]_{PbS}$ (Fig. 1.4). Each lillianite homologue can be denoted as ${}^{N1,N2}L$ where N_1 and N_2 are the (not necessarily equal) values of N for the two alternating sets of layers (Fig. 1.4). Its chemical formula is $Pb_{N-1-2x}Bi_{2+x}Ag_xS_{N+2}$ ($Z = 4$) where $N = (N_1 + N_2)/2$ and x is the coefficient of the Ag + Bi = 2Pb substitution (Fig. 1.5). If the trigonal coordination prisms of Pb cannot be substituted (which is very close to the real situation), $x_{max} = (N - 2)/2$. This structure type is quite frequent also outside the Pb–Bi–Ag compositional space, for example for a number of complex **lanthanide sulphides**; the general formula then becomes $M^{2+}_{N-1}M^{3+}_{2}S_{N+2}$. The rare instance of $TlSb_3S_5$ ($N = 3$) leads to the formula $M^{+}_{(N-1)/2}M^{3+}_{(N+3)/2}\ S_{N+2}$. A somewhat different generalized approach to the series is given by Hyde *et al.* (1979).

The existence of lillianite homologues (*sensu lato*) depends on the suitable sizes of coordination polyhedra (trigonal prisms versus octahedra), satisfactory local valence balance and feasibility of close-to-regular octahedral (i.e. ccp or PbS-like) arrays. Thus, the cases with $N = 1$ and 2 do not tolerate lone electron pairs of Bi^{3+} or Sb^{3+} that enlarge selected volumes of individual layers. Therefore, $PbBi_2S_4$ is not a lillianite homologue although about a half of its structure approximates the $N = 2$ configuration. Examples of lillianite homologues are in Table 1.1.

For the **MnS–Y_2S_3 system**, Bakker and Hyde (1978) found that the homologous pair **MnY_2S_4** ($N = 2$) and '**$MnYS_3$**' ($N = 1$) (which occurs only as a layer in MnY_4S_7) form a **combinatorial series** that comprises **MnY_4S_7** (${}^{1,2}L$), **$Mn_2Y_6S_{11}$** (${}^{1,2,2}L$), **$Mn_4Y_{10}S_{19}$**(${}^{1,2,2,2,2}L$), etc. Both the cases with ideal '**aristotype**' symmetry and those with subgroup symmetry (Table 1.1) are present for $N = 1$–3; the reduction of symmetry is caused either by distortions of coordination polyhedra or by the asymmetric position of cations in the trigonal coordination prisms. The tetrahedral voids on the mirror planes of unit cell twinning are occupied only in exceptional cases (**Eu_2CuS_3**, ${}^{1,1}L$).

Higher homologues start at $N = 4$ (exceptionally already at 3 as for $TlSb_3S_5$); they allow more pronounced departures from the galena-like array, especially in the form of

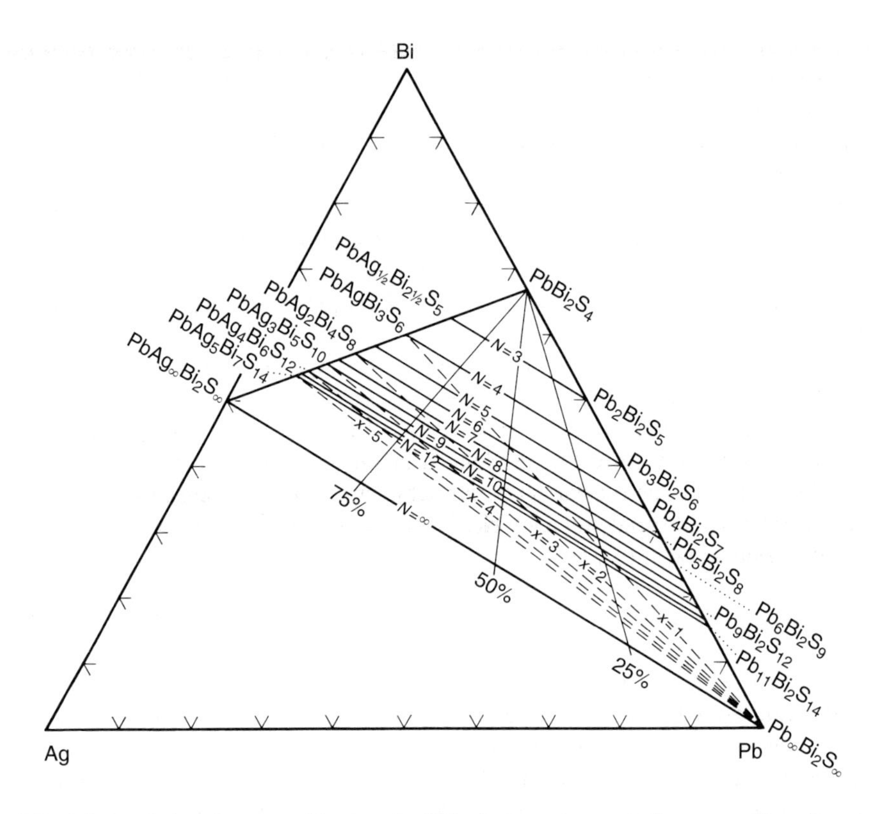

Fig. 1.5. Isopleths of constant N values for lillianite homologues in the composition diagram Ag–Bi–Pb with, respectively, Ag-free and maximally Ag–Bi substituted compositions indicated. Percentages of substitution and the x values in the formula $Pb_{N-1-2x}Bi_{2x}Ag_xS_{N+2}$ are shown.

locally 'inflated' interspaces that accommodate lone electron pairs of quasi-octahedral Bi or Sb (the common volume for lone electron pairs of several adjacent cations was named a **lone electron pair micelle**) by Makovicky and Mumme (1983) (Fig. 1.6). Only $^{4,4}L$ and $^{7,7}L$ are known for the Ag-free subsystem Pb–Bi–S. Reduction of symmetry from the usual orthorhombic to monoclinic, caused by different Pb/Bi ordering into the two mirror-related slabs, was observed for $^{4,4}L$ (Berlepsch *et al.* 2001*a*) and $^{7,7}L$ (Mumme and Makovicky, unpublished). With the Ag + Bi = 2Pb substitution active in the octahedral layers, also the cases $^{4,7}L$, $^{4,8}L$, $^{5,9}L$ and $^{11,11}L$ (Makovicky and Karup-Møller 1977*b*) are found, as well as the disordered combinations based on $N_1 = 4$ and $N_2 = 7$ in different proportions (*ibid.*, Skowron and Tilley 1990). The structures with close-to-ideal PbS-like arrays and those with extensive lone electron pair micelles either do not form continuous solid solutions inspite of the same N or they become separated by exsolution at low temperatures (e.g. $Pb_3Bi_2S_6$–$PbAgBi_3S_6$, Figs 1.4 and 1.6). The match of coordination polyhedra is better in the Pb–Ag–Bi sulphosalts and results in an extensive accretional series. In the parallel system Pb–Ag–Sb–S, the size and shape mismatch of Pb, Ag and Sb coordination polyhedra appears serious; only members with $N = 4$ were found and they are known only for ≥ 50 percent of (Ag + Sb) substitution for Pb; still, they form a string of intermediate (often mutually exsolved) phases with different spatial distribution of lone electron pair micelles in the structure and with different superperiods (2-, 4-, and 6-tuple) of the 4 Å dimension. For phases with substitution close to 50 percent, incorporation of smaller M^{2+} instead of octahedral Pb appears critical for the formation of viable unit layers; for example, $\mathbf{AgPb_3MnSb_5S_{12}}$ (Moëlo *et al.* 1984*b*).

The symmetry or asymmetry of the trigonal prismatic cation site varies widely for the lanthanide-based lillianite homologues with lower N values seemingly not having significant crystal-chemical role. However, the difference in the symmetry or asymmetry of this site (symmetrically coordinated Pb versus asymmetrically, sideways oriented Bi) becomes the substantial factor defining the crystal-chemically distinct **lillianite and pavonite homologous series** in the Pb–Bi–Ag–Cu system (Makovicky *et al.* 1977; Makovicky 1981, 1989).

1.4.4.2 *The pavonite accretional series*

The **pavonite homologous series** was defined by Makovicky *et al.* (1977) as an extensive series of complex sulphides from the Cu–Ag–(Pb)–Bi–S system. However, it has been a series with the most vigorous recent growth, acquiring all the time new members and chemical compositions. It includes, as new acquisitions, Cu–Bi selenides as well as sulphides of cadmium ± lead and indium.

The topology of the pavonite homologous series is similar to that of lillianite homologues with $N_1 \neq N_2$, that is, it is a classical case of contracted-set reflection twinning and, to certain degree also of a **heterochemical homologous series**. In the **pavonite homologues** (Fig. 1.7), all members have N_1; $N_2 = 1$; N_{pav} where the order of pavonite homologue, $N_{pav} = 2$–8, and possibly even higher (Table 1.2). The trigonal prisms on the planes of contracted-set unit-cell twinning are distorted and are occupied by square-pyramidal Bi and its lone electron pair. Bi is displaced towards those prism caps that form a part of the thin $N = 1$ layers. The extensive PbS-like portions (those with N_2) contain quasi-octahedral to octahedral Bi coordinations combined with Ag, Cu and some Pb, occasionally also Cd and In. The sole, skewed octahedron in the narrow portions ($N_1 = 1$) represents AgS_{2+4} or this octahedral column contains three- and four-coordinated Cu. Only for $N_{pav} = 2$, these octahedra are occupied by Pb; in the case of $\mathbf{CdBi_2S_4}$ with $N = 3$, they accommodate Cd.

Peculiar for synthetic, **Cu–Bi pavonite homologues** is the statistical substitution of Bi by Cu which is accommodated in the walls of, and adjacent to, the vacated Bi polyhedra. Synthetic phases (sulphides and selenides) with $N_{pav} = 3$ and 4 are built on this principle which is assumed to be active up to $N_{pav} = 8$ with the possible exception of $N_{pav} = 7$. For natural phases with $N_{pav} = 4$, 5, and 8, this substitutional/interstitial

Table 1.1 Selected homologues of lillianite[a]

Compound	Homologue	Unit cell parameters (Å and degrees)	Space group	Note	Reference
$NdYbS_3$	$^{1,1}L$	$\underline{a}$ 12.55 $\underline{b}$ 9.44 $\underline{c}$ 3.85	$B22_12$		Carré and Laruelle (1974)
$UFeS_3$	$^{1,1}L$	$\underline{b}$ 11.63 $\underline{c}$ 8.72 $\underline{a}$ 3.80	*Cmcm*		Noël and Padiou (1976)
$CuEu_2S_3$	$^{1,1}L$	$\underline{b}$ 12.86 $\underline{a}$ 10.35 $\underline{c}$ 3.95	*Pnam*	Cu tetrahedral	Lemoine *et al.* (1986*a*)
$MnEr_2S_4$[b]	$^{2,2}L$	$\underline{b}$ 12.60 $\underline{c}$ 12.75 $\underline{a}$ 3.79	$Cmc2_1$		Landa-Canovas and Otero-Diaz (1992)
$CrEr_2S_4$	$^{2,2}L$	$\underline{a}$ 12.56 $\underline{b}$ 12.48 $\underline{c}$ 7.54	$Pb2_1a$	$c' = c/2$	Tomas and Guittard (1980)
$FeHo_4S_7$	$^{1,2}L$	$\underline{a}$ 12.57 $\underline{c}$ 11.35 $\underline{b}$ 3.78 β 105.7	$C2/m$		Adolphe and Laruelle (1968)
$MnEr_4S_7$[b]	$^{1,2}L$	$\underline{a}$ 12.54 $\underline{c}$ 11.44 $\underline{b}$ 3.76 β 105.4	$C2/m$		Landa-Canovas and Otero-Diaz (1992)
$TlSb_3S_5$	$^{3,3}L$	$\underline{a}$ 7.23 $\underline{b}$ 15.55 $\underline{c}$ 8.95 β 113.6	$P2_1/c$	$c' = c/2$	Gostojič *et al.* (1982)
$Pb_3Bi_2S_6$ (lillianite)	$^{4,4}L$	$\underline{a}$ 13.54 $\underline{b}$ 20.45 $\underline{c}$ 4.10	*Bbmm*	Minor Ag	Takagi and Takéuchi (1972)
$Pb_3Bi_2S_6$ (xilingolite)	$^{4,4}L$	$\underline{a}$ 13.51 $\underline{c}$ 20.65 $\underline{b}$ 4.09 β 92.2	$C2/m$	Cation ordering	Berlepsch *et al.* (2002*c*)
$PbAgBi_3S_6$ (gustavite)	$^{4,4}L$	$\underline{a}$ 7.08 $\underline{b}$ 19.57 $\underline{c}$ 8.27 β 107.2	$P2_1/c$	$c' = c/2$	Harris and Chen (1975)
$Pb_8Ag_5Bi_{13}S_{30}$[c] (vikingite)	$^{4,7}L$	$\underline{a}$ 7.10 $\underline{c}$ 25.25 $\underline{b}$ 8.22 α 90.0 β 95.36 γ 106.8	$P\bar{1}$	$c' = c/2$	Makovicky and Karup-Møller (1977*b*), Makovicky *et al.* (1992)
$Pb_6Sb_{11}Ag_3S_{24}$ (ramdohrite)	$^{4,4}L$	$\underline{b}$ 13.08 $\underline{a}$ 19.24 $\underline{c}$ 8.73 β 90.3	$P2_1/n$	$c' = c/2$	Makovicky and Mumme (1983)
Uchucchacuaite $AgMnPb_3Sb_5S_{12}$	$^{4,4}L$	$\underline{a}$ 12.67 $\underline{b}$ 19.34 $\underline{c}$ 4.38[d]		Pseudoorthorhomb.	Moëlo *et al.* (1984*a*, *b*)
$Pb_{18}Ag_{15}Sb_{47}S_{96}$ (andorite IV = quatrandorite)	$^{4,4}L$	$\underline{a}$ 13.04 $\underline{b}$ 19.18 $\underline{c}$ 17.07 γ 90.0	$P2_1/a$	$c' = c/4$	Moëlo *et al.* (1988*b*)
$Pb_{24}Ag_{24}Sb_{72}S_{144}$ (andorite VI = senandorite)	$^{4,4}L$	$\underline{a}$ 13.02 $\underline{b}$ 19.18 $\underline{c}$ 25.48	$Pn2_1a$	$c' = c/6$	Moëlo *et al.* (1988*b*), Sawada *et al.* (1987)
$Pb_6Bi_2S_9$ (heyrovskyite)	$^{7,7}L$	$\underline{a}$ 13.71 $\underline{b}$ 31.21 $\underline{c}$ 4.13	*Bbmm*		Takéuchi and Takagi (1974)
$Pb_{5.92}Bi_{2.06}S_9$ (aschamalmite)	$^{7,7}L$	$\underline{a}$ 13.71 $\underline{c}$ 31.43 $\underline{b}$ 4.09 β 91.0	$C2/m$	Cation ordering	Mumme *et al.* (1989)
$Pb_{3.36}Ag_{1.32}Bi_{3.32}S_9$ (Ag-Bi-heyrovskyite)	$^{7,7}L$	$\underline{b}$ 13.60 $\underline{c}$ 30.49 $\underline{a}$ 4.11	*Cmcm*		Makovicky *et al.* (1991)
$Ag_7Pb_{10}Bi_{15}S_{36}$ (eskimoite)	$^{5,9}L$	$\underline{a}$ 13.46 $\underline{b}$ 30.19 $\underline{c}$ 4.10 β 93.4	$B2/m$		Makovicky and Karup-Møller (1977*b*)
$Pb_4Ag_3Bi_5S_{13}$ (ourayite)	$^{11,11}L$	$\underline{a}$ 13.49 $\underline{b}$ 44.17 $\underline{c}$ 4.05	*Bbmm*		Makovicky and Karup-Møller (1984)

[a] Synthetic homologues $^{8,8}L$, $^{7,8}L$, $^{4,5}L$ were observed at high temperatures in the system Ag_2S–Bi_2S_3-PbS prepared by HRTEM (Skowron and Tilley 1990).

[b] Complex homologues $Mn_2Er_6S_{11}$($^{2,2,2,1}L$) and $Mn_3Er_8S_{15}$($^{2,2,1,2,2,1}L$) occur in this system (Landa-Canovas and Otero-Diaz 1992); the equivalent phases, MnY_2S_4, MnY_4S_7, and $Mn_2Y_6S_{11}$ occur in the MnS–Y_2S_3 system (Bakker and Hyde 1978). Random complex intergrowths ^4L–^7L occur in the PbS-Bi_2S_3–Ag_2S system (‘schirmerite’, Makovicky and Karup-Møller 1977*b*).

[c] The 4 Å subcell of vikingite has $\underline{a}$ 13.60 Å, $\underline{b}$ 4.11 Å, $\underline{c}$ 25.25 Å and β 95.60°, space group $C2/m$.

[d] Determined from powder data.

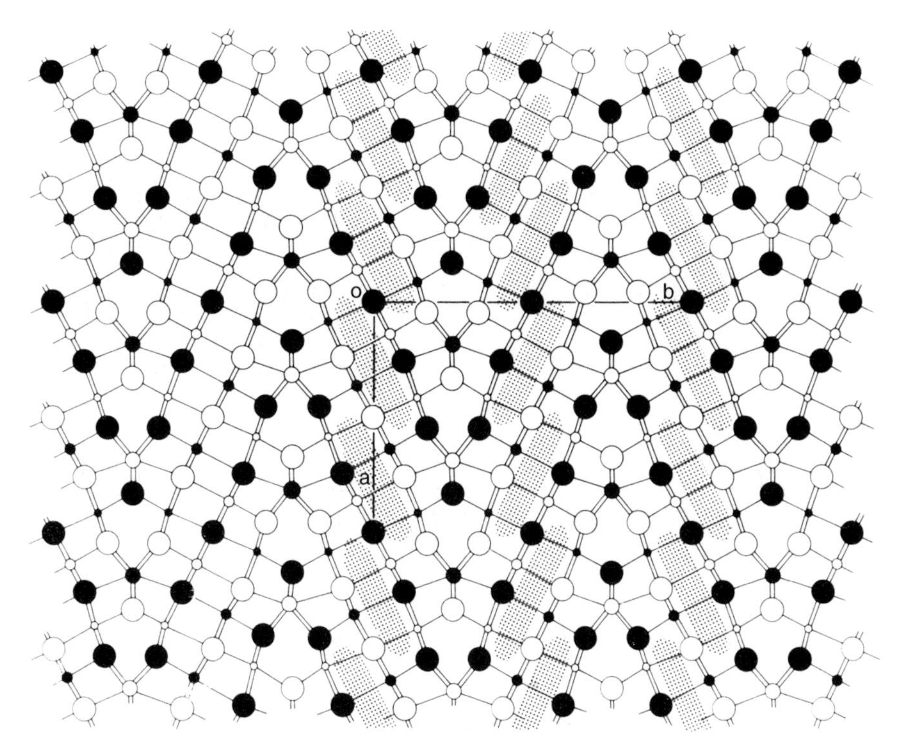

Fig. 1.6. The packing scheme for lone electron pair micelles of Bi in gustavite $PbAgBi_3S_6$ (Kupčík and Steins 1990) ($N = 4$). In order of decreasing size circles represent S, Pb, Bi and $(Ag_{0.5}Bi_{0.5})$. Lone electron pair micelles of flat-cylindrical cross-section are stippled.

solid solution exsolves into an intimate lamellar intergrowth of two, respectively, Cu–Pb poor and Cu–Pb highly substituted phases; the nature of the Cu–Pb-rich phase will be discussed further below. Natural pavonite homologues $N_{pav} = 5$ and 7 prefer Ag in the skewed sole octahedra; those with $N_{pav} = 3$ are only known with Cu or with Cd filling this position; $N = 8$ has mixed occupancy and $N = 4$ is known in both the Cu- and Ag varieties (Table 1.2).

Sb alone does not form pavonite homologues because of its inability to assume near-regular octahedral coordination in the interior of the thicker slabs. The only exceptions are $MnSb_2S_4$, triclinic $\mathbf{Li_{3x}Sb_{6-x}S_9}$ and a Cu–Sb selenide with pavonite-like structure. For low N values, the distinction between the pavonite and lillianite principles becomes blurred in the case of rare-earth sulphides with laterally displaced cations in the bi-capped trigonal prisms on the planes of unit-cell twinning.

Two ways of bridging the differences between the **lillianite and pavonite accretional series** have been observed. The first one is the **Phase V homologous series** summarized by Takéuchi (1997). We can describe this series, defined as

Table 1.2 Selected pavonite homologues

Mineral	Formula	Homologue	Unit cell parameters (Å)			β (degrees)	Space group	References
Synthetic V-I	$Pb_{1.46}Bi_{8.36}S_{14}$ (approx. $PbBi_4S_7$)	^{2}P	*a* 13.25	*b* 4.03	*c* 12.04	105	$C2/m$	Takéuchi *et al.* (1974, 1979)[a]
Synthetic	$CdBi_4S_7$	^{2}P	*a* 13.11	*b* 4.00	*c* 11.77	105.2	$C2/m$	Choe *et al.* (1997)
Synthetic	$Cd_{2.8}Bi_{8.1}S_{15}$	$^{2,3}P$	*a* 13.11	*b* 3.99	*c* 24.71	97.8	$C2/m$	Choe *et al.* (1997)
Synthetic	$Cd_2Bi_6S_{11}$	$^{2,2,3}P$	*a* 13.11	*b* 4.00	*c* 35.84	90.4	$C2/m$	Choe *et al.* (1997)
Synthetic	$CdBi_2S_4$	^{3}P	*a* 13.10	*b* 3.98	*c* 14.61	116.3	$C2/m$	Choe *et al.* (1997)
Synthetic	$Cu_{1.57}Bi_{4.57}S_8$	^{3}P	*a* 13.21	*b* 4.03	*c* 14.09	115.6	$C2/m$	Ohmasa and Nowacki (1973), Tomeoka *et al.* (1980)
Synthetic	$HgBi_2S_4$	^{3}P	*a* 14.17	*b* 4.06	*c* 13.99	118.3	$C2/m$	Mumme and Watts (1980)
Synthetic	$Cu_{3.21}Bi_{4.79}S_9$	^{4}P	*a* 13.21	*b* 3.99	*c* 14.81	100.2	$C2/m$	Ohmasa (1973)
Synthetic	$Cu_2Pb_{1.5}Bi_{4.5}S_9$	^{4}P	*a* 13.45	*b* 4.03	*c* 14.99	99.8	$C2/m$	Mariolacos *et al.* (oral communication)
Makovickyite	$Cu_{1.12}Ag_{0.81}Pb_{0.27}Bi_{5.35}S_9$	^{4}P	*a* 13.37	*b* 4.05	*c* 14.71	99.5	$C2/m$	Mumme (1990), Žák *et al.* (1994)
Cupromakovickyite	$Cu_{1.85}Ag_{0.60}Pb_{0.70}Bi_{4.40}S_9$	^{4}P[a]	*a* 13.40	*b* 4.01	*c* 29.93	100.07	$C2/m$	Topa *et al.* (in preparation)
Ag-rich makovickyite	$Cu_{0.11}Ag_{0.80}Pb_{0.86}Bi_{5.01}S_9$	^{4}P	*a* 13.83	*b* 4.04	*c* 14.72	97.50	$C2/m$	Mumme (1990), Žák *et al.* (1994)
Synthetic	$Li_{3x}Sb_{6-x}S_9 (x = 1/3)$	^{4}P	*b* 6.68	*a* 4.09	*c* 14.70	[b]	$P1$-	Olivier-Fourcade *et al.* (1983)
Synthetic	$AgBi_3S_5$	^{5}P	*a* 13.31	*b* 4.04	*c* 16.42	94.0	$C2/m$	Makovicky *et al.* (1977)
Pavonite	$Cu_{0.27}Ag_{0.78}Pb_{0.33}Bi_{2.78}S_5$	^{5}P	*a* 13.42	*b* 3.99	*c* 16.39	94.3	$C2/m$	Karup-Møller and Makovicky (1979)
Cupropavonite	$Cu_{0.9}Ag_{0.5}Pb_{0.6}Bi_{2.5}S_5$	^{5}P[a]	*a* 13.45	*b* 4.02	*c* 33.06	93.5	$C2/m$	Karup-Møller and Makovicky (1979)
Benjaminite	$Ag_3Bi_7S_{12}$	^{7}P	*a* 13.25	*b* 4.05	*c* 20.25	103.1	$C2/m$	Herbert and Mumme (1981)
Benjaminite	$Cu_{0.5}Ag_{2.3}Pb_{0.4}Bi_{6.8}S_{12}$	^{7}P	*a* 13.30	*b* 4.07	*c* 20.21	103.3	$C2/m$	Makovicky and Mumme (1979)
Mummeite	$Cu_{0.58}Ag_{3.11}Pb_{1.10}Bi_{6.65}S_{13}$	^{8}P	*a* 13.47	*b* 4.06	*c* 21.63	92.9	$C2/m$	Karup-Møller and Makovicky (1986), (1992), Mumme (1990)

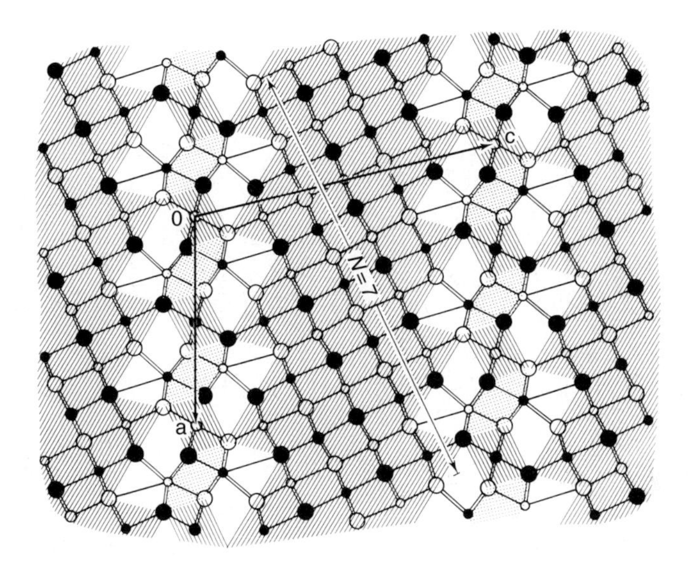

Fig. 1.7. The crystal structure of benjaminite, the $N = 7$ member of the pavonite homologous series (Makovicky and Mumme 1979). In order of decreasing size, circles denote S, Ag and Bi. Atoms at $z = 0$ and $z = 1/2$. Broad PbS-like slabs $(311)_{PbS}$ are ruled; homologous order $N = 7$ is defined.

$Pb_{1-x}Bi_{2x/3}S \cdot 2Bi_2S_3$ by Takéuchi, as a **combinatorial series** of ordered intergrowths of lillianite $^{2,2}L$ and $^{1,2}L$ modules; the latter are, in fact, pavonite-like modules, present as the sole component of the phase V-1 (Takéuchi *et al.* 1974). In this case, the lillianite and pavonite-like composition planes of unit-cell twinning are spatially separated.

The second way has recently been found in the structure of **cupromakovickyite** ($N_{1,2} = 1, 4$) (in preparation) and apparently relates also to that of **cupropavonite** ($N_{1,2} = 1, 5$). In **pavonite homologues**, the asymmetrically Bi-occupied trigonal prisms across the thin, $N = 1$, layer share edges of their caps, resulting in a typical column of paired Bi square pyramids. In the 'cupro-' varieties, one of these Bi polyhedra from a pair is exchanged by a symmetrically occupied trigonal prism with lead instead of bismuth. The occupation of the adjacent octahedral column in the thin layer by statistically distributed Cu atoms becomes oriented as well.

1.4.4.3 *Glide-plane twinned accretional series*

Glide-plane twinned accretional series have more individualized coordination polyhedra on composition planes of adjacent slabs. Such are the **meneghinite series** of sulphosalts of Cu, Pb, Sb or/and Bi (Fig. 1.8) and the **sartorite homologous series** (sulphosalts of Pb and As). These, respectively, represent twinning of SnS-like

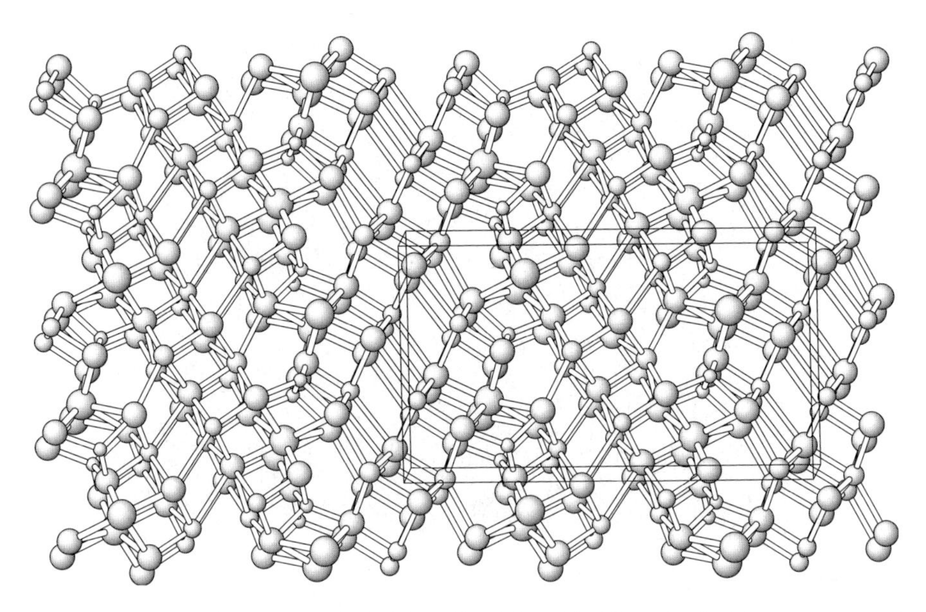

Fig. 1.8. The crystal structure of jaskolskiite $Cu_{0.2}Pb_{2.2}$ (Sb, Bi)$_{1.8}$ S_5, the $N = 4$ member of the meneghinite homologous series. In order of decreasing size, circles denote S, Pb, Bi and Cu. Atoms at $z = 0$ and $z = 1/2$.

arrays on $(501)_{SnS}$ and $(301)_{SnS}$ (Makovicky 1985*b*). Among silicates, the glide-plane twinned accretional series are, for example, **clinobiopyriboles**, in which the marginal cation positions of octahedral ribbons lie along the glide planes of unit-cell twinning (Figs 1.20 and 3.11). Both the meneghinite- and the biopyribole series were also described in a polysomatic ('slab-slicing') way, with the coordinations lining the glide planes defined as one kind of layers and the accreting slabs as the other, incrementally growing type of layers (Smith and Hyde 1983; Thompson 1978). Both ways of description are possible—in all these series, the marginal atoms conform to the configurations of the incrementally growing layers on the one side of their coordination polyhedron whereas their coordinations are altered on the side of the glide plane.

1.4.4.4 *Coherent intergrowths/polysomatic series*

The cuprobismutite homologues (Cu–Bi sulphosalts), the layered cuprates (several high-T superconductor families) and the pyroxenoid series (Kato *et al.* 1976) are typical examples of **coherent intergrowths** of two structure types. In the **cuprobismutite homologues**, incremental slabs of octahedral arrangement intergrow with slabs of trigonal prismatic arrangement. These homologues form a combinatorial triplet, the known phases being: N_1, N_2 = (1, 1), (2,2) and their combination (1,2)

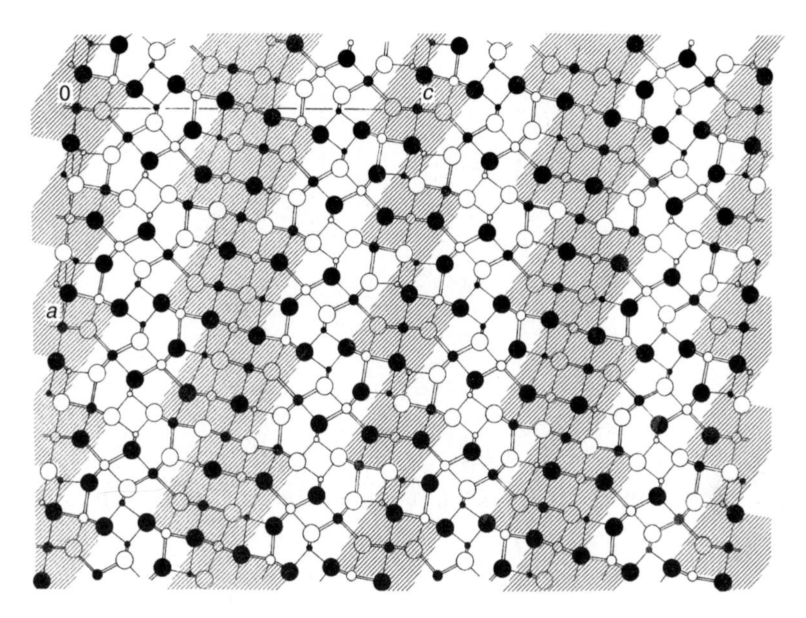

Fig. 1.9. Crystal structure of hodrushite (Kupčík and Makovicky 1968), $N_{1,2} = 1, 2$ homologue of the cuprobismutite combinatorial series. Circles in order of decreasing size indicate S, Bi and Cu. PbS-like portions $(331)_{\mathrm{PbS}}$ are ruled.

(Fig. 1.9). The **pyroxenoid series** is genuinely polysomatic with a number of intermediate members in which the number of pyroxene-like repeats P proportionally increases at the expense of wollastonite-like W repeats (WWW, WP, WPP, WPPP are known as macromembers, as well as pure P). Polysomatic description of pyroxenoids was given by Thompson (1978), a polyhedral one, for example, by Angel and Burnham (1991). Large number of unit cell intergrowth series were described among intermetallic compounds (Parthé *et al.* 1985; Grin *et al.* 1982) as 'linear heterogeneous series'.

1.4.4.5 *The brownmillerite homologous series*

Brownmillerite, $Ca_2(Fe^{3+}, Al)_2O_5$ is one of the principal components of cement clinker. Its structure consists of alternating layers of two kinds (Fig. 1.10):

(a) octahedral layers composed of corner-linked octahedra; these are slightly distorted unit layers (001) of perovskite structure;
(b) tetrahedral layers consisting of tetrahedron chains; these tetrahedra share corners with the adjacent octahedral layers.

The large Ca cations are situated between the (a) and (b) layers.

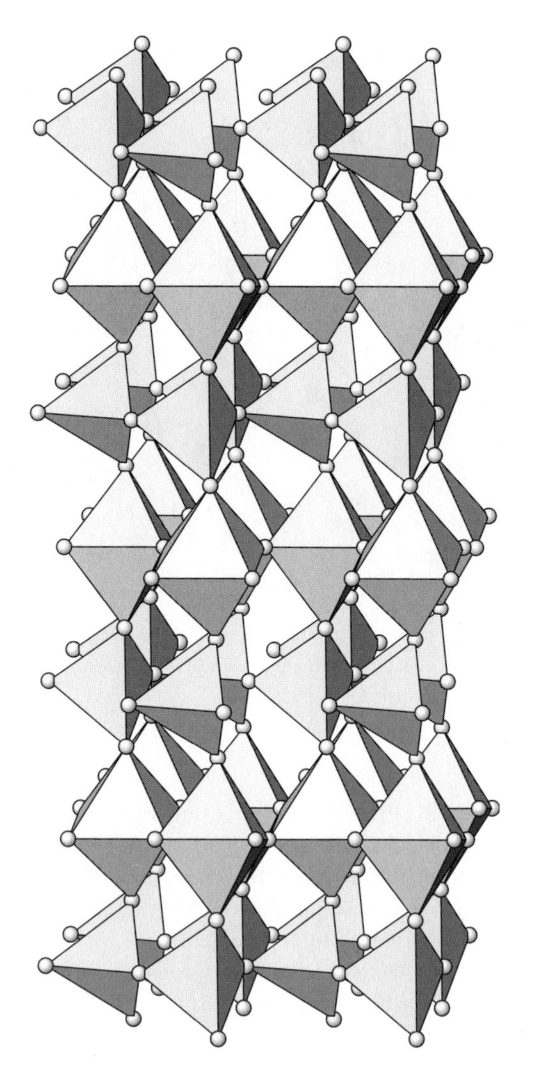

Fig. 1.10. Crystal structure of brownmillerite Ca_2FeAlO_5 (Colville and Geller 1978). Fe and Al are distributed in tetrahedral chains and octahedral layers; Ca in interspaces.

Brownmillerite is a type compound of a small homologous series in which the thickness of (a) layers changes by increments of one octahedron. Actually, brownmillerite-like compounds (orthorhombic $A_2B_2O_5$ compounds with A = Ca, Sr or Ba, B = Al, Sc, Cr, Fe^{3+}, Ga, and In) (Mitchell 2002) are basis for two parallel, closely related homologous sub-series: (A) a series based on *Ibm2* brownmillerite $Ca_2Fe^{3+}AlO_5$ (Colville and Geller 1971) and (B) a series based on *Pcmn* $Ca_2Fe^{3+}_2O_5$ (Bertaut *et al.* 1959). These two type compounds differ by the orientation of tetrahedral layers in the (010) planes. The same orientation for adjacent tetrahedral layers occurs in Fe–Al brownmillerite as well as in $Sr_2Fe_2^{3+}O_5$ and $Ba_2In_2O_5$ whereas their opposite orientation is observed in $\mathbf{Ca_2Fe_2O_5}$.

The homologous order may be defined as a number of octahedral layers in the octahedral slab of **brownmillerite homologues**. Members $N = 2$ have been synthesized: $\mathbf{Ca_3Fe_2TiO_8}$, $Ca_2LaFe_3O_8$ (Grenier *et al.* 1977) and $Ba_3In_2HfO_8$ (Goodenough *et al.* 1990); they belong to the second series, the stacking axis being about a triple of a single layer thickness. Those with $N = 3$ are $\mathbf{Sr_4Fe_4O_{11}}$ (Tofield *et al.* 1975) and $\mathbf{Ca_4Fe_2Ti_2O_{11}}$ (Gonzales-Calbert and Vallet-Regi 1987), belonging to the first series, have the stacking length doubled, that is, eight-tuple of the single layer thickness. Single four-tuple thickness was observed for $Ca_4Fe_2Ti_2O_{11}$; combinatorial members $N = 1$; 2 ($\mathbf{Ca_4YFe_4O_{13}}$; Bando *et al.* 1981) and $N = 1, 1, 2$ ($\mathbf{Ca_7Fe_6TiO_{18}}$; Rao and Raveau 1995) relate to the first family of structures.

Members with $N = 1$ have the general formula $A_2B_2O_5$, those with $N = 2$ are $A_3B_3O_8$, $N = 3$ $A_4B_4O_{11}$, that is, the general formula of the series is $A_{N+1}B_{N+1}O_{3N+2}$ where N is the number of octahedral layers per tetrahedral layer. This general formula suggests that the ratio of large and small cations cannot be used to determine the homologous order of a member of this series from microprobe analysis alone. The same is true about the ratios of diverse small, highly charged cations due to their readiness to assume both tetrahedral and octahedral positions.

1.4.4.6 *The humite and leucophoenicite homologous series*

These are nesosilicates (i.e. silicates with isolated SiO_4 tetrahedra) with Mg, Mn, Zn, Fe^{2+} and, rarely, also Ca in octahedral coordination. The monoclinic and orthorhombic structures of these series are based on hexagonal close-packed O arrays including subsidiary OH/F. All published descriptions of these compounds are based on the configuration of the serrated chains of edge-sharing octahedra. These descriptions mostly do not differentiate between the humite and leucophoenicite series; exceptions are Makovicky (1995) and Gaines *et al.* (1997). Olivine structure $(Mg,Fe)_2SiO_4$ (Fig. 1.11) is one end-member of both series and represents a point of their intersection.

Structures of these series can be divided along the incremental (~ chain) direction into layers with thickness from a breakpoint to the adjacent breakpoint of serrated chains. The octahedra in the breakpoints (tips) are divided into halves; straight sections are, respectively, two octahedra +2 halves in norbergite, one octahedron +2 halves in olivine, and only two halves in the oxide, **goethite**. The structures of the **humite series** can be understood as ordered intergrowths of these, respectively, three and two octahedra broad layers in changing proportions: 22 sequence forms olivine, 3222 sequence **clinohumite** (Fig. 1.12), 322 is **humite**, 32 **chondrodite** and 33 **norbergite** (Table 1.3). The leucophoenicite series starts with **olivine**, 22; continues with **jerrygibbsite** 22221 (Fig. 1.13), then **leucophoenicite** 2221, **ribbeite** 221, whereas the term 21 is thus far known as double layers in the crystal structures of the high-pressure 'superhydrous phase B' (Pacalo and Parise 1992) in which they are intercalated by

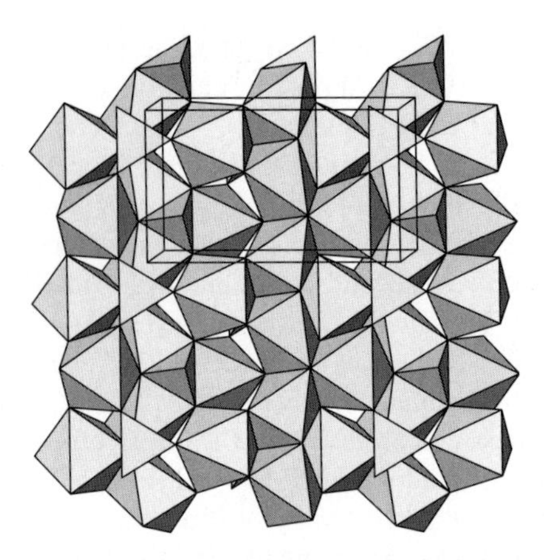

Fig. 1.11. Crystal structure of forsterite Mg_2SiO_4 (Ribbe 1982). Note the zig-zag pattern of Mg coordination octahedra composed of break-to-break slabs related by mirror reflection.

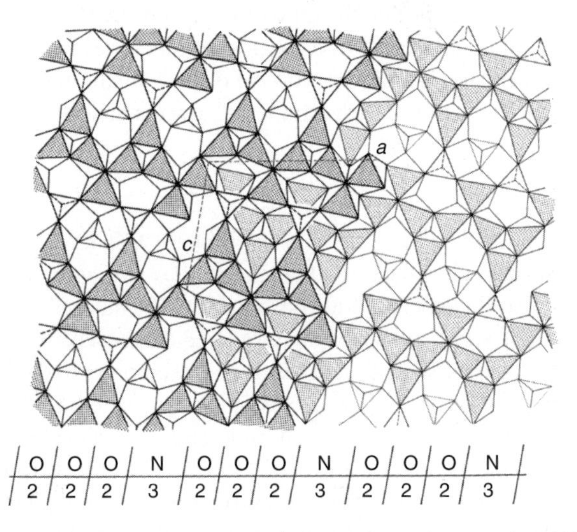

Fig. 1.12. The crystal structure of clinohumite $Mg_9(SiO_4)_4(F, OH)_2$, the 2223 member of the humite homologous series (polysome OOON where O are olivine-like and N norbergite-like slabs). Two octahedral–tetrahedral layers are shown.

layers with octahedral Mg and Si (Table 1.3). The member 11 is **goethite** FeOOH or **groutite** α-MnOOH.

Both series are **combinatorial series** with all members being a combination of two adjacent members N and $N + 1$ based on the same principle of serrate (i.e. unit cell twinned by reflection) octahedral chains. Members 44 and higher are not known, those with the 11 sequence are chemically very different, that is, no extensive series $N = 11, 22, 33, 44, \ldots$ is known to exist among silicates. Still, synadelphite

Table 1.3 Humite and leucophoenicite homologues; related high-pressure B phases

Mineral	Formula	Space group	a (Å)	b (Å)	c (Å)	α (°)	Homologue	Reference
Norbergite	$Mg_3SiO_4(F,OH)_2$	*Pbnm*	4.71	10.27	8.73	–	33	Gibbs and Ribbe (1969)
Chondrodite	$Mg_5(SiO_4)_2(F,OH)_2$	$P2_1/b$	4.75	10.27	7.80	109.2	32	Gibbs *et al.* (1970)
Alleghanyite	$Mn_5(SiO_4)_2(F,OH)_2$	$P2_1/b$	4.85	10.72	8.28	108.64	32	Rentzeperis (1970)
Humite	$Mg_7(SiO_4)_3(F,OH)_2$	*Pbnm*	4.74	10.24	20.72	–	322	Ribbe and Gibbs (1971)
Manganhumite	$(Mn,Mg)_7(SiO_4)_3(F,OH)_2$	*Pbnm*	4.82	10.58	21.45	–	322	Moore (1978)
Clinohumite[a]	$Mg_9(SiO_4)_4(F,OH)_2$	$P2_1/b$	4.74	10.23	13.58	100.9	3222	Robinson *et al.* (1973)
Sonolite	$Mn_9(SiO_4)_4(F,OH)_2$	$P2_1/b$	4.85	10.54	14.02	100.3	3222	Kato *et al.* (1989)
Forsterite	Mg_2SiO_4	*Pbnm*	4.76	10.20	5.98	–	22	Ribbe (1982)
Jerrygibbsite	$(Mn,Zn)_9(SiO_4)_4(OH)_2$	$Pbn2_1$	4.85	10.70	28.17	–	22221	Kato *et al.* (1989)
Leucophoenicite	$Mn_7(SiO_4)_3(OH)_2$	$P2_1/b$	4.83	10.84	11.32	103.73	2221	Moore (1970*a*)
Ribbeite	$Mn_5(SiO_4)_2(OH)_2$	*Pbnm*	4.8	10.73	15.67	–	221	Freed *et al.* (1993)
Anhydrous B	$Mg_{14}Si_5O_{24}$	*Pmcb*	5.87	14.18	10.05	–	−22	Finger and Hazen (1991*a*)
Hydrous B	$Mg_{12}Si_4O_{19}(OH)_2$	$P2_1/c$	10.59	14.1	10.07	104.1	−2221	Finger and Hazen (1991*a*)
Superhydrous B	$Mg_{10}Si_3O_{14}(OH)_4$	*Pnnm*[b]	5.09	13.97	10.08	120.34	−21	Pacalo and Parise (1992)

[a] Hydroxyloclinohumite is a fully hydroxylated member of this group (Ferraris *et al.* 2000).
[b] Axes were recalculated to match the rest of the series.

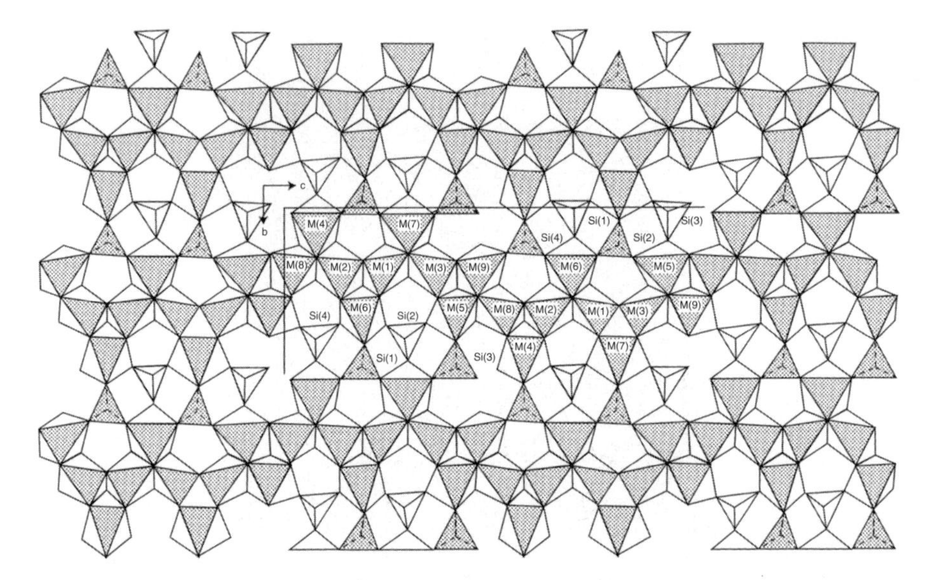

Fig. 1.13. The crystal structure of jerrygibbsite $(Mn, Zn)_9(SiO_4)_4(OH)_2$, the 22221 member of the leucophoenicite homologous series (polysome OOOOG). One octahedral/tetrahedral layer (100) is shown. The thin G slab comprises halves of the M(5) and M(9) octahedra.

$Mn_9(AsO_3)(AsO_4)_2(OH)_9 \cdot 2H_2O$ appears to approximate, with modifications in the position of (AsO_4) tetrahedron, the member 66 (a plesiotype).

The **humite series** *sensu stricto* was one of the series (the other being pyriboles) used by Thompson (1978) to define polysomatism. He described the humite series as a polysomatic series composed of 'norbergite-like' modules (N) interleaved by one to several 'olivine-like' modules (O). However, these slabs (defined above as three and two octahedra wide, respectively) resemble each other much more than the 'pyroxene-like' and 'mica-like' modules of the biopyribole structures and, in fact, one is the expanded version of the other. Thus, although satisfactory from a mathematical point of view (the compositions can be obtained in a simple way from the NO^n formula and the slab compositions), the polysomatic description obscures the combinatorial character of these series and it is not equivalent to such description of biopyribole structures.

1.4.4.7 *Two-dimensional homologous series*

Examples of **two-dimensional homologous series** of 'block types' are: (a) blocks of ReO_3 structure limited by two systems of crystallographic shear planes (Ijima 1975; Hyde *et al.* 1974) (Fig. 1.14), (b) triangular, incrementally growing blocks in the alloy/silicide homologous series $Ln_{n^2+3n+2}Ni_2(Ni,Si)_{2n^2}$ (Rogl 1992) or in a similar **zinckenite sulphosalt series** $Me_{6+x+N(N+5)}S_{12+N(N+7)}$ (Makovicky 1985*a*). Two indices, N_1 and N_2, are necessary for the block oxide structures; only

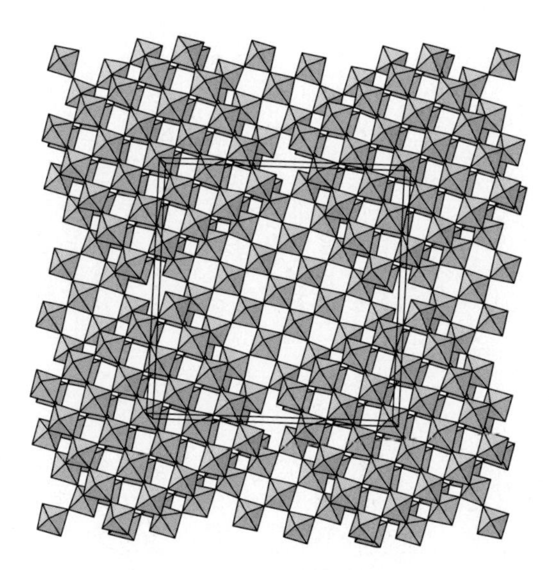

Fig. 1.14. The structure of $W_8Nb_{18}O_{69}$ (Roth and Wadsley 1965), an example of a two-dimensional homologous series. 5×5 columns of ReO_3 structure are separated by crystallographic-shear boundaries.

one, N, for the two latter series. Hollow blocks with channel cations occur in the **manganese oxide tunnel structures** (Turner and Buseck 1979). An example of a two-dimensional homologous series among silicates is given in Chapter 1.9 (**kalsilites**).

1.4.4.8 *Manganese oxides with tunnel structures*

Supergene oxides of manganese occur in continental and marine environments with oxidizing regime. Their complex mineralogy is complicated by fine grain size of their aggregates, often poor crystallinity and abundant defects in their crystal structure. Variable valence of Mn, as well as presence of solid solutions and, eventually, absence of simple stoichiometry further complicates their study.

Manganese tunnel oxides are one of the two principal structural families of this group, the other being **'phyllomanganates'**, that is, layered structures. The tunnel oxides (Table 1.4) consist of parallel chains of edge-sharing MnO_6 octahedra which either can stand alone, sharing only corners with adjacent such chains or they can polymerize via further edge sharing and form two, three, or more chains broad ribbons (Fig. 1.15).

Four adjacent chains/ribbons are joined by sharing vertices into intersecting walls which define (nearly) rectangular tunnels with size determined by the width of the ribbons involved. Therefore, we have to define two values of homologue order N, for the two wall systems at (nearly) right angles to one another. All these compounds contain narrow $N_{1,2} = 1, 1$ channels in the points of wall intersection; these are the

Table 1.4 Manganese oxides and selected related compounds with tunnel structures

Mineral	Formula	Space group	Unit cell parameters				Channels	Reference
			a	b	c	Angles		
Pyrolusite	β-MnO_2	$P4_2mnm$	4.398	–	2.873		(1,1)	Abrahams and Bernstein (1971)
Ramsdellite	γ-MnO_2	*Pbnm*	4.333	9.27	2.866		(1,2)	Byström (1949)
Hollandite	$(Ba,Pb,Na,K)\ (Mn,Fe,Al)_8O_{16}$	$I2/m$	10.026	2.878	9.729	β 91.03°	(2,2)[a]	Post *et al.* (1982)
Coronadite	$(Pb,Ba)\ (Mn,V,Al)_8O_{16}$	$I2/m$	9.938	2.868	9.834	β 90.39°	(2,2)[a]	Post and Bish (1989)
High-pressure	$K_2Al_2Si_6O_{16}$	$I4/m$	9.36	–	2.74		(2,2)[a]	Ringwood, Reid and Wadsley (1967)
High-pressure	$Na_2Al_2Ge_6O_{16}$	$I4/m$	9.648	–	2.856		(2,2)[a]	Reid and Ringwood (1969)
High-pressure	$K_2Al_2Ti_6O_{16}$	$I4/m$	10.04	–	2.94		(2,2)[a]	Bayer and Hoffman (1966)
Synthetic	$NH_4Mn_8O_{16}$	$I/4m$	9.865	–	2.849		(2,2)[a]	Botkovitz *et al.* (1994)
Romanéchite	$(Ba,H_2O)_2Mn_5O_{10}$	$C2/m$	13.929	2.846	9.678	β 92.39°	(2,3)[a]	Turner and Post (1988)
Todorokite	$(Na,Ca,K)\ (Mn,Mg)_6O_{12}\cdot nH_2O$	$P2/m$	9.764	2.842	9.551	β 94.06°	(3,3)[a]	Post and Bish (1988)
Synthetic	$Rb_{0.33}MnO_2$	$C2/m$	14.191	2.851	24.343	β 91.29°	(2,4)[a]	Rziha *et al.* (1992)
Synthetic	$Ba_{6.3}Mn_{24}O_{48}$[b]	$I4/m$	18.173	–	2.836		(2,2)[a]	Boullay, Herrieu and Raveau (1997)

[a] Also contain (1,1) channels in ratio 1 : 1.
[b] Two distinct types of (2,2) tunnels; plesiotype of hollandite.

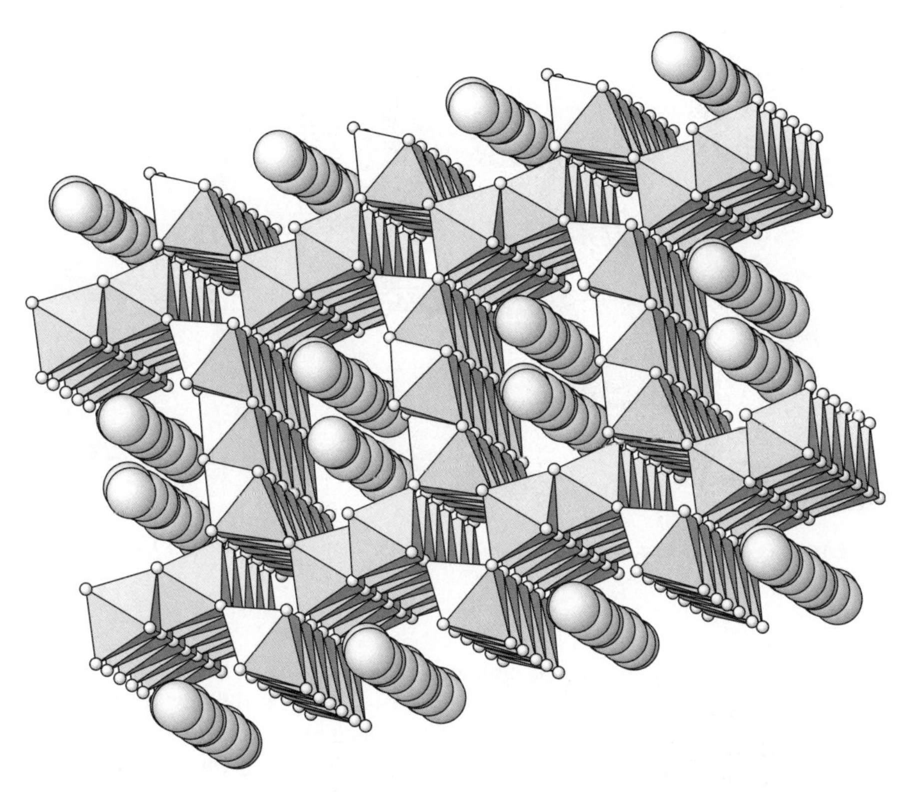

Fig. 1.15. Crystal structure of romanechite (Turner and Post 1988), the $N_{1,2} = 2, 3$ member of the manganese tunnel oxide series. Coordination octahedra of Mn form tunnel walls, Ba and H_2O occur in the tunnels.

only walls in the (1, 1) member of the series, pyrolusite MnO_2. Besides them, they contain (larger) tunnels (N_1, N_2), with the known combinations (1, 1) in **pyrolusite,** (2, 2) in **hollandite**, (2, 3) in **romanechite** (Fig. 1.15), (2, 4) in $\mathbf{Rb_{0.33}MnO_2}$, (2, 5) in $\mathbf{Rb_{0.27}MnO_2}$, and (3, 3) in **todorokite**. These are the macroscopic phases. However, the higher homologues are usually full of periodicity defects such as (2, 1), (2, 2), (2, 4), (2, 5), etc. in romanechite and especially (3, 2), (3, 4), (3, 5) etc. in todorokite; some of these were observed as domains.

Because all these minerals grow by accretion of subsequent layers of the structure, they generally preserve the ribbon (i.e. tunnel) widths tangential to the growing surface and the above ribbon-width variability is concentrated into the direction normal to the growing surface. Therefore, the refined classification by Chukhrov *et al.* (1989) distinguished these two directions. If we define N_1 parallel to the growth layers and N_2 normal to them, they distinguish a **pyrolusite family** ($1, N_2 = 1, 2, \ldots$), **hollandite family** ($2, N_2 = 2, 3, \ldots$), and **todorokite family** ($3, N_2 = 3, 4, \ldots$) of structures.

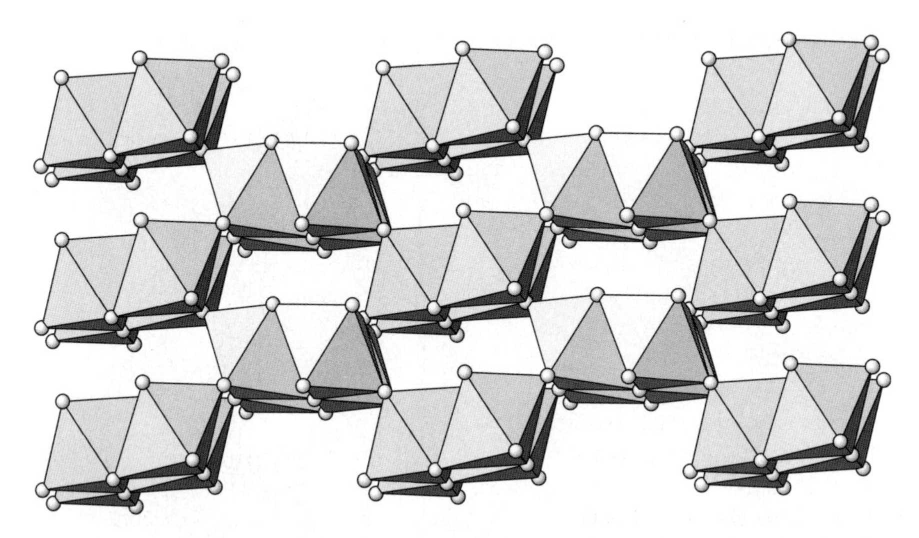

Fig. 1.16. Crystal structure of ramsdellite β-MnO_2 (Abrahams and Bernstein 1971) composed of double-octahedral ribbons.

A bit outside stands the crystal structure of **ramsdellite**, which lacks the (1,1) tunnels. It only has (1,2) tunnels; that is, double-octahedral ribbons in subparallel chess-board arrangement (Fig. 1.16). It can be joined to the rest of these structures by assuming that, instead of single-octahedral chains typical for **pyrolusite** (and pyrolusite-based sequence with $N_2 = 1, 2, 3, \ldots$), it has 'chains' with $N_1 = 1$ but two octahedra 'thick'. Chukhrov *et al.* (1989) build a hypothetical **'ramsdellite family'** with these double-chains combined with broader ribbons, resulting in channels (1, 2) combined with $(1, N_2 = 2, 3, \ldots)$. **Ramsdellite** MnO_2 forms regular intergrowths with pyrolusite MnO_2; these together (called 'nsutite') are the principal Mn ores in oxidation-zone deposits. Yet another exception is the tunnel structure of **$Ba_{6.3}Mn_{24}O_{48}$** (Boullay *et al.* 1997) with two ('single-Ba' and 'double-Ba') types of tunnels (2, 2) (the former of hollandite type) and much less frequent (1, 1) channels. Pronounced distortion of presumably Mn^{3+} sites is essential for this plesiotype of **hollandite**.

In all these structures, the framework stoichiometry yields MO_2 where M is the octahedrally coordinated cation. In MnO_2 (**pyrolusite**, **ramsdellite**), Mn is tetravalent and no additional cations can enter the narrow channels. Actually, a distorted monoclinic version of pyrolusite structure exists in manganite γ-$Mn^{3+}OOH$ with hydrogen bonds spanning the deformed (1, 1) tunnel with rhomb-shaped cross-section, and in the orthorhombic groutite α-MnOOH, the hydrogen bonds span the narrow (1, 2) channel of the **ramsdellite-type structure**.

Larger channels form only in the presence of large uni- and divalent cations. Variations in octahedron size and Jahn–Teller distortion indicate presence of Mn^{3+} or even of minor substituting cations, for example, Mg, Cu, Zn, Ni, Co in selected (especially 'corner') octahedra in the ribbons; these are compensating for the charge of large cations.

In hollandites (2,2), up to two large cations p.f.u. $A_{0-2}M_8O_{16}$ occur, like Ba (Miara 1986), Sr, Pb, K, less Na and even NH_4 (Table 1.4). The amount of cations is not only a function of valence but also of cation–cation interaction, resulting in complicated disordered sequences. In **romanechite** ('**psilomelanes**' in technical literature) $(Ba,H_2O)_2Mn_5O_{10}$, the tunnel components are ordered with tripled periodicity parallel to the tunnels. In the always poorly crystallized todorokite $(Na,Ca,K,Ba,Sr)_{0.3–0.7}$ $(Mn,Mg,Al)_6O_{12}\cdot 3.2–4.5H_2O$, four columns of partly disordered water molecules are supposed to support partly occupied cation sites around the channel centre (Post and Bish 1988). Thus, the channel species are variable and, in many instances, subject to own interactions and order–disorder. We may disregard them when classifying these phases as individual homologues.

These tunnel structures exist also for other octahedrally coordinated highly charged cations. Pyrolusite structure is the structure of TiO_2 (rutile) as well as of SnO_2, β-PbO_2, $TeTa_2O_6$, as well as GeO_2 and high-pressure SiO_2 (stishovite) (Strunz and Nickel 2001). The hollandite structure type has been synthesized with a number of octahedral cations ($V,Cr,Fe^{3+},Ti,Al\ldots$), especially in M^{4+}–M^{3+} combinations. These high-pressure phases include compositions like $\mathbf{K_2Al_2Ti_6O_{16}}$ (Bayer and Hoffman 1966) and high-pressure analogues of feldspars, for example, $\mathbf{K_2Al_2Si_6O_{16}}$ (Ringwood *et al.* 1967) and $Na_2Al_2Ge_6O_{16}$ (Reid and Ringwood 1969) (Table 1.4). Partly occupied channels occur, for example, in $Ba_{1.33}(Ti,Mg)_8O_{16}$ (Bursill and Grzinic 1980). Properties of these compounds were reviewed by Vicat *et al.* (1986) and Post and Burnham (1986).

1.4.5 *Pairs of homologues and combinatorial series*

An example for the **pair of homologues** is the pair $TlPbAs_3S_9$ (**hutchinsonite**, *Pbca*, Takéuchi *et al.* 1965)–$TlAs_5S_8$ (**bernardite**, $P2_1/c$, $TlAs_5S_8$; Pašava *et al.* 1989). Both structures are composed of alternating slabs (A) of SnS-like configuration, with As, Tl $\pm$ Pb, and (B) of complex spiral configuration, accommodating only As and its active lone electron pairs (Fig. 1.17). Similarity between the two structures, in nearly all coordination polyhedra and layer configurations, is striking. The only, and substantial, difference is that in bernardite the SnS-like slab is two coordination pyramids of As (respectively As and Tl in the marginal pyramids) broad, that is, a 2,2-homologue when considering the two faces of a tightly bonded fragment in the SnS-like slab, whereas in hutchinsonite this tightly bonded fragment is diagonally split, half of it rotated about [010], and an additional Pb pyramid is inserted into only one face of the fragment. This is the 2,3-homologue of the series.

It is easy to imagine a 3,3-homologue with two Pb pyramids inserted. However, it was not found, apparently because of the match problems between the large Pb polyhedra and the As pyramids. Homologues lower than 2,2 will have valence balance problems.

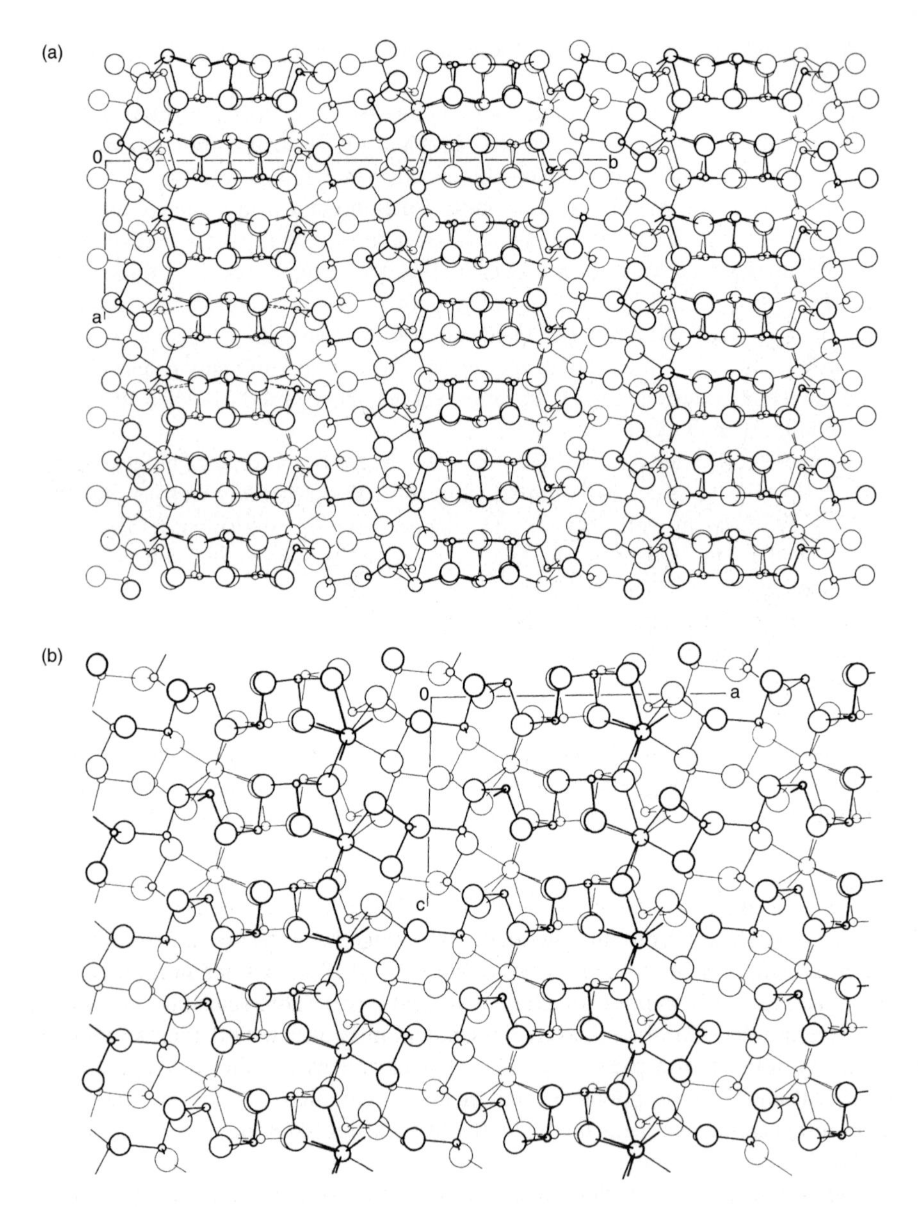

Fig. 1.17. Crystal structures of (a) hutchinsonite $TlPbAs_3S_9$ (Takéuchi *et al.* 1965) and (b) bernardite $TlAs_5S_8$ (Pašava *et al.* 1989), the $N = 2, 3$ and $N = 2, 2$ members of a homologous pair. Line thickness describes four height levels of the ~ 8.8 Å structures, in the order of decreasing circle size: S, Tl, Pb, As and S.

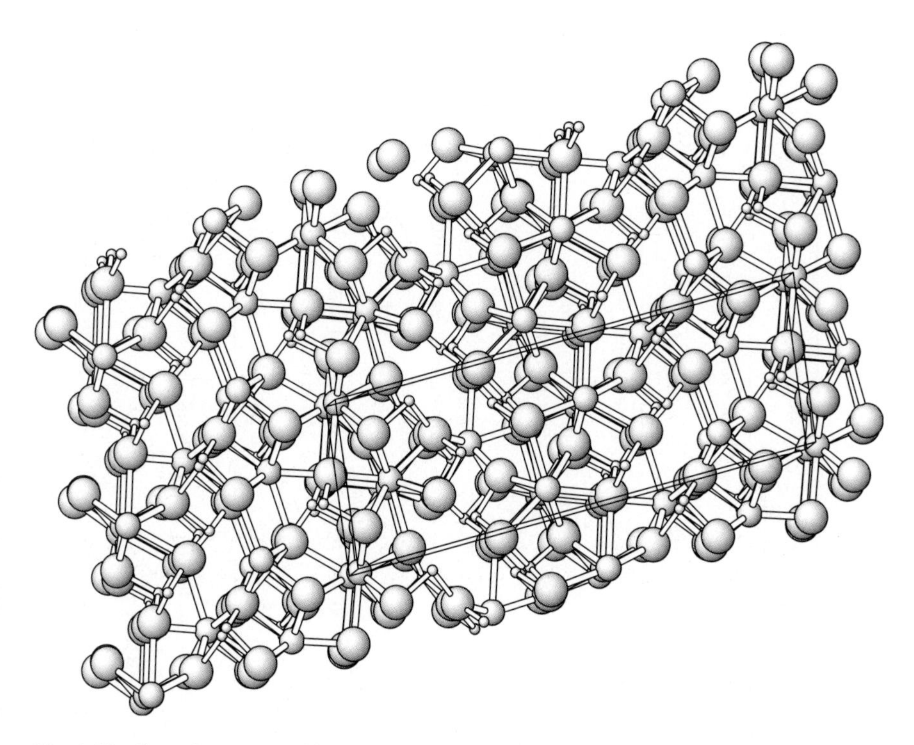

Fig. 1.18. Crystal structure of baumhauerite $Pb_{4.6}As_{15.7}Ag_{0.6}S_{36}$ (Engel and Nowacki 1969), the $N_{1,2} = 3, 4$ member of the combinatorial sartorite series. Cation count along a diagonal across the unit slabs includes both the As coordination polyhedra and those of Pb.

The same match problems between As pyramids and potentially inserted multiple Pb polyhedra limit the **sartorite homologous series** (simplified formula $Pb_{4N-8}As_8S_{4N+4}$) to combinations of $N = 3$ and $N = 4$ slabs in different proportions (Fig. 1.18); members with $N > 4$ would have entire slabs of Pb polyhedra inserted into their SnS-based layers, in apparent mismatch with the dimensions and geometry of these, primarily As-based configurations (Makovicky 1985b; Berlepsch *et al.* 2001*b*). Thus, the sartorite homologous series remains a combinatorial series. This is apparently valid also for its Ba–Sb analogues (Berlepsch *et al.* 2003*b*) (Table 1.5).

The structures of the small combinatorial cuprobismutite homologous series of Cu–Bi sulphides can be described as intergrowths of (A) slabs composed of double columns of BiS_5 coordination pyramids alternating with paired CuS_4 coordination tetrahedra, and (B) serrate slabs $(331)_{PbS}$ of PbS-like arrangement with two different thicknesses for $N = 1$ and $N = 2$, respectively (Fig. 1.9). The latter can also be described as Bi octahedra flanked on two sides either by a single Bi pyramid ($N = 1$) or a pair of pyramids ($N = 2$); trigonal bipyramidal Cu sites on the margins of (B) slabs reach into the (A) slabs. This homologous expansion has no meaningful

Table 1.5 Selected data on combinatorial sartorite homologues and related phases

Mineral	Chemical formula	Homologue	Unit cell parameters				Space group	Reference
Sartorite[a] (subcell)[b]	$PbAs_2S_4$	3,3	$\underline{a}$ 19.62	$\underline{b}$ 7.89	$\underline{c}$ 4.19	β 90.00	$P2_1/n$	Iitaka and Nowacki (1961)
Synthetic	$BaSb_2S_4$	3,3	$\underline{c}$ 20.60	$\underline{b}$ 8.20	$\underline{a}$ 8.99	$\underline{\beta}$ 101.4	$P2_1/c$	Cordier *et al.* (1984)
Synthetic	$BaSb_2Se_4$	3,3	$\underline{b}$ 20.76	$\underline{c}$ 8.55	$\underline{a}$ 9.23	$\underline{\beta}$ 91.2	$P2_1/n$	Cordier and Schäefer (1979)
Parapierrotite (synth.)	$TlSb_5S_8$	3,3	$\underline{b}$ 19.42	$\underline{a}$ 8.10	$\underline{c}$ 9.06	$\underline{\beta}$ 91.96	$P2_1/c$	Engel (1980)
Pierrotite	$Tl_2(As,Sb)_{10}S_{16}$	3,3	$\underline{a}$ 38.75	$\underline{c}$ 7.99	$\underline{b}$ 8.82		$Pna2_1$	Engel *et al.* (1983)
Synthetic	KSb_5S_8	3,3	$\underline{b}$ 19.50	$\underline{a}$ 8.13	$\underline{c}$ 9.06	$\underline{\beta}$ 91.93	Pn	Berlepsch *et al.* (1999)
Baumhauerite[a]-abc	$Pb_{4.6}As_{15.7}Ag_{0.6}S_{36}$	3,4	$\underline{a}$ 22.80	$\underline{c}$ 7.89	$\underline{b}$ 8.36		$P1$	Engel and Nowacki (1969)[c]
			α 90.05	β 97.26	γ 89.92			
Liveingite	$Pb_{18.5}As_{25}S_{56}$	4,3,4	$\underline{b}$ 70.49	$\underline{c}$ 7.91	$\underline{a}$ 8.37	$\underline{\beta}$ 90.13	$P2_1$	Engel and Nowacki (1970)
115 Å phase	$Pb_{32}As_{40}S_{92}$[d]	4,3,4,3,4,4, 3,4,3,4	$\underline{a}$ 115.75	$\underline{b}$ 7.90	$\underline{c}$ 8.37		–	Ozawa and Takéuchi (1993)
138 Å phase (rathite-IV)	$Pb_{38}As_{48}S_{110}$[e]	4,3,4,3,4,4, 3,4,3,4,3,4	$\underline{a}$ 138.6	$\underline{b}$ 7.92	$\underline{c}$ 8.46	$\underline{\beta}$ 90.33	$P2_1$ or $P2_1/m$	Ozawa and Tachikawa (1996)
Dufrenoysite[a]	$Pb_2As_2S_5$	4,4	$\underline{b}$ 25.74	$\underline{a}$ 7.90	$\underline{c}$ 8.37	$\underline{\beta}$ 90.35	$P2_1$	Marumo and Nowacki (1967)
Rathite I	$Pb_{11.1}Tl_{0.9}As_{17.6}Ag_{2.3}S_{40}$	4,4	$\underline{a}$ 25.16	$\underline{b}$ 7.94	$\underline{c}$ 8.47		$P1$ or $P\bar{1}$	Marumo and Nowacki (1965)
			α 90	β 100.47	γ 90		$(\sim P2_1/a)$	
Synthetic	$Ba_3Sb_{4.66}S_{10}$	4.4	$\underline{c}$ 26.76	$\underline{b}$ 8.23	$\underline{a}$ 8.96	$\underline{\beta}$ 100.29	$P2_1/c$	Choi and Kanatzidis (2000)

[a] Only these phases are known in the pure $PbS–As_2S_3$ system.
[b] Newest reference on superstructure of sartorite (Berlepsch *et al.* (2003*b*) suggests composition $Pb_{8.2}Tl_{1.4}As_{17.5}Sb_{0.5}S_{35}$ due to S vacancies and cation adjustments.
[c] Stacking polytypes of baumhauerite were studied by Pring and Graeser (1994) using HRTEM.
[d] Calculated from average $N = 3.6$ (Ozawa and Tachikawa 1996).
[e] Calculated from average $N = 3.58$ (Pring 2001). Graeser (in Pring 2001) gives $Pb_{32.3}Ag_{2.8}As_{49.2}Sb_{1.0}S_{110}$ from microprobe data.

polysomatic description. Contents of minor elements, Ag, Fe, or Pb, were ubiquitous in the phases of this structure series but their importance for the stability of individual homologues was not considered until recently.

Combined configurational and charge balance problems limit the cuprobismutite homologous series to only $N = 1$ and 2 homologues, as well as their 1:1 intergrowth, $N = 1, 2$ (Fig. 1.9). Synthetic cuprobismutite, with a formula far from ideal, $Cu_{10.4}Bi_{12.6}S_{24}$, was synthesized and its structure determined (Ozawa and Nowacki 1975). In this structure, two Bi sites out of the total of four sites are partly substituted by statistical copper distributed over more than one position. The homologue $N = 1$ was synthesized only once, by accident, with the formula $Cu_4Bi_5S_{10}$ (Mariolacos *et al.* 1975) requiring both Cu^+ and Cu^{2+} (Table 1.6).

Recent structure determinations on preanalysed crystals of natural phases (Topa *et al.* in preparation) clarified the situation and explained the limited character of the series. Pure copper-based $N = 1$ member shows lack of cation valence that in nature is compensated by partial Fe^{3+} for Cu^+ substitution in tetrahedrally distorted trigonal bipyramidal sites. Pure copper-based $N = 2$ member (cuprobismutite) has opposite problems. In synthetic cuprobismutite, they were compensated by Cu-for-Bi substitution (Ozawa and Nowacki 1975), in nature there is preference for Ag- and Pb-for-Bi substitution in octahedral Bi sites. Fe is absent. Hodrushite, $N = 1, 2$ has all these substitutions although also hodrushite low in Ag and with Cu-for-octahedral Bi is known. Substitution requirements for other N values than those found would be substantial; extensive PbS-like arrays of Bi polyhedra would need charge balance accommodation. Of equal importance, coordination problems arise for Bi arrays already at $N = 3$. Thus, this series remains limited to $N = 1$ and 2, respectively (Table 1.6).

1.5 Polysomatism

The concept of polysomatism (Thompson 1970) recognizes that any crystal structure can be sliced into slabs and in some cases the slabs of one structure can be combined with the slabs of another structure (Veblen 1991). The same concept was developed independently for intermetallics, under the name of **heterogeneous structural series** by Krypyakevich and Gladyshevskii (1972) and further perfected by Grin' *et al.* (1982) and Parthé *et al.* (1985). As mentioned above, both concepts evolved from **polytypy** (polytypes were called 'homogeneous structural series' by Krypyakevich and Gladyshevskii (1972) where slabs of the same structure occur in different combinations).

The choice of unit slabs in the **polysomatic description** is dictated by the need to obtain an additive general chemical formula for the described series. For each family, they are selected essentially in such a way that they give a simple mathematical description AB_n $(0 \leq n \leq \infty)$ where A and B are the two different **polysomatic slabs**. Stoichiometries of all polysomes are then linear combinations of the stoichiometries of the A and B slabs (Thompson 1978; Veblen 1991).

Table 1.6 Cuprobismutite homologues and plesiotypes

Mineral	Formula	$N_1; N_2$	Lattice parameters (Å)				Space group	Reference
Synthetic	$Cu_4Bi_5S_{10}$	1;1	$\underline{a}$ 17.54	$\underline{b}$ 3.93	$\underline{c}$ 12.85	β 108.0	$C2/m$	Mariolacos *et al.* (1975)
Kupčíkite	$Cu_{3.3}Fe_{0.7}Bi_5S_{10}$	1;1	$\underline{a}$ 17.51	$\underline{b}$ 3.91	$\underline{c}$ 12.87	β 108.57	$C2/m$	Topa *et al.* submitted
Hodrushite[a]	$Cu_{8.12}Fe_{0.29}Bi_{11.54}S_{22}$	1;2	$\underline{c}$ 17.58	$\underline{b}$ 3.94	$\underline{a}$ 27.21	β 92.15	$A2/m$	Kodera *et al.* (1970)
Synthetic[b]	$Cu_{10.4}Bi_{12.6}S_{24}$	2;2	$\underline{a}$ 17.52	$\underline{b}$ 3.93	$\underline{c}$ 15.26	β 100.2	$C2/m$	Ozawa and Nowacki (1975)
Cuprobismutite	$Cu_8AgBi_{13}S_{24}$	2;2	$\underline{a}$ 17.65	$\underline{b}$ 3.93	$\underline{c}$ 15.24	β 100.5	$C2/m$	Nuffield (1952)
Synthetic	$Cu_4Bi_4S_9$	Plesio[c]	$\underline{b}$ 11.66	$\underline{c}$ 3.97	$\underline{a}$ 31.68	–	*Pnam*	Bente and Kupčík (1984)
Synthetic	$Cu_4Bi_4Se_9$	Plesio	$\underline{c}$ 12.20	$\underline{b}$ 4.12	$\underline{a}$ 32.69	–	*Pnma*	Makovicky *et al.* (2002)
Paderaite	$Cu_{5.9}Ag_{1.3}Pb_{1.2}Bi_{11.2}S_{22}$	Plesio	$\underline{a}$ 28.44	$\underline{b}$ 3.90	$\underline{c}$ 17.55	β 106.0	$P2_1/m$	Mumme (1986)

[a] Hodrushite from Felbertal (Austria) is $Cu_{7.55}Ag_{0.39}Fe_{0.36}Bi_{11.58}S_{22.11}$ (Topa 2001).
[b] Synthetic cuprobismutite (Ozawa and Nowacki 1975).
[c] Plesiotypes of cuprobismutite homologues.

As a corrolary, several distinct choices of unit slabs may be possible, all of them equally valid for a given series. An example of this is given by the approaches of Ferraris *et al.* (1996) and Christiansen *et al.* (1999) to the **heterophyllosilicate polysomatic series** (Figs 4.20 and 4.21).

Contrary to this, unit slabs in the heterogeneous structural series of Grin' *et al.* (1982) and Parthé *et al.* (1985) are configurationally standardized, selected as a set of standard slabs from 'archetype' structures and are applicable to a wide range of intermetallic compounds (Fig. 3.1).

1.5.1 *Problems of application*

In the original field of application, the polysomatism of **biopyriboles** (Thompson 1970), the 'modular' choice of unit slabs in the accreting portions of the structure (the 'mica modules') coincides with the 'polyhedral' choice of unit (i.e. pyroxene-like) chains involved in the polymerization process (cf. Liebau 1985) (the modular choice also includes the octahedral 'backing'). A similar coincidence exists between the polyhedral and polysomatic descriptions for the **carlosturanite and antigorite accretional series**.

In the **olivine–norbergite** (= **humite**) **series** (Thompson 1978), the olivine and norbergite slabs were defined in a way conceptually different from biopyriboles—from a break-point to a break-point in the zig-zag chain of octahedra (Figs 1.11–1.13). Due to the combinatorial character of this series, mentioned elsewhere, they happen to be only one and two octahedra wide. In the combinatorial **sartorite–dufrenoysite series** of Pb–As sulphosalts (Makovicky 1985*b*), corresponding slabs are, respectively, three and four polyhedra wide (Fig. 1.18); they were selected as fundamental slabs 'S' and 'R', similar to Thompson's **olivine** and **norbergite** slabs, by Le Bihan (1962) and Pring (1990). In these two cases, the layer definition does not consider the fact that the layers are themselves products of polyhedral accretion. In this way, the polysomatic approach described obscures the difference between extensive and combinatorial series.

If Thompson's way of slicing devised for the humite structures is followed for the extensive series of **lillianite homologues**, we obtain a number of unit slabs that only rarely combine with one another (e.g. in vikingite, $N_{1,2} = 4, 7$). In a different approach, Veblen (1991) separated the (discontinuous) configurations on unit cell twin planes from the rest of the structure and sliced the 'galena-like' portions into (311) unit slices. The density of slicing is such that each cation-centred coordination octahedron spans three slices. Andersson and Hyde's (1974) model for these cases is different: the unit-cell twinned *hcp* and *ccp* structures are classified according to the number of $d(311)_{\text{ccp}}$- or $d(11\bar{2}1)_{\text{hcp}}$-spacings between the break-points (twin-planes), that is, the entire structure consists of slices of accretional type, disregarding the presence or absence, respectively, or coordination details of cations.

Detailed crystal chemistry of individual mineral groups requires analysis of coordination polyhedra, especially by cross-comparing those that assume analogous positions in the structures of different homologues (marginal, next-to-marginal, central, etc. polyhedra of the respective unit slabs). Schematic application of the concept of unit-cell twinning (or of other extended defects) or of the polysomatic slicing must, for this purpose, be abandoned in favour of polyhedral analysis: one follows the polyhedra from the surface into the interior of the accretionally obtained blocks and seeks site equivalence with analogous polyhedra in other homologues of the same series (see Makovicky and Norrestam 1985; Makovicky 1989; Veblen 1991; Berlepsch *et al.* 2001*a*). This analysis is important; no ideal homologues or polysomes with all accreting polyhedra (slabs) exactly identical exist in natural or synthetic compounds. The differences between them may be decisive for the series: for example, the majority of lillianite homologues exist because of the fine structural adjustments at the level of individual polyhedra (polyhedral groups), beyond the reach of simple concepts of unit-cell twinning, homology or polysomatism.

1.5.2 *The sapphirine polysomatic series*

Phases of this series (Merlino and Zvyagin 1998; Zvyagin and Merlino 2003), with ideal end-members (a) **Mg-clinopyroxene** close to ccp (typified, e.g. by $MgGeO_3$ Yamanaka *et al.* 1985) and (b) ideal **spinel** have been interpreted as polysomatic intergrowths of pyroxene-like P modules with **spinel-like S modules** in several distinct proportions. The **P modules** are one chain-width broad (010) slices of pyroxene structure whereas the S modules are one octahedron/tetrahedron broad (010) slices of spinel structure. The ccp anion framework of both slice types is coherent across the slab boundaries (Fig. 1.19).

Any P:S contact leads to the formation of a branched tetrahedral chain, with a lateral tetrahedron attached to the pyroxene chain. In the PS and PSS compounds (**sapphirine** and **SFCA-I**; Table 1.7), two-sided branch chains occur whereas in the PPS case (**surinamite**) one-sided branched chains are formed, building pyroxene-like slabs PP in the structure. Thus, this series, P_mS_n consists of two subseries, one with $m > n$ and pyroxene-like portions in the structure, and another one with $m < n$, in which only fully branched chains, more-or-less inserted into each other, occur.

In the latter case, only every second two-tetrahedra-long interval of the chain is branched (Fig. 1.19). This automatically leads to OD phenomena, with the boundaries of the OD layers centred on the pyroxene-like chains and the alternative positions of the $(n + 1)$ layer being one repetition period of the pyroxene chain apart. In the PPS case, a similar situation occurs on the P–P boundary, between the adjoining one-sidedly branched pyroxene chains. Details of this **polytypy** as well as further, hypothetical cases were worked out by Zvyagin and Merlino (2003).

Phases of this series are **silicates/aluminosilicates** with very variable percentage of Si in the total composition; in industrial products of the **SFCA type** (Table 1.7) Si may be absent, these being pure ferrialuminates. If the formulae of polysomatic modules are given as $P = M_4T_4O_{12}$ and $S = M_4T_2O_8$, the polysomatic formula of **sapphirine** PS is $M_8T_6S_{20}$, that of **SFCA-I**, PSS is $M_{12}T_8O_{28}$, whereas the

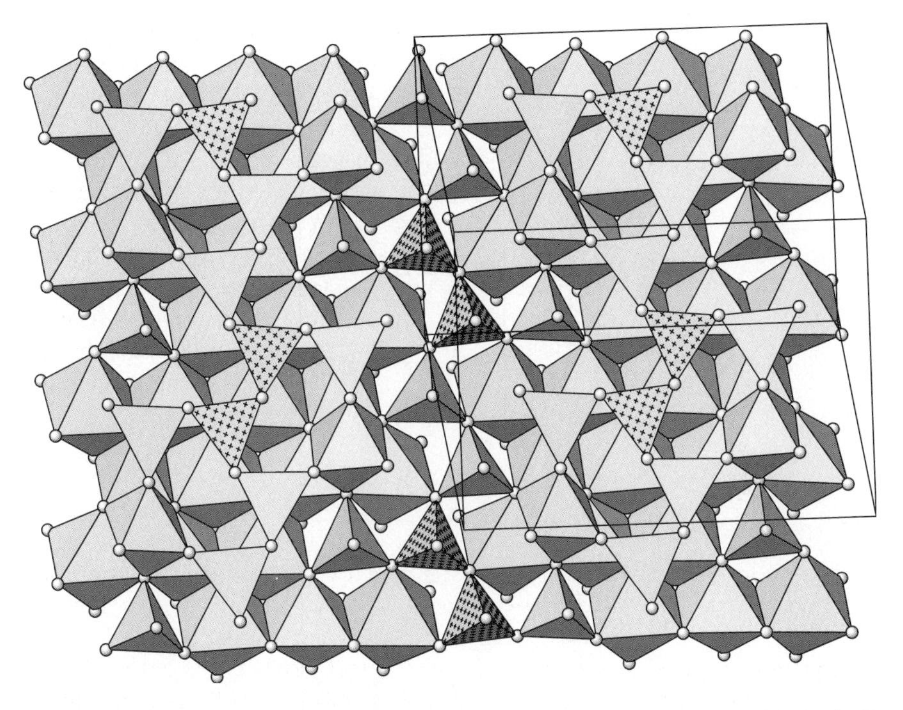

Fig. 1.19. The crystal structure of sapphirine $Mg_4Al_4(Al_4Si_2)O_{20}$ (Merlino 1980) the SPSP homologue of the sapphirine homologous series. The S modules (010) are at $y \sim 0.25$, the P modules are sandwiched between them.

formula of **surinamite**, PPS, is $M_6T_5O_{32}$, where T denotes tetrahedral sites and M the octahedral ones. Detailed treatment of these phases is given in Section 4.2.2.1.

1.5.3 *Biopyriboles*

The single-chain silicates **pyroxenes**, double-chain silicates **amphiboles**, and the phyllosilicates **micas** (typified by biotite) served as the basis of the concept of **polysomatic series** (Thompson 1970). Thompson realized that the usual concept of pyroxene-chain polymerization to amphiboles, etc. can be modified so that the pyroxene structure is sliced via median planes of (idealized) pyroxene chains and appropriate, coherent modules from an (idealized) **mica-like structure** are inserted in the planes of slicing. In this way, amphibole (a double-chain **inosilicate**) or more complicated structures are created, essentially with all the chemical and crystallographic properties observed in these minerals. The concept has been confirmed by the discovery of jimthompsonite (Table 1.23) with triple silicate chains (Fig. 1.20) or chesterite in which double- and triple chains and appropriate modules alternate.

Table 1.7 Selected phases of the sapphirine polysomatic series[a]

Mineral	Composition	Space group	Lattice parameters						Polysome	Reference
			a (Å)	b (Å)	c (Å)	α (Å)	β (Å)	γ (Å)		
Sapphirine	$(Mg_4Al_4)(Al_4Si_2)O_{20}$	$P\bar{1}$	9.85	8.62	9.97	90	110.4	90	PS	Merlino (1980)
		$P2_1/n$	9.814	14.438	9.957		110.39		PS	Higgins and Ribbe (1979)
Rhönite	$Ca_2(Mg,Fe^{2+}, Fe^{3+}, Ti)_6(Si,Al)_6O_{20}$	$P\bar{1}$	10.188	8.925	10.428	96.13	108.95	118.42	PS	Bonaccorsi *et al.* (1990)
Synthetic SFCA	$Ca_{2.3}Mg_{0.8}Al_{1.5}Si_{1.1}Fe_{8.3}O_{20}$	$P\bar{1}$	10.395	9.06	10.546	95.67	109.58	118.45	PS	Hamilton *et al.* (1989)
Synthetic SFCA-I	$Ca_{3.2}Fe^{2+}_{0.8}Fe^{3+}_{14.7}Al_{1.3}O_{28}$	$P\bar{1}$	10.431	11.839	10.610	94.14	110.27	111.35	PSS	Mumme *et al.* (1998)
Surinamite	$Mg_3Al_4Si_3BeO_{16}$	$P2/n$	9.631	11.384	9.916		109.3		PPS	Moore and Araki (1983)
Synthetic	$Mg_4Ga_4Ge_3O_{16}$	$A2/a$	10.073	23.733	10.324		110.29		PPS	Barbier (1998)

[a] Further details in Zvyagin and Merlino (2003).

Fig. 1.20. The crystal structure of jimthompsonite $(Mg, Fe)_5Si_6O_{16}(OH)_2$ (Veblen and Burnham 1977) the member $N = 3$ of the biopyribole homologous (polysomatic) series.

In the polysomatic model, structures consist of pyroxene-like P modules (halves of silicate chains with backing of M1 octahedra and with the larger, less regular M2 polyhedra) and mica-like M modules ('double-tetrahedral' slices of the stack of **T–O–T layers**, that is, of layers of six-fold silicate rings with the backing of coordination octahedra and the relevant OH groups). One P module can be combined with one or more M modules, yielding the PP, PM, PMM, ... MM members, that is, **pyroxenes**, **amphiboles**, **jimthomsonite**, ... **biotite**.

The chemical formula of the P module is $M_4T_4O_{12}$, that of M module $AM_3T_4O_{10}(OH)_2$, giving the combinations $AM_7T_8O_{22}(OH)_2$ for PM and $A_2M_{10}T_{12}O_{32}(OH)_4$ for PMM (the A site is often vacant). Compared to pyroxene (**enstatite**) the b parameter is doubled in **anthophyllite**, tripled in **jimthompsonite** and quintupled in **chesterit**e (Strunz and Nickel 2001).

All these compounds occur in clino- and ortho-families; this difference furthermore may or may not be connected with the presence of large cations (Ca, Na) in the polyhedra lining the glide plane positioned centrally in the P module. However, with the third or higher homologues being Mg,Fe-based compounds, the standing polysomatic classification is not concerned with these details.

A polyhedral description of this series is generally known and used, based on single (Si_2O_6), double (Si_4O_{11}), triple (Si_6O_{16}), etc. chains (Liebau 1985; Strunz and Nickel 2001). A homologous description is obtained by shearing the **mica-like archetype** by means of periodically spaced diagonal glide planes, disrupting the tetrahedral sheets into stripes of different thickness and modifying the marginal coordination octahedron of the T–O–T fragment in the process. The opposite approach is possible as well: Zvyagin and Merlino (2003) sliced the pyroxene archetype by periodic diagonal glide planes, creating blocks of mica-like modules of desired thickness.

1.6 Chemical composition series

A persistent, although not frequent, misconception is the misunderstanding when 'chemical composition series' are called homologous series. In reality, a series of compositions combining formally two oxide or sulfide compositions A_mX_n–B_oX_p may not be a single homologous (polysomatic) series but it may contain one or two homologous series and/or several structurally unrelated compounds. Thus, the Cu_2S–Bi_2S_3 series contains wittichenite as well as (idealized) members of the **cuprobismutite** and of the **pavonite homologous series.**The PbS–Sb_2S_3 series contains plagionite homologues (Takéuchi 1997) as well as a plesioseries of **rod-layer structures** and individual rod-based structures of other types, ending with the most complicated box-work type. **Plagionite homologues** are only a subset of all these compositions, that is, only those which obey a restricted formula $\mathbf{Pb_{3+2x}Sb_8S_{15+2x}}$ with $x = 0, 1, 2$ or 3 (Takéuchi 1997). Finally, the MgO–SiO_2–H_2O compositions contain two distinct series with the general formulae in which n is the number of olivine-like slabs between two consecutive slabs of another type: (a) the **norbergite–olivine series $\mathbf{Mg_{2n+3}Si_{n+1}O_{4n+4}(OH)_2}$** and (b) the **leucophoenicite–olivine series $\mathbf{Mg_{2n+1}Si_nO_{4n}(OH)_2}$**, as well as the biopyribole series $\mathbf{Mg_{3N+1}Si_{4N}O_{10N+2}(OH)_{2N-2}}$ (N = chain multiplicity), **serpentinite–carlosturanite series** and the **high-pressure B-phase series $\mathbf{Mg_{7n+3}Si_{2.5n+0.5}O_{12n+2}(OH)_4}$** besides others, less prominent groups (e.g. **Mg-chlorites**).

1.7 Variable-fit homologous series and series with a combined character

1.7.1 *Definitions and principal families*

Variable fit homologous series occur in crystal structures that are composed of two kinds of alternating, mutually non-commensurate layers (rarely of such columns or of a matrix/infilling combination). Each kind of layer has its own short-range (sub)periodicity and it takes m periods of one layer and n periods of the other layer before they meet in the same configuration as at the origin (Fig. 1.21). These are so-called semi-commensurate cases for which the two layer (sub)periodicities comprise a ratio of two not very large integers. Besides them, incommensurate cases exist for which these short-range periodicities comprise irrational fractions or a ratio of very large integers (these two cases are practically indistinguishable). Non-commensurability of layers may occur in one or two interplanar directions, the former case is much more common, the other direction being commensurate or semi-commensurate with one of the simplest ratios. With minor compositional changes in the cations, the m/n ratio and the M/S ratio vary within certain, rather narrow limits, leading to a series of closely related compounds (Makovicky and Hyde 1981, 1992). **Incommensurate layer structures** are also known as **vernier compounds** (Hyde *et al.* 1974).

Non-commensurate crystal structures require fairly weak interactions between the two types of regularly alternating component layers or flexible coordination polyhedra for the elements with coordination spheres spanning the noncommensurate interface. Although they come from most diverse groups of inorganic compounds, common to

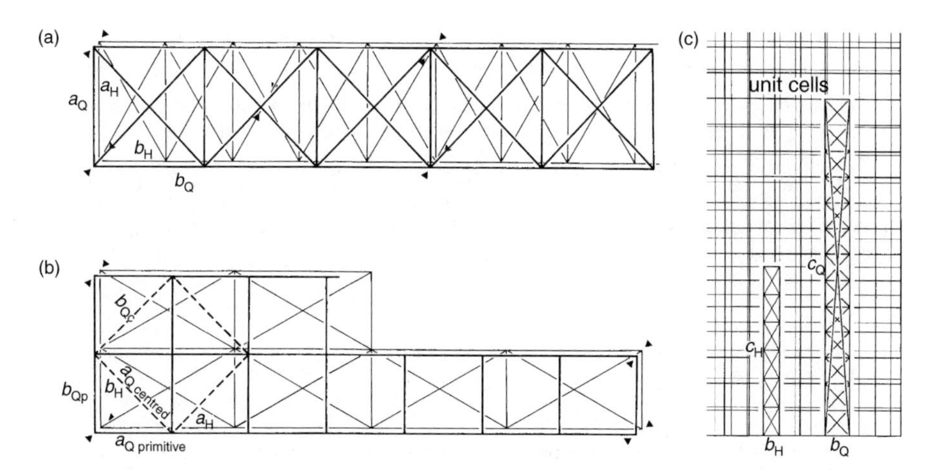

Fig. 1.21. Match of pseudotetragonal (Q) and orthohexagonal (H) subcells on non-commensurate interfaces in the crystal structures of (a) $\sim LaCrS_3$ and (b) $Er_9La_{10}S_{27}$ (an example of a 'cannizzarite-like' match) and (c) cylindrite-like match (not to scale). Note the centred Q submesh in (a) and (c), and the primitive Q submesh in (b).

most of them are polarizable anions such as S, OH, Cl, much less also F and O; and cations with adjustable coordination polyhedron such as Pb, REE, Zr and, with active lone electron pairs partaking in the interlayer configurations, also Bi^{3+}, Sn^{2+}, As^{3+} and Sb^{3+}.

The participating layers belong to fairly simple and mostly thin layer types:

1. The hexagonal 6^3 net in **graphite** and the triangular anion nets 3^6 in the lanthanide oxyfluoride group.
2. The fully or partly occupied **octahedral layer** $(111)_{ccp}$—mostly a one octahedron thick slice but also double and, possibly, triple octahedral layers are present.
3. The (100) slices of **PbS** (i.e. ccp) and **SnS-structure types**, two-to-six atomic planes thick.
4. The **'valleriite-type' trigonal layers** and the **'mackinawite-type' tetragonal layers** with (almost) all tetrahedral interstices of two-layer sulphur array filled by Fe, Cu and Ni.
5. A trigonal prismatic single-to-triple layer of coordination prisms of Nb and Ta in synthetic layer-misfit compounds.
6. A (100) slice of **fluorite-type structure**, in **lanthanide oxyfluorides**.

The spectrum of interlayer matches is fairly limited: most of them are hexagonal/hexagonal and (pseudo)hexagonal/(pseudo)tetragonal match. Three subtypes of the latter match are known: if the Q layers are described as (100) and the match is expressed in terms of centred pseudotetragonal and orthohexagonal ($c_H = b_H\sqrt{3}$ subcells, the first, **'cylindrite-like'** match implies (exactly or approximately)

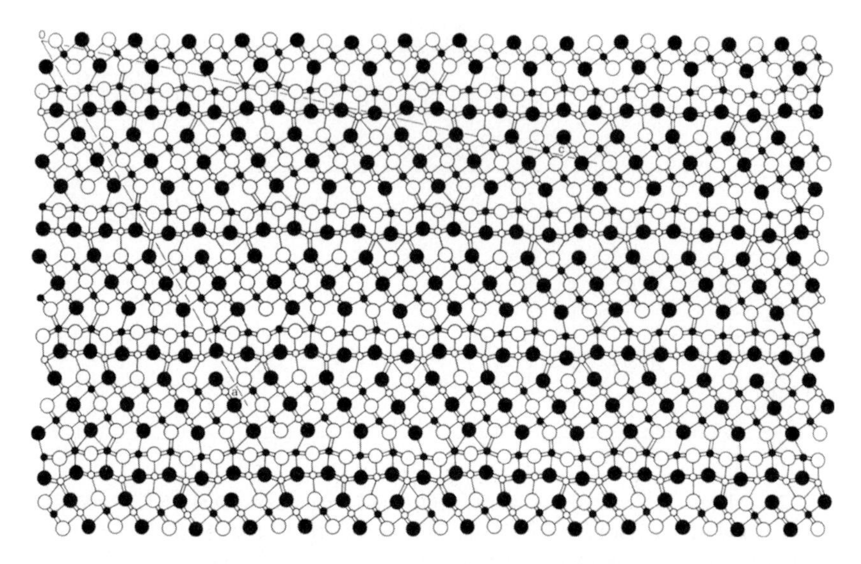

Fig. 1.22. The non-commensurate structure of cannizzarite, $Pb_{46}Bi_{54}S_{127}$ (Matzat 1979) projected upon (010). Large circles denote S, small circles undifferentiated metal atoms. Double octahedral H layers alternate with two-atoms thick Q, PbS-like layers. A total of 46 primitive Q subcells are commensurate with 27 centred H subcells.

$b_Q \approx b_H$ and $c_Q \approx c_H$ with $b_Q > b_H$ and $c_Q < c_H$ (e.g. in **cylindrite**—$FePb_3Sn_4Sb_2S_{14}$, $19b_Q \approx 30b_H$ and $13c_Q = 12c_H$, common modulation of both layers $\parallel c$). The second, **'$LaCrS_3$-like' match** implies $c_Q = c_H$ and $b_Q > b_H$ (b is the semi- or incommensurate modulation direction, for example, $3b_Q = 5b_H$ for ~ $LaCrS_3$). The third match involves Q layers turned by 45° so that $b_H \parallel$ (and equal to) $b_{Q\text{primitive}}$ ($= \frac{1}{2}$ diagonal of Q centred) and $c_H > c_{Q\text{primitive}}$ (e.g. **cannizzarite** ~ $Pb_{24}Bi_{28}S_{66}$ with $12c_Q \approx 7c_H$). The $LaCrS_3$-like match occurs for the largest Q/H cation radius ratio whereas, as the other extreme, the **cannizzarite match** is typical for about equally large average cations in the Q and H layers (Fig. 1.22).

Match ratios vary from incommensurate for chemically (occupationally) simple component layers with weak interactions to semicommensurate, 'lock-in' cases for chemically complex layers in which local stronger interactions can be assumed. The source of non-commensurate to semi-commensurate layer modulation will also vary accordingly.

The former category yields only **heterochemical** (configurational) **variable-fit homologous series**. For the 'ABX_3' compounds with trigonal prismatic layers (BX_6 prisms), the iso- and homeotypes of '$PbTaS_3$' show layer matches (i.e. stoichiometries) and a_Q/a_H ratios as follows:

$(PbS)_{1.13}TaS_2$ (5.803/3.306 Å), **$(SmS)_{1.19}TaS_2$** (5.562/3.292 Å), **$(SnS)_{1.15}TaS_2$** (5.739/3.311 Å), and **$(YS)_{1.23}NbS_2$** (5.393/3.322 Å). The other interlayer subcell axes, b, are equal for the two component layers (Table 1.8).

Table 1.8 Selected 'ABS_3' layer misfit sulfides (A = lanthanide)

Compound	Component[a]	Subcell data						Symmetry	Match Q/H	Reference
		a (Å)	b (Å)	c (Å)	α (Å)	β (Å)	γ (Å)			
$(LaS)_{1.20}CrS_2$[b]	Q = LaS	5.94	5.75	1.04	90.3	95.3	90	Triclinic	1.675	Makovicky and Hyde (1992)
	H = CrS_2	5.94	3.44	1.05	93.3	95.3	90	$C\bar{1}$		
$(CeS)_{1.14}NbS_2$[c]	Q = CeS	5.77	5.73	1.41	Orthorhombic			$Cm2a$	1.731	Kuypers *et al.* (1990)
	H = NbS_2	5.77	3.31	2.81	Orthorhombic			$Fm2m$		
$(LaS)_{1.13}TaS_2$[c]	Q = LaS	5.77	5.81	1.53	Orthorhombic			$Cm2a$	1.764	Makovicky and Hyde (1992)
	H = TaS_2	5.77	3.3	3.06				$Fm2m$		
$(SmS)_{1.19}TaS_2$[c]	Q = SmS	5.65	5.56	2.56	Orthorhombic			$Fm2m$	1.69	Kuypers *et al.* (1990)
	H = TaS_2	5.65	3.29	2.56				$Fm2m$		

[a] Q = pseudotetragonal component layer; H = pseudohexagonal component layer.

[b] Incommensurate $Ce_{1.203}CrS_{3.203}$, $Pr_{1.220}CrS_{3.220}$, $Nd_{1.231}CrS_{3.231}$, '$LaTiS_3$', '$LaVS_3$' and '$\sim ErCrS_3$' are analogous in the [001] zone. Octahedral H layers are present.

[c] a and b axes are interchanged against the original notation in order to match with '$LaCrS_3$' and others. Trigonal prismatic H layers are present; $(CeS)_{1.14}TaS_2$, $(LaS)_{1.14}NbS_2$ (Kuypers *et al.* 1990) and $(YS)_{1.23}NbS_2$ (Q/H = 1.623) were described as well. Synopsis in Makovicky and Hyde (1981, 1992).

Similar cases exist for octahedral B-based layers: here, for B = Cr, the b_Q/b_H ratio decreases with the radius of the lanthanide cation from $\mathbf{La_{1.200}CrS_{3.200}}$ to $\mathbf{Nd_{1.231}CrS_{3.231}}$ (Table 1.8).

It is the latter, chemically complex category that forms true variable-fit series. For cannizzarite, $\sim Pb_{48}Bi_{54}S_{127}$, Matzat (1979) found match of 46 subcells of the pseudotetragonal (Q) layer with 27 centred orthohexagonal subcells of the pseudo-hexagonal (H) layer (Fig. 1.22). There is a number of near-matches between the two layer sets in this model. Should these be fully materialized, the k match coefficient in the ratio $MeS \cdot kMe_2S_3$ (0.587 for the above case) will alter as follows: for 12Q/7H, the composition is $Pb_{12}Bi_{14}S_{33}$, that is, $k = 0.583$; for 17Q/10H, that is, $Pb_{17}Bi_{20}S_{47}$, $k = 0.588$. For the very simple match, perhaps not attainable by the structure, 5Q/3H, that is, $Pb_5Bi_6S_{14}$, $k = 0.600$. In **cannizzarite**, the H layer is two octahedra thick, in the synthetic phase of Graham *et al.* (1953), it is presumably three octahedra thick, whereas in the $\mathbf{Cr_2Sn_3Se_7}$ polymorph (Jobic *et al.* 1994), it is a single-octahedral layer. A Q layer of the same thickness occurs in all these phases (Table 1.9).

Variability in anions, instead of cations, creates the **yttrium oxyfluoride variable-fit series** $\mathbf{Y_nO_{n-1}F_{n+1}}$. The two types of alternating layers are the pseudotetragonal (100) layers of fluorite type and simple 3^6 nets of anions (Hyde *et al.* 1974).

Recognized matches vary from $\mathbf{Y_5O_4F_7}$ to $\mathbf{Y_8O_7F_{10}}$ (or $\mathbf{YX_{2.12}}$ to $\mathbf{YX_{2.22}}$; X = O, F), that is, n = 5–8, or combinations thereof. In these structures, n Q subcells match with $n + 1$ H subcells (Fig. 1.23). Similar variable-fit series $Ln(O,F)_x$ are known for other lanthanides with the exception of the largest lanthanides. Furthermore, they are also known for zirconium oxyfluorides, zirconium–uranium oxides and zirconium/uranium nitride–fluorides. The large structure of $\mathbf{Zr_{108}N_{98}F_{138}}$ (Jung and Juza 1973) with 27 Q subcells of the fluorite-like structure matching 32 H subcells is an example.

1.7.2 *Doubly non-commensurate example*

Non-commensurability in two dimensions is typical for the chemically complex structures of the **cylindrite–franckeite family** of Pb, Sn^{2+}, Sb, Fe sulfides, rather recently broadened to include similar compounds with Bi (and Cu), or As. In this family two- or, alternatively, four atomic layers thick layers based on the **SnS archetype** alternate regularly with essentially SnS_2 layers. Minor elements, Sb (Bi,As), and Fe (Cu) are believed to be distributed over both layer types.

The type compound, cylindrite, with average composition $FePb_3Sn_4Sb_2S_{14}$ (Table 1.10), has the coincidence mesh in terms of b and c parameters of pseudotetragonal subcells 19,0/0,13 whereas in terms of orthohexagonal subcells 30,0/0,12. Layers are stacked along a. Common geometrical and compositional modulation occurs along [001], resulting in the fixed semi-commensurate match 13 Q/12 H. Along [010], there is an incommensurate match, expressed as the empirical ratio 19 Q/30 H without modulation, that is, the layers only have the repetition periods 5.79 and 3.67 Å, respectively (Makovicky 1976).

Table 1.9 Selected cannizzarite homologues

Mineral	Formula	Cell type	a (Å)	b (Å)	c(Å)	β	Space group	Reference
Cannizzarite	$Pb_{46}Bi_{54}S_{127}$[a]	Unit cell	189.8	4.09	74.06	11.9	$P2_1/m$	Matzat (1979)
		Q subcell	4.13	4.09	15.48	98.6	$P2_1/m$	
		H subcell	7.03	4.09	15.46	98.0	$C2/m$	
Synthetic	$Pb_{38}Bi_{28}S_{80}$[a]	Q subcell	4.11	4.08	18.58	93.6	$P2/m$[b]	Graham *et al.* (1953)
		H subcell	7.03	4.08	27.16	93.6	$F2/m$	
		Unit cell not determined						
Synthetic	$Cr_2Sn_3Se_7$	Unit cell	12.77	3.84	11.79	105.2	$P2_1/m$	Jobic *et al.* (1994)

[a] The compositions are derived from the crystal structure.

[b] Probably $P2_1/m$ (Matzat 1979).

(a)

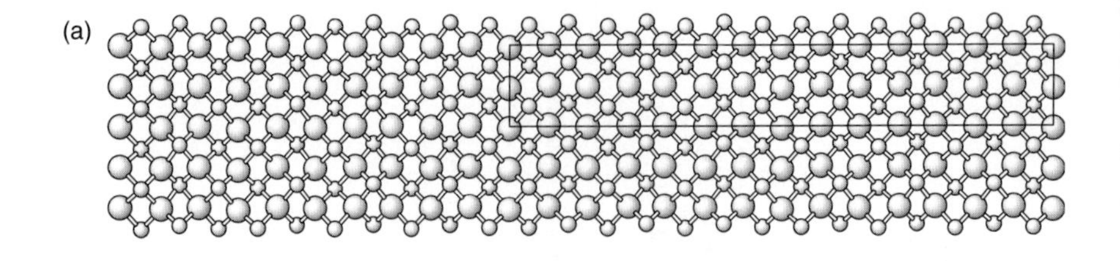

(b)

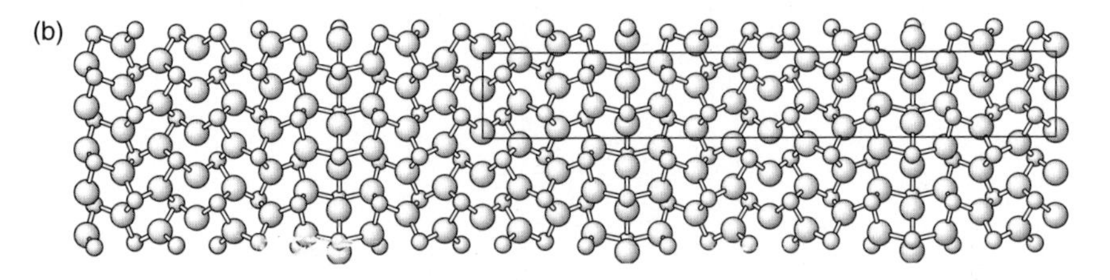

Fig. 1.23. Crystal structure of $Y_7O_6F_9$ (Bevan *et al.* 1990). Separate presentation of pseudotetragonal anion layers at $x = 0.5$ and pseudohexagonal layers at $x = 0.0$. In both cases, cations on both sides of an anion layer are included.

Pseudotetragonal layers are essentially $(Pb,Sn^{2+})S$ with substitutions of minor elements; with the change in the $Pb:Sn^{2+}$ ratio in these layers, a change in the c_Q/c_H ratio is connected, giving matches from 12 Q/11 H to 16 Q/15 H; variation in the ratio of b parameters was not systemized.

If the latter ratio is fixed in a simplified way as $7b_Q = 11b_H$, the 12/11 ratio means a composition $(12 \times 7)Me_4S_4 + (11 \times 11)Me_2S_4$, resulting in the Me_Q/Me_H ratio of 1.388 and Me/S ratio of 0.7048. In the same way, the ratio 16/15 gives Me_Q/Me_H ratio equal to 1.358 but the Me/S ratio of 0.7022, that is, a minuscule change in the resulting valence distribution in this variable-fit series (Makovicky and Hyde 1992).

Franckeite (Table 1.10) is closely related to **cylindrite** and displays all the above variations and a similar variable-fit series (Henriksen *et al.* 2002). It is an accretion homologue of cylindrite: the Q layers were two atomic layers thick in the latter, whereas they are four such layers thick in franckeite, stacked in a SnS-like fashion. **Lengenbachite**, approximating $Pb_{182}Ag_{45}Cu_{20}As_{117}S_{390}$ is a closely related structure with the semi-commensurate match fixed at 12 Q/11 H subcells in the c direction and 2 Q/3 H subcells in the b direction.

The match of the 13 Q subcells tall A-centred Q cell with two 6 H subcells tall primitive hexagonal cell in cylindrite results in indistinct OD character with two positions of the Q layer. This becomes pronounced for lengenbachite in which a similar situation occurs. It is caused by coincidences due to the combination of the 12 c_Q/11 c_H match with the 2 b_Q/3 b_H match. It results in a complete disorder of the H layer (Makovicky *et al.* 1994).

1.7.3 *Structures with a combination of accretional and variable-fit principles*

In layered misfit structures, bonds at layer surfaces will be strained or local valence balance is expected to fluctuate periodically (or quasiperiodically) along the

Table 1.10 Selected crystallographic data on the members of the cylindrite-franckeite series

Compound		Q component		H component		Coincidence data	Reference
		Subcell	Unit cell	Subcell	Unit cell		
Cylindrite	a	11.73 Å	11.73 Å	11.71 Å	11.71 Å	13Q/12H	Makovicky (1976)
~ $FePb_3Sn_4Sb_2S_{14}$	b	5.79 Å	5.79 Å	3.67 Å	3.67 Å		
	c	5.81 Å	75.53 Å	6.32 Å	37.92 Å		
	α	90°	90°	90°	90°		
	β	92.38°	92.38°	92.58°	92.58°		
	γ	93.87°	93.88°	90.85°	90.85°		
	Space group	$A1$	$A1$	$A1$	$P1$		
Potosiite	a	17.28 Å	17.28 Å	17.28 Å	17.28 Å	16Q/15H	Wolf *et al.* (1981)
~ $Pb_{24}Ag_{0.2}Sn_{8.8}$	b	5.84 Å	5.84 Å	3.70 Å	3.70 Å		
$Sb_{7.8}Fe_{3.74}S_{55.6}$	c	5.88 Å	188.06 Å	6.26 Å	188.06 Å		
	α	90°	90°	90°	90°		
	β	92.2°	92.2°	92.2°	92.2°		
	γ	90°	90°	90°	90°		
	Space group	$A1$ or $A\bar{1}$	$P1$ or $P\bar{1}$	$A1$ or $A\bar{1}$			
Franckeite natural	a	17.3 Å		17.3 Å		16Q/15H	Wang (1989)
	b	5.84 Å		3.68 Å			
	c	5.90 Å		6.32 Å			
	α	91°		91°			

	β	95°		96°			
	γ	88°		88°			
	Space group	$A\bar{1}$		$A\bar{1}$			
Lévyclaudite subcell	a	11.84 Å		11.84 Å		13Q/12H	Moëlo *et al.* (1990)
	b	5.83 Å		3.67 Å			
	c	5.83 Å		6.31 Å			
	α	90°		90°			
	β	92.6°		92.6°			
	γ	90°		90°			
	Space group	$A2/m$, $A2$ or Am		$A2/m$, $A2$ or Am			
Langenbachite	a	36.89 Å	36.89 Å	36.89 Å	36.89 Å	12Q/11H	Makovicky *et al.* (1994)[a]
	b	5.84 Å	11.68 Å	3.90 Å	11.68 Å		
	c	5.85 Å	70.16 Å	6.38 Å	70.16 Å		
	α	90°	90°	90°	90°		
	β	90°	90°	90°	90°		
	γ	91°	91°	91°	91°		
		A2/*m*, A2 or A*m*	A centred	A2/*m*, A2 or A*m*	A centred		

[a] Also other variants and stacking disorder of the H component are reported.

non-commensurate lattice direction. Layer mismatch will cause tension in one layer set and compression in the alternative layer set. In more complicated cases, each of the layer types will periodically or quasi-periodically experience regions of tension and compression.

When exceeding critical values, stresses and strains as well as the charge balance problems in the non-commensurate layer structures can be relieved by breaking them up by means of glide planes, antiphase or out-of-phase boundaries, or by exposing alternatively the Q and the H surfaces of the same layer. With respect to the rest of the structure, these planes of break-up may be composition-and-structure-conservative or non-conservative. In some cases, only the ratios of elements present may be altered; in other cases new, minor elements are introduced.

The possible situations are schematically illustrated in Fig. 1.24. Here, the non-commensurate structure has the fully outlined set of layers in compression and the alternating set of layers (dotted lines) in tension. The commensurate solutions are (i) composition-conservative breakup along antiphase boundaries or glide planes, (ii) composition non-conservative boundaries of the same type, with one type of layers overlapping in the boundary zone, (iii) the same with the composition/configuration non-conservative boundaries of a type different from any of the layer configurations, and (iv) alternation of two types of surfaces along each layer, alternating the sense of incommensurate interface in the process, and (v) formation of stepped out-of-phase boundaries by slip or crystallographic shear (Fig. 1.54).

A case with non-conservative boundaries having chemistry completely different from that of layer set is, for example, $Cu_3Bi_2S_4Cl$ (Lewis and Kupčík 1974); several further examples occur among rare-earth sulphides (Makovicky 1992*b*).

Those cases for which the pile of the two alternating non-commensurate layers is periodically sheared, kinked, or modified by antiphase (out-of-phase) boundaries or glide planes, represent combination of the accretional (separately for each layer) and variable-fit (for the interlayer match) principles. Therefore, the geometrical constraints in these structures are much more severe than in the pure variable-fit structures and result, with the exception of $(Nb/Ta)_2Zr_{n-2}O_{2n+1}$, in a very reduced number of homologues (usually only in pairs of homologues). These cases were not considered as variable-fit structures until the publication by Makovicky and Hyde (1981).

1.7.3.1 *Derivatives of cannizzarite*

Cannizzarite (Fig. 1.22) acts as a parent structure for the majority of such phases in the realm of sulphides (Makovicky and Hyde 1981; Makovicky 1981, 1992*a*) (Table 1.11). They represent piles of alternating Q- and H-type layers with different thicknesses (not always the layer thicknesses observed in cannizzarite itself); they have similar match modes as cannizzarite and similar average cation radii. The pairs **$PbBi_2S_4$** (Iitaka and Nowacki 1962)–**$Ag_{0.33}Pb_{5.33}Bi_{8.33}(S,Se)_{18}$** (Mumme 1980*b*) and **$PbBi_2S_4$–$Pb_4In_3Bi_7S_{18}$** (Krämer and Reis 1986) are interesting cases in which two potential homologous series intersect.

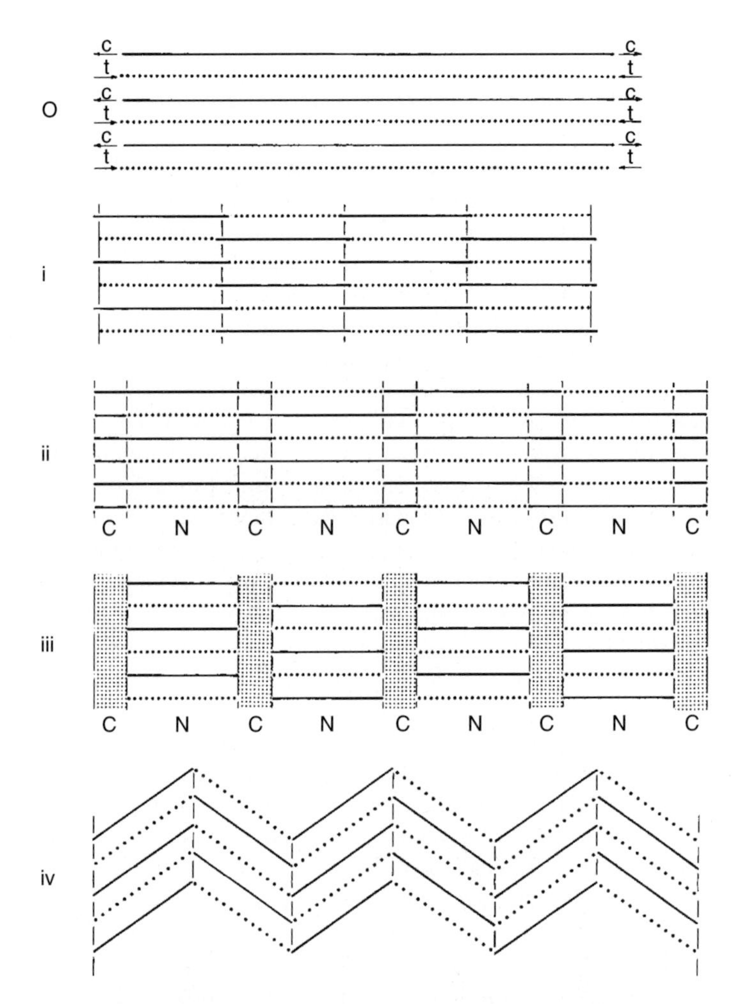

Fig. 1.24. Schematic representation of non-commensurate layer stacks (continuous and stippled lines, respectively) and their modification by (i) composition-conservative antiphase boundaries/glide planes, (ii) composition non-conservative such planes, (iii) non-conservative interlayers, and (iv) compensation of differences by alternation of pseudohexagonal and pseudotetragonal intervals in the same layer. For step-like modifications see Fig. 1.54.

The structure of **weibullite** $Ag_{0.33}Pb_{5.33}Bi_{8.33}(S,Se)_{18}$ (Mumme 1980*a*) is the most obvious derivative of cannizzarite by shearing by means of composition non-conservative glide planes (Fig. 1.25). That of $\mathbf{Pb_4In_3Bi_7S_{18}}$ (Krämer and Reis 1986) is based on double as thick pseudotetragonal layers and only single-octahedral layers; the out-of-phase boundaries are underlain by thicker non-conservative slabs (Fig. 1.26).

Table 1.11 Selected plesiotypes based on cannizzarite structure type

Mineral	Formula	Lattice parameters (Å)			β	Space group	Reference
Galenobismutite	$PbBi_2S_4$	*a* 11.79	*b* 14.59	*c* 4.10	–	*Pnam*	Iitaka and Nowacki (1962)
Weibullite	$Ag_{0.33}Pb_{5.33}Bi_{8.33}(S,Se)_{18}$	*a* 53.68	*c* 15.42	*b* 4.11	–	*Pnma*	Mumme (1980*b*)
Synthetic	$Pb_4In_3Bi_7S_{18}$	*a* 21.02	*b* 4.01	*c* 18.90	97.1	$P2_1/m$	Krämer and Reis (1986)
Junoite[a]	$Cu_2Pb_3Bi_8(S,Se)_{16}$	26.66	4.06	17.03	127.2	$C2/m$	Mumme (1975*b*)
Felbertalite[a]	$Cu_2Pb_6Bi_8S_{19}$	27.64	4.05	20.74	131.3	$C2/m$	Topa *et al.*(2000*b*)
Proudite	$Cu_xPb_{7.5}Bi_{9.67-0.33x}(S.Se)_{22}$	31.96	4.12	36.69	109.5	$C2/m$	Mumme (1976)
Nordströmite	$CuPb_3Bi_7(S.Se)_{14}$	17.98	4.11	17.62	94.3	$P2_1/m$	Mumme (1980*b*)
Neyite	$Ag_2Cu_6Pb_{25}Bi_{26}S_{68}$	37.53	4.07	43.70	108.8	$C2/m$	Makovicky *et al.* (2001*a*)
Synthetic	$Pb_3In_{6.67}S_{13}$[b]	*a* 38.13	*c* 3.87	*b* 13.81	γ 91.3	$B2/m$	Ginderow. (1978)
Synthetic	$Bi_3In_5S_{12}$	33.13	3.87	14.41	91.2	$C2/m$	Krämer (1980)

[a] Homologues $N = 1$ and $N = 2$, respectively, of the junoite homologous series.
[b] 1/3 *Me* is vacant in one metal site.

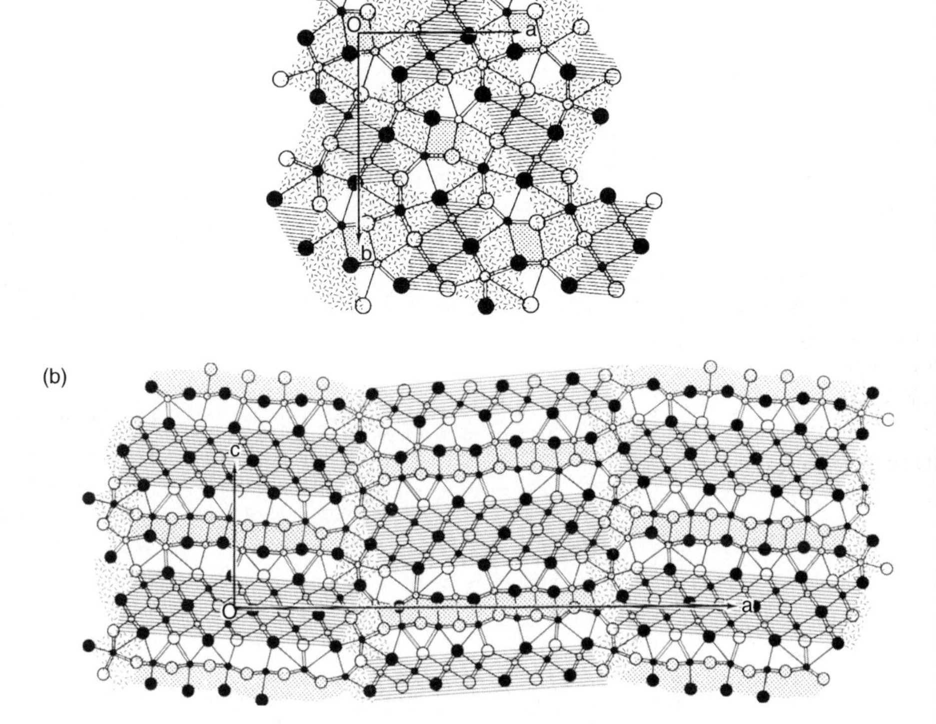

Fig. 1.25. The homologous pair $PbBi_2S_4$–$Ag_{0.33}Pb_{5.33}Bi_{8.33}(S, Se)_{18}$ (galenobismutite–weibullite). $(111)_{PbS}$ slabs are hatched, $(100)_{PbS}$ are stippled and configurations created by interface modulation dashed.

For both phases, the nearest lower homologue of the series with combined character is **galenobismutite** $PbBi_2S_4$. This structure combines pairs of distorted octahedra of Bi, those of 'lying' monocapped trigonal prisms of Bi and 'standing' bicapped trigonal prisms of Pb; it shows affinity to that of $CaFe_2O_4$ (Mumme 1980*a*). In the **'weibullite-like' expansion** (parallel to *a* of galenobismutite) the half subcell broad Q module expands to five-and-half subcells whereas the half H subcell (i.e. one octahedron) broad pseudohexagonal module expands to three-and-half H subcells long layer fragment (Fig. 1.25). In the 'Pb–In–Bi sulphide-like' expansion, perpendicular to $(\bar{1}10)$ of **galenobismutite** (Fig. 1.26), it is only the pairs of Bi octahedra that expand. One set of them expands from one Q subcell to four Q subcells, the alternative set from one H-subcell width to that of three subcells. In both cases, the width of pseudohexagonal modules exceeds somewhat that of pseudotetragonal ones: the differences are compensated by shearing of the structure and 'buffering' by means of non-conservative boundaries.

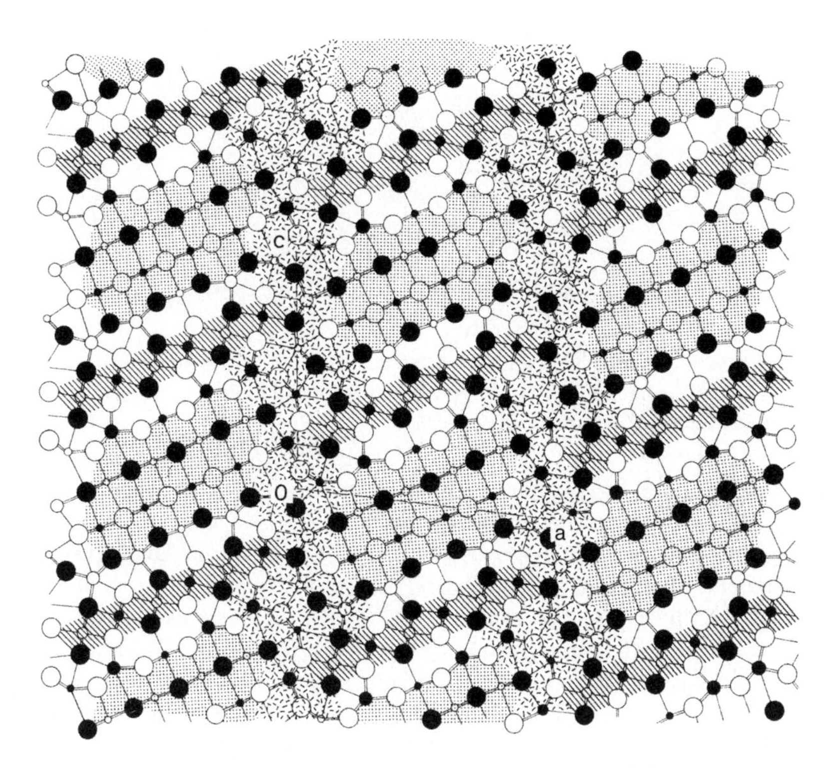

Fig. 1.26. The crystal structure of $Pb_4In_3Bi_7S_{18}$ (Krämer and Reis 1986), the higher homologue of galenobismutite. In the order of decreasing size, circles represent S, Pb, In and Bi. Pseudotetragonal slabs are stippled, pseudohexagonal layers ruled. They are separated by the chemically non-conservative boundaries (dashed).

1.7.3.2 *Sheared structures of lanthanide halides*

The non-commensurate structures of **lanthanide oxyfluorides**, $Y_nO_{n-1}F_{n+2}$ (Fig. 1.23) (Bevan and Mann 1975; Makovicky and Hyde 1981, 1992) are parents to a series of sheared structures of **lanthanide halides** (Bärnighausen 1976, for a synopsis see Makovicky and Hyde (1992)). All these structures represent piles of alternating **fluorite-type layers** and 3^6 nets. Only semi-commensurate cases Ln_nX_{2n+1} ($4 \leq n \leq 7$) are known for the halides. Examples are $\mathbf{Yb_5ErCl_{13}}$ (i.e. Ln_6Cl_{13}) (Lüke and Eick 1982), a homologue (Fig. 1.27) with pseudotetragonal strips between adjacent shear planes three Q subcells wide (and the matching pseudohexagonal strips three-and-half centred subcells wide), and the next lower homologue, $\mathbf{Dy_5Cl_{11}}$ (Bärnighausen 1976) in which these strips are two-and-half Q subcells and three H subcells wide. $\mathbf{Eu_4Cl_9}$ has only two Q subcell wide strips whereas Dy_7Cl_{15} has three-and-half Q subcells wide strips (listings of these and related phases are in Makovicky and Hyde (1992). Only **semi-commensurate structures** were ascertained in this family (even $Sm_{11}Br_{24}$, an intergrowth of two-and-half Q and three Q structure, is

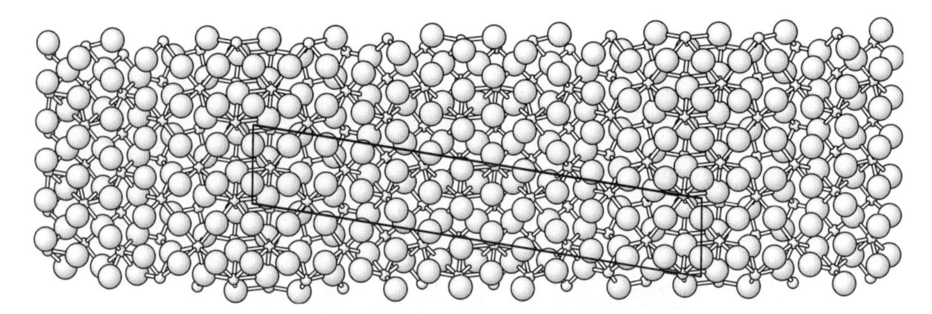

Fig. 1.27. Crystal structure of Yb_5ErCl_{13} (Lüke and Eick 1982). The hexagonal 3^6 net is in foreground in the central portion of the figure, the tetragonal fluorite-type net is in foreground in the portions about $x = 0$ and 1, respectively.

semi-commensurate) and two-phase regions exist between two adjacent members of this, to a certain degree configurational family.

The crystal structures of $\mathbf{Nb_2Zr_6O_{17}}$ (Galy and Roth 1973) and $\mathbf{Ta_2Zr_8O_{21}}$ (Galy, oral communication) typify the case with H strips overlapping in the shear regions. They are, respectively, the $n = 8$ and $n = 10$ members of the series $\mathbf{M_2Zr_{n-2}O_{2n+1}}$ with M = Nb or Ta. The former has Q strips three-and-half Q subcells wide, matching with four subcells of H strips. However, the latter strips are five H subcells wide and in each antiphase boundary region, there is an overlap of half H, with the commensurate H–H match. The Nb-based series is continuous in the range $7.1 < n < 10.2$; no two-phase regions have been detected in this range (Thompson *et al.* 1990).

1.7.3.3 *Structures with step-like shear*

The last type of the sheared non-commensurate structures are those modified by steps in pseudohexagonal layers, often with their partial overlap by the width of one or more octahedra; and by similar, more complex steps in the pseudotetragonal layers. Typical is junoite $Cu_2Pb_3Bi_8(S,Se)_{16}$ (Mumme 1975*a*) (Fig. 1.54), also dealt with in the Section 1.8.2.8 on 'sliding series'. Single-octahedral layers with five octahedra long intervals and one octahedron long overlap are matched by three Q subcells wide fragments of pseudotetragonal layers that are two atomic layers thick. Steps in the latter are occupied by paired Cu tetrahedra. Accretional homologue, **felbertalite** $Cu_2Pb_6Bi_8S_{19}$ (Topa *et al.* 2000*b*, 2001) has double octahedral layers. Pavonite homologue $N = 3$ (Tomeoka *et al.* 1980) is the only known lower homologue of the series with single octahedral layers; it can be described as three octahedra long octahedral fragments (with one octahedron broad overlap) combined with one Q subcell broad pseudotetragonal fragments; Cu coordination in the step portions of the Q layer corresponds only partly to that in junoite. The two shear principles—steps and composition non-conservative boundaries join in **neyite** $Ag_2Cu_6Pb_{25}Bi_{26}S_{68}$ (Makovicky *et al.* 2001*a*) with triple octahedral layers and two atomic layers thick pseudotetragonal slabs (Fig. 1.28).

Fig. 1.28. Crystal structure of neyite $Ag_2Cu_6Pb_{25}Bi_{26}S_{68}$ (Makovicky *et al.* 2001*a*) with two types of blocks indicated by ruling, stippling and the subvertical layers left unshaded. Circle fill indicates two levels of the 4 Å structure, in the order of decreasing size circles describe S, Pb, Bi, Ag (linear) and Cu.

1.8 Merotype and plesiotype series

1.8.1 *Definitions*

On a modular level, a number of structural relationships exist between recombination structures or between the distinct homologous series of such structures. These relationships are geometrically less accurate than the relationships between members of a homologous series. Sometimes, they have been treated by redefining homologous series on a 'higher level', that is, by removing some configuration/coordination features from the set of conditions valid for all its members. In other cases, these structures were attached to the series as an 'appendix' (Strunz and Tennyson 1978; Strunz 1993) or relegated to the category of 'related structures' (Makovicky 1989; Strunz and Nickel 2001). Such treatment does not allow construction of a proper multi-level **hierarchical structural classification**. Therefore, Makovicky (1997*a*) introduced two higher categories of structural relationships that address most of the classification problems we have encountered above the level of homologous series; their usefulness was tested, for example, by Ferraris (1997) and Christiansen *et al.* (1999).

Merotypic structures (merotypes; meros = part) are composed of alternating layers (blocks), in the same way as do the homologous series. However, one set of these building layers (blocks) are common to all merotypes (i.e. they are isotypic, homeotypic or they are mutually related via homologous expansion/contraction) whereas *the layers (blocks) of the other set(s) differ for different merotypes.* Homologous series may occur as subsets of merotypic families; members with the second layer set missing may sometimes exist as well.

Plesiotypic structures (plesiotypes; plesios = near, close) form a group that is built on the same overall principles. It means that (a) they contain fundamental structural elements (blocks, layers) of the same general type(s) and (b) mutual disposition/interconnection of these elements in all plesiotypes follows the same general rules. However, unlike the homologous series, (1) the plesiotypic structures may contain additional structural elements that differ from one member of the family to another, (2) details of fundamental elements may also differ between distinct members of one plesiotypic family; within this family, such elements may be interrelated by means of homologous or non-homologous expansion, truncation, slip planes slicing them or in their interior, etc., (3) details of the relationships defined in (2) may differ as well. Diagrams expressing various degrees and kinds of kinship can be constructed for a plesiotype family.

Merotypic and plesiotypic families rank higher than the homeotypic, ordering or chemical variations between individual real structures. Merotypes and homologous series (homologous pairs) can enter a plesiotype family as subsets. The lower limits of plesiotypism are fairly obvious—the next nearest level above homology and, eventually, merotypy—but its upper limits are open to practical interpretation in exactly the same way as for homeotypism (Lima-de-Faria *et al.* 1990). The upper cut-off level should depend on the purpose and level/degree of detail of a particular investigation/classification.

Similar to the homologous (polysomatic) series, the mero- and plesiotypic families can sometimes be defined not only at the level of quaternary and quinary configurations but also at the level of secondary and tertiary configurations, for example, the polyhedral classifications used for silicates.

Plesiotypic families can be constructed both for special *ad hoc* needs (e.g. for all Ca silicates or for all those oxides, sulphides, etc. that are variously related to the **lillianite** structure principle) and for the purpose of general crystal chemical classification of inorganic compounds where their natural tendency to transcend the classical chemical boundaries requires careful evaluation of each individual case. A number of merotypic/plesiotypic families exist in literature, hidden under the name of 'homologous' or 'polysomatic' series. In these cases, authors felt clearly that there was a classification category (a group of similar structures) to be defined but, for a lack of better name, they resorted to a loose definition of homology/polysomatism, thus endangering its definition and proper use.

Examples of merotype and plesiotype families abound among complex sulphides, oxides and hydroxides; among oxysalts they can be traced especially among silicates. Examples given below do not exhaust this variety, they only illustrate several typical cases of large families or only small interesting groups with such relationships.

1.8.2 *Examples of merotypy*

1.8.2.1 *Hutchinsonite merotypes*

Hutchinsonite merotypes (Makovicky 1997*b*) is a group of complex sulphides of As or Sb and large uni- and divalent cations (Tl^+, Pb^{2+}, Na^+, Cs^+, NH_4^+, and others, including organic cations (Table 1.12)). Their structures are regular 1 : 1 intergrowths of slabs A which can be described as $(010)_{SnS}$ cut-outs of different widths from the SnS-archetype or $(110)_{PbS}$ cut-outs from the **PbS-archetype**, with layers B of variable thickness and configuration, based primarily on MeS_3 pyramids (Me = As, Sb) with active lone electron pairs which are combined with coordination polyhedra of large (eventually organic) cations. Slabs A and B share certain S atoms in common.

Slabs A are built according to the above common principles in all these structures, whereas the B slabs may differ, being always adapted to the requirements of large cations. In rare cases, B slabs are reduced out as in edenharterite $TlPbAs_3S_6$ and jentschite $TlPbAs_2SbS_6$ (Balič Žunič and Engel 1983; Berlepsch 1996; Berlepsch *et al.* 2000). The width of SnS-like layers has been quantified by giving *N* values independently to the two opposing sides of the SnS-like tightly bonded strips (Makovicky 1989) (Fig. 1.17(*a*,*b*)). This scheme can equally well be applied to PbS-like tightly bonded strips, for example, $Rb_2Sb_8S_{12}(S_2) \cdot 2H_2O$ is $N_{1,2} = 1; 2$ (Berlepsch *et al.*, 2001*d*) (Fig. 1.29).

In the layers based on PbS-archetype, structural allowances are made for a more active role of lone electron pairs. In a number of these structures, the pattern of short Sb–S bonds forms chains composed of corner-sharing SbS_3 coordination pyramids. This type of chains is found in $\mathbf{(NH_4)_2Sb_4S_7}$ (Dittmar and Schäfer 1977), **gerstleyite** $Na_2(Sb, As)_8S_{12} \cdot 2H_2O$ (Nakai and Appleman 1981), **gillulyite** $Tl_2(As, Sb)_8S_{13}$ (Foit *et al.* 1995; Makovicky and Balič Žunič 1999), $\mathbf{Rb_2Sb_8S_{12}(S_2) \cdot 2H_2O}$ (Berlepsch *et al.* 2001*d*), $\mathbf{[C_4H_8N_2][Sb_4S_7]}$ (Parise and Ko 1994), and $\mathbf{[C_2H_8N]_2[Sb_8S_{12}(S_2)]}$ (Tan *et al.* 1996). In all these cases, via weak Sb(As)–S interactions, the same A slab configuration arises, that of distorted PbS-like motif $N = 1, 2$ with the $N = 2$ layers containing additional, periodically sideways inserted SbS_3 groups and with the coordination polyhedra of trapezoidal cross-section.

Tan *et al.* (1996) classified interconnections of A slabs across the B slabs as follows: the lateral Sb–S chains of A slabs can be (1) isolated as in $[C_4H_8N_2]Sb_4S_7$, (2) linked via common sulphurs $\mathbf{[C_2H_{10}N_2]Sb_8S_{13}}$ (Tan *et al.* 1994), (3) linked via disulphide groups $[C_2H_8N]_2Sb_8S_{12}(S_2)$, and $Rb_2Sb_8S_{12}$, or (4) via additional SbS_3 groups $[C_3H_{12}N_2]Sb_{10}S_{16}$ (Wang 1995).

Imhofite $Tl_3As_{7.66}S_{13}$, **hutchinsonite** $TlPbAs_5S_9$, **bernardite** $TlAs_5S_8$, **edenharterite** $PbTlAs_3S_6$, **jentschite** $PbTlAs_2SbS_6$, **pääkkönenite** $Sb_2(As,Sb)S_2$, and

Table 1.12 Selected hutchinsonite merotypes

Compound/ mineral	Homologue no.[a]	Chemical formula	Lattice parameters			Space group	References
Hutchinsonite	2;3	$TlPbAs_5S_9$	*a* 10.79	*b* 35.39	*c* 8.14	*Pbca*	Matsushita and Takéuchi (1994)
Bernardite	2;2	$TlAs_5S_8$	*c* 10.75	*a* 15.65	*b* 8.04	$P2_1/c$	Pašava *et al.* (1989)
				β 91.27			
Imhofite[b]	1;2	$Tl_{5.8}As_{15.4}S_{26}$	*c* 5.74	*b* 24.43	*a* 8.76	$P2_1/n$	Divjakovič and Nowacki (1976)
				β 108.28			
Gillulyite[c] (subcell)	1;2	$Tl_2(As,Sb)_8S_{13}$	*b* 5.68	*c* 21.50	*a* 9.58	$P2/n$	Foit *et al.* (1995), Makovicky and Balič Žunič (1999)
				β 100.07			
Synthetic	1;2	$Rb_2As_8S_{13} \cdot H_2O$	*b* 11.52	*c* 21.76	*a* 9.60	$P2_1/n$	Sheldrick and Kaub (1985)
				β 98.8			
Edenharterite[d]	2;3	$PbTlAs_3S_6$	*c* 5.85	*a* 47.45	*b* 15.48	*Fdd*2	Balič Žunič and Engel (1983)
Jentschite[d]	2;3	$TlPbAs_2SbS_6$	*c* 5.89	*b* 23.92	*a* 8.10	$P2_1/n$	Berlepsch (1996), Berlepsch *et al.* (2000)
				β 108.06			
Synthetic[c]	1;2	$(NH_4)_2Sb_4S_7$	*a* 11.33	*b* 26.25	*c* 9.94	*Pbca*	Dittmar and Schäfer (1977)
Synthetic[c]	1;2	$Rb_2Sb_2S_{12}(S_2) \cdot 2H_2O$	*a* 7.08	*b* 25.40	*a* 8.05	$P2_1/n$	Berlepsch *et al.* (2001*a*)
				β 97.84			
Synthetic	2;2	$Cs_2Sb_8S_{13}$	*b* 11.48	*a* 15.43	*c* 8.29[e]	$P\bar{1}$	Volk and Schäfer (1979)
			α 71.89	β 102.45	γ 95.16		
Synthetic	2;2	$[CH_3NH_3]_2Sb_8S_{13}$	*b* 11.58	*a* 15.87	*c* 8.30[e]	$P\bar{1}$	Wang and Liebau (1994)
			α 71.46	β 75.71	γ 82.25		
Synthetic	2;2	$[H_3N(CH_2)_3NH_3]Sb_{10}S_{16}$	*b* 10.93	*a* 18.36	*c* 17.39	$P2_1/n$	Wang (1995)
				β 111.44			

Kermesite	1;1	Sb_2S_2O	*a* 5.79	*b* 10.71	*a* 8.15	$P\bar{1}$	Bonazzi *et al.* (1987),
			α 102.78	β 110.63	γ 101.00		Kupčík (1967)
Pääkkönenite	2;2	$Sb_2(As_{0.84}Sb_{0.16})S_2$	*a* 10.75	*c* 12.49	*b* 3.60	$C2/m$	Bonazzi *et al.* (1995)
				β 115.25			
Synthetic	1;2	$(NH_3CH_2CH_2NH_3)Sb_8S_{13}$	*c* 11.34	*a* 22.87	*b* 10.06	$Cmc2_1$	Tan *et al.* (1994)
		ethylenediammonium sulphide					
Synthetic[c]		$(C_2H_8N)_2Sb_8S_{12}(S_2)$	*b* 11.65	*c* 25.98	*a* 9.97	*Cmca*	Tan *et al.* (1996)
Gerstleyite[c]	1;2	$Na_2(Sb,As)_8S_{13}2H_2O$	*c* 7.10	*b* 23.05	*a* 9.91	*Cm*	Nakai and Appleman (1981)
				β 127.85			

[a] Homologue order is determined by numbers of square coordination pyramids (SnS subcells) across the width of the SnS slab *on the two surfaces of the tightly bonded SnS-like ribbon in this slab.*

[b] OD character described by Balič Žunič and Makovicky (1993).

[c] PbS archetype.

[d] B slabs are reduced out (see text). Structures of edenharterite and jentschite respectively contain 2-fold axes and $\bar{1}$ as operators of unit-cell twinning.

[e] Tightly-bonded SnS-like ribbons are parallel to [011].

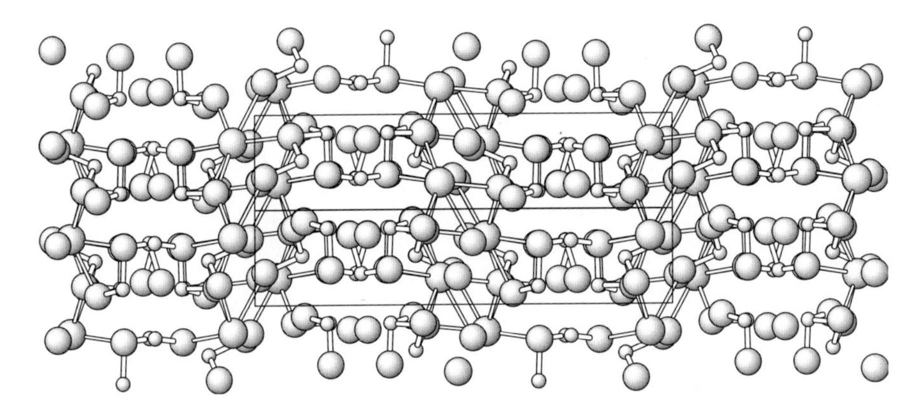

Fig. 1.29. Crystal structure of $Rb_2Sb_2S_{12} \cdot 2H_2O$ (Berlepsch *et al.* 2001*d*), a member of the hutchinsonite merotype series. SnS-like slabs $N_{1,2} = 1$; 2 house Sb polyhedra, complex interlayers contain Rb (black) and H_2O molecules (white).

kermesite Sb_2S_2O belong to the branch based on SnS-archetype; the configuration of A slabs are designed to accommodate active lone electron pairs, often with large cations (Tl^+) alternating with As polyhedra along the slab margins. These margins can sometimes be strongly modified relative to the archetype. In the case of $(NH_4)_2Sb_4S_7$ (Dittmar and Schäfer 1977), the distribution of bonds in A slabs is transitional between the PbS and SnS principles.

1.8.2.2 *Reyerite merotypes*

Reyerite merotypes (Ferraris 1997, Table 1.13) are phyllosilicates of Ca with eventual smaller presence of Na(K), all with hydroxyl groups and hydrated. Their structures consist of the same type of simple octahedral Ca–(O,OH) sheets paired via a 'simple' phyllosilicate sheet with tetrahedra alternately joining the two bounding Ca–(O,OH) sheets (Fig. 1.30). These 'Ca–Si–Ca' slabs are interleaved by 'complex silicate sheets' which differ between distinct members of the reyerite family.

The complex sheet of **minehillite** (Table 1.13) consists of a central, openwork sheet composed of elongated Zn tetrahedra and Al octahedra. At its upper and lower boundaries, Si_6O_{18} rings are attached, sandwiching between them potassium in a mica-like configuration (Dai *et al.* 1995). The complex sheet reminds one of a modified cordierite (or, better, osumilite) structure.

In **reyerite** (Table 1.13), the Si_6O_{18} rings are joined via $AlSiO_7$ groups instead of the above openwork sheet (Fig. 1.30). The resulting interspace is filled by large cations (Merlino 1988*a*). Large cations also fill interspaces outside the rings. The corresponding layer thickness increases from 9.7 Å in minehillite to 12.1 Å in reyerite.

Truscottite is nearly isostructural with **reyerite**; in gyrolite (Table 1.13), the thickest complex layer, 15.3 Å in thickness, is encountered (Lachowski *et al.* 1979; Merlino 1988*b*). Detailed treatment of this series, with an example of a different slab choice and structure prediction is given in the Section 4.3.5.2.

Table 1.13 Reyerite merotypes

Mineral	Formula	Space group	Lattice parameters (Å)			Note	Reference
Reyerite	$(Na,K)_2Ca_{14}Al_2Si_{22}O_{58}(OH)_8 \cdot 6H_2O$	$P\bar{3}$	a 9.764	c 19.07		–	Merlino (1988*a*)
Minehillite	$(K,Na)_2Ca_{28}Zn_5Al_4Si_{40}O_{112}(OH)_{16}$	$P\bar{3}c1$	a 9.777	c 33.29		OD twinning	Dai *et al.* (1995)
Gyrolite	$NaCa_{16}Si_{23}AlO_{60}(OH)_8 \cdot 14H_2O$	$C\bar{1}$ or $P\bar{1}$	a 9.74	b 9.74	c 22.40	Layer stacking	Merlino (1988*b*)
			α 95.7°	β 91.5°	γ 120°	disorder	
Truscottite	$(Ca,Mn^{2+})_{14}Si_{24}O_{58}(OH)_8 \cdot 2H_2O$	$P\bar{3}$	a 9.731	c 18.836			Lachowski *et al.* (1979)
Tungusite	$Ca_{14}Fe^{2+}_9Si_{24}O_{60}(OH)_{22}$	$P\bar{1}$	a 9.714	b 9.721	c 22.09	Structural model	Ferraris *et al.* (1995)
			α 90.13°	β 98.3°	γ 120.0°		

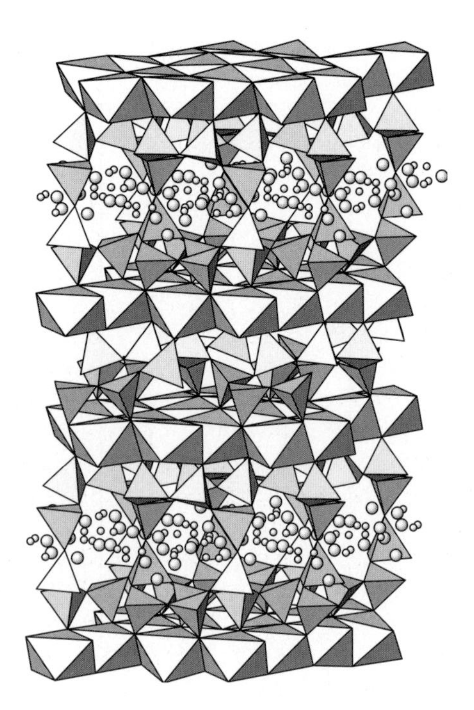

Fig. 1.30. The crystal structure of reyerite, $(Na, K)_2Ca_{14}Al_2Si_{22}O_{58}(OH)_8 \cdot 6H_2O$ (Merlino 1988*a*), the type structure of reyerite merotypic family. Thinner slabs composed of octahedral calcium layers sandwiching a tetrahedral layer are interleaved by an extensive interlayer with water molecules and channel cations that varies for various members of the series.

1.8.2.3 *Layered superconducting cuprates*

The immensely broad family of layered superconducting cuprates contains conducting CuO_2 sheets alternating with insulating separation layers. Several types of the latter layers occur in one compound. Complex symbolism has been devised (Shaked *et al.* 1994) to describe these sequences. All these compounds are layer stacks with tetragonal symmetry or with its orthorhombic distortion.

These compounds contain numerous homologous pairs or entire series based on homologous expansion of conducting blocks or of insulating layers. Conventional derivation of these series from perovskite, symbolism, and illustrations are given in Section 4.1.3.1; a somewhat different approach is adopted in this brief account, stressing the merotypic character of this family. The **$TlBa_2Ca_{n-1}Cu_nO_{2n+3}$ series** (Vainshtein *et al.* 1994) contains a one-cell thick, face-centred TlO structure intergrown with a conducting block; the latter is limited by layers of CuO_5 pyramids that are interspaced by $(n-1)$ Ca layers and $(n-2)$ flat-square CuO_2 layers (Fig. 4.4(a–c)). This is a classical homologous series with one kind of configuration in the thin layer of stable thickness and an incrementally growing layer with another, Ca–Cu composition. Vainshtein *et al.* (1994) described a parallel series in which the Tl layer is 1.5

body-centered cells thick whereas the conducting block resembles the previous one: $(n-1)$ Ca layers are interspaced by $(n-2)$ CuO_2 layers; these blocks again attach themselves to Tl-containing layers via vertices of CuO_5 pyramids (Fig. 4.5(b–d)). Members $N = 1$–3 are known. For $N = 1$, in both the former and the latter series, the marginal cations of the 'Tl' layer, on the level of pyramidal tips, can be substituted by Ba, giving the $Tl_nBa_2CuO_{n+4}$ ($n = 1, 2$) pair. In the $N = 1$ members, no Ca interlayers occur and the CuO_5 pyramids coalesce into CuO_6 octahedra.

A large group of cuprates can be derived from the terms $N = 2$ of the above series: The stable layer element is a pair of CuO_5 pyramids oriented outwards, with Ba or Sr situated among their tips, and with edge-sharing YO_8 or GdO_8 etc. coordination cubes in the interior of the layer. The alternating interlayers can be absent as in $YBaCuFeO_5$ (Vaughney and Poeppelmeier 1991), or they represent single RuO_6 octahedral layers sharing vertices with the CuO_5 pyramids ($RuSr_2GdCu_2O_8$; Tang *et al.* 1997). They can also be omission derivatives of octahedral layers, in form of CuO_4 squares as in $CuBa_2YCu_2O_7$ (Jorgensen *et al.* 1950), or of linearly coordinated Cu ($CuBa_2YCu_2O_6$; Raveau *et al.* 1991).

Doubling of Cu interlayers, resulting in ribbons of edge-sharing squares gives a higher homologue, $Cu_2Ba_2YCu_2O_8$ (Fischer *et al.* 1992). In numerous other compounds, with varying types of interlayers, the above mentioned cubes of rare-earth elements comprise a double layer (distorted **CaF_2 archetype**)—these are Gd, Pr or NdO_8 cubes, $N = 2$ for this layer.

Thus, the just outlined complex of cuprates is a series of merotypes with one set of layers, those between the bases of CuO_5 pyramids, having MO_8 coordinations that are interleaved in some cases by horizontal square planar Cu coordinations. Homologous expansion can occur with or without the horizontal square planar Cu groups, reduction to zero (i.e. coalesced Cu pyramids) as well. The alternating set of layers comprises layers of various types: oxygen-free bismuth layers in Bi_2Sr_2(Gd, Ce) Cu_2O_{10} (Shaked *et al.* 1994), Pb–O layers in (Pb,Cu) (Eu, Ce)$_2$ Cu_2O_9 (Shaked *et al.* 1994), Ru–O octahedral layers and various Cu-based layers mentioned above.

1.8.2.4 *Layered organic–inorganic perovskites*

Layered organic halide perovskites with variously thick slices (100), (110) or (111) of perovskite structure, intercalated by chain-like organic molecules that have NH_4 groups, form typical merotypic series in which the intercalating molecules vary widely. Detailed description of these merotypes will be given in Section 4.1.3.3.

1.8.2.5 *Phyllomanganates and other layer structures with decorated punctured octahedral layers*

Structures of this mero- to plesiotypic family are composed of octahedral layers with a regular-to-random pattern of vacant octahedra that are decorated on both sides (rarely on one side) by coordination octahedra or coordination tetrahedra which share three anions with the layer. Layers are separated by a single-to-multiple layer of H_2O molecules $\pm$ anions that take part in coordination of decorating cations. Interlayer cation polyhedra are present in lawsonbauerite and related compounds.

The first subfamily are layered Mn oxides, known as **phyllomanganates**. In the stable structure of **chalcophanite** (Table 1.14, Fig. 1.31), decoration proceeds by means of $ZnO_3(H_2O)_3$ octahedra; their external triangular faces form the H_2O interlayer so that Zn alternately is above and below this layer. Every seventh Mn octahedron is vacated.

Birnessite, an important and frequent secondary manganese mineral has changeable octahedral layers, with different amounts of only partly ordered vacancies. In Na-saturated birnessite, only very few vacancies occur. Na is in a single interlayer together with water molecules (the same in **K-birnessite**) and about 25 percent of Mn is trivalent instead of tetravalent (Post and Veblen 1990) (Fig. 1.31). In hydrogenated birnessite with Na leached away, part of Mn migrates to cover the vacancy sites. This is primarily Mn^{3+}; some Mn^{2+} is absorbed from solution as well. The migration and absorption of Mn onto vacancy sites depends on pH. Both Mn in H-birnessite and Zn in Zn-birnessite cover mostly only one side of each octahedral vacancy, the other side being compensated for by H^+ (Fig. 1.31) (Silvester *et al.* 1997).

The '7 Å phyllomanganates' of the birnessite type are parallelled by '10 Å phyllomanganates' of **buserite type** for which a double water layer is assumed, pushing layers and decorating octahedra apart. According to Chukhrov *et al.* (1989), so-called **buserite I** has only isolated vacancies and decorating octahedra, and is capable of collapsing to a single-layer birnessite on dehydration whereas **buserite II** has small clusters of such octahedra, and the collapse is made impossible.

In the high-pressure phase D, $MgSi_2H_2O_6$ (Yang *et al.* 1997), the Si-containing dioctahedral layers have octahedral voids capped by Mg octahedra on both sides. However, each Mg octahedron caps voids in two adjacent octahedral layers, interlinking them into a solid three-dimensional framework. The remaining interspace is occupied by hydrogen bonds between the adjacent octahedral layers.

The other subfamily are the Zn–Cu hydroxysalts of the **simonkolleite family** (Hawthorne and Sokolova 2002), several examples of which are in Table 1.14. In hydroxychloride simonkolleite, every fourth Zn octahedron is vacated and covered by Zn tetrahedra from both sides; a single layer of H_2O with Cl as the tetrahedral tips separates octahedral layers (Fig. 1.32). In Mn-Mg-Zn hydroxide cianculllite (Grice and Dunn 1991) every fourth octahedron is vacant and covered by Zn tetrahedra. In gordaite (Table 1.14), every seventh octahedron is vacant and again covered by Zn tetrahedra; the SO_4 groups are attached to the octahedral sheet as well; interlayer is formed by H_2O molecules. No mention was found of their behaviour on hydration/dehydration.

A special subgroup of this family are the structures of **lawsonbauerite** (Table 1.14) and related Mg-dominant torreyite and mooreite (Treiman and Peacor 1982). In lawsonbauerite, 2/9 of the octahedral sheet sites are vacant, capped by Zn tetrahedra. These, however, share a vertex with interlayer $(Mn,Mg)(OH)_2(H_2O)_4$ octahedra; SO_4 tetrahedra complete the interlayer. In the closely related mineral mooreite, 2/11

Table 1.14 Selected structures of the mero/plesiotypic family with decorated punctured octahedral layers

Mineral	Formula	Space group	Unit cell parameters (Å) a	b	c	Angles	Decoration and interlayer	Reference
Chalcophanite	$ZnMn_3O_7 \cdot 3H_2O$	$R\bar{3}$	7.533	–	20.794	–	$ZnO_3(H_2O)_3$ octahedra; H_2O molecules	Post and Appleman (1988)
Na-birnessite	$Na_{0.6}Mn_2O_4 \cdot 1.5H_2O$[a]	$C2/m$	5.175	2.85	7.337	103.18°	less vacancies; interlayer Na + H_2O	Post and Veblen (1990)[b]
H-birnessite	H-exchanged	Hexagonal	2.845	–	7	–	H^+ and $MnO_3(H_2O)_3$ octahedra; interlayer H_2O	Silvester *et al.* (1997)
Buserite I	birnessite + nH_2O[c]	Hexagonal	2.83	–	10	–	Na, Mn^{3+}, etc. octahedra; two H_2O layers	Chukhrov *et al.* (1987)
Buserite II		Hexagonal	2.84	–	9.7	–	Ca, Mn^{3+}, etc. octahedra; two H_2O layers	Chukhrov *et al.* (1987)
Phase D	$MgSi_2H_2O_6$	$P\bar{3}1m$	4.745	–	4.345	–	Mg octahedra	Yang *et al.* (1997)
Simonkolleite	$Zn_5Cl_2(OH)_8H_2O$	$R\bar{3}m$	6.341	–	23.646	–	$ZnCl(OH)_3$ tetrahedra; Cl + H_2O molecules	Hawthorne and Sokolova (2002)
Gordaite	$NaZn_4(SO_4)Cl(OH)_6(H_2O)_6$	$P\bar{3}$	8.356	–	13.025	–	$Zn(OH)_4$ tetrahedra and (SO_4) groups; H_2O layer	Adiwidjaja *et al.* (1997)
Chalcophyllite	$Cu_9Al(AsO_4)_2(OH)_{12}(H_2O)_6(SO_4)_{1.5}(H_2O)_{12}$	$R\bar{3}$	10.756	–	28.678	–	one-sided (AsO_4) tetrahedra; (SO_4) and H_2O layer	Sabelli (1980)
Bechererite	$Zn_7Cu(OH)_{13}[SiO(OH)_3SO_4]$	$P3$	8.319	–	7.377	–	$Zn_2(OH)_7$ pyrogroups; $SiO(OH)_3$ and SO_4 groups	Hoffmann *et al.* (1997)
Lawsonbauerite	$(Mn,Mg)_9Zn_4(SO_4)_2(OH)_{22} \cdot 8H_2O$	$P2_1/c$	10.50	9.64	16.41	95.21°	$Zn(OH)_4$ tetrahedra; $Mn(OH)_2(H_2O)_4$ octahedra, (SO_4) tetrahedra	Treiman and Peacor (1982)
Phase X	$(Na, K)_{2-x}Mg_2Si_2O_7H_x$	$P\bar{3}1m/P6_3cm$	4.98–5.08	–	6.44/13.22	–	Si_2O_7 groups; K, Na cations; OH by IR	Yang *et al.* (2001)

[a] Formula refined by Silvester *et al.* (1997).

[b] Synthetic birnessites. Mg-exchanged birnessite $Mg_{0.3}Mn_{1.9}O_4 \cdot 1.6H_2O$ is monoclinic (*c* 7.05 Å) and has $MgO_3(H_2O)_3$ octahedra; K-substituted birnessite $K_{0.5}Mn_2O_4 1.4H_2O$ (K and H_2O coincide in one layer, *c* 7.18 Å (Post and Veblen 1990).

[c] Formulae of buserite I and II are not given by Chukhrov *et al.* (1987). Giovannoli *et al.* (1975) gives approximate formula $Na_4Mn_{14}O_{27} \cdot 21H_2O$. For natural birnessite, a mixture of Na, Ca and K as well as partial substitution of Mn by Mg is indicated.

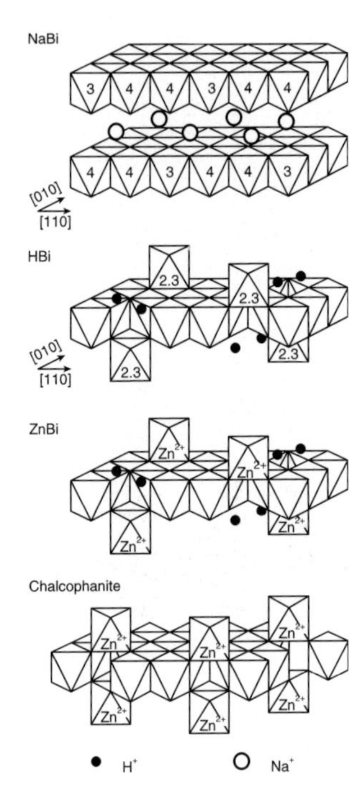

Fig. 1.31. Schematic representation of the crystal structures of Na-, H^+-, and Zn-substituted birnessite as well as chalcophanite. The decorating H^+ ions and Mn/Zn octahedra flank the vacancies in the octahedral layers. 2, 3, and 4 denote preferred valence for Mn at the site. Redrawn with permission from Silvester *et al.* (1997).

of octahedral sites are vacant, suggesting a spectrum of possible vacancy arrangements in these phases.

In **bechererite** $Zn_7Cu[SiO(OH)_3](SO_4)(OH)_{13}$, the Zn tetrahedra polymerize to $Zn_2(OH)_7$ pyrogroups and the interlayer space is filled by 'free' SO_4 tetrahedra and by Si tetrahedra attached by a single vertex to the adjacent octahedral layer. One-seventh of the octahedral sites are void.

Yang *et al.* (2001) synthesized **Phase X**, a high-pressure alkali-rich silicate of (primarily) magnesium, with compositions spanning $Na_{1.8}Mg_{1.9}Al_{0.1}Si_2O_7$ ('**anhydrous phase X**') $K_{1.5}Mg_{1.9}Si_{1.9}O_7H$ ('**hydrous phase X**'). This trigonal phase (Table 1.14) has punctured layers with 1/3 of the octahedra vacated and decorated from both sides by Si tetrahedra, and with Na/K in interlayer. Silicon tetrahedra form Si_2O_7 dimers; H is present as OH groups in the octahedral layers.

Structures with cation-decorated punctured sheets and an interlayer of water associated with additional ions form a large plesiotypic family with those of the hydrotalcite family (**hydrotalcite** $Mg_6Al_2(OH)_{16}(CO_3) \cdot 4H_2O$, with octahedral, undecorated hydroxide layers alternating with interlayers of water plus various

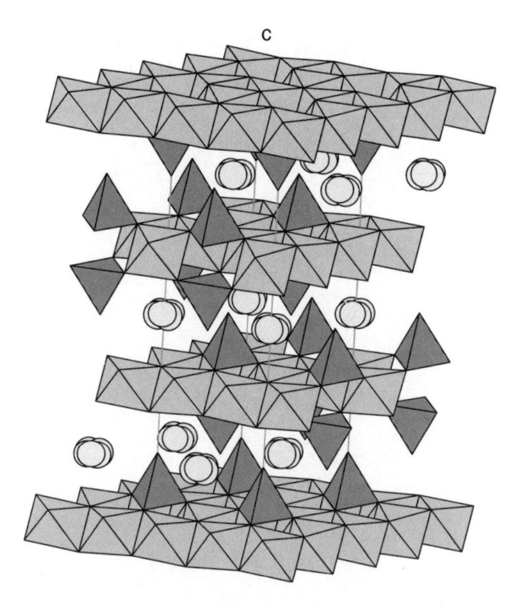

Fig. 1.32. The crystal structure of simonkolleite $Zn_5Cl_2(OH)_8 \cdot H_2O$ (Hawthorne and Sokolova 2002). Zinc in octahedral and tetrahedral positions; chlorine forms tetrahedral apices; interlayer water molecules. A type structure of the merotypic family with decorated punctured octahedral layers.

anions), and with the composite layer structures in which octahedral oxide/hydroxide layers alternate with hydroxide layers based on different cations (e.g. **lithiophorite** $(MnO_2)(Al,Li)(OH)_2$; Post and Appleman 1994), with chloride layers (**koenenite** $Na_4(Ca, Mg)_2Cl_{12} \cdot Mg_7Al_4(OH)_{22}$, Allmann *et al.* 1968) or even with sulphide layers (e.g. **valleriite** 1.562 $Mg_{0.7}Al_{0.3}(OH)_2 . Fe_{1.1}Cu_{0.9}S_2$; Evans and Allmann 1968).

1.8.2.6 *Thalcusite–rohaite series*

The **thalcusite–rohaite series** (Makovicky *et al.* 1980) with only one known intermediate member, chalcothallite, allows two alternative interpretations. As a merotypic series, it is composed of regularly occurring layers of Tl coordination cubes that alternate either with quadratic, tetrahedral sulphide layers fully occupied by (Cu, Fe) in $TlCu_3FeS_4$ or with thicker, complex sulphide–antimonide layers in $Tl_2Cu_{8.67}Sb_2S_4$ (Fig. 1.33). In **chalcothallite** $Tl_2(Cu, Fe)_{6.35}SbS_4$, these two-layer types alternate regularly. In $TlCu_4Se_3$ and $TlCu_6S_4$, the tetrahedral, Cu-filled layers between two consecutive Tl layers undergo homologous expansion giving double- and triple-tetrahedral layers (Berger 1987). Another modification of the thalcusite principle takes place by introducing ordered vacancies in tetrahedral layers, for example, in $Rb_2Mn_3S_4$ (Bronger and Böttcher 1972).

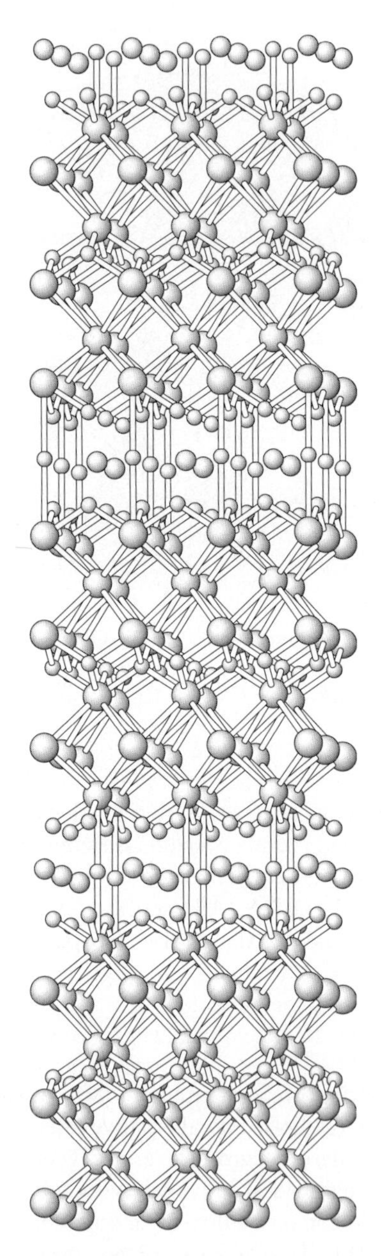

Fig. 1.33. Crystal structure of chalcothallite $Tl_{1.7}K_{0.2}Cu_{5.5}Fe_{0.7}Ag_{0.1}SbS_4$ (Makovicky *et al.* 1980). Tl in eight-fold coordination; Cu in tetrahedral mackinawite-like layers. Cu also forms triple 'antimonide' layers with a partial, $\frac{1}{3}$ Cu site and Sb in their central row.

1.8.2.7 *Mixed-layer clay minerals*

An extension of merotype series are mixed-layer clay minerals (Drits 1997). If we limit our description to the most common pair smectite-illite, both end-members contain topologically identical layers composed of three sheets: tetrahedral–octahedral–tetrahedral (**T–O–T**). They differ in the 'interlayer', which for smectites consists of defect Ca, Na layers with coordinating water molecules (amount and configuration is a function of partial H_2O pressure) whereas for illite, it consists of a more-or-less complete K layer. These differences are reflected in the interlayer spacing, about 12–14 Å for smectites and about 10 Å for illite. Illite and smectite layer modules may combine in variable ratios and in an ordered or disordered fashion. Thus, illite–smectite is an intergrowth of two members of the merotype series. These intergrowths alone can also be described as a polysomatic series with two slabs, A and B, in all possible proportions and sequences. Such an attitude will reflect the fact that the appropriate sides of **T–O–T layers** have to adjust compositionally (by element substitutions) to the interlayer contents. However, it separates the illite–smectite pair from other mixed-layer clay structures, in which other modules (**tobellite**, **vermiculite**, **kaolinite**, ...) are present mixed to one or both of the above mentioned ones.

1.8.2.8 *Manganese oxide subsilicates*

A merotypic **bixbyite–braunite series**, described by de Villiers and Buseck (1989) is another example of intergrowths of two merotypes. All structures of this series, often understood as defect derivatives of CaF_2 type, contain characteristically deformed octahedral A layers (with Mn, Fe) which regularly alternate with configurationally distorted A′ layers (Mn, Fe) or with B layers that contain (Ca, Mn^{2+}) in deformed cubic coordination together with tetrahedral Si and octahedral (Mn^{3+},Fe). A pure AA′AA′ sequence characterizes **bixbyite** $(Mn,Fe)_8O_{12}$; a pure ABAB sequence is in **braunite** and **neltnerite** (Mn^{2+} or Ca)$Mn_6^{3+}SiO_{12}$. The 1 : 1 merotypic combination AA′ABAA′ABA forms **braunite-II** $Ca_{0.5}Mn_7^{3+}Si_{0.5}O_{12}$; a number of longer-repeat, more complex A′ and B combinations in ratios different from 1 : 1 were observed, especially by HRTEM. Omitting A layers from the symbol, e.g. A′BBA′BB, i.e. assigning halves of A layers to the immediately adjacent A′ and B layers, leads to an interpretation of this series as a simple combinatorial polysomatic series; this is only possible because just two alternative types, A′ and B, of the second layer have so far been described in these structures.

1.8.2.9 *Covellite merotypes*

Covellite, CuS, is a sulphide with a layer structure that is composed of consecutive sheets (0001) of tetrahedrally coordinated copper, trigonally coordinated copper and those of the S–S groups that are parallel to the hexagonal *c* axis (Fig. 1.34). The same structure is adopted by CuSe (**klockmannite**).

For the purpose of merotypy studies, the **covellite** structure can be divided into composite layers comprising the sulphur atoms of the trigonal planar sheet, the adjacent

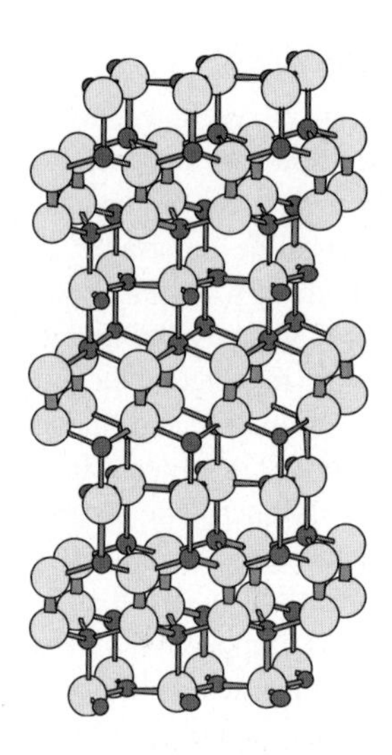

Fig. 1.34. Crystal structure of covellite (Evans and Konnert 1976). Alternating layers of 3- and 4- coordinated Cu, and S_2 groups.

Table 1.15 Covellite merotypes

Mineral	Formula	Space group	Lattice parameters a	Lattice parameters c	Interlayer	Reference
Covellite	CuS	$P6_3/mmc$	3.79	16.34	$Cu^{[3]}$	Evans and Konnert (1976)
Klockmannite	CuSe	$P6_3/mmc$	3.94	17.25	$Cu^{[3]}$	Effenberger and Pertlik (1981)
Nukundamite	$Cu_{3.39}Fe_{0.61}S_4$	$P\bar{3}m1$	3.78	11.2	$(Cu,Fe)^{[4]}$	Sugaki *et al.* (1981)
Synthetic	Cu_4SnS_6	$R\bar{3}m$	3.74	32.94	$Cu^{[3]}$, $Sn^{[6]}$	Chen *et al.* (1999)

tetrahedral sheets and the central S–S layer. In covellite, these layers share the apical S atoms; the changeable interlayer is formed only by the trigonal planar Cu atoms.

The composite layer is unchanged in other known merotypes (Table 1.15). In $Cu_{3.39}Fe_{0.61}S_4$, **nukundamite** (Sugaki *et al.* 1981), the interlayer has all tetrahedra filled with (Cu, Fe) in a valleriite-like arrangement (Fig. 1.35); the interlayer in $\mathbf{Cu_4SnS_6}$ (Chen *et al.* 1999) contains trigonal planar copper on both surfaces, and octahedrally coordinated Sn^{4+} inside the sandwich. Occupancy of the Sn^{4+} position is $\frac{2}{3}$, that of limiting planar Cu is $\frac{1}{3}$ (Fig. 1.36).

$\mathbf{Cu_4Sn_7S_{16}}$ (Chen *et al.* 1998) preserves substantial features of Sn-interlayer whereas the 'principal' composite layer is destroyed, replaced by tetrahedral/trigonal planar Cu positions reminding us of **valleriite**, with interspersed Sn^{4+} octahedra.

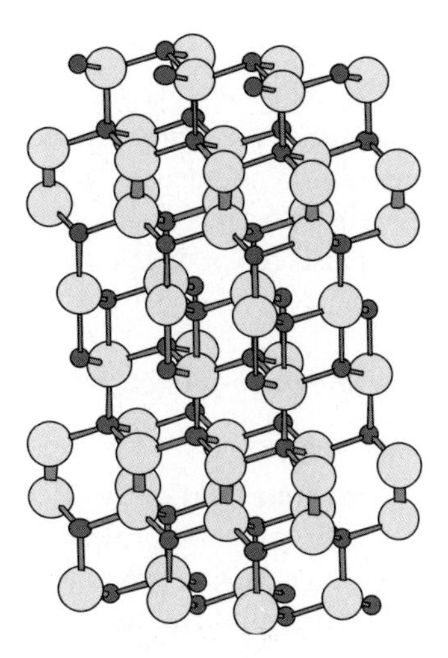

Fig. 1.35. Crystal structure of nukundamite $Cu_{3.4}Fe_{0.6}S_4$ with alternating layers of tetrahedrally coordinated Cu, of edge-sharing (Cu, Fe) tetrahedra and of S_2 groups.

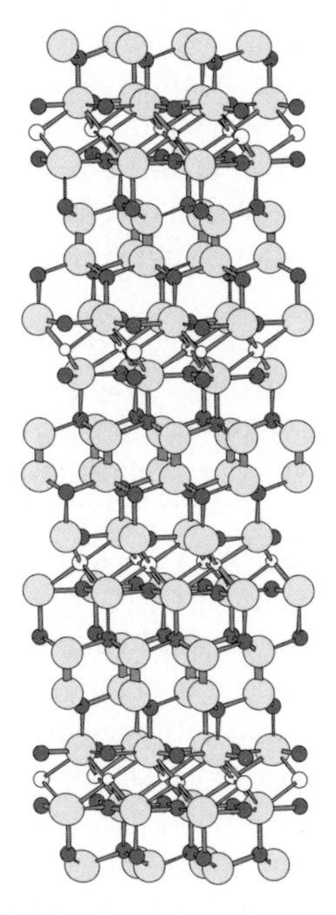

Fig. 1.36. Crystal structure of Cu_4SnS_6 (Chen *et al.* 1999). Small white circles: partially occupied Sn^{4+} positions flanked by partly occupied triangular Cu sites; fully occupied Cu tetrahedra and S_2 groups. Merotype of covellite.

1.8.2.10 *Further examples of merotypy*

An elegant merotypic pair is formed by $Pb_3In_{6.67}vacancy_{0.33}S_{13}$ (Ginderow 1978) and $Er_9La_{10}S_{27}$ (Carré and Laruelle 1973). Three atomic layers thick PbS-like layers are preserved in both structures (are homologous), but the sheared SnS_2-like layers in the former are replaced by very complex layer stacks in the latter (Makovicky 1992*a*); the original interlayer match is preserved in this process.

1.8.3 *Examples of plesiotypy and mero-plesiotypy*

1.8.3.1 *Sursassite–clinozoisite plesiotypes*

The **sursassite–pumpellyite–ardennite OD family** (Merlino 1990*a*) and the **zoisite–clinozoisite OD family** (Ito 1950) form together a plesiotypic family (Table 1.16). The structural basis of the former group is an O–D alternation of continuous 'L_1' layers (Merlino 1990*a*), composed of octahedral columns surrounded and interconnected by SiO_4 tetrahedra, with thinner 'L_0' layers of octahedra and tetrahedra; these are interconnected with the tetrahedra of the L_1 layer (Fig. 1.37(a–d)).

Plesiotypic relationship between **sursassite** $Mn_2Al_3[SiO_4][Si_2O_7](OH)_3$ and **clinozoisite** $Ca_2Al_3[SiO_4][Si_2O_7]O(OH)$ preserves the L_0 layer and one half of octahedra in the L_1 layer intact. A half of the large cation sites surrounding the octahedra of the L_0 layer are, in a *trans*-configuration, replaced by smaller, octahedral configurations (Fig. 1.38). Those octahedra in the L_1 layer that would find themselves squeezed between these new octahedra become strongly deformed, vacated and replaced by two larger cations placed eccentrically outside their volume. These sites are centres of pronounced distortion of the clinozoisite structure when compared to that of sursassite (Fig. 1.37(a)).

In zoisite, the clinozoisite structure is sheared on a new set of OD boundaries at 116° to the original O–D slabs (Fig. 3.16). The relation is that of OD structure composed of two unit layers (Merlino 1990*a*), one comprising octahedral columns whereas the other the principal parts of the structure with flanking octahedra, Ca polyhedra and silicate groups.

Ardennite $Mn_4(Mg,Al,Fe)_2Al_4[(As,V)O_4][SiO_4]_2[Si_3O_{10}](OH)_6$, **pumpellyite** $Ca_2Al_2(Fe,Mg,Al)[SiO_4][Si_2(O,OH)_7](OH,O)_3$ and sursassite are three distinct, maximally ordered polytypes of the **sursassite family**. They all contain the L_0 and L_1 layers and are, because of the compositional differences, configurational polytypes.

Ardennite or, when additional unit-cell twinning is involved, pumpellyite is in a merotypic relationship with **lawsonite** $CaAl_2[Si_2O_7](OH)_2 \cdot H_2O$ and its isotypes: the L_0 layers from the former are eliminated in the latter and the L_1 layers are in mutual contact via unit-cell reflection. Sites for Mn/Ca atoms on the L_0/L_1 boundaries coalesce into a single large site (Fig. 1.39). A corresponding phase with L_0 layers related via two-fold rotation axes or symmetry centres was not found.

A structure based on alternation of L_0 and reduced L_1 layers (reduction by means of crystallographic shear applied to the L_1 layers) from sursassite is **fornacite** $CuPb_2(CrO_4)(AsO_4)(OH)$ (Cocco *et al.* 1967) and **törnebohmite**

Table 1.16 Selected phases of the sursassite-clinozoisite family of plesiotypes

Mineral	Formula	Space group	Unit cell parameters (Å)			Angles	Reference
			a	b	c		
Sursassite	$Mn_2Al_3(SiO_4)(Si_2O_7)(OH)_3$	$P2_1/m$	8.719	5.808	9.813	109.00°	Allmann (1984)
Pumpellyite	$CaMgAl_2(SiO_4)(Si_2O_7)(OH)_2 \cdot H_2O$	$A2/m$	8.812	5.895	19.116	97.41°	Yoshiasa and Matsumoto (1985)
Ardennite	$(Mn,Ca,Mg)_4(Al,Fe)_5Mg(AsO_4)(SiO_4)_2(Si_3O_{10})(OH)_6$	$Pnmm$	8.713	5.811	18.521	–	Donnay and Allmann (1968)
Clinozoisite[a]	$Ca_2Al_3(SiO_4)(Si_2O_7)O(OH)$	$P2_1/m$	8.879	5.583	10.155	115.50°	Dollase (1968)
Zoisite	$Ca_2Al_3(SiO_4)(Si_2O_7)O(OH)$	$Pnma$	16.212	5.559	10.036	–	Dollase (1968)
Lawsonite[b]	$CaAl_2(Si_2O_7)(OH)_2 \cdot H_2O$	$Ccmm$	8.795	5.847	13.142	–	Baur (1978)
Fornacite	$Pb_2Cu(AsO_4)(CrO_4)OH$	$P2_1/c$	7.91	5.91	17.46	109.50°	Cocco *et al.* (1967)
Törnebohmite	$Ce_2Al(SiO_4)_2OH$	$P2_1/c$	7.383	5.673	16.937	112.04°	Shen and Moore (1982)
Gatelite	$CaCe_3Al_2(Al,Mg,Fe)_2(Si_2O_7)(SiO_4)_3(O,OH,F)_3$	$P2_1/a$	17.77	5.651	17.458	116.18°	Bonazzi *et al.* (2003)

[a] Epidote $Ca_2(Fe,Al)_3(SiO_4)(Si_2O_7)(O,OH)_2$ and allanite $Ce(Ce,Y,Ca)_2FeAl_2(SiO_4)(Si_2O_7)O(OH)$ are isotypes.
[b] Hennomartinite $SrMn_2(Si_2O_7)(OH)_2 \cdot H_2O$ is an isotype.

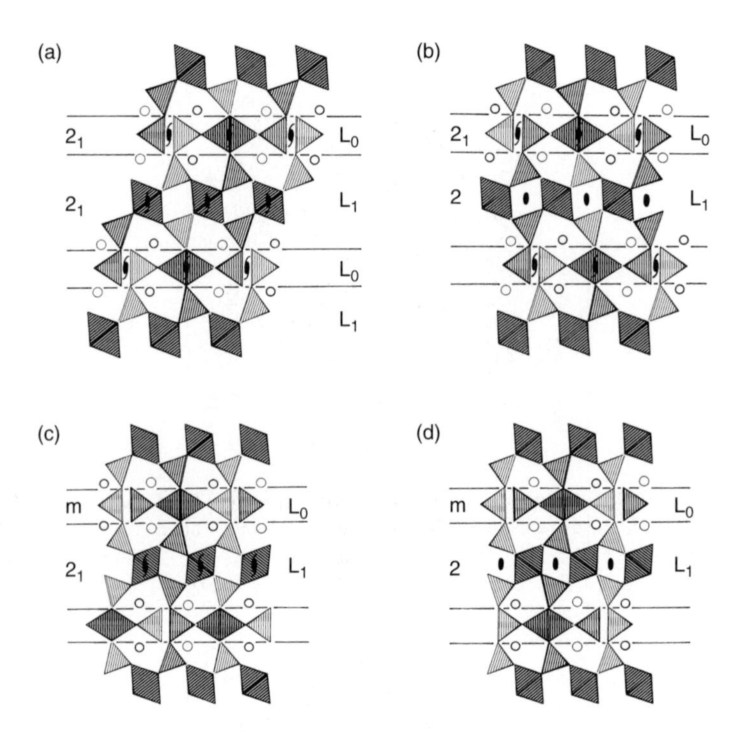

Fig. 1.37. Idealized crystal structures of (a) sursassite $Mn_2Al_3[SiO_4][Si_2O_7](OH)_3$, (b) pumpellyite $Ca_2Al_2(Fe, Mg, Al)[SiO_4][Si_2(O, OH)_7](OH, O)_3$, (c) ardennite $Mn_4(Mg, Al, Fe)_2Al_4[(As, V)O_4][SiO_4]_2[Si_3O_{10}](OH)_6$, and (d) a still hypothetical member of a family of configurational polytypes (Merlino 1990*a*) composed of two types of unit layers, L_0 and L_1. Intralayer symmetry elements relating adjacent layers of opposite kind are indicated. Octahedral columns in both kinds of layers are interconnected by single, paired or triple Si tetrahedra; positions of large cations are given as circles.

$Ce_2Al(SiO_4)_2(OH)$ (Shen and Moore 1982) (Table 1.16). Crystallographic shear creates rectangular channels situated between two octahedral columns from two adjacent L_0 layers, positioned at about 45° to these layers, and populated by two columns of large cations (Fig. 1.40). The resulting configuration, and the sequence of channels and octahedra along $(\bar{1}02)$ of törnebohmite corresponds to the above described channels with two columns of Ca atoms in clinozoisite. This opens a possibility for intergrowths of clinozoisite and törnebohmite structures, discovered and described as polysomatic by Bonazzi *et al.* (2003): Ca–Ce neso-sorosilicate **gatelite** (Table 1.16) is a 1 : 1 intergrowth of unit 'E' slabs (010) of epidote (i.e. clinozoisite structure type) with unit 'T' slabs $(\bar{1}02)$ of **törnebohmite**.

The phases closest to pure L_0 layers are **tsumcorite** $Pb(Zn,Fe)_2[AsO_4]_2(OH,H_2O)_2$ (Tillmans and Gebert 1973) and **medenbachite** $Bi_2Fe\ (Cu,Fe)\ (O,OH)_2\ (OH)_2\ (AsO_4)_2$ (Krause *et al.* 1996) as well as **staurolite** (Section 1.8.2.5). Figure 1.41 expresses relationships in the **sursassite–clinozoisite family** of plesiotypes.

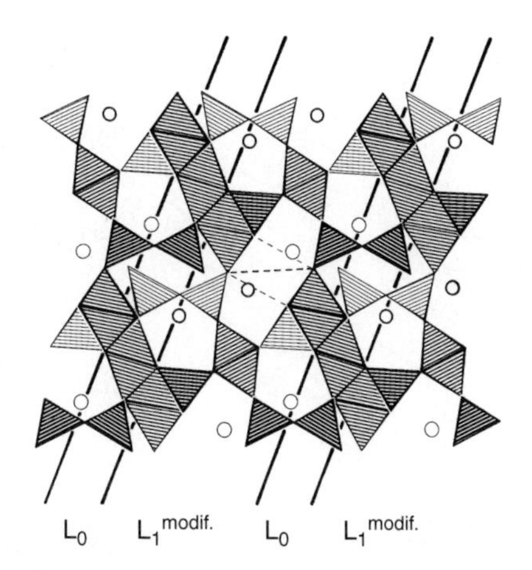

Fig. 1.38. Crystal structure of clinozoisite $Ca_2Al_3[SiO_4][Si_2O_7]O(OH)$ (Dollase 1968). Octahedral columns in L_0 layers are flanked by additional octahedra which replace some of the Mn atoms in sursassite (Fig. 1.37), dotted lines indicate an octahedral column from the latter structure that was replaced by a pair of Ca sites (Makovicky 1997*a*).

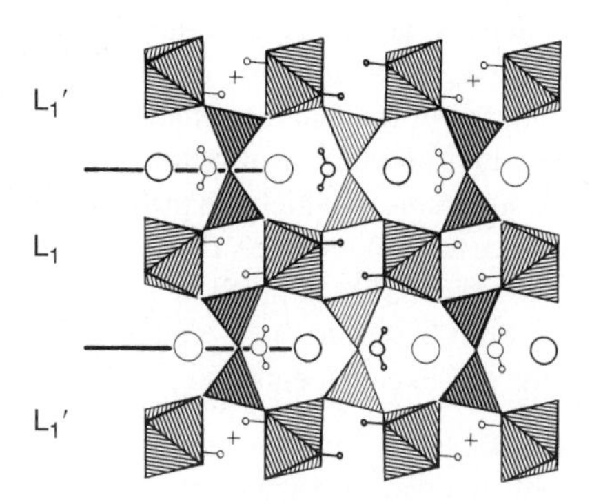

Fig. 1.39. The crystal structure of hennomartinite $SrMn_2[Si_2O_7](OH)_2 \cdot H_2O$ (Libowitzky and Armbruster 1996), an isotype of lawsonite $CaAl_2[Si_2O_7](OH)_2 \cdot H_2O$. This is a merotype of ardennite with L_0 layers left out, triple and single tetrahedra reduced to paired tetrahedra and marginal Mn sites of L_0 layers in ardennite coalesced into single sites.

1.8.3.2 *Rod-based complex sulphides*

Plesiotype family of rod-based sulphides includes complex sulphides of Pb and Sb, Pb and Bi, Sn and Sb, as well as a number of synthetic sulphides of alkalies, alkaline earths, and lanthanides combined with Sb or Bi.

Structures of rod-based sulphides (Fig. 1.3) contain rods of simpler configuration, based on PbS or SnS archetype, generally with lozenge cross-section, several PbS/SnS subcells (i.e. cation polyhedra) wide and two or more atomic layers thick. Their surfaces are alternatively of pseudotetragonal and (sheared or non-sheared)

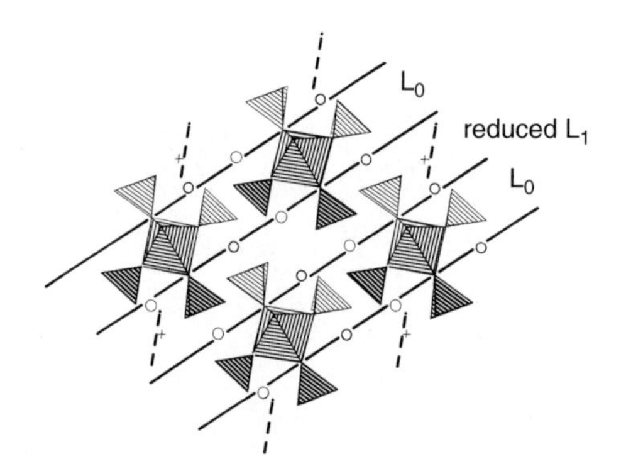

Fig. 1.40. Crystal structure of fornacite $CuPb_2$ $(CrO_4)(AsO_4)OH$ (Cocco *et al.* 1967). Octahedral columns and tetrahedra compose L_0 layers whereas L_1 layers of sursassite type were reduced by crystallographic shear. Stippled lines indicate planes of alternating octahedra and clinozoisite-like rectangular channels.

pseudohexagonal type. Typically the former surfaces face the latter in a complicated non-commensurate match across rod interfaces.

All these structures show distinct accumulation of Sb and Bi in rod interiors, forming **lone electron pair micelles**, and of Pb and the above-mentioned large cations on the pseudotetragonal surfaces. The interiors of larger rods contain mixed Pb–Sb and Sn–Sb positions, as means of maintaining charge balance and minimizing size differences of polyhedral arrays. A number of fairly recent structure determinations enlarged this spectrum by structures in which the presence of univalent large cations is outweighed by the presence of trivalent ones, for example, $KBi_{6.33}S_{10}$ (Kanatzidis *et al.* 1996) is isostructural with $Pb_4Sb_4Se_{10}$ (Skowron and Brown 1990*c*).

The plesiotypic family of rod-based structures can be divided into four subfamilies according to the pattern the rod-like elements compose:

1. **Rod-layer structures** (Table 1.17) in which rods are arranged in parallel rows and joined via common polyhedra in their acute corners (Fig. 1.42). There are up to 12 distinct ways of this interconnection into a rod-layer. Adjacent rod-layers indent each other by their obtuse corners and non-commensurate interfaces. There exist both the structures with all rod-layers composed of the same rod type and those in which the length of surfaces of opposite kind which are supposed to match requires alternation of rod-layers of two kinds. Individual structures differ both in rod dimensions and in the modes of their interconnection (Makovicky 1993).
2. Chess-board structures in which no rod-layers are present, but the rods form a chess-board pattern with non-commensurate interfaces. In the **kobellite homologous series** of Pb–Sb–Bi sulphides (Table 1.18) that belongs to this category, rods can have complicated cross-sections (Fig. 1.43). The chess-board family of **kobellite plesiotypes** offers a wealth of non-homologous and homologous relationships (Makovicky 1993).
3. Structures with a cyclic (six-fold) arrangement of rods; interfaces can either be of non-commensurate kind or they resemble contacts of archetype slabs in the **lillianite series** (Table 1.19, Fig. 1.44).

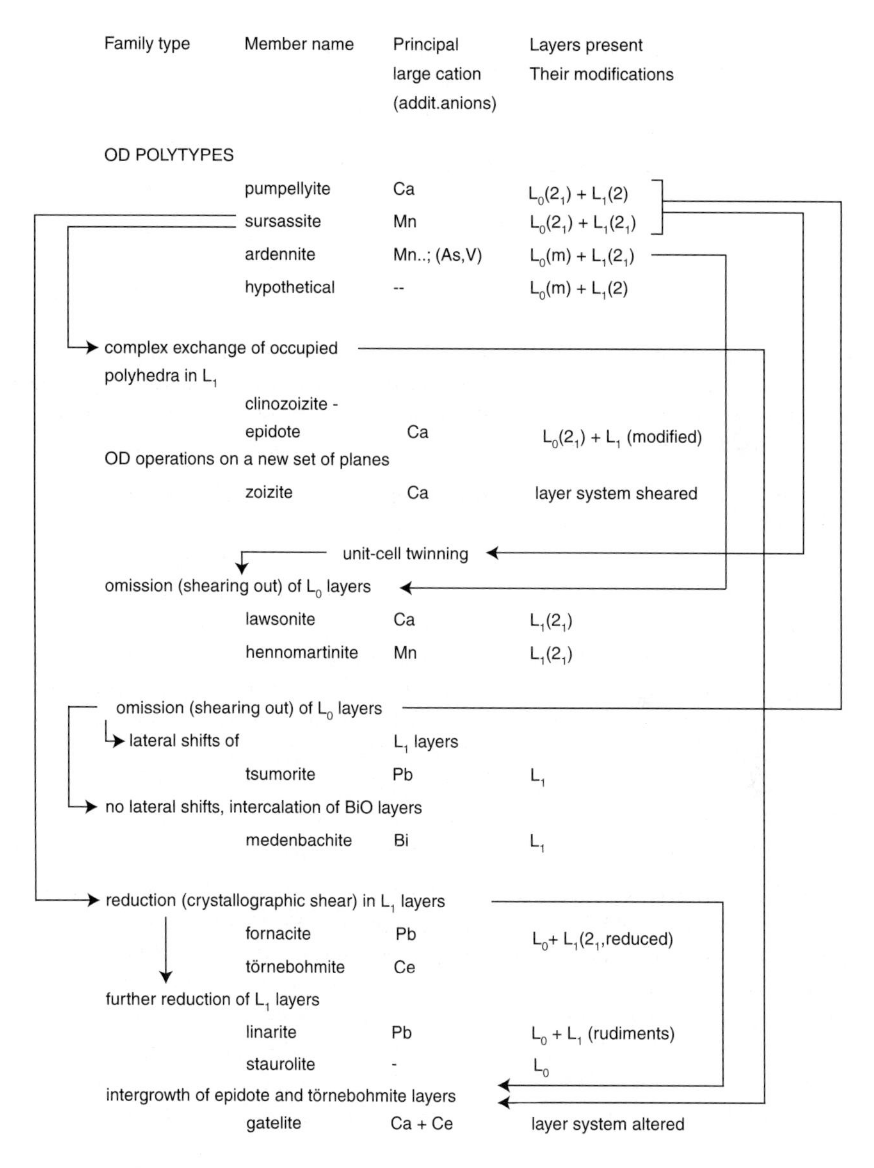

Fig. 1.41. Polytypic, merotypic and plesiotypic relationships in the sursassite–clinozoisite family.

4. Large-scale, doubly non-commensurate structures of a **box-work type** (Makovicky *et al.* 2000). They contain three types of rods, two of which form a box-work with pseudotetragonal surfaces, and the third one occurs inside the channels of the box-work, and displays only pseudohexagonal surfaces (Table 1.20, Figs 1.28 and 1.45).

Close similarity of rod types found in these categories and identity of match modes justifies joining all these plesiotypic categories into a unique, higher plesiotype family. In all categories, mostly only homologous pairs are found because of their combined, accretional and variable-fit character. The pairs $FePb_4Sb_6S_{14}$ (**jamesonite**)–$Pb_4Sb_4S_{11}$ or the $Ba_9Bi_{18}S_{36}$–$Ba_{12}Bi_{24}S_{48}$ can be quoted as examples.

In the most populated of these categories, the plesiotype subfamily of **rod-layer structures**, a wealth of non-homologous relationships exists interconnecting various phases (Makovicky 1993): 12 non-homologous ways of interconnecting individual rods into layers, two types of homologously expanding noncommensurate interfaces, non-homologous expansion of rods changing their contacts and leading to the **sliding series** described elsewhere in this book and combination of distinct rod types, with compatible interfaces, in one structure.

1.8.3.3 *Pyrochlore derivatives*

A large plesiotypic family are the oxysalts based on intergrowths of slices of **pyrochlore-like structure** (Fig. 1.46) with slabs based on complex tetrahedral anions. For example, one- and two octahedra thick (111) **pyrochlore layers** alternate with layers of $Si(Ge)_4O_{12}$ rings in **K-nenadkevichite** $(Na,K)(Nb,Ti)_2[Si_4O_{12}](O,OH)_2 \cdot 1.6\,H_2O$ (Rastsvetayeva *et al.* 1994) and $\mathbf{Cs_4Sb_4O_8[Si_4O_{12}]}$ (Pagnoux *et al.* 1992), respectively; one- to three octahedra thick pyrochlore (001) layers alternate with slabs of PO_4 tetrahedra (plus interlayer water), Ge_2O_7 groups or of Si_3O_9 groups in $K_3Sb_3O_6(PO_4)_2\ xH_2O$, $Cs_3Sb_3O_6[Ge_2O_7]$ and $\mathbf{Cs_8Nb_{10}O_{23}[Si_3O_9]_2}$ (Fig. 1.47) (Pagnoux *et al.* 1993; Tournoux *et al.* 1992). Interestingly, the recently descibed structure of **Na-komarovite** $Na_{5.5}Ca_{0.8}La_{0.2}Ti_{0.5}Nb_{5.5}Si_4O_{26}F_2 \cdot H_2O$ (Balič Žunič *et al.* 2002*a*) is a configurational homologue of $Cs_4Sb_4O_8[Si_4O_{12}]$, with all Si_4O_{12} groups oriented the same way whereas in the Cs–Sb compound their layers had alternating orientations; **pyrochlore-like layers** are three octahedra thick in komarovite (Fig. 1.48) whereas only two octahedra thick in the cesium compound.

Single **pyrochlore-like octahedral layers** occur in the structures of the alunite–jarosite–crandallite family $K(Al/Fe)_3(SO_4)_2(OH)_6$–$Ca(Al/Fe)_3(P/AsO_3OH)(P/AsO_4)(OH)_6$ (Strunz and Nickel 2001). Triplets of inclined octahedra are decorated by (SO_4) or (P/AsO_4) tetrahedra; these decorated octahedral layers (0001) are intercalated by large cations.

1.8.3.4 *Serpentinite-like structures*

The family of **serpentinite-like structures** (Mg-based 1 : 1 **phyllosilicates**) that includes lizardite as an 'aristotype', the chrysotile varieties, the antigorite homologous

Table 1.17 Selected sulphosalts with rod-based layer structures

Mineral	Chemical formula	Layer type	Rod type	N[a]	N′[a]	Lattice parameters[b] (Å)				Space group	Reference
SnS archetype											
Jamesonite	$FePb_4Sb_6S_{14}$	2	[001]SnS	3	4	*a* 15.57	*b* 18.98	*c* 4.03	β 91.8°	$P2_1/a$	Niizeki and Buerger (1957)
Synthetic	$Pb_4Sb_4S_{11}$	5	[001]SnS	2	4	*b* 15.56	*a* 15.01	*c* 4.03	–	*Pb*am	Petrova *et al.* (1979)
Boulangerite	$Pb_5Sb_4S_{11}$	1	[001]SnS	3	6	*b* 23.51	*a* 21.24	*c* 4.04	–	*Pbnm*[c]	Petrova *et al.* (1978)
						b 23.54	*a* 21.61	*c* 8.08	β 100.71°	$P2_1/a$;	Mumme (1989)
Synthetic	$Sn_3Sb_2S_6$	4	[001]SnS	5	6	*c* 34.91	*a* 23.15	*b* 3.96	–	*Pn*ma	Smith (1984); Parise *et al.* (1984)
Robinsonite	$Pb_4Sb_6S_{13}$	1	[001]SnS	3	4	*b* 17.69	*a* 16.56	*c* 3.98		*Pn*am	Makovicky *et al.* (in press).
		4	[001]SnS	2	4	*a* 96.5°	β 97.8°	γ 91.1°			
Synthetic	$Sn_4Sb_6S_{13}$	1	[001]SnS	3	4	*a* 24.31[d]	*c* 23.49	*c* 3.92	β 94.05°	*I2/m*	Jumas *et al.* (1980)
		4	[001]SnS	2	4						
Synthetic	$Pb_{12.65}Sb_{11.35}S_{28.35}Cl_{2.65}$	1	[001]SnS	2	6	*c* 35.13	*a* 19.51	*b* 4.05	β 96.34°	*I2/m*	Kostov and Macíček (1995)
		5	[001]SnS	2	4						
Dadsonite	$Pb_{10+x}Sb_{14-x}S_{31-x}Cl_x$	3	[001]SnS	3	4	*c* 17.33	*a* 19.04	*b* 8.23	β 96.30°	monoclinic,	Cervelle *et al.* (1979)
		5	[001]SnS	3	4					triclinic;	Mumme and Makovicky (in preparation)
Synthetic	$Pb_5Sb_6S_{14}$	6	[001]SnS	4	4	*a* 28.37[d]	*c* 22.04	*c* 4.02	β 92.28°	*I*1	Skowron and Brown,
		4	[001]SnS	2	4						(written communication)
Synthetic	$Pb_7Sb_4S_{13}^{f}$	6	[001]SnS	4	6	*a* 23.67	*b* 25.55	*c* 4.00	–	*Pn*am	Skowron and Brown, (written communication)
Cosalite	$Pb_2Bi_2S_5^{e}$	3	[011]PbS	4	4	*b* 23.89	*a* 19.10	*c* 4.06	–	*Pb*nm	Srikrishnan and Nowacki (1974)

Synthetic	$KBi_{6.33}S_{10}$	3	[011]PbS	4	4	*a* 24.05	*c* 19.44	*b* 4.10	–	*P*nma	Kanatzidis *et al.* (1996)
Synthetic	$Pb_4Sb_4Se_{10}$	3	[011]PbS	4	4	*a* 24.59	*b* 19.58	*c* 4.17	–	*P*nam	Skowron and Brown (1990*c*)
Synthetic	$Pb_4In_9S_{17}$	3	[011]PbS	4	3	*a* 22.76	*b* 15.20	*c* 3.86	–	*P*bam	Ginderow (1978)
Synthetic	$Sn_7In_{18}S_{34}$	3	[011]PbS	4	3	*b* 22.70	*a* 15.12	*c* 3.83	–	*P*bam	Likforman *et al.* (1987)
Synthetic	$Ce_{1.25}Bi_{3.78}S_8$	4	[011]PbS	3	4	*c* 21.52	*a* 16.55	*b* 4.05	–	*P*nam	Ceolin *et al.* (1977)
Synthetic	$KLa_{1.28}Bi_{3.72}S_8^g$	4	[011]PbS	3	4	*c* 21.59	*a* 16.65	*b* 4.07	–	*P*nma	Iordanidis *et al.* (1999)
Moëloite	$Pb_6Sb_6S_{14}(S_3)$	10	[001]SnS	3	4	*c* 23.05	*a* 15.33	*b* 4.04	–	$P2_122_1$	Orlandi *et al.* (2002)
Synthetic	$BaBiSe_3$	10	[011]PbS	2	4	*a* 17.24	*b* 16.00	*c* 4.37	–	$P2_12_12_1$	Volk *et al.* (1980)
Synthetic	$BaBiTe_3$	10	[011]PbS	2	4	*a* 18.09	*b* 16.94	*c* 4.64	–	$P2_12_12_1$	Volk *et al.* (1980)
Synthetic	$SrBiSe_3$	10	[011]PbS	4	4	*a* 33.55	*b* 15.76	*c* 4.26	–	$P2_12_12_1$	Cook and Schäfer (1982)
Synthetic	$Sr_6Sb_6S_{17}$	10	[001]PbS	3	4	*c* 22.87	*b* 15.35	*c* 8.29	–	$P2_12_12_1$	Choi and Kanatzidis (2000)

[a] *N* denotes number of coordination polyhedra along, *N′* number of atom planes across the rod-like element.

[b] When not indicated otherwise, the first parameter (or the relevant vector *d*) is parallel to the periodicity of the rod-layer which has rods infinite along the 4 Å direction.

[c] Presumed ordering variants for boulangerite. Structure refinements were published for $Pb_4Sb_6S_{13}$ in P1 (Petrova *et al.* 1978) and $I2/m$ (Skowron and Brown 1990*a*) and for $Pb_5Sb_4S_{11}$ in *Pnam* (Skowron and Brown 1990*b*, Makovicky *et al.* in press).

[d] Unit-cell vectors were selected diagonal to the direction and stacking of rod-layers.

[e] Minor elements (especially Cu^{2+}) present. According to Sugaki *et al.* (1987) synthetic cosalite represents $CuPb_7Bi_8S_{20}$.

[f] Structure proposed on the basis of unit-cell parameters and weighted reciprocal lattice (compared with boulangerite).

[g] Isostructural are $RbCe_{0.84}Bi_{4.16}S_8$, $RbLa_{1\pm x}Bi_{4\pm x}S_8$, $KCe_{1\pm x}Bi_{4\pm x}S_8$, $KPr_{1\pm x}Bi_{4\pm x}S_8$ and $KNd_{1\pm x}Bi_{4\pm x}S_8$. All these compounds are homeotypes of $Ce_{1.25}Bi_{3.78}S_8$.

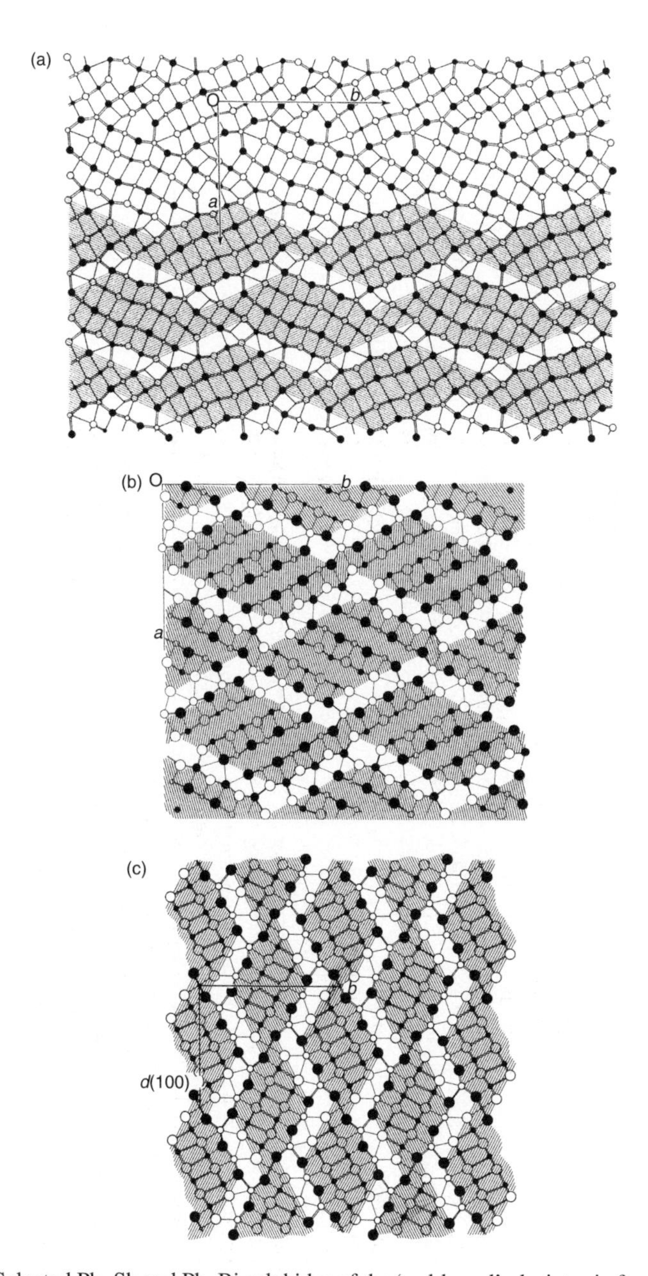

Fig. 1.42. Selected Pb–Sb and Pb–Bi sulphides of the 'rod-based' plesiotypic family: (a) cosalite $Pb_2Bi_2S_5$, (b) boulangerite $Pb_5Sb_4S_{11}$, (c) jamesonite $FePb_4Sb_6S_{14}$. References to individual structures are in Table 1.17. Rods based on PbS (for cosalite) and SnS (for the remaining structures) archetypes are indicated by shading, the non-commensurate interspaces are left void.

Table 1.18 Kobellite homologues, their derivatives and related phases with rods $[001]_{SnS}$

Mineral	Chemical formula[a]	No. of rod types	Rod[b] type	N	N'	Lattice parameters (Å)	Space[c] group	Reference
Kobellite	$(Cu,Fe)_2Pb_{12}(Bi,Sb)_{14}S_{35}$	2	[011]PbS	6[t]	3(+1)	a22.58 b34.10 c4.04	$P2_1/n$	Miehe (1972)
			[001]SnS	2	4	Monoclinic angle α90.0°		
Tintinaite	$Cu_2Pb_{10.7}Sb_{15.3}S_{35}$	2	[011]PbS	6[t]	3(+1)	a22.30 b34.00 c4.04	*Pnnm*	Harris *et al.* (1968)
			[001]SnS	2	4			Moëlo *et al.* (1984*a*)
Izoklakeite	$(Cu,Fe)_2Pb_{26.5}(Sb,Bi)_{19.5}S_{57}$	2	[011]PbS	6[t]	5(+1)	b37.98 a34.07 c4.07	*Pnnm*	Makovicky and Mumme (1986)
			[001]SnS	4	4			Armbruster and Hummel (1987)
Giessenite	$(Cu,Fe)_2Pb_{26.4}(Bi,Sb)_{19.6}S_{57}$	2	[011]PbS	6[t]	5(+1)	b38.05 a34.34 c4.06	$P2_1/n$	Makovicky and Karup-Møller 1986
			[001]SnS	4	4	Monoclinic angle β90.3°		
Eclarite	$(Cu,Fe)Pb_9Bi_{12}S_{28}$	2	[011]PbS	4[t]	3(+1)	c22.75 a54.76 b4.03	*Pnma*	Kupčik (1984)
			[011]PbS	2	4			
Synthetic	$Pb_4In_2Bi_4S_{13}$[d]	2	[011]PbS	3	3	c26.49 a21.34 b4.00	*Pcma*	Krämer (1986)
			[011]PbS	3	3			
Owyheeite	$Pb_{10}Ag_3Sb_{11}S_{28}$	2	[011]PbS	4	4	c22.93 b27.34 a4.11	$P2_1/c$	Olsen *et al.* (in preparation)
			[011]PbS	2	4			
Synthetical	$SnSb_2Se_4$[e]	2	Both [001] SnS[f]	2	4	a26.610 b21.066 c4.042	*Pnmm*	Smith and Parise (1985)

N denotes number of coordination polyhedra along, and N' number of atom layers across the rod element.

[a] Idealized formulae. Variations in the Pb/Me^{3+} ratio occur, connected with the variations in the $Cu^+/(Cu^2 + Fe)$ ratio.

[b] Independent rods.

[c] Bi-rich members of the kobellite homologous series have symmetry $P2_1/n$, Sb-richer members (tintinaite, Sb-kobellite and izoklakeite) are orthorhombic, *Pnnm* (Miehe 1972; Makovicky and Mumme 1986; Zakrzewski and Makovicky 1986).

[d] Both octahedrally and tetrahedrally coordinated In is present.

[e] $SnSb_2S_4$ (Smith and Parise 1985) and $PbSb_2Se_4$ (Skowron *et al.* 1994) are isostructural.

[f] An additional large cation is situated outside the rods.

[t] Rod cross-sections are truncated lozenges.

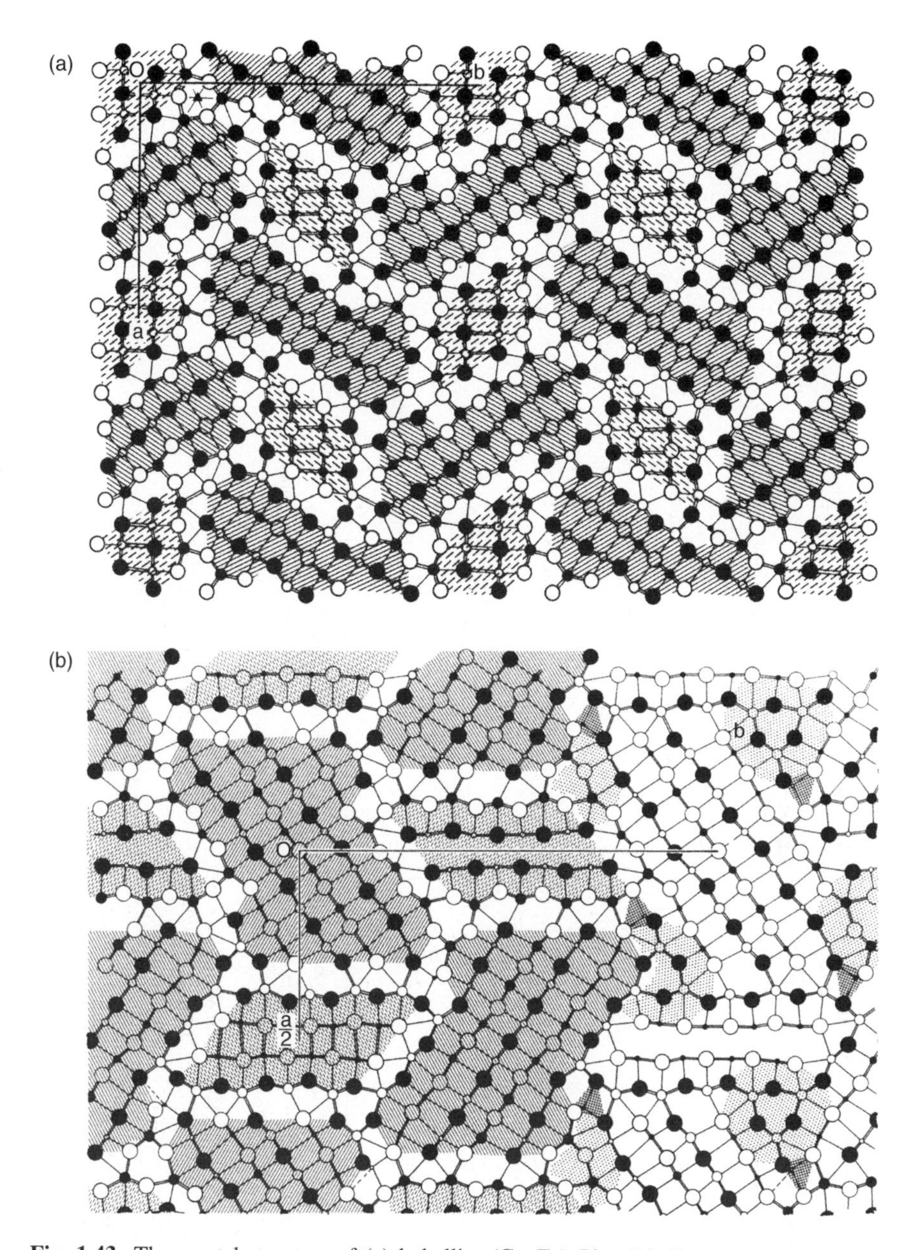

Fig. 1.43. The crystal structure of (a) kobellite $(Cu, Fe)_2Pb_{12}(Bi, Sb)_{14}S_{35}$ (Miehe 1971), a rod-based structure $N = 2$ with a chess-board arrangement of rods and (b) of its higher accretional homologue, izoklakeite $(Cu, Fe)_2Pb_{26.5}(Sb, Bi)_{19.5}S_{57}$ (Makovicky and Mumme 1986). In order of decreasing size, circles indicate S, Pb, mixed (Pb, Bi, Sb) positions, Bi or Sb and (Cu, Fe). PbS-based rods are hatched, SnS-based ones dashed; rings of Pb coordination prisms are stippled.

Table 1.19 Cyclically twinned rod-based sulpho-salt structures

Mineral	Chemical formula	Rod type	N^a	N'^a	Lattice parameters		Space group (Å)	Reference
Homologues of zinckenite								
Synthetic	$Bi_{0.67}Bi_{12}S_{18}I_2$	$(100)^b$ PbS	1	2	*a* 15.63	*c* 4.02	$P6_3^c$	Miehe and Kupčik (1971)
Synthetic	$Bi_{0.67}Bi_{12}S_{18}Br_2$	$(100)^b$ PbS	1	2	*a* 15.55	*c* 4.02	$P6_3$	Mariolacos (1976)
Zinckenite[d]	$Pb_9Sb_{22}S_{42}$–$CuPb_{10}Sb_{21}S_{42}$	[110] PbS	3^t	4	*a* 22.15	*c* 8.66	$P6_3^c$	Portheine and Nowacki (1975)
Homologues of $BaBi_2S_4^e$								
Synthetic	$Ba_9Bi_{18}S_{36}$	[110] PbS	2^t	4	*a* 21.71	*c* 4.16	$P6_3/m$	Aurivilius (1983)
Synthetic	$Ba_{12}Bi2_4S_{48}^f$	[011] PbS	3^t	4	*a* 25.27	*c* 4.18	$P6_3/m$	Aurivilius (1983)

[1] All these structures contain ring walls of coordination polyhedra of (primarily) Sb or Bi.

[a] For explanation of N and N', see Table 1.18.

[b] Segments of layers $(100)_{PbS}$.

[c] Also refined as $P6_3/m$.

[d] Moëlo (1982), see also Makovicky (1985*a*) for discussion.

[e] Expanded low-symmetry derivatives of $Ba_{12}Bi_{24}S_{48}$ (by means of intercalation with a rod-layer principle)–scainiite $Pb_{14}Sb_{30}S_{54}O_5$ (Moëlo *et al.* 2000) and pillaite $Pb_9Sb_{10}S_{23}ClO_{0.5}$ (Meerschaut *et al.* 2001) are in Table 1.20

[f] Isostructural are $Sr_{12}Bi_{24}S_{48}$ (Aurivilius 1983) and $Eu_{13.2}Bi_{24}S_{48}$ (Lemoine *et al.* 1986*b*).

[t] Rod cross-sections are truncated lozenges.

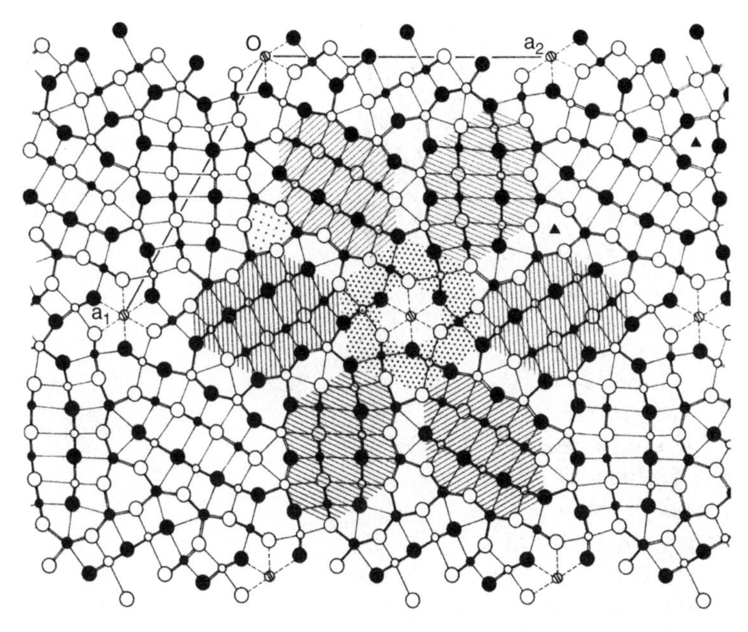

Fig. 1.44. The crystal structure of $Ba_{12}Bi_{24}S_{48}$ (Aurivilius 1983), homologue $N = 3$ of a series of rod-based structures with cyclic arrangement of PbS-based structural rods (ruled). In order of increasing size, circles indicate Bi, Ba and S. Channels contain statistically occupied cation sites.

Table 1.20 Doubly non-commensurate structures of box-work type

Mineral	Formula	Unit cell parameters (Å)				Space group	Reference
Neyite	$Ag_2Cu_6Pb_{25}Bi_{26}S_{68}$	a 37.53	b 4.07	c 43.7	β 108.80	$C2/m$	Makovicky *et al.* (2001*a*,*b*)
Pillaite[a]	$Pb_9Sb_{10}S_{23}ClO_{0.5}$	a 49.49	b 4.13	c 21.83	β 99.62	$C2/m$	Meerschaut *et al.* (2001)
Scainiite[a]	$Pb_{14}Sb_{30}S_{54}O_5$	a 52.00	b 8.15	c 24.31	β 104.09	$C2/m$	Moëlo *et al.* (2000)
Synthetic	$Er_9La_{10}S_{27}$	c 21.83	b 3.94	a 29.71	β 122	$C2/m$	Carré and Laruelle (1973)

[a] Quasi-homologues; related to cyclically twinned structures.

series and the carlosturanite series, is an outstanding plesiotypic family with a number of structural features in common. Section 4.3.4 contains detailed polysomatic treatment of these two groups.

2 : 1 phyllosilicates with tetrahedral sheets variously curled up and reversed in order to alleviate layer misfit (Guggenheim and Eggleton 1988) form another such family with several recognizable tiers of kinship.

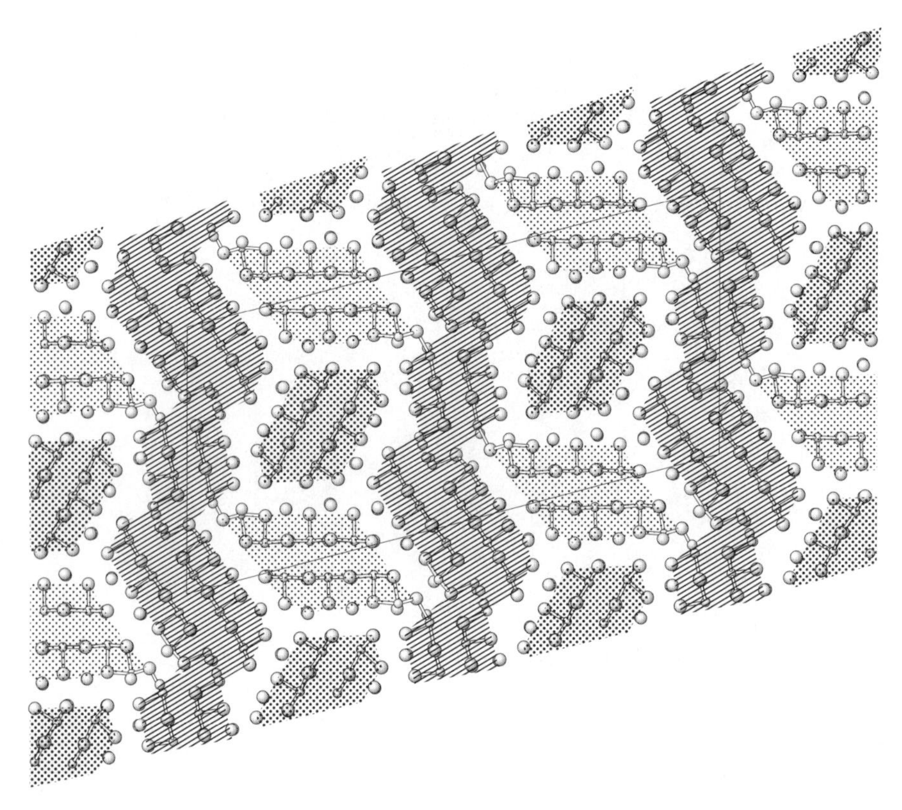

Fig. 1.45. Crystal structure of scainiite $Pb_{14}Sb_{30}S_{54}O_5$ (Moëlo *et al.* 2000). Complex wavy layers (100), consisting of two types of amalgamated rods are interconnected by $N = 3$ rods at $z = 0$, forming a box-work arrangement that encloses another type of $N = 3$ rods at $z = 1/2$. Central portions of rods are occupied by Sb, their surfaces primarily by Pb. Oxygen is shown as small circles situated very close to some Sb sites.

1.8.3.5 *Phyllosilicates with modular tetrahedral layers*

This group of plesiotypic structures was called by Guggenheim and Eggleton (1991) 'modulated layer silicates' although the changes in them are far from simple modulation. They form two subgroups, the 1 : 1 phyllosilicates and 2 : 1 phyllosilicates, that is, those with tetrahedral–octahedral layers and those with tetrahedral–octahedral–tetrahedral layers. Fe^{2+} and/or Mn^{2+} are the principal octahedral cations in these compounds and the size of their coordination octahedra is the reason of all complications in the tetrahedral layers with which Nature attempts to solve the resulting layer mismatch problems. In this distinctly plesiotypic group, we can include those structures for which the octahedral layer is continuous; cases in which it is not so are included among silicates with I-beam based structures (Section 1.8.2.9).

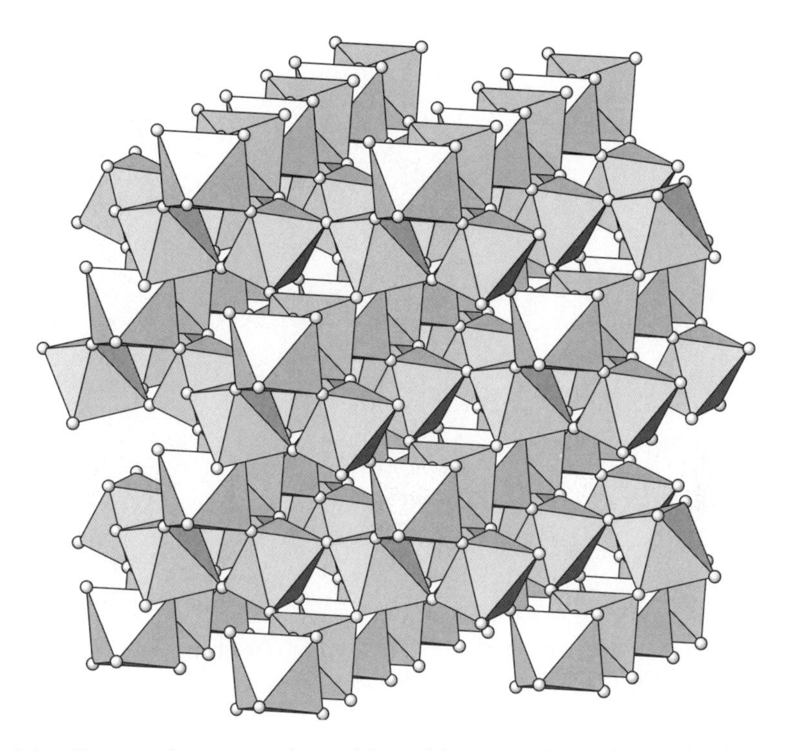

Fig. 1.46. The crystal structure of pyrochlore with non-octahedral ions omitted and (111) planes accentuated.

In the 1 : 1 subgroup, **greenalite** (Fe^{2+}) contains islands of 'normal' 1 : 1 arrangement, about three hexagonal rings in diameter which are out-of-phase with adjacent rings and the border areas contain inverted three- and four-tetrahedral rings attached to the adjacent octahedral layer (Guggenheim and Eggleton 1991). Related Mn-based **caryopilite** has reduced diameter of tetrahedral patches, in agreement with the mismatch of layers increasing from Fe^{2+} to Mn.

In the exceedingly complex and not yet quite understood **pyrosmalite group** $(Mn,Fe,Mg,Zn)_{16}(Si,Fe,Al,As)_{12}O_3(OH,Cl)_{20}$ (Guggenheim and Eggleton 1991), the tetrahedral layer attached to one octahedral layer is reduced to single six-fold rings of tetrahedra, which are, in respect to a continuous tetrahedral layer configuration, out-of-phase in such a way that they alternatively accomodate inverted six-fold rings and large 'empty' islands surrounded by 12 tetrahedra (Kato and Takéuchi 1983). Guggenheim and Eggleton (1991) propose for related **bementite** a different structure model, that of two rings broad strips of alternatively inverted tetrahedral layer composed of five, six and seven-fold rings.

Among the 2 : 1 phyllosilicates, in **zussmanite** $K(Fe,Mg,Mn,Al)_{13}(Si,Al)_{18}O_{42}(OH)_{14}$ (Lopes-Vieira and Zussman 1969) the 2 : 1 principle is limited to single rings that on each side of the octahedral layer again are out-of-phase in respect to the continuous layer and are interconnected by three-fold rings of tetrahedra, shared with the

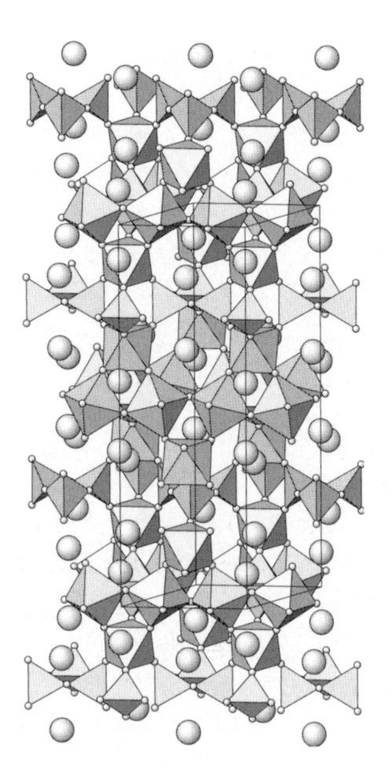

Fig. 1.47. Crystal structure of $Cs_8Nb_{10}O_{23}(Si_3O_9)_2$ (Tournoux *et al.* 1992) with triple-octahedral slabs (111) of pyrochlore structure separated by Si_3O_9 rings.

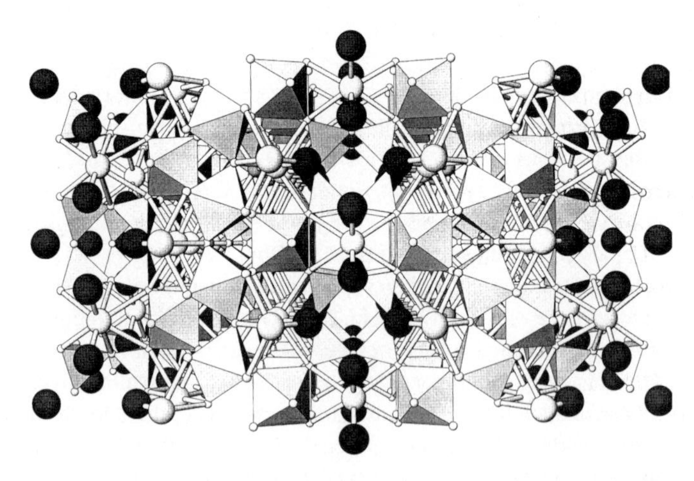

Fig. 1.48. The crystal structure of sodium komarovite $Na_{5.5}Ca_{0.8}La_{0.2}Ti_{0.5}Nb_{5.5}Si_4O_{26}F_2 \cdot 4H_2O$ (Balič Žunič *et al.* 2002*a*). Pyrochlore-like portions are three octahedra thick and interconnected by Si_4O_{12} groups.

opposing 2 : 1 layer. In the **stilpnomelane group** (Guggenheim and Eggleton 1991), typified, for example, by **ferrostilpnomelane** $K_{0.6}Fe_6(Si_8Al)(O,OH)_{27} \cdot 2H_2O$, islands of seven six-fold rings are not out-of-phase but every three such islands are connected by an inverted ring (a more complex form of 'rotational stacking fault' of Hyde and O'Keeffe (1973)) which, together with a corresponding ring from the opposing 2 : 1 layer, forms a ditrigonal double ring, pillaring the phyllosilicate structure.

Strips of tetrahedral layer, interconnected periodically across the interlayer space occur in **minnesotaite**, an Fe^{2+} analogue of talc (the interconnecting suture is a row of single tetrahedra) and **ganophyllite** $(K,Na,Ca)_6(Mn,Fe,Mg)_{24}(Si,Al)_{40.5}O_{96}(OH)_{16}\cdot 21H_2O$ (interconnection by a row of pillaring Si_4O_{12} rings) (Guggenheim and Eggleton 1991). Again, the preserved portions of a regular tetrahedral sheet are out-of-phase with each other.

Modelling of these plesiotypic structures has been done by Liebau (1985) and Guggenheim and Eggleton (1991); additional considerations are found in Makovicky and Hyde (1992). Guggenheim and Eggleton (1991) stress the possibility of various homologous series due to increasing/decreasing layer mismatch on substitution; the importance of the varying out-of-phase character of the patchwork of preserved tetrahedral sheets islands/strips has not been stressed as yet.

1.8.3.6 *Plesiotype family of W, Re and Mo bronzes*

O'Keefe and Hyde (1996) define 'bronzes' as compounds of variable composition involving in most cases alkali metal atoms combined with an early transition metal oxide. Coordination octahedra of W, Re, and Mo are corner-sharing, only rarely edge sharing. They outline trigonal, four-sided, pentagonal, hexagonal and even heptagonal channels along the *c* direction; in about all cases, several types of channels are intimately combined in one structure.

Perovskite-type bronzes contain only four-sided channels (Fig. 4.1). **Tetragonal tungsten bronzes** combine pentagonal, four-sided and trigonal channels (Fig. 1.49) whereas **hexagonal tungsten bronzes** combine hexagonal and trigonal channels (Fig. 1.50). The first group may be typified by ReO_3 $Pm\bar{3}m$ and Li_xWO_3 ($0.3 < x < 0.6$) (Rao and Ravenau 1995); the tetragonal bronzes by $\mathbf{K_{0.37}WO_3}$ (*P*4/*mbm*; Kihlborg and Klug 1973); the hexagonal ones by $\mathbf{Cs_{0.6}W_{10}O_{30}}$ ($P6_3/mcm$; Kihlborg and Hussain 1979); cation occupancies vary broadly (Hyde and O'Keefe 1973, Mitchell 2002). The three basic types are in fact archetypal structures that give rise to a plethora of structures built form columnar fragments of them in different combinations and proportions. Some of these **bronzoids** (Mitchell 2002) contain the same configurations as the tetragonal bronzes (i.e. the pentagonal, four-sided and trigonal channels), like $\mathbf{Ba_{0.15}WO_3}$ (*Pbmm*; Michel *et al.* 1984), and $\mathbf{KCuTa_3O_9}$ (*Pnc*2; Harneit and Müller-Buschbaum 1992). These two structures are crystallographic-shear derivatives of a tetragonal bronze structure, obtained by periodic CS reduction on planes (110) of the latter. $Ba_2NaWb_5O_{15}$ (*Pba*2; Foulon *et al.* 1996) is a true tetragonal bronze. Intergrowth of larger portions (two octahedra broad strips) of perovskite-like (i.e. 'framework') bronze with single strips of hexagonal bronze is

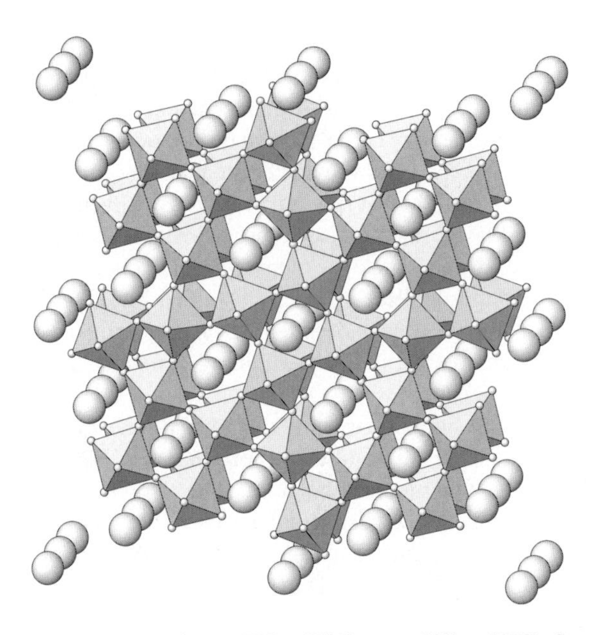

Fig. 1.49. The crystal structure of $K_{0.37}WO_3$ (Kihlborg and Klug 1973). Structure combines triangular, square, and pentagonal tunnels containing potassium sites.

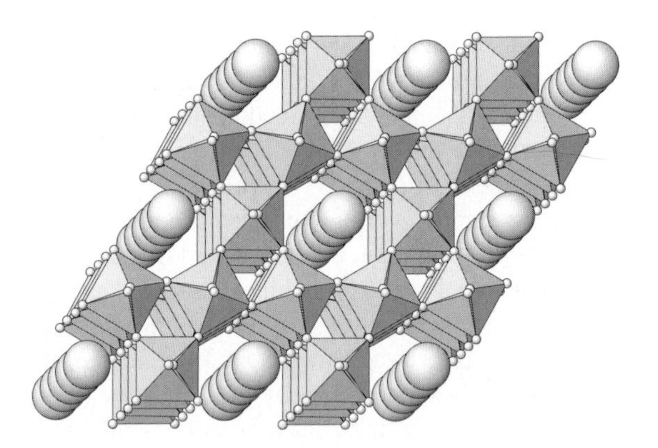

Fig. 1.50. Crystal structure of $Cs_{0.6}W_{10}O_{30}$ (Kihlborg and Hussain 1979), an archetypal hexagonal tungsten bronze. Cesium ions fill the hexagonal channels parallel to the *c* axes.

typified by $Ca_2TlTa_5O_{15}$ (Ganne *et al.* 1979). This compound is a member of a potential homologous series. Sheared strips of hexagonal bronze with ∼(110) strips of perovskite-like bronze occur in $W_{18}O_{49}$ (*P2/m*; Lamire *et al.* 1987).

A modular systemization of more complex bronzoid structures can best be achieved by recourse to Hyde and O'Keefe's (1973) derivation of a 'rotational stacking fault' in an ideal $\mathbf{ReO_3}$ framework (Fig. 1.51). The resulting cluster composed of a central

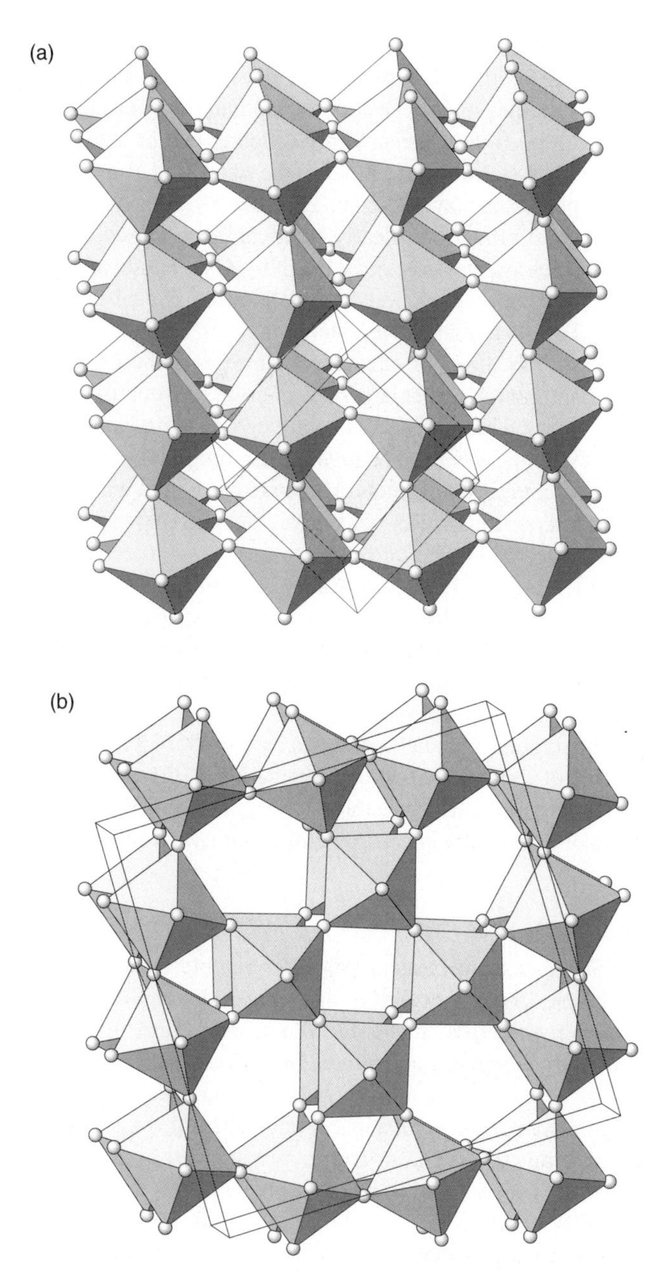

Fig. 1.51. Rotational stacking fault in perovskite array creating an element of 'tungsten bronze' structure. Cut-outs of real structures without and with the fault.

rotated square channel, four surrounding pentagonal channels, and intermeshed trigonal channels will be the basic module of reference. In the structure of tetragonal bronze, these modules intermesh and overlap without contradiction. Thus, such a cluster is, in fact, a portion of the structure of tetragonal bronze. In $\mathbf{Nb_{18}W_{16}O_{93}}$ (*Pbam*; Stephenson 1968), a distorted motif of the same kind as in a tetragonal bronze is present, with a part of pentagonal channels filled by Nb. Symmetry of the latter channels changes from bilateral to nearly rotational as a consequence of Nb occupancy. In the pseudotetragonal bronzoid $\mathbf{Tl_{3.84}Nb_{28}O_{73}}$ (Bhide and Gasperin 1979), these clusters are disjoint and separated by walls of combined hexagonal and heptagonal channels. The change in cluster interconnection again is caused by the shape of Nb-occupied channels. If a pair of rotation centres based on adjacent columns of ReO_3 structure is imagined, an 'enlarged' version of a group of two adjacent pentagonal channels with intervening and surrounding octahedra is obtained. Structures of $W_{18}O_{49}$ (Lamire *et al.* 1987) and $Mo_{17}O_{47}$ (Kihlborg 1963) contain this enlarged version of a pair of pentagonal channels. These are combined with open or collapsed six-fold channels and portions of ReO_3-like structure. All these structures and further examples are illustrated in Mitchell (2002).

Thus, the complex bronzoid structures can be described as a patchwork of diverse portions from the perovskite-like, tetragonal and hexagonal bronzes, that is, as intergrowth of variously large and variously shaped rods of these archetypal structures, with or without involvement of crystallographic shear, that is, as a family of plesiotypes in its broadest sense.

Regularity of pentagonal channels occupied by Nb, W, or Mo, as distinct from the bilateral symmetry of those occupied by large cations (Ba), or even left empty, has substantial influence on rod interconnection. It is a major contribution to the formation of complex bronzoid structures instead of simple tetragonal bronzes.

1.8.3.7 *Kyanite plesiotypes*

Kyanite, Al_2SiO_5, is a dense structure based on cubic close packing of oxygens (Winther and Ghoose 1979). Ten percent of tetrahedral and 40 percent of octahedral voids are, respectively, filled by Si and Al. Al octahedra form zig-zag chains parallel to [001], isolated tetrahedra are concentrated into rows parallel to the same direction. Zig-zag chains of octahedra are loosely arranged into one-to-two octahedra thick layers (100), separated by zig-zag interspaces of octahedral voids and SiO_4 tetrahedra.

Kyanite is a desymmetrized OD structure, as witnessed by its twinning on (100). Unit layers can be defined as bounded by (100) planes at $x = 0$ and $x = 1.0$ of Winter and Ghose's (1979) structure (Fig. 1.52(a)). Idealized, they have layer symmetry *P(b)cm*, with 2 positions of the consecutive layer $1/2b + 1/2c$ apart.

There exists a $(1/2t)$ shear (or slip) derivative of kyanite structure. It is **yoderite** $Mg(Al,Fe)_3O(OH)[SiO_4]_2$, the structure of which was determined by Fleet and Megaw (1962) who disregarded satellite reflections included in the study by Higgins *et al.* (1982) (Fig. 1.52(b)). Space group is $P2_1/m$, i.e. higher than for kyanite, $P\bar{1}$. The structural formula of yoderite is ${}^{VI}(Mg,Al,Fe^{3+})_4{}^{V}(Mg,Al,Fe^{3+})_4O_2(OH)_2[SiO_4]_4$,

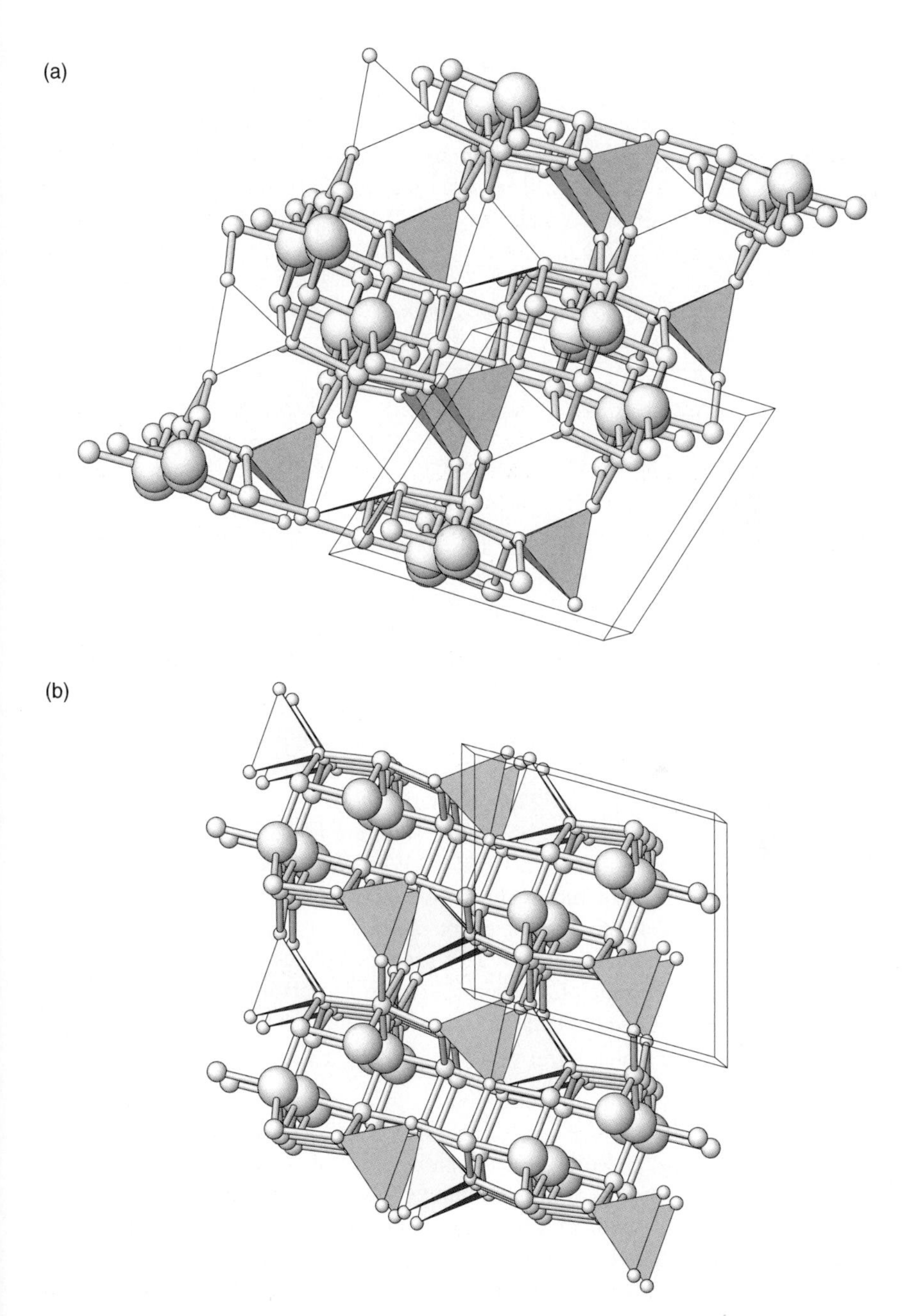

Fig. 1.52. Crystal structures of kyanite Al_2SiO_5 (Winter and Ghose 1979) and yoderite $Mg(Al,Fe)_3O(OH)[SiO_4]_2$ (Higgins *et al.* 1982), two structures composed of layers with the same configuration but related by non-OD polytypy. The boundary polyhedra have CN = 6 and 5, respectively.

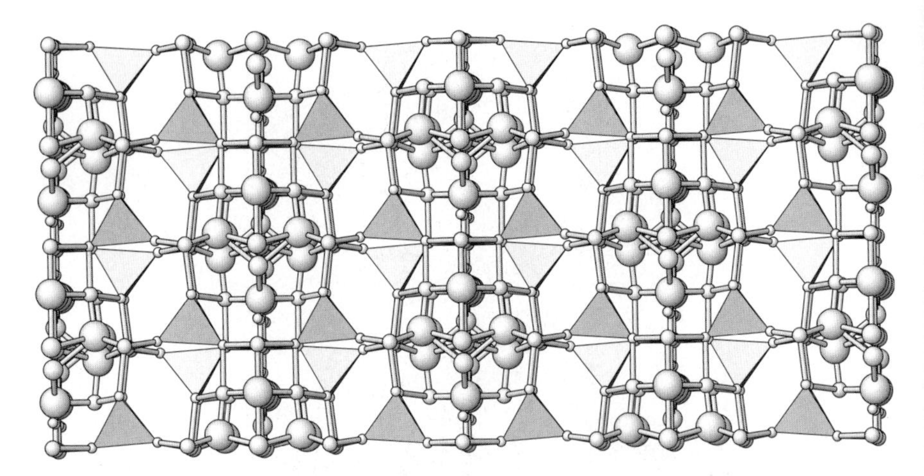

Fig. 1.53. The crystal structure of staurolite $(Fe,Mg)_4Al_{17}O_{13}[(Si,Al)O_4]_8(OH)_3$ (Tagai and Joswig 1985), a structure with kyanite-like slabs (vertical) separated by layers of Fe tetrahedra and Al octahedra; it is a merotype of kyanite.

the element exchange against kyanite is Mg + OH forAl + O. One half of octahedral sites in kyanite are altered to five-coordinated, trigonal bipyramidal, by a half-period slip along *c* on the same planes [(001) in yoderite orientation matrix] that served as boundaries of OD unit layers in kyanite. This structure, modified by shear planes with zero volume reduction, can also be described as a configurational non-OD polytype to kyanite. Nature recognizes this by producing frequent oriented macroscopic intergrowths of these two phases (Higgins *et al.* 1982).

A striking merotype extension of kyanite structure is the structure of **staurolite** (Náray-Szabó 1929; Takéuchi *et al.* 1972a; Griffen and Ribbe 1973, among others). In principle, the kyanite structure is sliced on the same (100) planes [(010) planes of staurolite] and intercalated by a single cation–anion sheet $Al_{0.7}Fe_2O_2(OH)_2$ (Náray-Szabó and Sasváry 1958) (Fig. 1.53). This sheet completes coordination octahedra of Al and represents, in principle, a chemically non-conservative shear of kyanite on $(100)_{kyanite}$ with *extension* of the structure by a width of half-octahedron on every crystallographic shear plane. Shear vector is $[\bar{2}21]_{kyanite}$.

1.8.3.8 *Sliding series*

The crystal structures of a **sliding series** (a particular kind of plesiotype series) consist of structure slabs of constant type (configuration) and stoichiometry that are limited on both sides by layer fragments arranged en échelon. These, together with the rest of the slab can slide along each other. Surfaces of this boundary layer and its mode of attachment to the rest of the slab remain unchanged in the sliding process. It, however, alters the length of overlap intervals for the en échelon stacking and generates new volume and new coordination polyhedra (although analogous to the existing ones) along the plane of sliding. As a consequence, chemical composition of individual sliding derivatives changes by constant increments in this process (Makovicky *et al.* 2000).

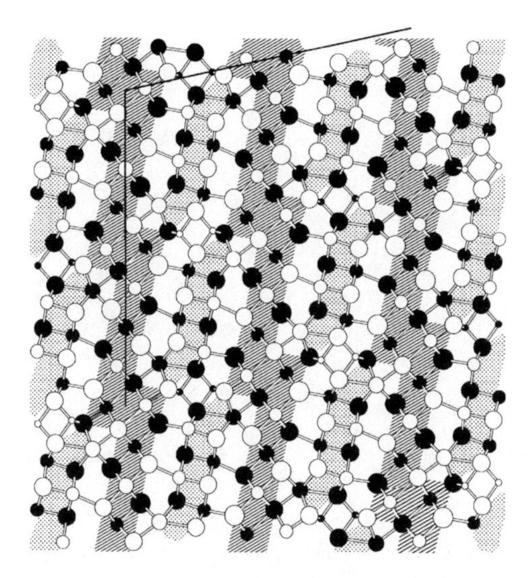

Fig. 1.54. The crystal structure of junoite $CuPb_3Bi_7(S,Se)_{14}$ (Mumme 1980), the $N = 1$ member of the junoite–felbertalite homologous pair. In the order of increasing size, circles indicate Cu, Bi, Pb, and S. Sheared pseudohexagonal slabs are ruled, pseudotetragonal ribbon-like fragments with pairs of Cu tetrahedra in the step regions are stippled.

A simple example of this series is based on a pair **junoite** $Cu_2Pb_3Bi_8(S,Se)_{16}$–**felbertalite** $Cu_2Pb_6Bi_8S_{19}$, for which structure determinations were performed by Mumme (1975*a*) and Topa *et al.* (2000*b*). These structures are composed of fragments of $(100)_{PbS}$ layer, two atomic layers thick and three coordination polyhedra broad (stippled in Figs. 1.54 and 1.55) alternating with *en échelon* fragments of $(111)_{PbS}$ octahedral layers which have a thickness of one octahedron (Fig. 1.54) or two octahedra (Fig. 1.55). The former fragments form the interior, and the latter the contact surfaces of the above defined mutually sliding slabs.

Owing to their difference in the thickness of octahedral layers, these two compounds were recognized as the first and second homologue of an accretional homologous series (Topa *et al.* 2000*b*). From each of them a sliding series can be derived, most of it still awaiting to be discovered among complex sulphides.

The a, b, and d_{001} parameters are, in principle, constant for each of these monoclinic series, the β and c parameters vary. For the modelling purposes, the large di- and trivalent cations are lumped together in Tables 1.21 and 1.22. Figures 1.56–1.58 show the three sliding derivatives of each series; for the case with single-octahedral layers even the aberrant case of 'minus one octahedron' shift was derived (Fig. 1.57).

Distinct sliding derivatives will satisfy different configurational criteria such as the local valence balance or coordination details. Zero shift for felbertalite and plus-one-octahedron shift for junoite were both adopted by these sulphosalts in order to create a lone electron pair micelle, that is, an extended common space for lone electron pairs of bismuth in the overlapped, kinked portions of the $(111)_{PbS}$ layer. For octahedral elements without active lone electron pairs, other sliding derivatives might be formed instead.

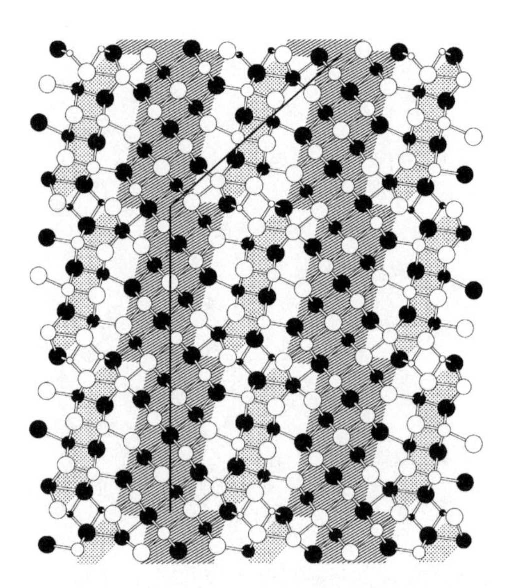

Fig. 1.55. The crystal structure of felbertalite $Cu_2Pb_6Bi_8S_{19}$ (Topa *et al.* 2000*a*,*b*), the $N = 2$ homologue of the junoite–felbertalite pair. These structures form the basis of a sliding series defined in the Section 1.8.2.8.

Table 1.21 Sliding derivatives of junoite $Cu_2Pb_3Bi_8(S,Se)_{16}$

Overlap n oct. heights	Formula	Note
3	$Cu_2M_{13}S_{18}$	Q fragments unchanged; H layer changes
2	$Cu_2M_{12}S_{17}$	–
1	$Cu_2M_{11}S_{16}$	Junoite
0	$Cu_2M_{10}S_{15}$	–
'−1'	$Cu_2M_9S_{14}$	–
n	$Cu_2M_{10+n}S_{15+n}$	Single-octahedral H layers

Table 1.22 Sliding derivatives of felbertalite $Cu_2Pb_6Bi_8S_{19}$

Overlap n oct. heights	Formula	Note
3	$Cu_2M_{17}S_{22}$	Q fragments unchanged; H layer changes
2	$Cu_2M_{16}S_{21}$	–
1	$Cu_2M_{15}S_{20}$	–
0	$Cu_2M_{14}S_{19}$	Felbertalite
n	$Cu_2M_{14+n}S_{19+n}$	Double-octahedral H layers

Note: Junoite and felbertalite preserve the lone electron pair micelle in the kinks of the H layer.

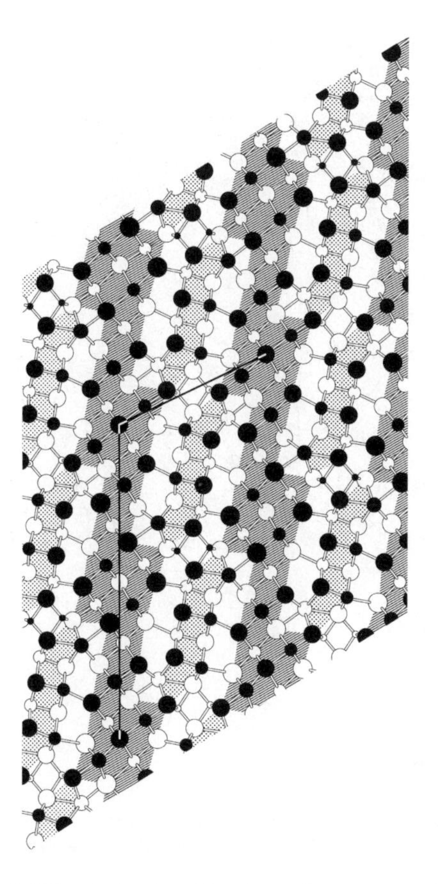

Fig. 1.56. A hypothetical sliding derivative of junoite with an overlap of two octahedral lengths in the $(111)_{PbS}$ layers. Compare with Fig. 1.54.

Other examples of sliding series are pairs $Pb_4Sb_6S_{11}$ (**robinsonite;** Skowron and Brown 1990)–$\mathbf{Pb_5Sb_6S_{14}}$ (synthetic; Skowron 1991) and $Pb_5Sb_4S_{11}$ (**boulangerite;** Petrova *et al.* 1978; Mumme 1989)–$\mathbf{Pb_7Sb_4S_{13}}$ (Skowron 1991). In the first pair (Fig. 1.59(a,b)), the structures consist of two kinds of layers, both based on a periodical series of interconnected structural rods of SnS-like archetype. The thinner layers and their interfaces with the thicker layers form a constant configuration (= slab) whereas the details of the internal structure of the thicker layers differ: the rods are three pseudotetragonal subcells broad in $Pb_4Sb_6S_{11}$ whereas they are four subcells broad in $Pb_5Sb_6S_{14}$. This difference arises formally from adding a diagonal slice with one additional Pb and Sb pair to the larger rod of the $Pb_4Sb_6S_{11}$ structure. The zig-zag interface to the thinner layer moves by a unit increment in this process, together with the entire slab, changing the phase composition as well. In the second pair, only one kind of rod-layers is present but again their interfaces form an unchanging configuration and the difference dwells in adding a diagonal slice to each rod, and the subsequent shift of adjacent interface by a unit increment.

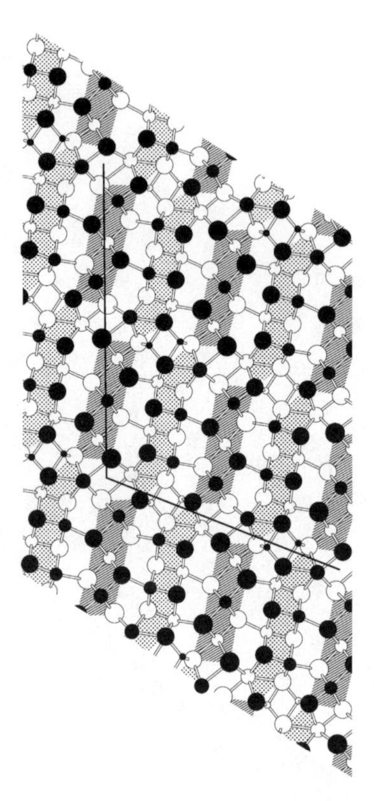

Fig. 1.57. A hypothetical sliding derivative of junoite with $(111)_{PbS}$ layers dismembered by a shift of minus one octahedron length. Compare with Figs 1.54 and 1.56.

1.8.3.9 *Silicate rod-layer plesiotypes (I-beam plesiotypes and related structures)*

The basic element of this family are 'TOT' rods formed by a ribbon of edge-sharing cation octahedra sandwiched on both surfaces by single-to-multiple silicate chains. Rods are parallel with each other, in the closest-packed arrangements forming a chess-board pattern in a section perpendicular to the rods, and to their layer-like arrangement. In the voids (channels) of the chess-board pattern, between 'free' sides of tetrahedral chains, large cations and/or water molecules may be accommodated. Marginal polyhedra on the sides of the octahedral strip can accommodate larger cations, eventually also coordinating to some of the water molecules in the channels.

Two large subdivisions of this family are related by local crystallographic shear—the common bases of tetrahedral chains can either be in the same plane, forming phyllosilicates with a periodic reversal of tetrahedral orientations—or collapsed by a height of an octahedron, all tetrahedral bodies being in the same layer, resulting in **inosilicates** with separated ribbons (Table 1.23). This family was defined by Egorov-Tismenko *et al.* (1996), its phyllosilicate branch was treated as plesiotypes by Ferraris (1997) and its plesiotypic aspects were discussed for $N = 1$ chains by Merlino and

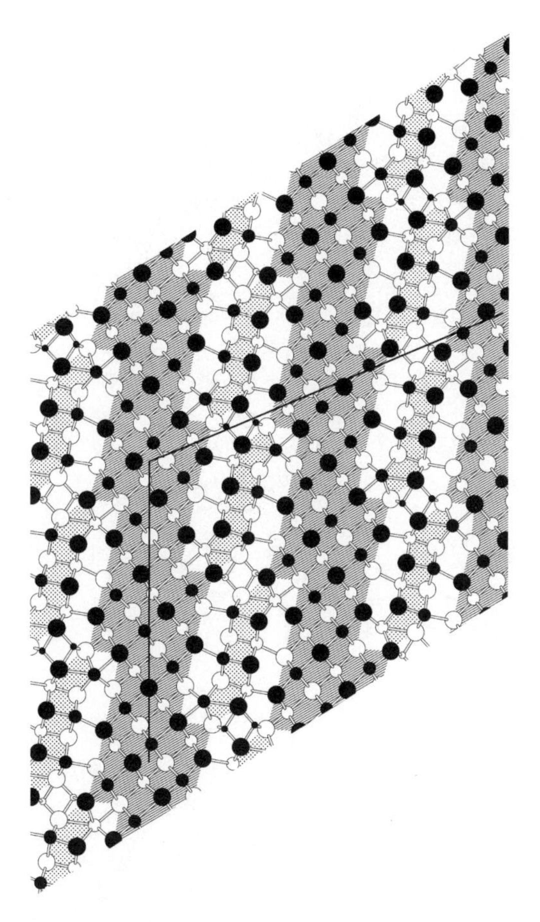

Fig. 1.58. A hypothetical sliding derivative of felbertalite with an overlap of two octahedral lengths in the $(111)_{PbS}$ layers. Compare with Fig. 1.55.

Pasero (1997). With the exception of inosilicates, **carpholite** and **nchwaningite** the phases enumerated here are treated in detail in the Sections 4.3.3 and 4.3.5.

The width of octahedral slabs, and that of silicate chains can vary from the single-chain **silinaite** (Grice 1991) to the triple-chain palygorskite (Artioli and Galli 1994) in the case of phyllosilicates, and from single to at least triple-chain structures in the inosilicate subfamily (Veblen and Burnham 1978). Although complying with the general structural principles for this family, **intersilite** (Egorov-Tismenko *et al.* 1996) widens its scope by introducing intergrowths of six-, eight-, and fivefold rings in the silicate sheet.

Combinations of these two subfamilies in one structure justify their treatment as a family of plesiotypes. In the members of the **silinaite** $Na_2Li_2Si_4O_{10} \cdot 4H_2O$– **lorenzenite** $Na_4Ti_4(Si_2O_6)_2O_6$ series (Merlino and Pasero 1997) (Fig. 4.29(a,b)),

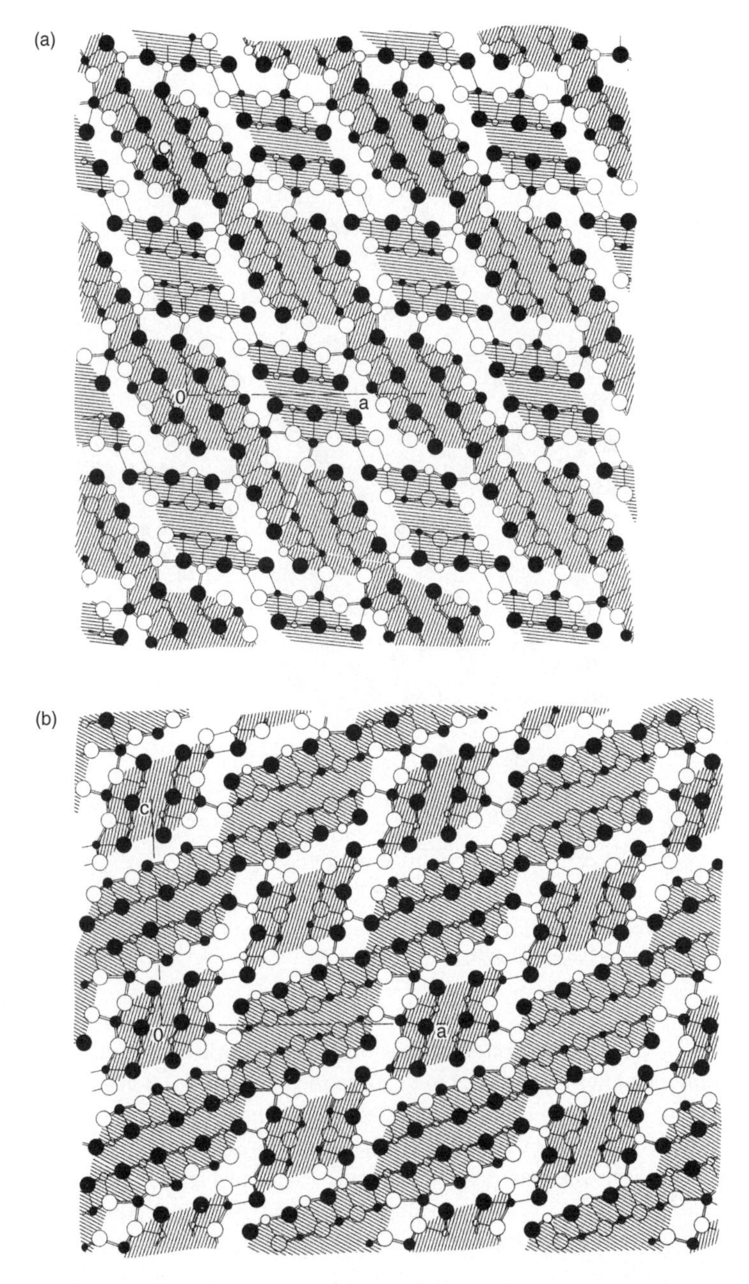

Fig. 1.59. The crystal structures of (a) robinsonite $Pb_4Sb_6S_{11}$ (Skowron and Brown 1990 *a*) and (b) the synthetic sulphosalt $Pb_5Sb_6S_{14}$ (Skowron 1991), a sliding-series pair. Details in the text.

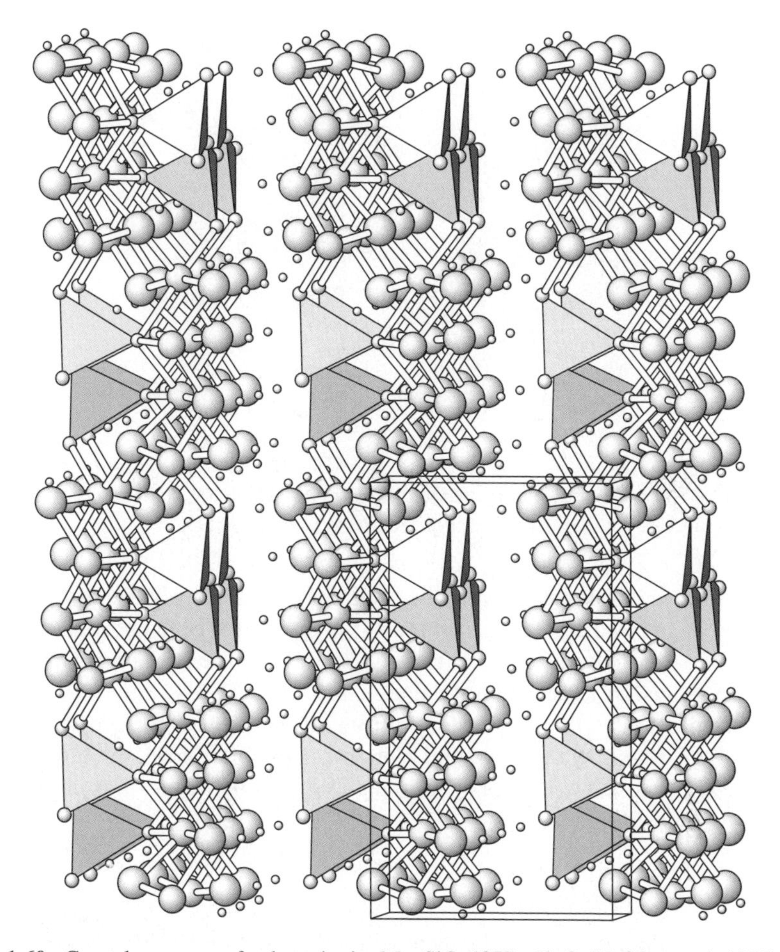

Fig. 1.60. Crystal structure of nchwaningite $Mn_2SiO_3(OH)_2{\cdot}H_2O$ (Nyfeler *et al.* 1995) with pyroxene-like chains and staggered octahedral ribbons.

the sole intermediate phase, **lintisite** $Na_3LiTi_2[Si_2O_6]_2O_2 \cdot 2H_2O$ is an intergrowth of pyroxene-like slabs limited by octahedral ribbons rather than tetrahedral surfaces with a truncated modified **palygorskite-like structure** ($N = 1$), which misses the inverted parts of the tetrahedral layer. Another modified structure is **nchwaningite** $Mn_2^{2+}SiO_3(OH)_2 \cdot H_2O$ (Nyfeler *et al.* 1995). In this Mn silicate, crystallographic shear starts again from the palygorskite principle but goes by one polyhedron width further than in pyroxenes, so that the octahedral strips are now on the level of adjacent pyroxene-like silicate chains. Octahedral strips share corners forming a double-layer. The double-layers consisting of 'half-I beams' bind via hydrogen bonding (Fig. 1.60).

Carpholite-group minerals (**carpholite** $MnAl_2(Si_2O_6)(OH)_4$, with ferro-, magnesio-, and Li–Ba analogues) are another branch of this plesiotypic family: I-beams are formed by Si_2O_6 chains with periodicity 2, sandwiching Al octahedral

Fig. 1.61. The crystal structure of karpholite $MnAlSi_2O_6(OH)_4$ (Lindemann *et al.* 1979). Pyroxene-like I-beam modules are separated by Mn–Al octahedral walls defining channels [001]. A member of the plesiotype family of rod-based silicate structures.

ribbons. However, the voids between the back sides of silicate ribbons that face each other are enlarged by the pillaring walls of Al polyhedra; the latter are interconnected by Mn in the space between adjacent ribbons (Fig. 1.61). The cavities generated may be partly occupied by large cations (Gaines *et al.* 1997).

Lorenzenite $Na_4Ti_4(Si_2O_6)O_6$ differs from this arrangement: the pyroxene-like chains build I-beams with octahedral ribbons of double thickness (Merlino and Pasero 1997), securing again enlarged, pillared spaces between—now chess-board arranged—beams (Fig. 4.29(b)).

1.8.3.10 *Heterophyllosilicates*

Heterophyllosilicates are crystal structures consisting of (a) so-called HOH silicate layers where O is the central octahedral sheet and H ('heterogeneous sheet') consists of ribbon-like strips of a mica-like tetrahedral sheet joined via rows of coordination octahedra (respectively half-octahedra), and (b) intermediate layers of widely different thickness, composition and configuration, which include the apical parts of the above rows of octahedra. The intermediate layers vary from single atomic layers to six to seven atomic layers in thickness and vary from purely cationic to complex structure slices comprising cations, simple and complex anions (PO_4, SO_4) and H_2O molecules.

This family displays polysomatic expansion in the HOH layers, cases of homologous expansion in the interlayers, extensive *merotypy* with a variety of layers intermediate to the HOH layers, plesiotypism in the development of octahedral, half-octahedral, or more complex polyhedra in the H layer and a spectrum of *polytypes* of various kinds. Principal reviews were prepared by Ferraris (1997), Egorov-Tismenko (1998) and Christiansen *et al.* (1999). It will be dealt with extensively in Sections 4.3.1 and 4.3.2.

Table 1.23 Selected rod-layer (I-beam) silicate structures

Mineral/phase	Formula	Space group	Lattice parameters				Note	Reference
			a (Å)	b (Å)	c (Å)	$\beta(^\circ)$		
Enstatite	$Mg_2Si_2O_6$	*Pbca*	18.235	8.818	5.179		Pyroxene N = 1	Gaines *et al.* (1997)
Diopside	$CaMgSi_2O_6$	*C2/c*	9.739	8.913	5.253	106.02	Pyroxene N = 1	Gaines *et al.* (1997)
Anthophyllite	$(Mg,Fe)_7Si_8O_{22}(OH)_2$	*Pnma*	18.554	18.026	5.282		Amphibole N = 2	Gaines *et al.* (1997)
Tremolite	$Ca_2Mg_5Si_8O_{22}(OH)_2$	*C2/m*	9.84	18.02	5.27	104.7	Amphibole N = 2	Gaines *et al.* (1997)
Jimthompsonite	$(Mg,Fe)_5Si_6O_{16}(OH)_2$	*Pbca*	18.626	27.23	5.297	–	Biopyribole N = 3	Gaines *et al.* (1997)
Carpholite	$MnAl_2(Si_2O_6)(OH)_4$	*Ccca*	13.718	20.216	5.132		$N = 1$	Gaines *et al.* (1997)
Balipholite	$LiBaMg_2Al_3(Si_2O_6)_2(OH)_4F_4$	*Ccca*	13.587	20.164	5.144		$N = 1$	Gaines *et al.* (1997)
Lorenzenite	$Na_2Ti_2O_3(Si_2O_6)$	*Pbcn*	8.713	5.233	14.487		$N = 1$	Gaines *et al.* (1997)
Silinaite	$NaLiSi_2O_52H_2O$	*A2/n*	5.061	8.334	14.383	96.67	$N = 1$ tetrahedral I-beams	Chao *et al.* (1991)
Nchwaningite	$Mn_2SiO_3(OH)_2{\cdot}H_2O$	$Pca2_1$	12.672	7.217	5.341		$N = 1$	Nyfeler *et al.* (1995)
Lintisite	$Na_3LiTi_2O_2(Si_2O_6)_2 \cdot 2H_2O$	*C2/c*	28.583	8.60	5.219	91.03	$N = 1$	
Sepiolite	$Mg_8Si_{12}O_{30}(OH)_4 \cdot 12H_2O$	*Pncn*	13.40	26.80	5.28		$N = 3$	Brauner and Preisinger (1995)
Palygorskite	$Mg_5Si_8O_{20}(OH)_2 \cdot 8H_2O$	*C2/m*	13.27	17.868	5.279	107.38	$N = 2$	Artioli and Galli (1994)
Palygorskite	$Mg_5Si_8O_{20}(OH)_2 \cdot 8H_2O$	*Pbmn*	12.763	17.842	5.241	–	$N = 2$	Artioli and Galli (1994)
Kalifersite	$(K,Na)_5Fe^{3+}_7Si_{20}O_{50}(OH)_6 \cdot 12H_2O$	$P\bar{1}$	14.86	20.54	5.29		$N = 3, 2$	Ferraris *et al.* (1997)
			α 95.6°	β 92.3°	γ 94.4°			
Raite	$Na_3Mn_3Ti_{0.25}Si_8O_{20}(OH)_2 \cdot 10H_2O$	*C2/m*	15.10	17.60	5.29	100.5	$N = 2$	Pluth *et al.* (1997)
Intersilite	$(Na,K)Mn(Ti,Nb)Na_5(O,OH)(OH)_2Si_{10}O_{23}(O,OH)_2 \cdot 4H_2O$	*I2/m*	13.033	18.717	12.264	99.62	N is modified	Egorov-Tismenko *et al.* (1996)

1.8.3.11 *The datolite–melilite plesiotypes*

These are structures with tetrahedral sheets linked together by layers of, primarily, eight-coordinated calcium, sometimes combined with Mn octahedra (Table 1.24). The tetrahedral sheets consist of four- and eight-membered rings in **datolite** (and **herderite**), exclusively of five-membered rings in melilite, of their combination in harstigite, semenovite and samfowlerite (Table 1.24). **Aminoffite** contains four- and six-membered rings (Fig. 1.62). A portion of tetrahedra always point up and down, and represent single tetrahedra or tetrahedral groups T_2O_7, whereas the tetrahedra interconnecting them have their $\bar{4}$ axes perpendicular to the layers.

Leucophanite (Grice and Hawthorne 1989), a chain silicate, and melanophanite (Grice and Hawthorne 2002), a three-membered sorosilicate, represent substitutional variants of **melilite** (Ca,Na) $(Mg,Al)Si_2O_7$, a sorosilicate with Si_2O_7 groups; in these minerals certain silicate and non-silicate tetrahedral sites are exchanged (i.e. they are homeotypes).

The plesiotypic relationships (i.e. intergrowths of islands with different ring configurations) are in a sense a two-dimensional analogy to tungsten bronzes mentioned elsewhere in this volume. The 'winged tetrahedra' of melilite structures are fully preserved in aminoffite, although differently interconnected, but their pairs underwent crystallographic shear in datolite, and partly also in **harstigite** and **samfowlerite**. Harstigite topology can be obtained applying crystallographic shear to samfowlerite. Graph treatment of these relationships, by means of decorated connectivity nets is in Huminicki and Hawthorne (2002).

1.8.3.12 *Zeolites*

Several of the universally recognized zeolite families are very good examples of plesiotypic relationships. Zeolites of the **mordenite family** (Gottardi and Galli 1985; Armbruster and Gunther 2001) are alternatively described as composed of 5–1 tetrahedral groups (rings of five tetrahedra with an attached additional tetrahedron) on the secondary level, and as built of puckered 6^3 sheets interconnected by additional species-specific groups on quinary level. In this family, orthorhombic **mordenite** $(Na_2, K_2, Ca)_4[Al_8Si_{40}O_{96}]\cdot 28H_2O$ approximates a twice unit-cell twinned version of monoclinic **dachiardite**, $(Na, K, Ca_{0.5})_4[Al_4Si_{20}O_{48}] \cdot 18H_2O$. **Ferrierite** and mordenite are two homologues, the four-member double rings connecting the puckered 6^3 layers in mordenite are reduced to half of their thickness, altering the relevant unit cell parameter from 18.1 to 14.1 Å in ferrierite.

The **heulandite family** (*ibid.*) structures are on secondary level composed of $(Si,Al)_{10}O_{20}$ configurations (or 4–4–1 units; Alberti 1979) arranged into parallel chains that yield only weakly interconnected layers (010). In **heulandite/clinoptilolite/stilbite**, the above units which represent $4^2 \cdot 5^4$ clusters of tetrahedra, are corner-sharing; in **brewsterite**, the same groups are stacked *en échelon*, and edge-sharing. The (010) layers arise by interconnecting these chains of $4^2 \cdot 5^4$ elements by 4.5^2, $4^2.6$ and 4.6 chains, respectively, that is, some Si–O bonds

Table 1.24 Datolite–melilite plesiotypes

Mineral	Formula	Space group	Lattice parameters (Å) a	b	c		Note	Reference
Åkermanite	$Ca_2MgSi_2O_7$	$P\bar{4}2_1m$	7.84		5.01		Melilite str.	Kimata and Ii (1981)
Gugiaite	$Ca_2BeSi_2O_7$	$P\bar{4}2_1m$	7.43		5.00		Melilite str.	Yang *et al.* (2001)
Synthetic	$CaNaAlSi_2O_7$	$P\bar{4}2_1m$	7.63		5.05		Melilite str.	Louisnathan (1970)
Leucophanite	$CaNaBeSi_2O_6F$	$P2_12_12_1$	7.40	7.41	9.99		Inosilicate	Grice and Hawthorne (1989)
Meliphanite	$Ca_4Na_4Be_4AlSi_7O_{24}F_4$	$I\bar{4}$	10.51		9.89		Sorosilicate	Grice and Hawthorne (2002)
Aminoffite	$Ca_3Be_2Si_3O_{10}(OH)_2$	$P4_2/n$	9.86		9.93		Sorosilicate	Huminicki and Hawthorne (2002)
Harstigite	$Ca_6MnBe_4(SiO_4)_2(Si_2O_7)_2(OH)_2$	*Pnam*	9.79	13.64	13.83		Neso-sorosilicate	Hesse and Stümpel (1986)
Semenovite	$(Ce,La)_2(Ca,Na)_8(Fe,Mn)Be_6(Si(O,OH)_4)_2Si_{12}O_{32}F_6(OH)_2$	*Pnnm*	13.88	13.84	9.94		Inosilicate	Mazzi *et al.* (1979)
Samfowlerite	$Ca_{14}Mn_3(Zn,Be)_4Be_6(SiO_4)_6(Si_2O_7)_4(OH, F)_6$	$P2_1/c$	9.07	17.99	14.59	β 104.9°	Neso-sorosilicate	Rouse *et al.* (1994)
Herderite	$CaBe(PO_4)(F,OH)$	$P2_1/a$	9.82	7.70	4.81	β 90.1°	Neso-phosphate	Pavlov and Belov (1959)
Datolite	$CaB(SiO_4)OH$	$P2_1/c$	4.83	7.61	9.64	β 90.4°	Nesosilicate	Foit *et al.* (1973)

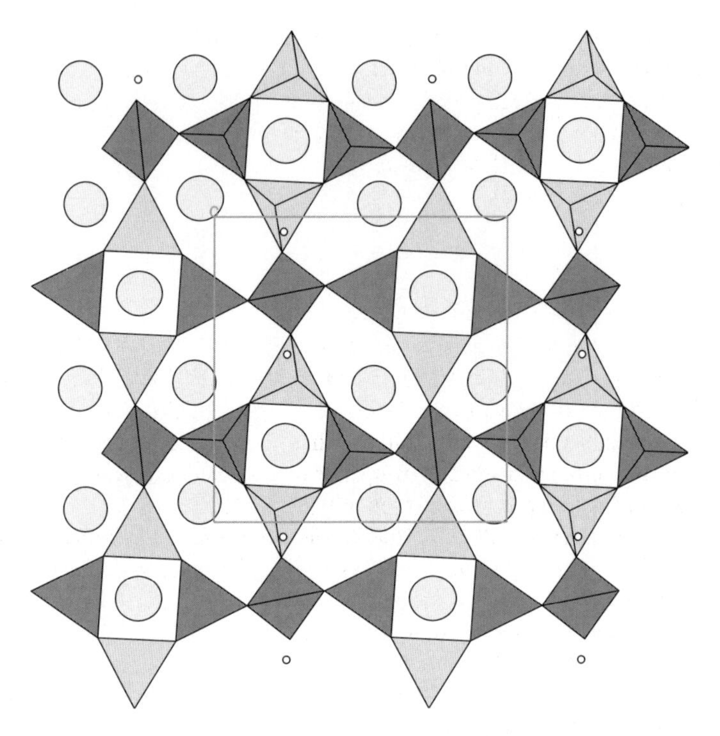

Fig. 1.62. A tetrahedral layer from the crystal structure of aminoffite $Ca_3Be(OH)_2Si_3O_{10}$ (Huminicki and Hawthorne 2002). Si-based tetrahedra are dark, Be-based ones light; Ca atoms are situated above and below the silicate layer; H sites are indicated by small circles.

differ by their structural role in different members of this family whereas their bulk build clusters identical for all structures.

Structures of the broad **chabasite–sodalite family** represent ABC and ABAB stackings of single- and double hexagonal rings in different combinations (Smith 1988). If these three-dimensional structures are considered as stacking of sheets composed of two types of rings and the associated cation sites, mixed-layer structures (polysomes and polytypes) are formed.

A nice example of an *ad hoc* plesiotypic family are zeolite structures constructed from **sodalite-like cages** (Meier *et al.* 1992). The 'sodalite' cage consists of 24 SiO_4 tetrahedra, with eight faces formed by six-fold rings and six faces by four-fold rings of tetrahedra.

In sodalite, all of these rings are shared with adjacent cages, forming a cubic aristotype ('topological') structure (space group $Im\bar{3}m$). Sharing four-fold *double rings* yields the structure of **Linde Type A** (LTA) zeolite, in which the I-centring sodalite cage was enlarged and represents an intersection of eight-fold channels. In LTA, the four-fold double rings are shared by two cages only; if they are shared by six sodalite cages, the latter are brought close together and share all hexagonal rings with neighbours: the cubic $Fm\bar{3}\,m$ structure of **AST zeolite** results.

Sharing of hexagonal double rings leads either to a cubic, very open structure of **faujasite** (FAU) with sodalite cages in ABC sequence (space group $Fd\bar{3}\,m$) or to a hexagonal open structure of **EMT zeolite** (AB sequence of cages) with space group $P6_3/mmc$ (Meier *et al.* 1992).

1.8.3.13 *Further examples of plesiotypy*

Hexagonal structures of the plesiotypic fluoborite family—**fluoborite** $Mg_6[BO_3](OH,F)_2$—**jaffeite** $Ca_6\ [Si_2O_7](OH)_6$, **yeremeyevite** $Al_6[BO_3]_5F_3$, and **painite** $Al_9CaZr[BO_3]O_{15}$ (Yamnova *et al.* 1993)—are topologically close to each other. In these structures, trigonal channels situated in a framework of octahedral walls are occupied by BO_3 groups, disilicate groups, or by combinations BO_3/F and BO_3/Zr (in trigonal coordination prisms), respectively (Fig. 1.63).

The crystal structures of $CaTi_2O_4$ (Bertaut and Blum 1956; a **lillianite-homologue** $N = 2$) and $\mathbf{CaFe_2O_4}$ (Apostolov and Bassi 1971) represent a plesiotypic pair. In common, they have bicapped trigonal prisms of Ca in anti-parallel rows, separated by partitions composed of double columns of coordination octahedra of Ti, respectively, Fe. The difference between the two structure types is the orientation of these octahedra: the same type of orientation for all paired columns of octahedra in $\mathbf{CaTi_2O_4}$ and a chess-board of two opposite orientations in $CaFe_2O_4$.

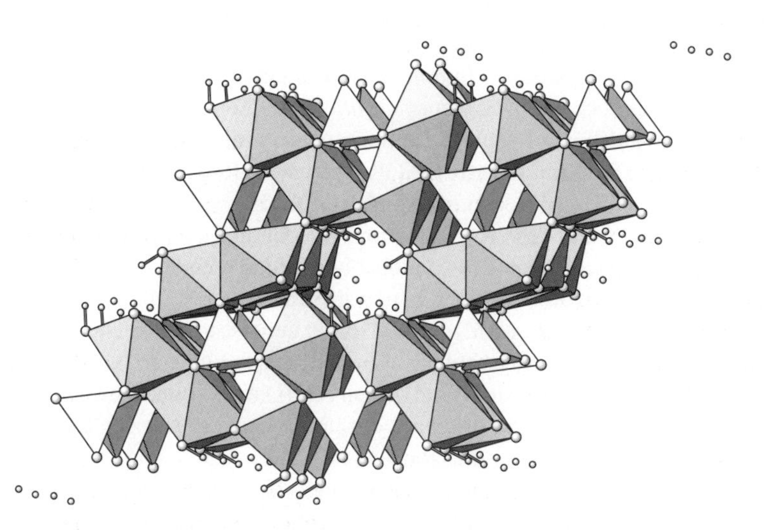

Fig. 1.63. Crystal structure of jaffeite $Ca_6[Si_2O_7](OH)_6$ (Yamnova *et al.* 1993). Double-octahedral ribbons form hexagonal channels lined by OH groups and trigonal channels housing Si_2O_7 groups. A member of the plesiotypic fluoborite family.

Inclusion of **paderaite** (Mumme 1986), **$Cu_4Bi_4S_9$** (Bente and Kupčík 1984) and **$Cu_4Bi_4Se_9$** (Makovicky *et al.* 2002) widens the **combinatorial cuprobismutite series** of unit-cell intergrowths into a plesiotypic family (Table 1.6) by relaxing the strictly homologous conditions imposed on the PbS-based layers or modifying the composite layers of Cu tetrahedra plus Bi pyramids.

1.9 Ordered derivatives of solid solutions as modular structures

Series of ordered derivatives of (usually) high-temperature solid solutions are by their nature series of modular structures—they contain modules of definite widths and their intergrowths or even form homologous series. We shall examine three examples of distinct nature.

Individual ordered structures of the **bismuthinite–aikinite** (Bi_2S_3–$CuPbBiS_3$) **solid solution** series were constantly described using three types of Me_4S_6 ribbons which in them combine in an ordered fashion, two at a time: **bismuthinite-like Bi_4S_6 ribbons**, **krupkaite-like $CuPbBi_3S_6$ ribbons**, and **aikinite-like $Cu_2Pb_2Bi_2S_6$ ribbons** (Ohmasa and Nowacki 1970; Mumme *et al.* 1976). Instead of this classification, we can use the observation that the above Cu atoms occupy slightly warped planes spaced at different intervals, that is at different submultiples of the Bi_4S_6 subcell. For the well-known central part of the series, three different modules have been defined:

(1) gladite-like modules (G) with copper-occupied planes $1\frac{1}{2}$ subcell period apart;
(2) krupkaite-like modules (K) with such planes 1 subcell period apart; and
(3) aikinite modules (A), with these planes $\frac{1}{2}$ subcell period apart.

In the structures from which they derive names, these modules occur in pure sequences, in those positioned intermediate to them, $1:1$ and $1:2$, or more complicated sequences occur (Makovicky *et al.* 2000; Topa *et al.* 2002*b*). Thus, in **salzburgite** (Fig. 1.64 and Table 1.25) the sequence of modules is gladite–gladite–krupkaite, that is, GGK, in **paarite** GKGK, whereas in **lindströmite** it is KKAKKA, in **emilite** KKAKAKKAKA and in **hammarite** KAKA.

In this scheme, we disregard small amounts of Cu-and-Pb present in the interiors of some modules, incorporated most probably in order to equalize the dimensions of different modules that are combined in one structure.

A different example is the series of **n-kalsilites**, superstructures of nepheline. These structures lie compositionally between the 'pure' nepheline pole $NaAlSiO_4$ and the kalsilite pole $KAlSiO_4$ (Merlino 1984). Each (0001) net of alternatively upward and downward oriented tetrahedra of their tectosilicate network can contain three kinds of six-fold rings: hexagonal or ditrigonal about K and oval, 'collapsed', housing Na. Hexagonal rings form corners of the hexagonal subcell, oval ones outline the triangular net of the hexagonal cell whereas the trigonal rings alone or in triplets occupy the sites on three-fold axes (Fig. 1.65). Absence of the latter rings gives **nepheline** (Table 1.26), single rings occur in **trikalsilite**, and ring triplets in **tetrakalsilite**. Edges of the cell have one, two, and three oval rings, respectively, intervening between

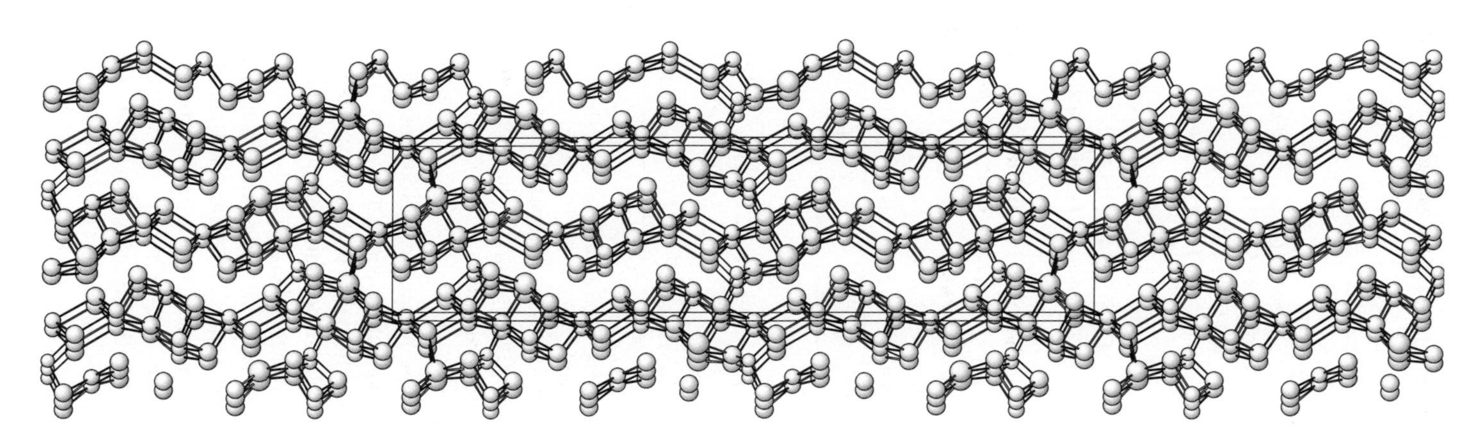

Fig. 1.64. The crystal structure of salzburgite $Cu_{1.6}Pb_{1.6}Bi_{6.4}S_{12}$, a fourfold superstructure of the bismuthinite–aikinite solid solution series (Topa *et al.* 2000*a*,*b*). In the order of decreasing size, spheres denote Pb, S, Bi and Cu (tetrahedrally coordinated). Partly occupied Cu positions have been omitted in order to accentuate division of the structure into krupkaite–gladite–gladite modules.

Table 1.25 Selected members of the bismuthinite–aikinite series

Mineral	Formula	Lattice parameters			Space group	Reference[a]
		a	b	c		
Bismuthinite	Bi_2S_3	3.985	11.163	11.314	*Pmcn*	Topa *et al.* (2002*a*)
Gladite	$CuPbBi_5S_9$	4.004	33.575	11.48	*Pmcn*	Topa *et al.* (2002*b*)
Salzburgite	$Cu_{1.6}Pb_{1.6}Bi_{6.4}S_{12}$	4.007	44.81	11.513	$Pmc2_1$	Topa *et al.* (2000*a*)
Paarite	$Cu_{1.7}Pb_{1.7}Bi_{6.3}S_{12}$	4.007	55.998	11.512	*Pmcn*	Makovicky *et al.* (2001*b*)
Krupkaite	$CuPbBi_3S_6$	4.013	11.208	11.56	$Pmc2_1$	Topa *et al.* (2002*b*)
Lindströmite	$Cu_3Pb_3Bi_7S_{15}$	4.018	56.141	11.578	*Pmcn*	Topa *et al.* (2002*a*)
Emilite	$Cu_{10.7}Pb_{10.7}Bi_{21.3}S_{48}$	4.029	44.986	11.599	$Pmc2_1$	Balič Žunič *et al.* (2002*b*)
Hammarite	$Cu_2Pb_2Bi_4S_9$	4.025	33.773	11.595	*Pmcn*	Topa *et al.* (2002*a*)
Aikinite	$CuPbBiS_3$	4.042	11.339	11.652	*Pmcn*	Topa *et al.* (2002*a*)

References relate to the newest refinements of unit cell parameters.

Table 1.26 Nepheline–kalsilite family[a]

Mineral	Composition	Space group	Lattice parameters (Å)		Homologue	Reference
			a	c		
Kalsilite	$KAlSiO_4$	$P6_3$	5.161	8.693	[0^b]	Perrotta and Smith (1965)
Nepheline	$Na_3KAl_4Si_4O_{16}$	$P6_3$	9.989	8.38	1	Hahn and Buerger (1955)
Trikalsilite	$K_2NaAl_3Si_3O_{12}$	$P6_3$	15.339	8.501	2	Bonaccorsi *et al.* (1988)
Tetrakalsilite	$K_3NaAl_4Si_4O_{16}$	$P6_3$	20.513	8.553	3	Merlino *et al.* (1985)[c]

[a] Plesiotypes to this family are kaliphylite $KAlSiO_4$ (Cellai *et al.* 1992) and megakalsilite (Khomyakov *et al.* 2002) in which at least some sixfold rings of tetrahedra deviate from the up–down UDUDUD topology of the nepheline-kalsilite group.

[b] Ditrigonalized rings of the average structure are in subsequent layers rotated by 180°; not a perfect analogue to the subsequent structures.

[c] Natural 'panunzite'.

hexagonal rings. No oval rings are present in kalsilite in which the layer consists of (ditrigonalized) hexagonal rings only; consecutive nets are rotated 'averaging' the ditrigonalization. The configurational change is accompanied by a change in the Na:K ratio. The **nepheline–tetrakalsilite family** is by its nature a two-dimensional accretional homologous series with members $N = 1$–3. Plesiotypes to this family of 'stuffed tridymite derivatives' have the up–down sequence of tetrahedra in the (0001) sheets different from the tridymite UDUD... scheme. Some are mentioned in Table 1.26; they have been summarized by Liebau (1985) and Palmer (1994) and discussed in part by Khomyakov *et al.* (2002).

Three structures of the **natrolite family** of fibrous zeolites also offer a polysomatic development for the ordered derivative of a solid solution. In this case, the substitution scheme for channel cations is $Ca + H_2O \leftrightarrow 2Na$; the Ca-filled channel

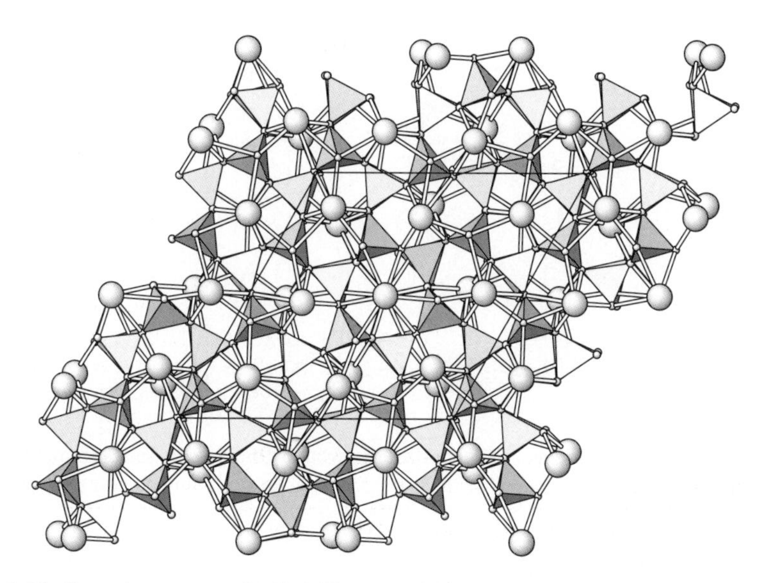

Fig. 1.65. Crystal structure of trikalsilite $K_2NaAl_3Si_3O_{12}$ (Bonaccorsi *et al.* 1988), the $N = 2$ homologue of the nepheline-kalsilite series. Note hexagonal, trigonal and distorted six-membered rings in the silicate skeleton.

is asymmetrically occupied by the above pair (Armbruster and Gunter 2001). **Natrolite** $Na_{16}[Al_{16}Si_{24}O_{80}] \cdot 16H_2O$ has symmetrically occupied channels (orthorhombic symmetry, space group *Fdd*2), **scolecite** $Ca_8[Al_{16}Si_{24}O_{80}] \cdot 24H_2O$ is monoclinic, space group $F1d1$ whereas the—so far—only known intermediate member, **mesolite** $Na_{16}Ca_{16}[Al_{48}Si_{72}O_{240}] \cdot 64H_2O$ is a polysome in which channels are arranged *en échelon* into layers NSSNSS... parallel to (010) where the infill of the adjacent scolecite double-layers 'points' in the opposite directions of the interlayer *a* axis and the total *b*-period is 56.65 Å instead of the original 18.64 and 18.98 Å for natrolite and scolecite (Artioli *et al.* 1986).

1.10 Principles of prediction of modular structures

1.10.1 *Exploiting accretional homology*

In principle, the modular structures can be predicted by extrapolation of known structures by means of homologous or plesiotypic relationships. In this process, one has to remember that in real structure families, many compounds will only have subgroups of the ideal space group derived from the idealized framework for the family. Often, one can profit from the existence of **configurational homologues**, built on the same homology principles but with fairly different chemical compositions (all such members forming together a heterochemical homologous series).

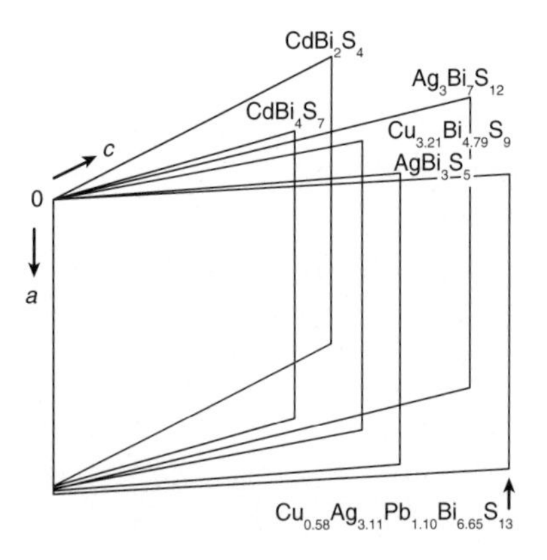

Fig. 1.66. Unit cells of selected members of the pavonite homologous series. In the order of increasing d_{001}, the homologues $N = 2, 3, 4, 5, 7, 8$ are shown. Sulphide homologue $N = 6$ is not known.

The incremental growth of moduli by accretion of new polyhedra brings about incremental growth of one (suitably defined) unit cell edge (more exactly of one of the fundamental interplanar spacings). The other two unit cell edges stay constant. For many types of accreting slabs, the accreting unit cell edge may be 'flapping', with the corresponding unit cell angles differing widely for adjacent homologues, and without any simple progression of angles observed with increasing homologue order (Fig. 1.66).

Having recognized the exact mechanism of accretion, a mathematical model of the accreting structure can be constructed, the results of which can be compared with the chemistry, unit cell parameters and space group of the candidate compound.

As an example, these operatios will be sketched for the case of **lillianite homologues**. The average octahedra of the PbS-like slabs in these homologues are the principal geometric element of the structure (Fig. 1.67). They determine the geometry of the idealized trigonal co-ordination prisms and the lattice parameters of the phase. In the (001) projection of the structure, we define the length l and the width w of an octahedron in a way shown in Fig. 1.67. A constraint must be introduced in order to make the idealized calculations possible, such as the perpendicularity of the l and w directions. It is well maintained in almost all known **lillianite homologues**.

Under this assumption

$$a_0 = \sqrt{(3w)^2 + l^2} \tag{1.1}$$

or

$$\cos\varepsilon = \frac{3w}{a_0} \tag{1.2}$$

and

$$1/2b_0 = \cos\varepsilon \cdot ql \tag{1.3}$$

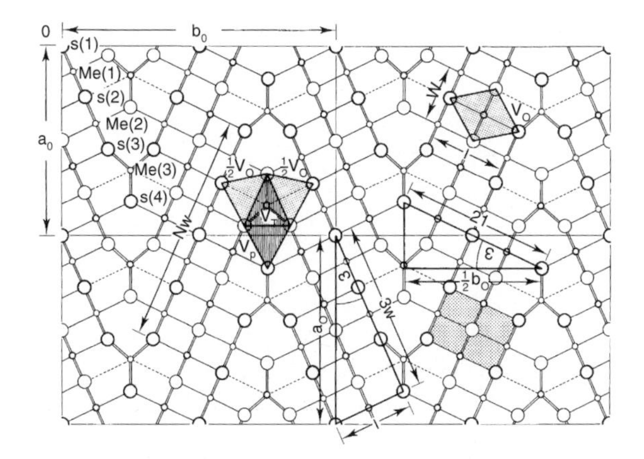

Fig. 1.67. Atomic positions and geometry of a generalized lillianite structure ($N = 4$). V_o, octahedral site; V_T, trigonal prismatic site; V_p, pyramidal/tetrahedral vacancies; w and l, 'width' and 'length' of a coordination octahedron; w and l also are dimensions of the I-centred subcell that characterizes the PbS-like submotif; N, no. of octahedra in a unit chain; ε, angle between a_o and the direction of the unit chain.

where $q = (N + 2)/3$ is the number of l dimensions (i.e. the number of octahedra) in a straight part of the zig-zag chain [010] (e.g. $q = 2$ octahedra for $N = 4$ in Fig. 1.67).

From these relationships, we obtain the equation

$$324q^2w^4 - 36q^2a_0^2w^2 + b_0^2a_0^2 = 0 \tag{1.4}$$

from which

$$w = \sqrt{\frac{1}{18}a_0\left(a_0 + \sqrt{a_0^2 - \frac{b_0^2}{q^2}}\right)} \tag{1.5}$$

and

$$l = \frac{b_0a_0}{6qw} \tag{1.6}$$

For the monoclinic cases, b_0 is to be replaced by d_{010}(c-setting). Search for the right N value proceeds by trial-and-error, trying out different q, that is, N values until realistic l and w dimensions are obtained. It should be remarked, however, that small differences in l and w values arise when the cation sites are occupied by different elements or the ratios of cations involved changes across a **solid-solution** range. For the monoclinic $N_1 \neq N_2$ cases, the result represents N_{aver}. Experience shows that even for large values of N (e.g. equal to 11), an erroneous estimate of N will result in unrealistic values for l and w.

Except for the position of trigonal prismatic Pb at the ends of the chain, atomic positions in one diagonal chain of octahedra in the idealized linkage pattern (space group *Bbmm*) can be described by the equations

$$x = \frac{1}{a} Qw \cos \varepsilon \tag{1.7}$$

$$y = \frac{1}{b} Qw \sin \varepsilon \tag{1.8}$$

where $z =$ either 0 or 1/2, differing for the two principal species (Me and S) involved and the value of the 'parity coefficient' Q is defined as follows. There are $(N/2 + 1)$ independent metal positions (including the trigonal prismatic site) and $(N/2 + 2)$ independent sulphur positions in a lillianite homologue with even order N and ideal symmetry. In the homologue with odd N there are $(N/2 + 1\frac{1}{2})$ independent metal or sulphur positions. In both cases, Q is equal to $(i - 1 + z)$ where z is the z coordinate in the 4 Å direction and i the order number of the atom in question starting from the centre of the chain which we place in the unit cell origin. $i = 1, 2, \ldots, i_{max}$ where i_{max} assumes the above values $(N/2+k)$ as dictated by the parity of N and the atomic species.

In the **lillianite homologues** of either parity, the position of the trigonal prismatic cation is defined as

$$x = x_{\text{S with imax}} - w/2a_0 \cos \varepsilon \tag{1.9}$$

$y = 0.25$; $z = 1/2$ for $N = 2n$ whereas $z = 0$ for $N = 2n + 1$.

Equation (9) holds for an idealized trigonal prismatic coordination with six equal Me–S bonds.

The above equations allow to build octahedral chains for desired values of N. Once a fitting value of $N_{\text{aver.}}$ was found for the observed parameter d_{010}(c-setting), the values of l and w attain realistic values and various combinations of N_1 and N_2 can be tried. The resulting γ_{calc} angles are compared with the γ_{obs} value and the right combination selected. The second, N_2, chain of octahedra has to be reflected on (010), translated and attached to the end of the first one, N_1, in the process of these calculations.

This procedure predicted correctly the crystal structure of **vikingite** $Pb_{11}Ag_{3.5}Bi_{11.5}S_{30}$ as N_1, $N_2 = 4, 7$ (average 5.5). For this combination, $\gamma_{\text{calc}} = 95.1°$ versus $\gamma_{\text{obs}} = 95.5°$. Other calculated combinations were 5,6 ($\gamma_{\text{calc}} = 98.6°$); 3,8 (102.1°); 2,9 (91.5°); 1,10 (105.5°) and 0,11 (i.e. two rows of Pb coordination prisms mutually indented by caps in direct contact) (92.13°). At the periphery of the **lillianite group**, both the structure of **pavonite** $AgBi_3S_5$ (N_1,N_2 = 1, 5) and that of **benjaminite**, idealized formula $Ag_3Bi_7S_{12}$ (N_1, $N_2 = 1, 7$) (Fig. 1.7) were correctly predicted in this way (Makovicky and Karup-Møller 1977*a*,*b*; Makovicky 1997*c*).

More complex structures are equally amenable to homologous expansion or interpolation, that is, prediction. In this way, the structure of **jaskolskiite** $Cu_{0.2}Pb_{2.2}(Sb, Bi)_{1.8}S_5$ (Makovicky and Norrestam 1985) (Fig. 1.8) was obtained by homologous contraction of the structure of **meneghinite** $CuPb_{13}Sb_7S_{24}$ (Euler and Hellner 1960) based on the homologous increments of the b parameter:

Table 1.27 Unit cell data for meneghinite homologues

Structure	N	a (Å)	b (Å)	c (Å)	Space group
Meneghinite	$N = 5$	a 11.36 Å	b 24.06 Å	c 4.13 Å	(subcell *Pbnm*)
Jaskolskiite	$N = 4^*$	a 11.43 Å	b 19.81 Å	c 4.08 Å	*Pbnm*
Aikinite	$N = 2$	a 11.32 Å	b 11.64 Å	c 4.04 Å	*Pbnm*

* Derived.

This is the best place to stress that, unlike the simple model of accretional expansion, a more realistic and detailed picture of bonding situation is obtained when the incremental slice is inserted in the centres of accreting slabs (blocks) and not at one of their margins. Detailed bonding patterns of the marginal, submarginal, and progressively deeper slices of polyhedra are inherited from the previous, lower homologues and the new (slightly different) bonding pattern is inserted in the central slab. Situation along central seams will differ in principle for even and odd values of N.

The largest structure we predicted before a corresponding phase was even known is **izoklakeite** $(Cu,Fe)_2Pb_{26.5}(Sb,Bi)_{19.5}S_{57}$ (Makovicky and Mumme 1986). It was obtained by homologous expansion of **kobellite** $(Cu,Fe)_2Pb_{12}(Bi,Sb)_{14}S_{35}$ (Fig. 1.43). The SnS-based rods were lengthened from $N = 2$ to $N = 4$ whereas the PbS-like moduli were extended from three to five full-length rows of atoms in order to maintain and widen their noncommensurate match.

The reader is invited to check the constancy and incremental growth of module interfaces under these changes.

A number of similar relationships were recognized *a posteriori*, for already determined structures, for example, those between **bernardite** $TlAs_5S_8$ (Pašava *et al.* 1989) and **hutchinsonite** $TlPbAs_3S_9$ (Takéuchi *et al.* 1965) (Fig. 1.17(a,b)). This example illustrates one of the dilemmas facing prediction of modular structures: homologous series will often find their upper or lower limit due to local crystal-chemical factors usually not contained in their general structure schemes. In this case, it is the potential incorporation of several large Pb polyhedra in the SnS-based slabs of the homologues higher than hutchinsonite. Some compositions in the lillianite homologous series will be absent either because of the problems connected with the incorporation of large divalent atoms in the central rows of slabs for high N values or because of the difficulties in incorporating lone electron pairs into PbS-like arrays for low N values. Volume requirements of lone electron pairs (their common micelles) are illustrated in Fig. 1.6.

1.10.2 *Exploiting plesiotypic relationships*

Besides homologous expansion/contraction, many modular structures can be modified by non-homologous truncation of the motif, additions to the motif, slip, or crystallographic shear along selected planes, by additional unit-cell twinning or by rotation of certain units. Such structures build plesiotypic pairs or entire groups and, as long as the previous research has not pre-emptied the field by finding all viable combinations, exact or at least general predictions can be made of new structural varieties, plesiotypic to known structures.

In the complex sulfides a 'free' interchange of moduli respectively based on the PbS archetype and on the SnS archetype, under preservation of the same shape and size, is rather frequent. Differences are compensated for by changes in the bonding scheme on the module interfaces. Moreover, in these sulphides from the extensive Pb–Sb and Pb–Bi plesiotypic family of **rod-based structures**, quite distinct types of layers composed of lozenge-shaped moduli can interchange in the structure when the incommensurate boundary between them and the underlying layer remains unchanged or is only modified in the above, permissible way. Such a predicted intergrowth between the layers of the **$Pb_2Bi_2S_5$type** (Srikrishnan and Nowacki 1974) and those of the **$Pb_5Sb_4S_{11}$type** (Petrova *et al.* 1978) (see Fig. 1.42b) has not yet been found but other hypothetical predictions by Makovicky (1993) were later found as existing compounds (e.g. $Pb_{12.65}Sb_{11.35}S_{28.35}Cl_{2.65}$; Kostov and Macíček 1995).

These predictions were made by combining twelve possible ways of interconnection of rod-like moduli into layers with the choice of rods with different dimensions. Layers thus created force a definite interlayer configuration for which a matching layer is to be found. In addition, the **chess-board and cyclic arrangements of rod-like moduli** open another spectrum of possibilities. Full treatment of these combinations, together with attempted geometric derivations, is in Makovicky (1993), a brief outline was given earlier in this section.

The structures of **kobellite** and **izoklakeite** can be reduced by truncation of an octahedral layer (parallel to (010) in kobellite), or of such an octahedral double-layer, off the ends of PbS-based 'dumb-bells'. An 180° rotation of one of the resulting structure halves may be required for joining the two now separated structure portions. A singly truncated derivative (only in principle, because it has $N = 3$) is $Pb_4In_2Bi_4S_{13}$ (Krämer 1983), a doubly truncated derivative (based on SnS-like modules only) is $SnSb_2S_4$ (Smith and Parise 1985) and, on a more complicated level, **eclarite** $(Cu,Fe)Pb_9Bi_{12}S_{28}$ (only PbS-like modules; Kupčík 1984) which is a direct **kobellite derivative** with alternating contacts which show respectively double truncation and zero truncation (Table 1.18).

1.10.3 *Distortion of archetypes*

In the majority of cases, the structures of the interior of the selected moduli are fairly, and recognizably close to an archetypal structure. In the marginal cases, this is no more true and the distortions may render the assignment of an archetype to a module problematic. For example, the rods in the structure of **zinkenite** $Pb_9Sb_{22}S_{42}$ (Portheine and Nowacki 1975) or nuffieldite $Cu_{1.4}Pb_{2.4}Bi_{2.4}Sb_{0.2}S_7$ (Moëlo *et al.* 1997) appear to be transitional between the PbS and SnS types. A sequence of structures from the (modified) SnS archetype towards PbS archetype is observed for $PbAs_2S_4$–$Tl_2(Sb,As)_{10}S_{16}$–$TlSb_5S_8$ (Berlepsch *et al.* 2001*b*).

A global geometric analysis of deviations by a least-squares fitting of diverse model archetypes to the structure of the module is difficult and appears still some way off in the future. Instead, after recognizing the general topological identity with the archetype, we can analyse quantitatively the configurational characteristics of all individual cations and anions (polyhedral distortion, sphericity of ligands, eccentricity of the central atom) and compare them with those in the archetypal structure. Relevant methods for obtaining these measures of deformation were derived by Balič Žunič and Makovicky (1996) and Makovicky and Balič Žunič (1998).

1.10.4 *Examination of reciprocal lattice*

Modules of archetypal structures will be reflected as a pronounced submotif in the weighted reciprocal lattice of the modular structure. The orientation of the submotif reveals the orientation of archetype blocks. Presence of several reciprocal submotifs reveals presence of several orientations (and presumably also of types) of modules. Maxima of the submotif are generally broad and overlap several reflections of the 'sampling reciprocal lattice'. From the sharpness of these maxima, information on the extent (and shape) of modules can be drawn: the sharper and more limited they are, the larger are modules of the archetypal structure.

If there is only a limited number of structural families known for a given combination of elements and lattice parameters alone are insufficient to indicate to which of these families the structure belongs, the reciprocal-lattice information will often allow an unambiguous assignment.

The PbS-like slabs in the crystal structure of **lillianite homologues** produce two submotifs in the reciprocal space. They are Fourier transforms of the two I-centred pseudo-orthorhombic subcells which can be recognized in these slabs in Fig. 1.67. The *hk*0 section of the reciprocal lattice of lillianite (the Pb-rich pole of the solid solution series with $N = 4$) is in Fig. 1.68.

In the reciprocal lattices of **pavonite** and **benjaminite** (which in terms of **lillianite homologues** have the N_1, N_2 combinations 1,5 and 1,7, respectively), only one extensive direct-space submotif exists, and only one reciprocal submotif is developed (Figs 1.69 and 1.70(a,b)). Among the structures with rod-like modules, the layered ones (Fig. 1.42) have two submotif orientations whereas the cyclic ones and the 'degenerated cyclic cases' such as **kobellite** and **izoklakeite** (Figs 1.43 and 1.68) have three orientations resulting in an approximately hexagonal pattern of very broad, composite maxima. Such observations were supportive, for example, in recognition of izoklakeite structure. It is important not to mix submotif and sublattice in this connection: submotif does not have to be commensurate with the lattice, its islands/slabs being separated by another structure motif. Sublattice describes a submotif with a specific orientation in respect to the lattice and being an integer submultiple of the lattice periodicity.

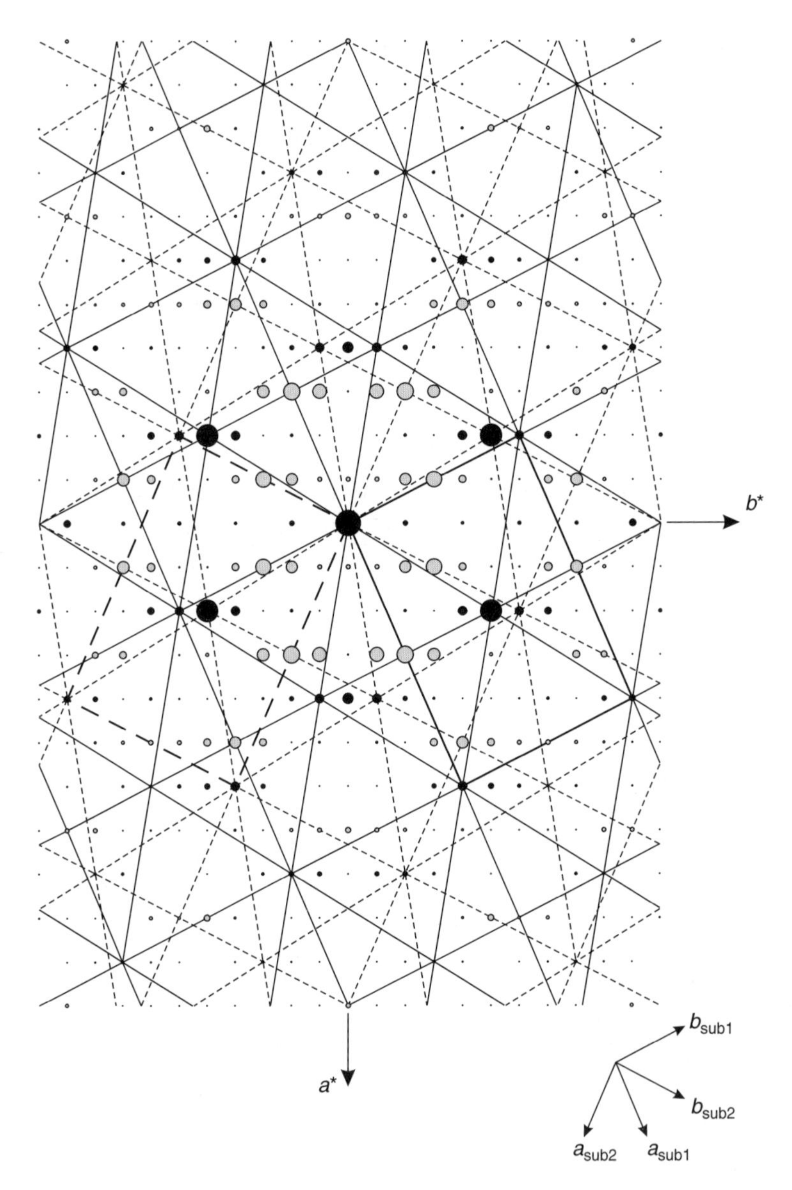

Fig. 1.68. Sections ($hk0$) (black) and ($hk1$) (grey) of the weighted reciprocal lattice of lillianite $Pb_3Bi_2S_6$ (Takagi and Takeuchi 1972). The 'F-centred' reciprocal submotifs (indicated by continuous and dashed lines, respectively) correspond to two orientations of the I-centred PbS-like submotif in Fig. 1.67. Their reciprocal cells are drawn in bold outlines (Ilinca and Makovicky 1999).

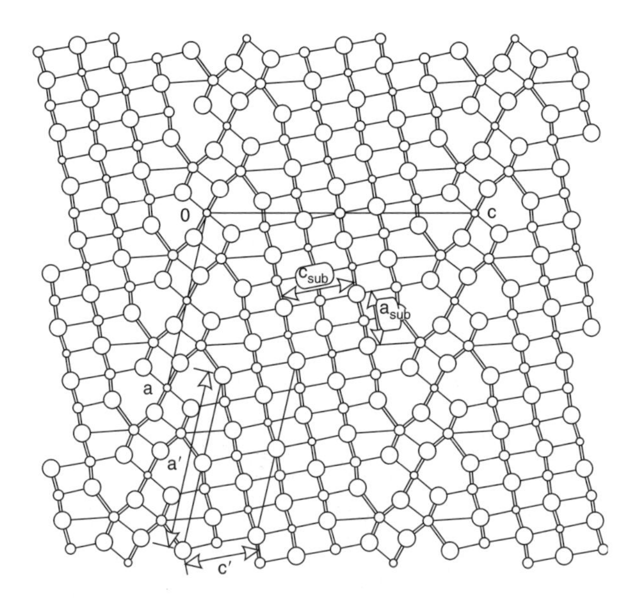

Fig. 1.69. Submotif cells in the crystal structure of benjaminite (Makovicky and Mumme 1979). The quasi-orthorhombic I-centred subcell (a_{sub}, b_{sub} = b, and c_{sub}) yields a strong reciprocal-space signature; the monoclinic C-centred subcell does not.

1.10.5 *Use of HRTEM*

Use of HRTEM in combination with modular models has been eminently successful for biopyriboles and interface modulated layer silicates (e.g. Veblen *et al.* 1977; Guggenheim and Eggleton 1988). In the world of complex modular sulphides, the structures of $\mathbf{SnSb_2S_4}$ and $\mathbf{SnSb_2Se_4}$ as well as that of $Sn_3Sb_2S_6$ were determined in this way (Smith and Parise 1985; Smith 1984). Several such studies were performed on the lillianite homologous series, especially for clarification of structure defects (slabs with wrong *N* values).

Maximum information content obtained in these studies is on the level of very basic models for individual series which are then refined using the crystal chemical information available from similar phases (e.g. Smith (1984) for $\mathbf{Sn_3Sb_2S_6}$). Electron diffractograms can yield excellent information about the orientation of the submotifs in the moduli (*ibid.*). Still, the current studies (Pring *et al.* 1999) on lillianite homologues confirm that great caution should be exercised in the interpretation of even very slightly misaligned specimens.

1.10.6 *Exploiting chemical information*

A series of structures related by homologous expansion will have a general structural formula, in terms of cation and anion sites that must involve the order *N*. In most cases,

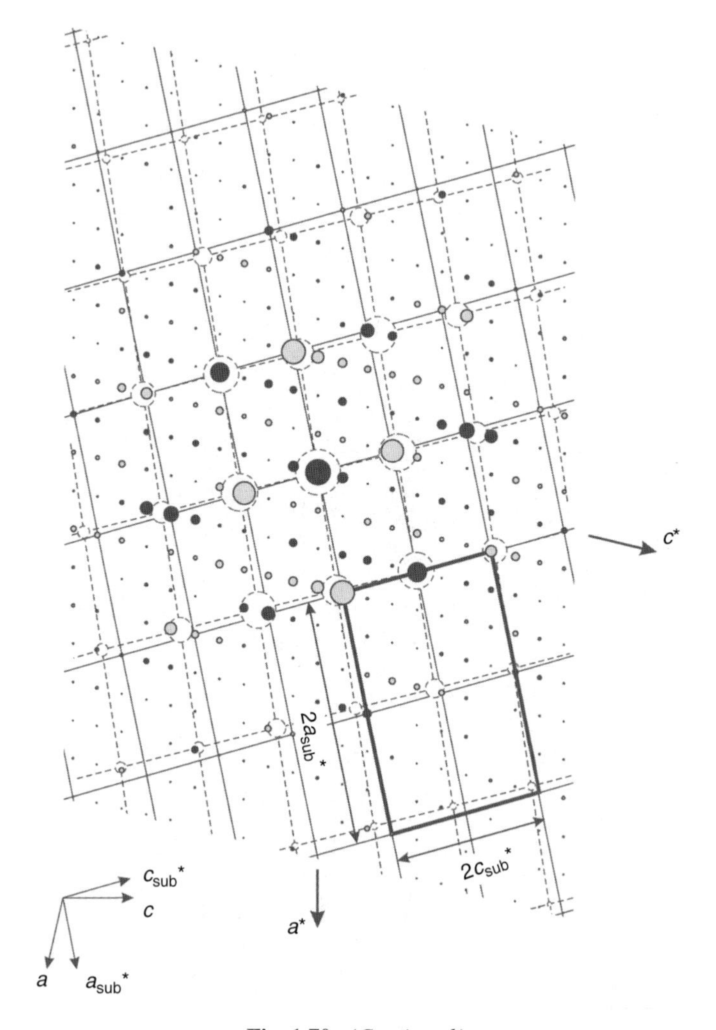

Fig. 1.70. (*Continued.*)

also a crystal-chemical formula can be derived that takes care of the known element substitutions which may take place in this series without changing the respective values of N.

For example, for lillianite homologues the latter formula is

$$\mathbf{Pb_{N-1-2x}Bi_{2+x}Ag_xS_{N+2}}$$

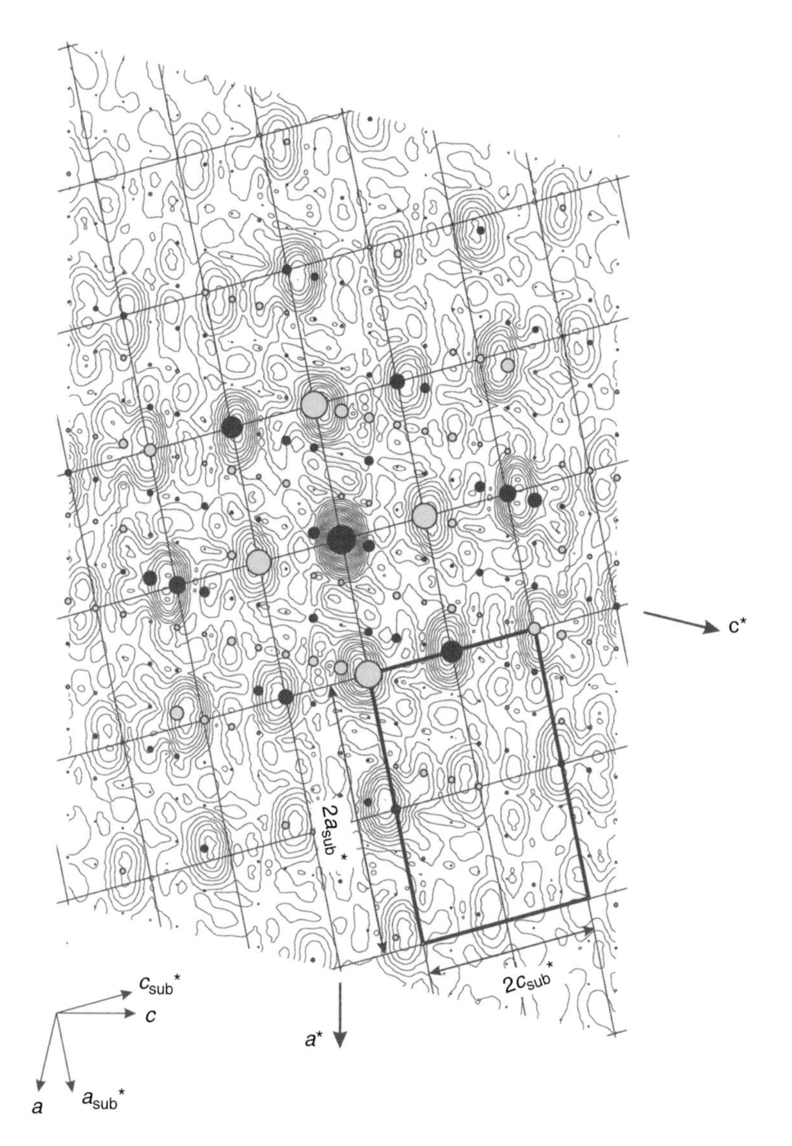

Fig. 1.70. Sections $(h0l)$ (black circles) and $(h1l)$ (grey circles) of the reciprocal lattice of benjaminite (Ilinca and Makovicky 1999). In (a) stippled sublattice describes diffraction maxima of the I-centred submotif; in both (a) and (b) the fully drawn sublattice interconnects 'sampling nodes' of the reciprocal lattice superimposed on the submotif. Part (b) contains Fourier transform (in projection on (010)) of the full structure of benjaminite illustrating the presence and nature of the reciprocal submotif. Sampling subcell is drawn in bold outlines.

where N is the order of the homologue and x the degree of Ag+Bi = 2Pb substitution (Fig. 1.5). For sartorite homologues, it is

$$\mathbf{Pb_{4N-8-2x}(As,Sb)_{8+x}Me^{+}_{x}S_{4N+4}}$$

where $Me+$ is Ag and/or Tl whereas for meneghinite homologues it reads

$$\mathbf{Cu_xPb_{2N+x-4}(Sb,Bi)_{4-x}S_{N+2}} \text{ (Fig. 1.71).}$$

Working with such formulae, the value of N can be extracted from the results of chemical analyses. In all cases, it will be the average N value, the arithmetic average of N_1 and N_2. The general formulae, which also take care of substitutions can be derived (Makovicky and Karup-Møller 1977), such as

$$N = -1 + \frac{1}{\mathrm{Bi} + \mathrm{Pb}/2 - 1/2} \tag{1.10}$$

for the lillianite homologues for which the cation contents have first been normalized as follows:

$$\mathrm{Ag} + \mathrm{Pb} + \mathrm{Bi} = 1.0. \tag{1.11}$$

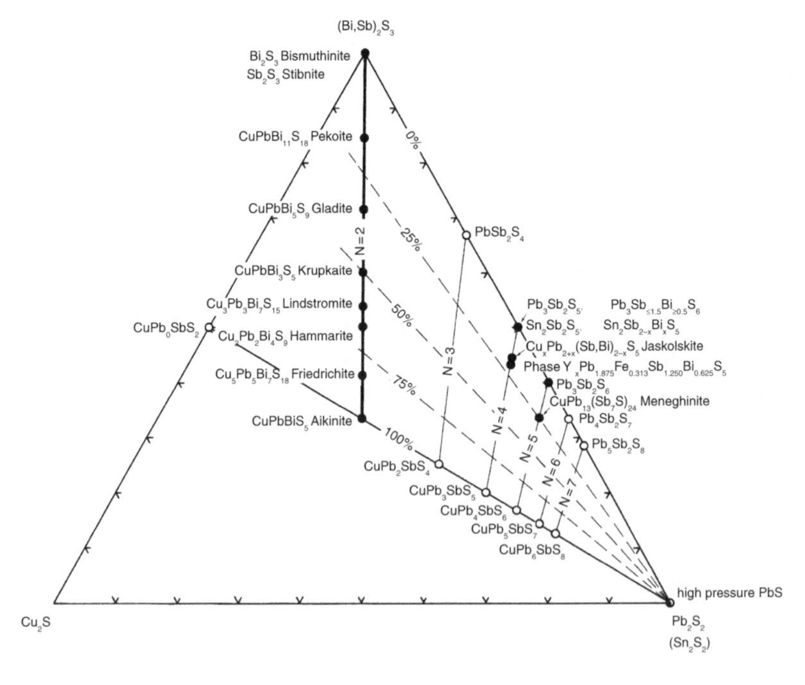

Fig. 1.71. The compositional plot (mol%) for the members of the meneghinite homologous series Cu_xPb_{2N+x-4} $(Sb,Bi)_{4-x}S_{N+2}$. Grid of order numbers N versus the Cu + Pb substitution percentage for vacancy + (Sb,Bi). Only some of the bismuthinite derivatives are plotted.

In this formula, Ag, Pb and Bi denote element contents (atomic proportions) obtained by means of chemical analysis. Approximate estimate of N can be obtained by plotting the cation contents in Fig. 1.5 (or, analogously, in Fig. 1.71).

Instead, we can (a) 'eliminate substitutions', for example, by converting nCu+nPb in the meneghinite-group formula into n(Bi+Sb)+vacancy and (b) use the 'corrected' cation contents for extracting N from the atomic ratios. Thus, for the meneghinite homologues after the conversion sub (a) we obtain

$$\mathrm{Pb}_{\mathrm{corr}}/(\mathrm{Sb}+\mathrm{Bi})_{\mathrm{corr}} = (2N-4)/4, \qquad (1.12)$$

giving

$$N_{\mathrm{average}} = 2\mathrm{Pb}_{\mathrm{corr}}/(\mathrm{Sb}+\mathrm{Bi})_{\mathrm{corr}} + 2 \qquad (1.13)$$

A very general formula for obtaining the homologue order N and cation substitution percentages can be devised for all accretional series of complex sulphides, requiring only insertion of specific subscripts for individual series. However, because of the presence of some cations with two valencies, such as Cu or Fe, it relies upon a precise determination of S in analyses what is currently a weak point of microprobe measurements. It will not be pursued further here.

The same procedures are valid for the homologous/polysomatic series of oxysalts. For the well known series of Mg-based **biopyriboles** (chain- to phyllosilicates) (Veblen *et al.* 1977) the individual homologues are as follows:

The general formula of the series is

$$\mathbf{Mg_{3N+1}Si_{4N}O_{10N+2}(OH)_{2N-2}}.$$

In microprobe analyses of these phases, the measurable elements are Mg and Si (Fe^{2+} is added to Mg). The value of N_{aver} is then obtained by converting the general formula to

$$N = \mathrm{Si}/(4\mathrm{Mg}-3\mathrm{Si}) \qquad (1.14)$$

where Mg and Si are the atomic proportions obtained from the microprobe analysis. This approach is important because of frequent non-integral N values stemming from

Table 1.28 Ideal composition of magnesium biopyriboles

Orthopyroxene	$Mg_4Si_4O_{12}$	$N = 1, 1, 1, 1, \ldots$
Orthoamphibole	$Mg_7Si_8O_{22}(OH)_2$	$N = 2, 2, 2, 2, \ldots$
Jimthompsonite	$Mg_{10}Si_{12}O_{32}(OH)_4$	$N = 3, 3, 3, 3, \ldots$
Chesterite	$Mg_{17}Si_{20}O_{54}(OH)_6$	$N = 2, 3, 2, 3, \ldots$
...		
Talc	$Mg_3Si_4O_{10}(OH)_2$	$N =$ infinite

intimate homologue intergrowths on a unit cell scale. Fitting the analytical results to the formula of a presumed integral homologue will lead to wrong conclusions and models in such a case.

Whenever we have a combination of a substitutional and an interstitial solid solution for the same substituting element in one phase, the above formulae break down. The analysis will, with equal probability, represent several adjacent homologues with different degrees of apparent substitution. Such cases are Cu–Bi sulphosalts with Cu-for-Bi substitution (the pavonite homologues and the cuprobismutite homologues) for which only single-crystal data are a reliable criterion. The same situation arises in **biopyriboles** with the partially or fully occupied A sites. Here the Ca + A-site vacancy configuration changes either into the $Na + Na_{A\ site}$ situation or in the Na + A vacancy one. The latter situation, connected with other element substitutions in the entire biopyribole framework, gives uncertain site assignment and the identity of the phase has to be ascertained by optical or diffraction means.

2

OD structures

2.1 Introduction

The concept of order/disorder **(OD) structures** was introduced in the crystallographic literature by Dornberger-Schiff (1956, 1964, 1966, 1979) and developed through the work of Dornberger-Schiff herself and her coworkers into a wide and dynamic field of research, resulting in a comprehensive theory and leading to a revision and generalization of basic concepts in 'classical crystallography', including the redefinition of the notion of 'crystal' itself. 'The introduction of the notion of OD structures was, in the first place, motivated by the task of determining the structure of a crystal exhibiting one-dimensional disorder. Considerations of the causes leading to the tendency to disorder of some crystalline substances led, furthermore, to the conviction that the classical definition of a crystal as a body with three-dimensional periodicity may have its shortcomings; properly speaking, it is not truly atomistic, although it refers to atoms: periodicity, by its very nature, makes statements on parts of the body which are far apart, whereas the forces between atoms are short range forces, falling off rapidly with distance. A truly atomistic crystal definition should, therefore, refer only to the neighbouring parts of the structure, and contain conditions understandable from the point of view of atomic theory' (Dornberger-Schiff 1979). Actually, each normally ordered structure—a crystal in the classical sense—may be described in terms of two-dimensionally periodic layers: if a definite relative arrangement of subsequent layers is energetically favoured (short range forces of interaction), this arrangement will occur throughout, which eventually results in periodicity (long range property of the structure). Pairs of adjacent layers, wherever taken in the structure, are obviously geometrically equivalent. 'This equivalence seems the most important feature of crystals, and may be used for a revision of the crystal definition' (Dornberger-Schiff 1979). However, whereas the equivalence of pairs of adjacent layers is a necessary condition for periodicity, it is not a sufficient condition. When adjacent layers may follow each to the other in two (or more) geometrically equivalent ways, then infinite possible sequences, ordered as well as disordered sequences, may result: the whole set builds up a **family of OD structures**; it is appropriate to call each of them 'crystal', within a wider definition of this concept.

OD structures may often be recognized through some peculiar features displayed by their diffraction patterns:

- presence of sharp spots and diffuse streaks;
- **non-space-group absences**;

- diffraction symmetry higher than that corresponding to the point group and Friedel law (**diffraction enhancement of symmetry**);
- a set of common reflections in diffraction patterns from different crystals in a family (**family reflections**);
- evidences for twinning, often polysynthetic twinning;
- polytypism.

All these features have been studied independently without any reference to the OD nature of the material. A lot of papers by Templeton (1956), Pabst (1959, 1961), Fischer (1961) discussed the 'non-space-group absences' (**Templeton effect**). Similarly, a lot of papers were devoted to the problem of 'diffraction enhancement of symmetry': Ramsdell and Kohn (1951) first described the structure of a polytype 10H of SiC with stacking sequence 3223 (Zhdanov notation), a polytype with space group symmetry $P3m1$ and diffraction Symmetry *6/mmm*; Ross *et al.* (1966) showed that a triclinic polytype of mica presented a diffraction pattern with monoclinic symmetry. The 'diffraction enhancement of symmetry' has been discussed, in general terms, by several authors: Sadanaga and Takeda (1968), Marumo and Saito (1972), Iwasaki (1972), Perez-Mato and Iglesias (1977). The cases in which this phenomenon is displayed by the whole diffraction pattern are probably rare; more frequently, the 'diffraction enhancement of symmetry' is shown by part of the reciprocal space (**partial diffraction enhancement of symmetry**) (Iwasaki 1972): it may happen, for example, that the diffraction pattern of a triclinic crystal presents monoclinic symmetry in the reciprocal lattice planes corresponding to $k = 2n$. As regards the presence of sharp spots and diffuse streaks in one diffraction pattern, the number of known examples in all classes of compounds (mineral phases as well as inorganic and organic products) is very large.

2.2 OD character of wollastonite

To introduce the reader to the terminology and the procedures of OD theory, we shall first discuss a single example, namely **wollastonite**, $CaSiO_3$, which appears the most appropriate for historical and didactical reasons. It was just the investigation of disorder phenomena and related diffraction features in wollastonite (as well as in Maddrell's salts and sodium polyarsenates) which gave the first hint for the development of OD theory (Dornberger-Schiff *et al.* 1955). Moreover, as it will soon appear, wollastonite displays the whole set of peculiar features we have previously listed.

In fact, a paper by Jeffery (1953) was just devoted to '*Unusual X-ray diffraction effects from a crystal of wollastonite*'. In that paper, the author recalled that wollastonite presents two modifications, triclinic and monoclinic [their unit cell parameters, taken from a paper of Ohashi (1984) are reported in Table 2.1], and presented the unusual diffraction pattern obtained with wollastonite from Devon, a pattern showing sharp spots in reciprocal lattice planes corresponding to $k = 2n$ (here and henceforth indices, vectors, and parameters are referred to the unit cell of the monoclinic

Table 2.1 Unit cell dimensions of the two 'main' modifications of wollastonite (Ohashi 1984)

Cell parameter	Wollastonite	
	$1A$	$2M$
a	7.926[a]	15.424[a]
b	7.302[a]	7.324[a]
c	7.065[a]	7.069[a]
α	90.06[b]	90.00[b]
β	95.22[b]	95.37[b]
γ	103.43[b]	90.00[b]
Space group	$P\bar{1}$	$P2_1/a$

[a] Units in Å.
[b] Units in degrees.

form) and diffuse streaks parallel to $\mathbf{a}^*$ in reciprocal lattice planes corresponding to $k = 2n + 1$.

Moreover, various other polytypes were subsequently found and studied with different techniques (X-ray diffraction, electron diffraction, high resolution transmission electron microscopy) by various authors (Wenk 1969; Takéuchi 1971; Jefferson and Bown 1973; Wenk *et al.* 1976) who confirmed the occasional occurrence of diffuse streaks along $\mathbf{a}^*$ for $k = 2n + 1$, streaks which were obviously indicative of structural disorder. It seems proper to stress that reflections corresponding to even k values are in the same positions and present the same intensity in all the structures, the ordered and the disordered ones (**family reflections**); they always display monoclinic symmetry (**partial diffraction enhancement of symmetry**) and are present only for $2h + k = 4n$ (**non-space-group absences**).

All these features may be comprehensively discussed and neatly explained on the basis of OD theory. In fact, all the structures we are discussing about belong to one family of OD structures built up with equivalent layers. A common feature of all the members of the family is that not only single layers, but also pairs of adjacent layers are equivalent, no matter whether taken in one member or in different members of the family. Figure 2.1 shows the structures of the two main polytypes of wollastonite, as seen along **c**.

The single layer (Fig. 2.2) has translation vectors **b** and **c**, which are the common vectors of the triclinic and monoclinic polytypes; the third basic vector $\mathbf{a}_0 = \mathbf{a}/2$, namely one half of the translation period of the monoclinic polytype, is not a translation vector; the symmetry of the layer is $P2_1/m$ or, more precisely, $P12_1/m1$.

It seems proper to recall here that this space group symmetry had been already indicated by Ito (1950) in his study on polymorphism in wollastonite, as the symmetry of the hypothetical form 'protowollastonite' from which, through 'gliding' (Ito 1950), the structures of wollastonite (triclinic form) and parawollastonite (monoclinic form) may be derived.

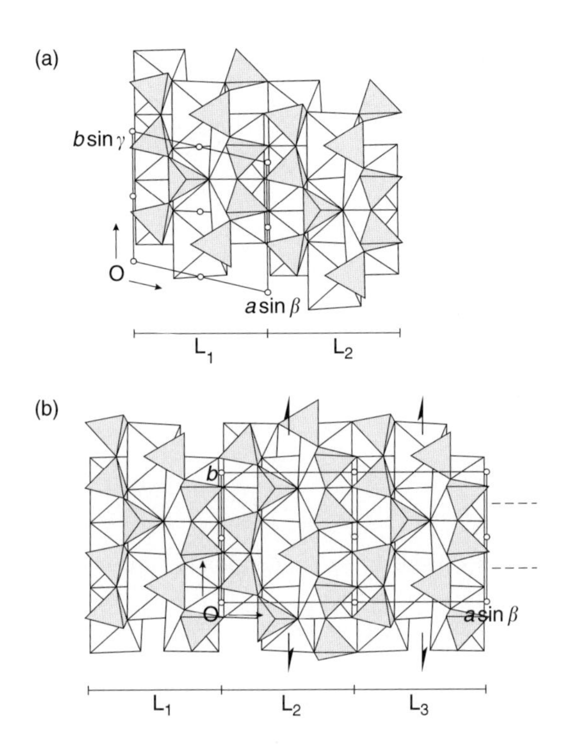

Fig. 2.1. The structures of the two main polytypes of wollastonite, as seen along **c**, with indication of the succeeding structural layers, as well as of the symmetry elements corresponding to the space groups $P\bar{1}$ and $P2_1/a$ of the triclinic (a) and monoclinic polytypes (b), respectively. In this figure, as well as in Figs 2.2 and 2.3, calcium polyhedra and silicon tetrahedra are drawn in light grey and dark grey, respectively.

Adjacent layers may be related through a twofold screw rotation with translational component $+\mathbf{b}/4$ $(2_{1/2})^1$ and a glide reflection perpendicular to **b** with translational component $\mathbf{a_0}$ $(a_2)^1$. Because of the mirror plane m in the single layer, adjacent layers may be related also through a screw rotation with translational component $-\mathbf{b}/4$ $(2_{-1/2})^1$. Pairs of adjacent layers related in both ways (Fig. 2.3) are geometrically equivalent.

Different sequences of operations $2_{1/2}$ and $2_{-1/2}$ give rise to different structures: an infinite number of polytypes, as well as of disordered structures is possible, corresponding to ordered or disordered sequences of those operations. The symmetry of each possible polytype may be derived from the symmetry properties of the single layer and the operations relating adjacent layers in the polytype.

If the operations $2_{1/2}$ and $2_{-1/2}$ regularly alternate (Fig. 2.1(b)), the first and third layer are at the same level, and the 2_1 operational element of each single layer is

[1] The symbols for the screw and glide operations will be explained in the following.

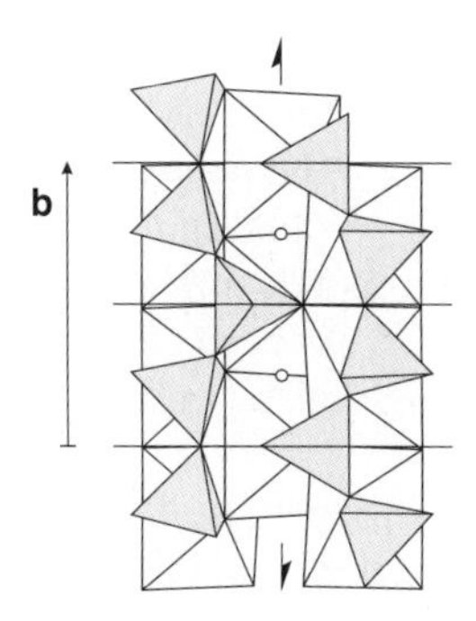

Fig. 2.2. The structure of the single layer, as seen along **c**, with indication of the elements of symmetry, corresponding to the layer group $P(1)2_1/m1$.

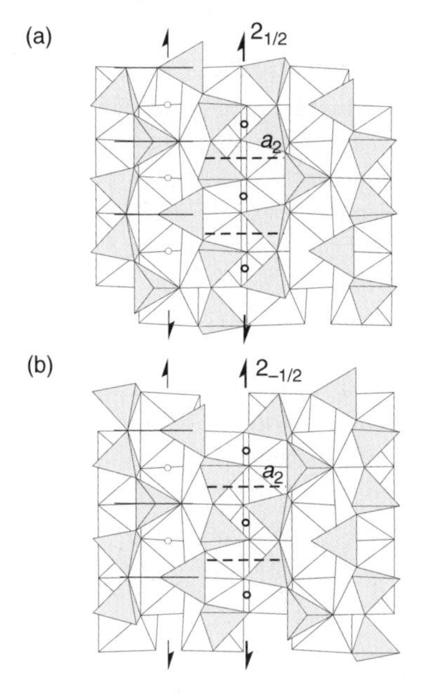

Fig. 2.3. Pairs of adjacent layers related through the operation $2_{1/2}$ (a) and $2_{-1/2}$ (b). The operations relating adjacent layers are indicated with heavy marks, whereas the operations which bring a layer into itself are indicated with light marks.

now valid for the whole structure; moreover, the glide operation a_2 is continued, becoming a true a glide valid for the whole structure, which has a parameter $a = 2a_0$. The overall space group symmetry of the resulting structure is therefore $P2_1/a$, which is just the space group of the polytype $2M$. On the other hand, if $2_{1/2}$ is constantly followed by $2_{1/2}$ (Fig. 2.1a), the operational elements 2_1 of the single layers are no longer total symmetry elements; moreover the operation a_2 is not continued from one layer to the other, but systematically transposed by **b**/4; only inversion centres now act as total symmetry elements and the overall space group symmetry is therefore $P\bar{1}$, which is just the space group of the polytype $1A$. It seems proper to recall that the structure obtained when $2_{-1/2}$ is constantly followed by $2_{-1/2}$ is not different from the preceding one and corresponds to its twinned (100) counterpart. Multiple (100) twinning on a very fine scale has been observed by Wenk *et al.* (1976) and by Thomas *et al.* (1978) in wollastonite and by Müller (1976) in the isostructural mineral **pectolite**, $Ca_2NaSi_3O_8(OH)$.

In a similar way, the symmetry properties of any other polytype in the family may be obtained from the symmetry properties of the single layer and the knowledge of the particular stacking sequence of that polytype. We have considered only the structures corresponding to wollastonite $1A$ and wollastonite $2M$, because of their significance and widespread occurrence. In all the families, a small number of 'main' polytypes exists, generally those most frequently found, which are called 'simple', 'standard', or 'regular'. OD theory presents neatly defined geometrical criteria by which the particular status of a small number of polytypes, named polytypes with maximum degree of order (**MDO structures**), may be recognized. These are the members in which not only pairs, but also triples (and quadruples, ... n-tuples) of consecutive layers are geometrically equivalent, as far as possible [a more precise definition will be given in the following Section 2.3.6]. In the family we are now considering only two MDO structures exist: MDO_1, corresponding to wollastonite $1A$ (Fig. 2.1a) and MDO_2, corresponding to wollastonite $2M$ (Fig. 2.1b).

The special stability observed for MDO structures is in keeping with the general assumptions behind OD theory. '... Even when the longer-range forces are much weaker than those between adjacent layers, they may not be negligible and, therefore, under given crystallization conditions either the one or the other kind of triples becomes energetically more favourable, it will occur again and again in the polytype thus formed, and not intermixed with the other kind. Needless to emphasize, however, that such structures are—as a rule—very sensitive to crystallization conditions; small fluctuations of these may reverse the energetical preferences, and this results in the occurrence of stacking faults, twinning, and other kinds of disorder' (Ďurovič and Weiss 1986).

These considerations, when applied to the case we are discussing about, explain not only why the most frequent polytypes are just $1A$ and $2M$, but also the occurrence of multiple twinning in the case of polytype $1A$. In fact, layers related by the operations $2_{1/2}$ and $2_{-1/2}$ are translationally equivalent and related by stacking vectors $\mathbf{t}_1 = \mathbf{a}_0 - \mathbf{b}/4$ and $\mathbf{t}_2 = \mathbf{a}_0 + \mathbf{b}/4$, respectively. Therefore, any adjoining layer may assume two possible positions, with displacements at left or right, according to the simple scheme represented in Fig. 2.4. Only two kinds of triples exist, 'stretched' and 'bent', corresponding to MDO_1 and MDO_2 respectively. As indicated in Fig. 2.4, the sequences $\mathbf{t}_1\mathbf{t}_1\mathbf{t}_1 \ldots$ and $\mathbf{t}_2\mathbf{t}_2\mathbf{t}_2 \ldots$ correspond to structures twinned on (100). If the 'stretched' triples are favoured in the crystallization process, the polytype could form as a polysynthetic twin with MDO_1 and MDO_1' domains due to the possible occurrence of stacking faults with local production of 'bent' triples. It seems appropriate

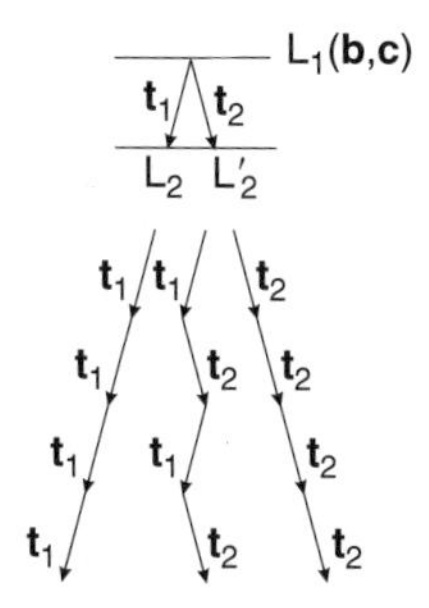

Fig. 2.4. Schematic representation of the MDO structures in the family of wollastonite.

to recall that the stretched triples, the bent triples, and the polysynthetic twinning closely correspond to the 'echelon', 'alternate', and 'complex' gliding, respectively, described by Ito (1950) in his study on polymorphism in wollastonite.

2.3 OD structures of equivalent layers

The family we have just described is an example of a family of OD structures built up with equivalent (namely, congruent or enantiomorphous) layers, characterized by two-fold periodicity and stacked one after the other in such a way to have the two translation vectors in common. In these structures, all pairs of adjacent layers are equivalent; this statement requires a specification, which will be given in the following (Section 2.3.2) after defining the categories of OD structures built up with equivalent layers.

2.3.1 *Partial operations*

Due to the equivalence of the layers, repeating operations exist which transform any layer in itself or into any other layer. These operations, which are generally not valid for the whole structure, are called **partial operations (POs)**. After numbering the layers according to their positions, the POs will be denoted through small letters with prefixes indicating the starting and the resulting layers: ${}_{p,q}a$ will denote a PO transforming L_p into L_q; ${}_{q,p}a^{-1}$ is the inverse operation.

It is important to observe that the complete set of POs does not form a group, as two POs ${}_{p,q}a$ and ${}_{r,s}b$ cannot be combined in this order unless $q = r$, namely, unless the resulting layer of the first transformation is identical with the starting layer of the second transformation. The POs do form a Brandt groupoid (Dornberger-Schiff 1964; Fichtner 1977a).

The POs ${}_{p,p}a$, which transform a layer L_p into itself are called **λ-POs**. The set of ${}_{p,p}a$ for a particular L_p forms a group, namely one of the **layer** (or **two-sided plane**) **groups**, which are the groups of symmetry operations of a structure built up with three-dimensional objects, but with two-dimensional lattice. The 80 layer groups were derived more than seventy years ago, through different approaches, by Alexander and Hermann (1929), Weber (1929), Heesch (1930) and reintroduced in the crystallographic literature by Holser (1958*a*, 1958*b*) in two papers about relation of symmetry to structure in twinning and by Dornberger-Schiff (1956, 1959) for the description of the symmetry of a structural layer in OD theory.

A PO ${}_{p,p+1}a$ or ${}_{p,p-1}a$ transforming a layer into an adjacent one is called a σ-**PO**. The complete set of POs of an OD structure may be generated from all λ- and σ-POs (as a matter of fact, it would be sufficient to give all λ-POs and one particular σ-PO, as the other σ-POs may be obtained by combining the whole set of λ-POs with the given σ-PO). Therefore, the presentation of both λ- and σ-POs fully describes the common symmetry properties of a whole OD family.

It is useful to distinguish the POs (both λ and σ) in two types; (i) the so called τ-**POs**, which do not turn the layer upside down, with reference to the direction in which the layers stack each after the other, and (ii) the so called ρ-**POs** which do turn the layer upside down. The multiplication rules for τ- and ρ-POs may be expressed as: $\tau \cdot \tau = \rho \cdot \rho = \tau; \tau \cdot \rho = \rho \cdot \tau = \rho$.

It is also useful to introduce the concept of **continuation**: if ${}_{p,q}a$ and ${}_{r,s}a$ are characterized by the same transformation applied to different layers, we call ${}_{p,q}a$ a continuation of ${}_{r,s}a$ and *vice versa*. This concept allows us to express the equivalence of pairs of adjacent layers as it follows: two pairs of adjacent layers, (L_p; L_{p+1}) and (L_q; L_{q+1}) are equivalent if there exists either a PO ${}_{p,q}\tau$ with continuation ${}_{p+1,q+1}\tau$ and/or a PO ${}_{p+1,q}\rho$ with continuation ${}_{p,q+1}\rho$.

2.3.2 *Categories of OD structures with equivalent layers*

Due to the equivalence of the OD layers, they are all **polar** or all **non-polar**, with reference to the stacking direction. The layer group symmetry of non-polar layers is characterized by the presence of both λ–τ-POs and λ–ρ-POs; consequently both σ–τ- and σ–ρ-POs exist. The layer group symmetry of polar layers presents only λ–τ-POs; consequently, the σ-POs are all of τ-character or all of ρ-character. Accordingly, there are three distinct **categories of OD structures of equivalent layers**.

- Category I: Character of λ-POs: τ and ρ; character of σ-POs: τ and ρ.
- Category II: Character of λ-POs: τ; character of σ-POs: τ.
- Category III: Character of λ-POs: τ; character of σ-POs: ρ.

For a simple description of the sequences of layers in the three categories, the layers are denoted as follows: a layer transformed into itself by a ρ-PO (namely a layer non-polar with respect to the stacking direction) is indicated with the symmetric letter A; a layer which is polar with respect to the stacking direction is indicated by a letter b or d; layers related to the previous ones through τ-POs or ρ-POs are indicated with the same letters (b, d) or the opposite pair (d, b), respectively. These letters are used in Fig. 2.5, which represents schematic examples for the three categories of OD structures.

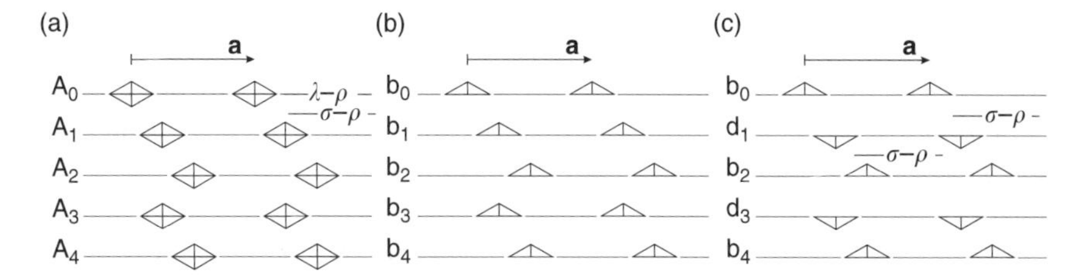

Fig. 2.5. Schematic examples of the three categories of OD structures built up with equivalent layers. The **a**,**b** layers are normal to the plane of the drawing, with **a** horizontal. The position of the $\lambda-\rho$-and $\sigma-\rho$-planes for category I and of both kinds of $\sigma-\rho$-planes for category III are indicated. In each scheme pairs of adjacent layers are equivalent and characterized by shifts of $\pm\mathbf{a}/4$.

In category I, the OD layers are non-polar and, as indicated in Fig. 2.5(a), the operational elements corresponding to the λ–ρ-POs lie in one plane (layer plane); similarly, the operational elements corresponding to the σ–ρ-POs lie in a plane which is located halfway between successive layer planes.

In category II, represented in Fig. 2.5(b), there are only τ-operations converting a layer into itself or into the adjacent one; all the layers are polar and presents always the same sense of polarity.

In OD structures of the first two categories, all pairs of adjacent layers are equivalent. This is no more valid for OD structures of the third category, where the layers are all polar and succeed each other with opposite sense of polarity. A look at Fig. 2.5(c) immediately shows that two distinct kinds of pairs of adjacent layers occur in OD structures of this category. In fact, according to the definition of equivalence of pairs of layers given at the end of the preceding chapter, the pairs of layers $(\mathrm{L}_p; \mathrm{L}_{p+1})$ and $(\mathrm{L}_{p+1}; \mathrm{L}_{p+2})$ cannot be equivalent, as neither a ${}_{p,p+1}\tau$ (σ-operation) nor a ${}_{p+1,p+1}\rho$ (λ-operation) exist.

In conclusion, $(\mathrm{L}_p; \mathrm{L}_{p+1})$ is equivalent to $(\mathrm{L}_q; \mathrm{L}_{q+1})$ for arbitrary values of p,q in the case of OD structures in categories I and II, and for values of p,q both even or both odd in the case of OD structures of category III. In the following, we shall refer to this statement as the **vicinity condition** (VC) for structures built up with equivalent layers.

2.3.3 *OD-groupoid families: notations for λ- and σ-POs*

As we have already said, the presentation of both λ- and σ-POs fully describes the common symmetry properties of a whole OD family. This is generally done with the so-called **OD-groupoid family** symbol, a symbol such as:

$$\begin{array}{cccc} P & m & m & (2) \\ & \{n_{s,2} & n_{2,r} & (2_2)\} \end{array} \tag{2.1}$$

In it, the first line presents the λ-operations, namely, the symmetry operations of the single layer corresponding to one of the 80 layer groups. The OD notation for the layer groups (Dornberger-Schiff 1959), derived from that proposed by Holser (1958*b*), follows the international notation, using a four-entry symbol (six-entry and eight-entry symbols for square and hexagonal nets, respectively) giving the type of lattice and the symmetry operations corresponding to the x, y, and z directions, and explicitly indicating the direction of missing periodicity, by placing in parentheses the symmetry operations corresponding to that direction.

The second line presents the σ-operations. The parentheses in the third position of each line indicate that only the basis vectors **a** and **b** are translation vectors corresponding to the periodicities of the single layer, whereas the third vector $\mathbf{c}_0$ is not a translation vector. In the example, the single layer has primitive lattice, mirror planes normal to **a** and to **b** and a twofold axis parallel to $\mathbf{c}_0$. The direction of $\mathbf{c}_0$ (in general the direction of the third vector, apart from the two translation vectors of the OD layer) is chosen so as to correspond to the point group isogonal[2] to the OD-groupoid family and its length spans the width of one OD layer for categories I and II, of two OD layers for category III. Therefore, in the present case, $\mathbf{c}_0$ is orthogonal to **a** and to **b** and its length corresponds to the width of the single layer. The σ-operations (the **symbols** for these POs are obtained as generalizations of the international space group symbols) of the second line are: n glide normal to **a**, with translational component $s\mathbf{b}/2 + \mathbf{c}_0$; n glide normal to **b**, with translational component $\mathbf{c}_0 + r\mathbf{a}/2$; a twofold screw axis with translational component $\mathbf{c}_0$. There is also a translation $\mathbf{t}(r, s, 1)$ $(= r\mathbf{a}+s\mathbf{b}+\mathbf{c}_0)$: a translation, not explicitly given in the symbol, is always present when both a λ-PO and a σ-PO correspond to the same point group symmetry. It seems proper to recall that the most general glide operation is denoted $n_{r,s}$: the order of the indices in the symbol is chosen in such way that the direction to which n and the two indices refer follow each other in a cyclic way. The indices r and s are only of interest *modulo* 2, so that r may be replaced by $r + 2$ or $r - 2$ and similarly s by $s + 2$ and $s - 2$.

It may be observed that three-line symbols are necessary for OD-families of category III (Ďurovič 1997*b*): the line with the λ-POs of the single OD layer (e.g. layer b_0); the line of the σ-POs transforming b_0 into d_1; the line of the σ-POs transforming d_1 into b_2.

In the case of square lattices (as well as in the case of trigonal/hexagonal lattices), more than three positions are necessary to indicate the operations related to the various directions. With the square lattice, the operations corresponding to the direction **a**, **b**, **c**, **a**+**b**, **a**−**b** have to be specified; with trigonal/hexagonal lattices the operations corresponding to the direction $\mathbf{a}_1, \mathbf{a}_2, \mathbf{a}_3, \mathbf{c}, \mathbf{a}_2 - \mathbf{a}_3, \mathbf{a}_3 - \mathbf{a}_1, \mathbf{a}_1 - \mathbf{a}_2$ have to be specified. An example of a symbol of an OD-groupoid family with POs isogonal to

[2] 'A symmetry operation is called **isogonal** to another, if it results from the composition of the latter with a translation. A point group is isogonal to a space group or an OD-groupoid, if its symmetry operations are isogonal to the symmetry operations of the space group or the POs (λ, σ or connecting further apart layers) of the groupoid, respectively. Space groups and/or OD-groupoids isogonal to the same point group are also called isogonal' (Dornberger-Schiff and Fichtner 1972).

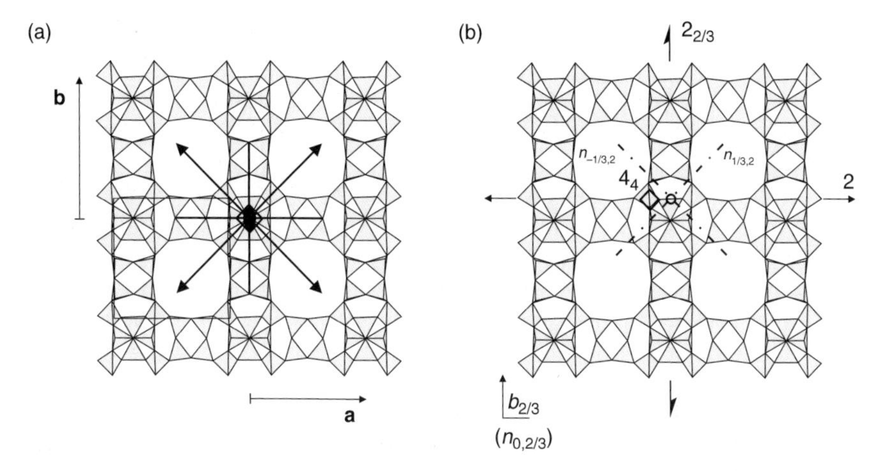

Fig. 2.6. Schematic drawing of the single layer in zeolite beta and tschernichite, with indication of the λ-POs (a) and σ-POs (b). The layer is built up by connection of (Si,Al) tetrahedra in a three-storied arrangement: in the upper level alignments of upwards-pointing tetrahedra run along **b**; in the lowest level alignments of downwards-pointing tetrahedra run along **a**; the two levels are connected through additional tetrahedra placed in the intermediate level.

4-fold rotation is here presented:

$$\begin{array}{llllll} P & m & m & (\bar{4}) & 2 & 2 \\ & \{2 & 2_{2/3} & (4_4/n_{0,2/3}) & n_{-1/3,2} & n_{1/3,2}\} \end{array}$$

This symbol refers to the OD family of **zeolite beta** (and its natural counterpart **tschernichite**). In fact, this family is built up with layers presenting $P\bar{4}m2$ layer group symmetry (Fig. 2.6(a)), layers which follow each to the other through the σ-operations listed in the second row and indicated in Fig. 2.6(b) (this OD-family will be more deeply discussed in Section 2.3.6).

2.3.4 *Possible OD-groupoid families*

In Section 2.3.3, we have presented a couple of OD-groupoid families. The derivation of all the OD-groupoid families has been one of the most important results of OD theory and a complete table of those families, which includes 400 members (Dornberger-Schiff 1964; Fichtner 1977*b*), is extremely useful to describe and discuss OD structures. In what follows, we shall assume, as it is usual, that **a** and **b** are translation vectors of the two-dimensional lattice of the OD layer, and $\mathbf{c}_0$ corresponds to the direction of 'missing periodicity'. For each of the five types of two-dimensional lattice (general, primitive rectangular, centred rectangular, square, and hexagonal), all the possible compatible λ- and σ-POs are considered. Table 2.2 presents these POs for the first three types of lattice (a complete table has been presented by Dornberger-Schiff and Grell-Niemann 1961 and by Dornberger-Schiff 1964). The

Table 2.2 λ- and σ-POs compatible with general and rectangular nets of the single layer

Description	λ–τ	λ–ρ	σ–τ	σ–ρ
		$\bar{1}$	$\boldsymbol{t}_{r,s,1}$	$\bar{1}$
General net	1 1 (2)	1 1 (m)	1 1 (2_2)	1 1 ($n_{r,s}$)
		1 1 (n)		
		$\bar{1}$	$\boldsymbol{t}_{r,s,1}$	$\bar{1}$
Rectangular	1 1 (2)	1 1 (m)	1 1 (2_2)	1 1 ($n_{r,s}$)
primitive, or	m 1 (1)	1 1 (n)	$n_{s,2}$ 1 (1)	
centred net *	b 1 (1)	1 1 (a)		
		2 1 (1)		
		2_1 1 (1)		2_r 1 (1)

* In addition, all the other operations must be considered in which the **a** and **b** directions are exchanged; for example, 1 m (1) as well as m 1 (1), or 1 $n_{2,r}$ (1), as well as $n_{s,2}$ 1 (1), etc.

λ-POs correspond, as it has been already said, to those occurring in the 80 layer groups; translational components parallel to **c** are obviously excluded. The possible σ-POs compatible with a given net differ from the λ-POs compatible with the same net only as regards the translational components: whereas the translational components of POs for x and y directions have values 0 or $\frac{1}{2}$ in λ-POs, they may have arbitrary values in σ-POs.

The groupoid may be derived by combining the POs in the following way:

(1) as regards the λ operations, the procedure is very simple, as they should build up one of the 80 layer groups;
(2) after defining the layer group, those σ operations are chosen (for the various directions) which are compatible with the given lattice (listed in Table 2.2 as regards the general and both rectangular lattices) and which are also compatible with the λ operations according to the concepts we have previously discussed.

For example, there are two distinct OD-groupoid families corresponding to the layer group $Pmm(2)$. All the λ-POs (the order of the group is 4) have character τ. Therefore, the four σ operations are all of τ (category II) or all of ρ character (category III). In the first case, the operational element relative to the **c** direction indicated in Table 2.2 is (2_2), and the operational elements relative to **a** and **b** are $n_{s,2}$ and $n_{2,r}$, respectively. In this way, the OD-groupoid family symbol (2.1) is obtained. In the second case, the operational elements of ρ character, relative to **c**, **a**, and **b** are $(n_{r,s})$, 2_r, and 2_s, respectively. The symbol of the corresponding OD-groupoid family is therefore:

$$\begin{matrix} P & m & m & (2) \\ & \{2_r & 2_s & (n_{r,s})\} \\ & \{2_{r'} & 2_{s'} & (n_{r',s'})\} \end{matrix}$$

As a second example, we shall derive the OD-groupoid family symbol for **wollastonite**. The layer group is $P(1)2_1/m1$; the lattice has orthogonal translations **b** and **c** (therefore rectangular lattice). It is proper to observe the direction of 'missing periodicity' is now $\mathbf{a}_0$, not $\mathbf{c}_0$ as assumed in Table 2.2. Among the σ operations compatible with the rectangular lattice, there are [- 2_s -] and [- $n_{r,2}$ -][3] (obtained from those of Table 2.2, by changing the direction of missing periodicity). Therefore, we obtain the following OD-groupoid family symbol:

$$\begin{matrix} P(1) & 2_1/m & 1 \\ \{(1) & 2_s/n_{r,2} & 1\} \end{matrix}$$

As the present OD-groupoid is isogonal with the point group $2/m$, with a proper choice of the direction of $\mathbf{a}_0$ r is 0 and the symbol becomes:

$$\begin{matrix} P(1) & 2_1/m & 1 \\ \{(1) & 2_s/a_2 & 1\} \end{matrix}$$

With $s = 1/2$, the symbol for the OD-groupoid family of wollastonite is obtained.

The first derivation of the OD-groupoid families has been carried on by Dornberger-Schiff (1964). Subsequently, a new procedure for listing all the OD-groupoid families has been presented (Dornberger-Schiff and Fichtner 1972; Fichtner 1977b), together with corrections to the previous list. The table given in Appendix 2.1, taken from the paper of Dornberger-Schiff and Fichtner (1972), presents all the OD-groupoid families not containing operations corresponding to three-, four-, or six-fold axes.

2.3.5 *Possible distinct positions of adjacent layers: NFZ formula*

In the preceding chapter, we have presented some examples of OD-groupoid family symbols. It is proper to observe that the set of σ–POs presented within braces transfer the starting layer to one particular among the various possible Z positions in keeping with the 'vicinity condition'. The σ–POs corresponding to the various possible positions differ as regards the sign of the parameters in the translational components ($r, s, \ldots$) or as regards the translation vectors to which they refer. We now shall discuss how to obtain the value of Z, the number of the possible distinct positions of a layer L_{p+1}, relative to the fixed position of an adjacent layer L_p. The knowledge of Z is important not only to determine whether the given arrangement is actually a **fully ordered structure** ($Z = 1$), but also to find whether the set of polytypes of a given type (e.g. MDO polytypes) is complete. It will be shown that the value of Z is dependent on the order N of the group G of λ–τ-POs and on the order F of the subgroup of λ–τ-POs ${}_{p,p}\tau$, which present a continuation ${}_{p+1,p+1}\tau$, namely those λ–τ-POs, which are valid both for the single layer and for the pair of adjacent layers.

[3] We have introduced here a convenient notation to indicate the operational elements of symmetry and their relations with the directions x, y, z.

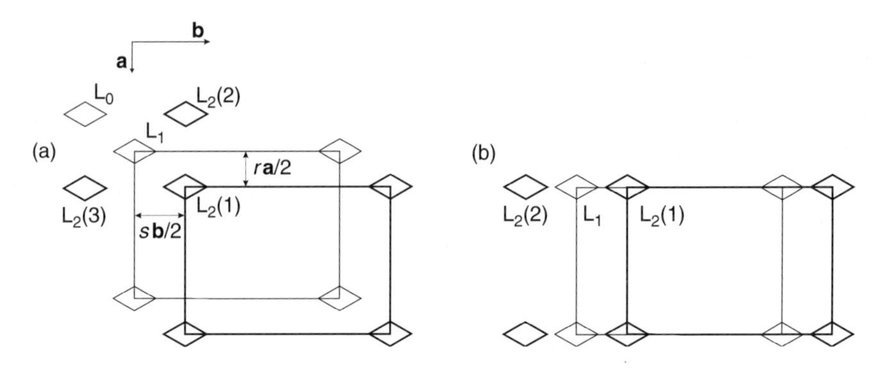

Fig. 2.7. Schematic representation of the layer stacking in OD-groupoid family

$$\begin{matrix} P & m & m & (2) \\ & \{n_{s,2} & n_{2,r} & (2_2)\} \end{matrix}$$

for general values of the s and r parameters (a) and for $r = 0$ (b). The 'lozenge' figures are drawn with light marks for layer L_0, with normal marks for layer L_1 and with heavy marks for layer L_2.

We shall start with a schematic example corresponding to the OD-groupoid family (2.1)

$$\begin{matrix} P & m & m & (2) \\ & \{n_{s,2} & n_{2,r} & (2_2)\} \end{matrix}$$

and represented in Fig. 2.7, where the layers stack in the direction normal to the drawing plane. This OD family belongs to category II, as explained beforehand. All the λ-POs have character τ; therefore $N = 4$ (the order of the layer group $Pmm2$). As it is usual in the schematic drawings for representing the layer groups (black–white groups), any of the 17 **polar layer groups** (in the present case, the polarity is with respect to the direction normal to the **a**, **b** plane) is represented with figures all looking white (they would look black if viewed on the other side)[4].

There are four distinct positions for the layer L_2, relative to the position of the layer L_1, all giving rise to geometrically equivalent pairs of layers, all compatible with the VC. The first position, $L_2(1)$, is obtained through the set of the σ-POs listed in braces (corresponding to the values $s = r = +1/2$ for the parameters appearing in the translational components), the other three positions are obtained from the first one through the action of the λ-POs of L_1 and correspond to three more sets of σ-POs, depending on the following values of the s and r parameters: $s = 1/2, r = -1/2$ for

[4] The other 63 layer groups are non-polar. Each of them is represented with a sheet of figures, half presenting the white side up, half presenting the black side up; 17 of these non-polar groups, corresponding to the 17 polar groups, present a mirror parallel to the **a**, **b** plane; both sides are equivalent and the figures are 'grey'.

$L_2(2)$; $s = -1/2, r = 1/2$ for $L_2(3)$; $s = r = -1/2$ for $L_2(4)$ (this last position is not indicated as it is superposed to the position of L_0).

The four triples $[L_0; L_1; L_2(n)]$ are not geometrically equivalent one to the others. With each of them, a full structure may be built as it will be shown in Section 2.3.6.

We shall now consider the same OD-groupoid family, but with special values for the s and r parameters. Figure 2.7(b) presents a pair of adjacent layers, obtained assuming $r = 0$. Now, there are only two distinct positions for the layer L_2 compatible with the VC. The λ-PO corresponding to the mirror plane normal to **a** is a symmetry operation not only of the single layer L_1, but also of the pair of layers $(L_1; L_2)$. Consequently, the subgroup of the λ–τ-POs presenting a continuation in the adjacent layers has order $F = 2$, whereas in the general case (Fig. 2.7(a)) $F = 1$, as in that case only the identity has a continuation. Consequently, in the present example and in all the OD-groupoid families of category II, we may easily calculate the value of Z through the relation: $Z = N/F$. It is easy to see that $F = 2$ also for $r = 1$; similarly $F = 2$ when s is equal to 0 or to 1. In the very special cases in which both r and s are equal to 0 or 1, $F = 4$, and consequently, $Z = 1$. In such cases, we obtain fully ordered structures: each operation has now continuations, becoming a 'true' symmetry operation valid for the whole structure.

Let us now consider the addition of a horizontal symmetry plane to the λ-POs of the OD-groupoid family we have just discussed, thus passing from the layer group $Pmm(2)$ to $Pmm(m)$. Due to the additional symmetry plane, the lozenges are now grey with equivalent upper and lower sides. The OD-groupoid family symbol is now

$$\begin{matrix} P & m & m & (m) \\ & \{n_{s,2} & n_{2,r} & (n_{r,s})\} \end{matrix} \tag{2.2}$$

and the family belongs to category I. The λ–τ-POs are the same of the preceding case, therefore both N and F have the same values. The schematic representation of the arrangement of succeeding layers is the same as in Fig. 2.7, with the substitution of grey lozenges to the white ones. Consequently, also in this case, for general values of s and r, for example for $s = r = 1/2$, Z is 4 and coincides with the ratio N/F. No additional positions for L_2 relative to L_1 result from the additional ρ-POs and the NFZ relationships given for category II remains valid.

This is not generally true for category I, as it will be shown by the next example, which presents the OD-groupoid family:

$$\begin{matrix} P & 2 & m & (m) \\ & \{2_r & n_{2,r} & (n_{r,s})\} \end{matrix} \tag{2.3}$$

Also, in this family, the layers are non-polar and therefore it belongs to category I. Figure 2.8 schematically illustrates the relative positions of layers L_0, L_1, and L_2.

L_0 transforms into L_1 through the action of the σ–ρ-PO $[- - n_{r,s}]$ ($r = s = +\frac{1}{2}$ in the example); L_1 transforms into $L_2(1)$ through the action of the same σ–ρ-PO; the position indicated as $L_2(2)$ is obtained by applying the λ–τ-POs of the L_1 layer, namely $[- m -]$ (Fig. 2.8(a)). The two positions of L_2 relative to L_1 are in keeping with

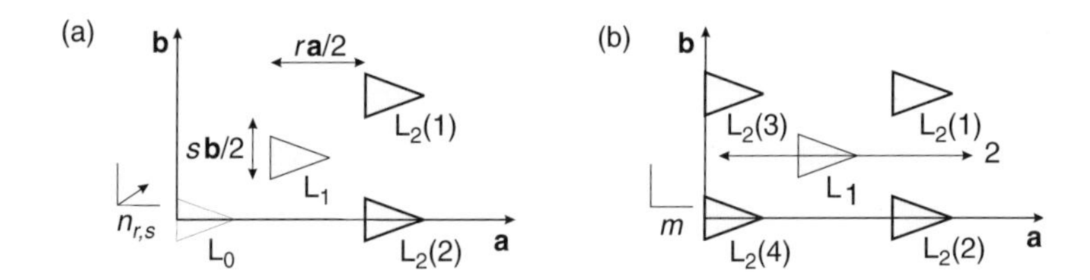

Fig. 2.8. Schematic representation of the layer stacking in OD-groupoid family

$$\begin{array}{cccc} P & 2 & m & (m) \\ & \{2_r & n_{2,r} & (n_{r,s})\} \end{array}$$

a) Relative positions of the layers L_0, L_1, and $L_2(1)$, $L_2(2)$.
b) Additional positions of the L_2 layer obtained from the position of L_0 through the action of $[- - m]$ and $[2 - -]$ in L_1.

the relationship $Z = N/F$. However, if we consider the action of the λ–ρ-POs in L_1, in particular the action of the mirror parallel to the **a**, **b** plane and of the twofold axis parallel to **a**, we find that additional positions for L_2 [$L_2(3)$ and $L_2(4)$, respectively] are obtained from the positions of L_0 (Fig. 2.8(b)). Therefore, four distinct positions for L_2 relative to L_1 are possible and the correct NFZ formula in the present case is: $Z = 2N/F$.

What distinguishes the two OD families of category I just discussed is the presence in the first one (2.2) of a σ–ρ-PO converting any layer into the adjacent one and *vice versa*; a σ–ρ-PO of such kind is said to have a **reverse continuation** and it is represented by an inversion centre in (2.2). No similar operation is present in (2.3). Consequently, OD structures in category I are subdivided in two distinct subcategories, Ia and Ib, according to the presence or absence, respectively, of σ–ρ-POs with reverse continuation.

As regards the OD structures in category III, two distinct types of OD layers are present and two values of Z, namely $Z_{2n,2n+1}$ and $Z_{2n+1,2n+2}$, have to be determined. In category III, all the layers are polar and related by σ–ρ-POs. Three distinct situations occur: (a) there is at least one ${}_{2n,2n+1}\rho$ and one ${}_{2n+1,2n+2}\rho$ with reverse continuation; (b) there is no ${}_{2n,2n+1}\rho$ and no ${}_{2n+1,2n+2}\rho$ with reverse continuation; and (c) there is at least one ${}_{2n,2n+1}\rho$ and no ${}_{2n+1,2n+2}\rho$ with reverse continuation, or *vice versa*. The three distinct situations give rise to three distinct subcategories: IIIa, IIIb, IIIc. The presence or absence of σ–ρ-POs with reverse continuation has in category III the same consequences as found in category I, as shown in Table 2.3, which comprehensively presents the formulas for obtaining Z values in the various categories and subcategories.

2.3.6 *MDO Structures*

The results obtained in the preceding chapter are useful in the derivation of the MDO structures in the various categories of OD families. In the presentation of the OD

Table 2.3 Z values in the various categories and subcategories of OD structures

Category	Description	Z	$Z_{2n,2n+1}$	$Z_{2n+1,2n+2}$
Ia	At least one ${}_{p,p+1}\rho$ has a reverse continuation	N/F		
Ib	No ${}_{p,p+1}\rho$ has a reverse continuation	$2N/F$		
II		N/F		
IIIa	At least one ${}_{2n,2n+1}\rho$ and one ${}_{2n+1,2n+2}\rho$ have reverse continuations		$N/F_{2n,2n+1}$	$N/F_{2n+1,2n+2}$
IIIb	No ${}_{2n,2n+1}\rho$ and no ${}_{2n+1,2n+2}\rho$ have reverse continuations		$2N/F_{2n,2+1}$	$2N/F_{2n+1,2n+2}$
IIIc	At least one ${}_{2n,2n+1}\rho$ and no ${}_{2n+1,2n+2}\rho$ have reverse continuations		$N/F_{2n,2n+1}$	$2N/F_{2n+1,2n+2}$

family of wollastonite, we introduced the concept of MDO structures indicating the physical reason of their special stability and widespread occurrence. Actually, as maintained by Dornberger-Schiff (1982), 'behind the OD theory and the concept of MDO polytypes, there is the basic idea of decreasing interatomic forces with increasing distance, thus leading to a preference of polytypes with a minimum number of layer pairs (principle of OD structures) or even a minimum number of kinds of n-tuples of layers, for any number n (principle of MDO structures)'. In the particular example of wollastonite, we emphasized the role played by the 'layer triples' and we have presented a rough definition of MDO structures as those members of an OD family in which 'triples (and quadruples, ... n-tuples) of consecutive layers are geometrically equivalent as far as possible'. This definition is actually satisfactory not only for the single case of wollastonite, but also for most of the OD families of structures built with equivalent layers, occurring in natural and synthetic compounds. As a matter of fact, the concretely realized examples of OD structures of equivalent layers mostly belong to categories Ia, II, and IIIa for which the definition based on the equivalence of triples, quadruples, etc. is satisfactory. However, to include all the categories and subcategories of OD structures built up with equivalent layers, it is necessary to clearly specify the second part of the definition, namely, to explain what '... as far as possible' actually means.

We shall first discuss three examples in categories II, Ia, and IIIa, and afterwards present a concrete example in category Ib, which will be useful to introduce a more comprehensive definition of MDO structures.

Let us consider the OD family (category II) described by the symbol (2.1) and examine the corresponding Fig. 2.7. We observed that assuming (L_0; L_1) as the starting layer pair four distinct positions for the L_2 layer were possible. The four positions and the corresponding triples of layers are realized when the four σ-POs ${}_{0,1}\tau(n)$ have continuations σ-POs ${}_{1,2}\tau(n)$. Of the four σ-POs ${}_{0,1}\tau(n)$, three are given in braces and the fourth is a translation; in fact, as we have already said, a translation, although not given in the symbol, is always present when both a λ-PO and a σ-PO correspond to the same point group symmetry. If the four σ–τ-POs have continuations $_{p,p+1}\tau(\mathrm{n})$ for all the values p, they become total operators for the polytypes thus obtained and are consequently defined **generating operations**.

The structure of the first polytype, MDO_1, characterized by the glide $[-\,n_{2,r}\,-]$ as generating operation and layer triples of type $(\mathrm{L}_0; \mathrm{L}_1; \mathrm{L}_2(1))$, has space group $P1c1$, with lattice translations $\mathbf{a}$, $\mathbf{b}$, $\mathbf{c}_1 = 2\mathbf{c}_0 + r\mathbf{a}$.

The structure of the second polytype, MDO_2, characterized by the glide $[n_{s,2}\,-\,-]$ as generating operation and layer triples of type $(\mathrm{L}_0; \mathrm{L}_1; \mathrm{L}_2(2))$, has space group $Pc11$, with lattice translations $\mathbf{a}$, $\mathbf{b}$, $\mathbf{c}_2 = 2\mathbf{c}_0 + s\mathbf{b}$.

The structure of the third polytype, MDO_3, characterized by the translation $\mathbf{t} = r\mathbf{a}/2 + s\mathbf{b}/2 + \mathbf{c}_0$ as generating operation and layer triples of type $(\mathrm{L}_0; \mathrm{L}_1; \mathrm{L}_2(3))$, has space group $P1$, with lattice translations $\mathbf{a}$, $\mathbf{b}$, $\mathbf{c}_3 = \mathbf{t}$.

The structure of the fourth polytype, MDO_4, characterized by the screw $[-\,-\,(2_2)]$ as generating operation and layer triples of type $(\mathrm{L}_0; \mathrm{L}_1; \mathrm{L}_2(4))$, has space group $P112_1$, with lattice translations $\mathbf{a}$, $\mathbf{b}$, $\mathbf{c}_4 = 2\mathbf{c}_0$. In the four polytypes, all the triples are of one kind, as are the quadruples, n-tuples, etc. These polytypes constitute the four possible MDO structures of the OD family described by the symbol (2.1).

It is useful in this context to discuss the problem of **twinning** which so frequently occurs in OD structures and is, in some way, connected to the MDO concept. We have already explained that each of the MDO polytypes is realized when the crystallization conditions favour the particular triple corresponding to that polytype. If the conditions change, other triples may be realized and that particular MDO occurs as a limited domain inside the whole crystalline arrangement. It may also happen that the conditions remain substantially unchanged during the crystallization process, but some stacking fault occasionally occurs. Let us consider the polytype MDO_3, characterized by the generating operation $\mathbf{t} = r\mathbf{a}/2 + s\mathbf{b}/2 + \mathbf{c}_0$ and consider that one of the other σ–τ-POs (glide or screw) present continuation and after this 'fault' again the σ-translation $\mathbf{t}'$ be continued, thus building again the structure-type of the third polytype. However, $\mathbf{t}'$ is distinct from $\mathbf{t}$ and the new MDO_3' domain is twin related with the previous one. Four distinct σ-translations $\mathbf{t} = r\mathbf{a}/2 + s\mathbf{b}/2 + \mathbf{c}_0$ are possible, depending on the sign of the r and s parameters. They are related each to the other by the λ–τ-POs, which are also the twin-elements of the possible twins: (100), (010) planes, [001] axis.

Twinning may occur also in MDO_1 and MDO_2 polytypes. In these cases, only one parameter occurs in the generating operation and only one twin element exists, (100) and (010) plane in MDO_1 and MDO_2, respectively. No twinning occurs in the case of MDO_4: in that case, the faults give rise to parallel-growth arrangements of MDO_4 domains.

We have already presented the OD family of **wollastonite** (Fig. 2.1),

$$\begin{array}{lll} P(1) & 2_1/m & 1 \\ \{(1) & 2_{1/2}/a_2 & 1\}. \end{array}$$

It is a family of category Ia. The λ–τ-POs are identity and mirror [– m –] ($N = 2$); only identity is valid for a pair of adjacent layers ($F = 1$) (Fig. 2.3). Therefore, each layer has two distinct positions relative to the preceding one ($Z = 2$). There are also two σ–τ-POs, the glide [– a_2 –] and the translation $\mathbf{t}_1 = \mathbf{a}_0 - \mathbf{b}/4$ (the presence of a σ-translation is guaranteed by the presence of a λ-PO and a σ-PO corresponding to the same point group symmetry). Starting with the pair of layers (L_0, L_1), the layer L_2 may assume two positions $L_2(1)$ and $L_2(2)$, related each to the other by the λ–τ-PO $_{1,1}$[– m –]. The two distinct positions and the corresponding triples of layers, the 'stretched' T_1 and the 'bent' T_2 (Fig. 2.4), are obtained when the two σ–τ-POs, $_{0,1}\mathbf{t}$ and $_{0,1}$[– a_2 –], have continuations $_{1,2}\mathbf{t}$ and $_{1,2}$[– a_2 –], respectively. As the two σ–τ-POs have general continuations $_{p,p+1}\mathbf{t}$ and $_{p,p+1}$[– a_2 –] for any value of p, they become the 'generating operations' of two MDO polytypes.

The structure of the first polytype MDO_1, characterized by the translation as generating operation and 'stretched' triples of layers, has space group $P\bar{1}$, with lattice translations $\mathbf{a}_0 - \mathbf{b}/4$, $\mathbf{b}$, $\mathbf{c}$, and corresponds to the polytype 1A already described. All the triples T_1, as well as all the quadruples, ... n-tuples, are equivalent. As explained in the previous example, twinning may occur with (010) as twin plane and formation of distinct domains with the same structure type but with different σ translations as generating operations, namely $\mathbf{t}_1 = \mathbf{a}_0 - \mathbf{b}/4$ and $\mathbf{t}_2 = \mathbf{a}_0 + \mathbf{b}/4$, related each to the other by the λ–τ-PO [– m –].

The structure of the second polytype, characterized by the glide [– a_2 –] as generating operation and 'bent' triples of layers, has space group $P2_1/a$, with lattice translations $2\mathbf{a}_0$, $\mathbf{b}$, $\mathbf{c}$, and corresponds to the polytype 2M. All the triples T_2, as well as all the quadruples, ... n-tuples, are equivalent.

An example of OD structure of category III is offered by **foshagite**, $Ca_4Si_3O_9(OH)_2$. The structure (Gard and Taylor 1960) is built up by walls of calcium 'octahedra', four octahedra large, connected through chains of silicon tetrahedra of wollastonite-type; hydroxyl anions occupy octahedral vertices not shared with silicate groups.

Its OD features are displayed in Fig. 2.9, which presents the structure as seen along $\mathbf{b}$, and shows that it may be described in terms of OD layers with translation vectors $\mathbf{b}$ and $\mathbf{c}$ ($b = 7.36$, $c = 14.07$ Å) and third vector $\mathbf{a}_0$ (not a translation vector, spanning—as usual in category III—two OD layers; $a_0 = 10.32$ Å, $\beta = 106.4°$), layer symmetry $A(1)m1$ and OD-groupoid family symbol

$$\begin{array}{ccc} A(1) & m & 1 \\ \{(1) & 2_{1/2} & 1\} \\ \{(1) & 2_{1/2} & 1\} \end{array}$$

In Fig. 2.10, we shall use a schematic representation of the structural layers, as seen along $\mathbf{c}$ ($\mathbf{b}$ horizontal). The family belongs to category IIIa. Each layer has four λ–τ-POs, namely identity, [– m –], translation $(\mathbf{b} + \mathbf{c})/2$, [– c –] ($N = 4$); only the identity and the translation are operations valid for pair of adjacent layers ($F = 2$). Consequently, each layer may be placed in two distinct positions relative to the preceding one ($Z = 2$). Let us assume b_1d_2, where the layers are related by the

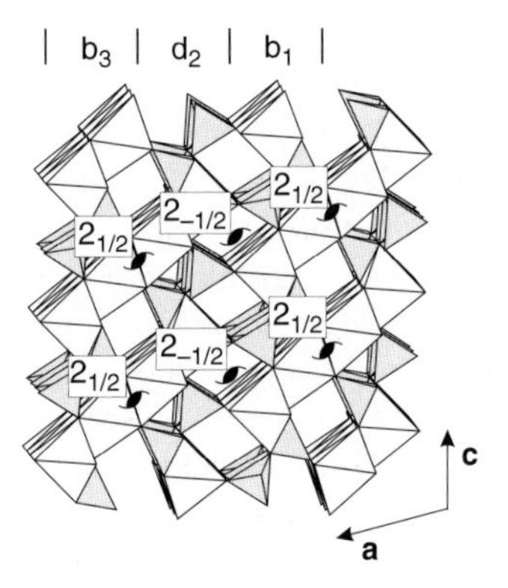

Fig. 2.9. The crystal structure of foshagite, $Ca_4Si_3O_9(OH)_2$ (polytype MDO_1, with space group symmetry $A\bar{1}$, $a = 10.32$, $b = 7.36$, $c = 14.07$ Å, $\alpha = 90°$, $\beta = 106.4°$, $\gamma = 90°$), as seen along **b**. The drawing is slightly rotated about the horizontal axis for a better appreciation of the polyhedral connections. Calcium 'octahedra' and silicon tetrahedra are drawn in light grey and dark grey, respectively. The sequence of layers $b_1d_2b_3$ is indicated, as well as the sequence of $2_{1/2}$ and $2_{-1/2}$ operators which alternate in the **a** direction.

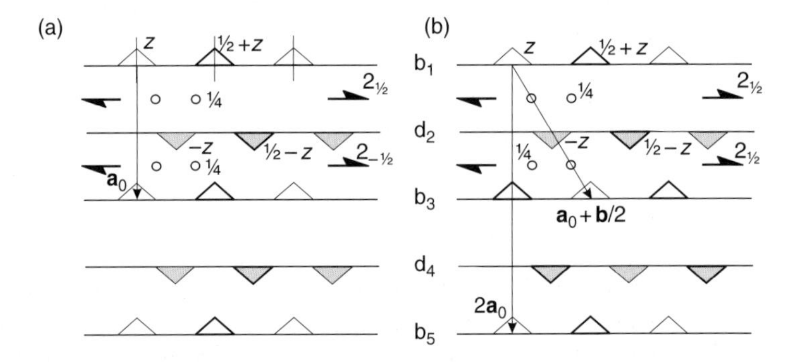

Fig. 2.10. Schematic representation of the sequence of the structural layers in foshagite, as seen along **c** (**b** horizontal), in MDO_1 (a) and MDO_2 (b) polytypes. The different z levels of the white triangles (b layers) and dark triangles (d layers) are indicated with light ($\pm z$) and heavy rims ($1/2 \pm z$). In (a) the position of the mirror plane [$-\ m\ -$] in b_1 is indicated.

operation $2_{1/2}$, as the starting pair. The other possible position of the layer d_2' relative to b_1 (operation $2_{-1/2}$) may be obtained from the preceding one by applying the mirror plane [$-\ m\ -$] in b_1: the following presentation would be valid also assuming b_1d_2' as the starting pair.

As our aim is to obtain the possible MDO structures of this family, we shall proceed according to the following steps:

1. We shall first derive the possible layer triples $b_1d_2b_3$; the *NFZ* relation indicates that two possible layer triples exist, as represented in Fig. 2.10(a) and (b).

2. Then, for both possible triples, we shall determine that position of d_4 for which $b_1d_2b_3$ and $d_2b_3d_4$ are geometrically equivalent; this position is obtained through the σ–ρ-POs which convert d_2 in b_3 and b_3 in d_2 (inversion centres at $z = 0$ and $z = 1/4$). By operating with both kind of inversion centres we obtain, in the case of the first triple, the position of d_4 given in Fig. 2.10(a) and, in the case of the second triple, the position of d_4 given in Fig. 2.10(b).
3. Last, for each resulting quadruple we shall look for those τ-operations which transform the pair b_1d_2 into the pair b_3d_4, and will use them as 'generating operations' to build the MDO structures. These τ-operations are the translations $\mathbf{t}_1 = \mathbf{a}_0$ (Fig. 2.10(a)), giving rise to MDO_1, with space group symmetry $A\bar{1}$, and translation vectors $\mathbf{a} = \mathbf{a}_0$, $\mathbf{b}$, $\mathbf{c}$; and $\mathbf{t}_2 = \mathbf{a}_0 + \mathbf{b}/2$ (Fig. 2.10(b)), giving rise to MDO_2, with space group symmetry $A\bar{1}$, and translation vectors $\mathbf{a} = \mathbf{a}_0 + \mathbf{b}/2$, $\mathbf{b}$, and $\mathbf{c}$. This structure may also be referred to space group $F\bar{1}$, with translation vectors $\mathbf{a} = 2\mathbf{a}_0$, $\mathbf{b}$, and $\mathbf{c}$.

The structural study of Gard and Taylor (1960) indicates that all the natural and synthetic crystals of foshagite are highly disordered as evidenced by their diffraction patterns, which present continuous streaks running along $\mathbf{a}^*$ for odd k values. The diffraction patterns of the natural crystals present maxima on the streaks in positions corresponding to the MDO_1 polytype, whereas the diffraction patterns of the synthetic crystals present maxima pointing to the presence of both MDO_1 and/or MDO_2 domains.

A very interesting example of OD structure is presented by **decaborane**, $B_{10}H_{14}$, a polytypic compound with OD-groupoid family

$$\begin{array}{cccc} P & 1 & (1) & 2/a \\ & \{2_{1/2}/n_{2,1} & (2_2/n_{1,-1/2}) & 1\} \end{array}$$

The structure-type of decaborane has been found by Kasper *et al.* (1950) who described it in a paper which is remarkable for various aspects: it presented the first successful solution of an unknown structure through inequalities relationships, it opened the way to the understanding of the structural features of boranes, it successfully dealt with the order–disorder problems of decaborane, just applying, *ante litteram*, an OD approach. The molecular structure of decaborane which was eventually found by Kasper *et al.* (1950) is represented in Fig. 2.11.

The molecules build up OD layers, with translation vectors $\mathbf{a}$, $\mathbf{c}$ ($a = 14.45$, $c = 5.68$ Å) and third vector (not a translation vector) $\mathbf{b}_0$ ($b_0 = 5.22$ Å). We shall now discuss the derivation of the MDO structures in that family with the help of the representation of Fig. 2.12, in which the decaborane molecules are drawn, for the sake of simplicity, with their skeleton of boron atoms, omitting the hydrogen atoms.

The layers are non polar; as none of the σ–ρ-POs, $[2_{1/2} - -]$ and $[- \, n_{1,-1/2} \, -]$, has reverse continuation, the family belongs to category Ib. Both λ–τ-POs, namely, identity and $[- - a\ ]$ ($N = 2$), are valid also for a pair of adjacent layers ($F = 2$); therefore $Z = 2N/F = 2$. If the position of the starting pair (A_1; A_2) is that indicated in Fig. 2.12(a), two generating operations occur, the screw $[- \, 2_2 \, -]$ and the

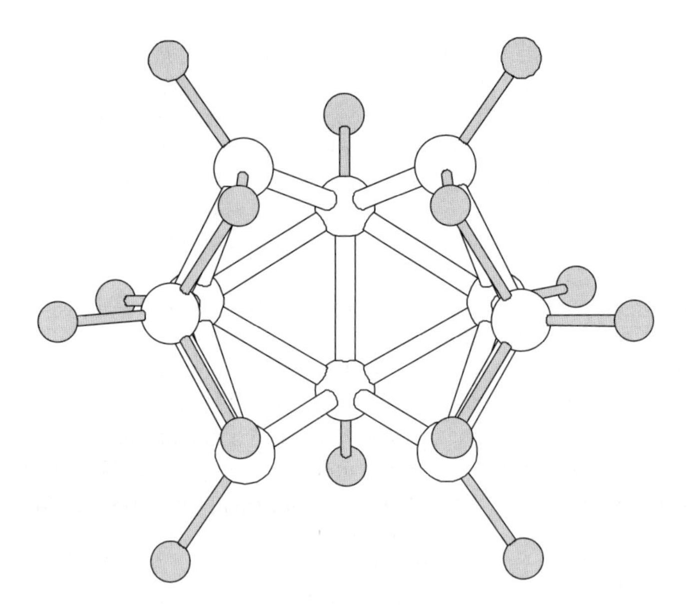

Fig. 2.11. The molecular structure of decaborane $B_{10}H_{14}$. The large white circles represent B atoms; the smaller grey circles represent H atoms. The molecule displays $2mm$ symmetry. Only the twofold axis appears in the layer group symmetry $P112/a$.

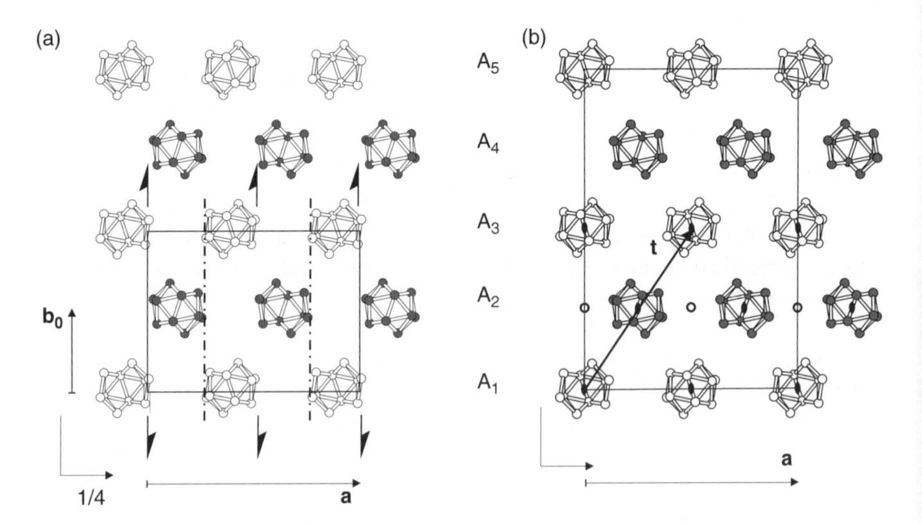

Fig. 2.12. The two MDO structures of the OD family of decaborane, as seen along **c** (**a** horizontal): (a) MDO_1; (b) MDO_2. In each row parallel to **a,** the molecules have their concavity alternately upwards and downwards. 'Open' and 'grey' circles represent boron atoms in molecules at $\pm z$ and $(1/2) \pm z$, respectively. In both (a) and (b), the symmetry operations of the structures and the generating operations are indicated.

glide $[n_{2,1} - -]$; their continuations to total operations give rise to the same structure, MDO_1, indicated in Fig. 2.12(a), with space group $Pn2_1a$ and lattice parameters $\mathbf{a}, \mathbf{b} = 2\mathbf{b}_0, \mathbf{c}$. As indicated by the *NFZ* relation, an alternative position exists for the layer A_3 relative to A_1, in keeping with the equivalence of $(A_1; A_2)$ and $(A_2; A_3)$ pairs. It may be obtained through the λ–ρ-POs transforming the layer A_2 into itself and A_1 into A_3. The two ρ-operations of such kind, namely the inversion $\bar{1}$ and the twofold rotation $[- - 2]$, lead to the same position of the A_3 layer (Fig. 2.12(b)). Two possible generating operations are now the b glide normal to $\mathbf{c}$ with translational component $2\mathbf{b}_0$ and the translation $\mathbf{t} = 2\mathbf{b}_0 + \mathbf{a}/2$ (only this one is represented in Fig. 2.12(b)): both operations lead to the same MDO_2 structure with space group $P112/a$ and lattice parameters $\mathbf{a}, \mathbf{b} = 2\mathbf{b}_0 - \mathbf{a}/2, \mathbf{c}$. It may also be described as $C112/a$, with $\mathbf{a}, \mathbf{b} = 4\mathbf{b}_0, \mathbf{c}$. In MDO_1, all the layers succeed each to the other through the constant application of the generating operation, which guarantees the equivalence of all the T_1 triples, as well as of all quadruples, ... n-tuples. In MDO_2, the generating operation transforms the pair of layers $(A_n; A_{n+1})$ into $(A_{n+2}; A_{n+3})$, which does not guarantee that $(A_n; A_{n+1}; A_{n+2})$ be equivalent to $(A_{n+1}; A_{n+2}; A_{n+3})$. In fact, they are not equivalent and MDO_2 presents two distinct kinds of triples, T_2 and T_3, exemplified by $(A_1; A_2; A_3)$ and $(A_2; A_3; A_4)$, respectively. On the other hand, all the quadruples are equivalent: the equivalence of $(A_n; A_{n+1}; A_{n+2}; A_{n+3})$ and $(A_{n+1}; A_{n+2}; A_{n+3}; A_{n+4})$ is ensured by the λ–ρ-POs (inversion centre and twofold rotation) transforming the A_{n+2} layer into itself, as well as A_n and A_{n+1} into A_{n+4} and A_{n+3}, respectively. In general, we shall find two kinds of n-tuples for n odd and one kind of n-tuple for n even.

In the structures of the OD family just discussed, three distinct kinds of triples have been found, namely T_1, T_2, T_3. If the triple T_1 is energetically favoured, the structure MDO_1 will be formed. Let us suppose that, in particular conditions of crystallization, the triples T_2 and T_3 be more stable than T_1. It is now important to remark that no structure may be built containing only T_2 or only T_3 triples; however, an ordered structure where T_2 and T_3 triples regularly alternate may form in those particular conditions. Moreover, in this arrangement, the MDO_2 structure, all the quadruples are geometrically equivalent. Although it presents two kinds of triples, it is proper to include it among the MDO structures.

We may now better specify, as it was our aim, the definition of structures with maximum degree of order (MDO). A structure containing M distinct kinds of triples is defined as an MDO structure if no other member of the family exists which is built only by a selection of the M types of triples, and if the same is valid for quadruples, ... n-tuples.

As regards the last three examples (wollastonite, category Ia; foshagite, category IIIa; and decaborane, category Ib), in which succeeding layers may be related by σ–ρ-POs, the two MDO structures in each family may be quickly obtained also considering the positive and negative signs of the translational parameters of those operations and deriving the two regular sequences.

In wollastonite, the MDO_2 polytype occurs when the σ-operations $2_{1/2}$ and $2_{-1/2}$ regularly alternate, whereas the MDO_1 polytype occurs when the σ-operation $2_{1/2}$ is constantly followed by $2_{1/2}$; when $2_{-1/2}$ is always followed by $2_{-1/2}$, the structural arrangement MDO'_1, twin-related through (010) to MDO_1, is realized.

By looking at Fig. 2.10 (schematic drawing for foshagite), it may be realized that MDO_1 and MDO_2 polytypes are obtained when the σ-operations $2_{1/2}$ and $2_{-1/2}$ regularly alternate, or when $2_{1/2}$ is always followed by $2_{1/2}$, respectively. When $2_{-1/2}$ is always followed by $2_{-1/2}$, a structural arrangement MDO'_2, twin-related through (010) to MDO_2, is realized: the generating operations is now $\mathbf{t}'_2 = \mathbf{a}_0 - \mathbf{b}/2$, related to $\mathbf{t}_2$ by the λ–τ-PO $[- m -]$.

Finally, in decaborane, the regular alternation of the σ-operations $2_{1/2}$ and $2_{-1/2}$ (parallel to **a**) (or the regular alternation of $[- n_{1,-1/2} -]$ and $[- n_{1,1/2} -]$) gives rise to the MDO_2 polytype with space group symmetry $Pn2_1a$; the MDO_1 polytype, with space group symmetry $C112/a$ occurs when $2_{1/2}$ is constantly followed by $2_{1/2}$, [the sequence $2_{-1/2}/2_{-1/2}/2_{-1/2}\ldots$ corresponds to the twin-related MDO'_1 arrangement, with (100) as twin plane].

The OD families discussed in the present chapter do not include examples from subcategories IIIb and IIIc. For these more difficult cases, and for a comprehensive treatment of the procedures to derive the MDO structures in all the six subcategories listed in Table 2.3, the interested reader may refer to Dornberger-Schiff (1982).

2.3.7 *Diffractional features of OD structures*

The OD character of a given compound generally fully reveals itself through its peculiar diffraction pattern. In a 'fully ordered' structure (for this kind of structure, $Z = 1$, according to our previous treatment), the single-crystal pattern presents only sharp spots: the intensity distribution of the pattern and the systematic absences conform to the point group (taking due account of the Friedel law) and to the space group, respectively, of the crystal under study.

When the positions of adjacent layers in keeping with the vicinity conditions are $Z \geq 2$, then random or regular sequences of the kinds of layer stacking may occur. The corresponding disordered and ordered structures display diffraction patterns with common reflections (**family reflections**: they present the same position and intensities in all the OD structures of the family), and can be distinguished for the position and intensities of the other reflections (**characteristic reflections**). In Appendix 2.2, we shall present the derivation of the diffractional features of the OD family of wollastonite: it will be clear that the family reflections are independent on the particular sequence of the layers and are always sharp, whereas the other reflections may be sharp, as well as more or less diffuse, sometimes appearing as continuous streaks running in the direction normal to the layers, depending on the more or less ordered sequences of the layers. The family reflections correspond to a fictitious structure, periodic in three dimensions, closely related to the structures of the family and called **family structure**. It may be obtained from a general polytype of the family, by superposing Z copies of it translated by the vector (or vectors) corresponding to the possible positions of each OD layer. In the case of **wollastonite**, the vector relating the two possible positions of the OD layer is **b**/2; consequently, the family structure may be easily obtained from any polytype of the family by applying the superposition vector $\mathbf{b}/2$.

Two basis vectors of the family structure are always chosen collinear with the translation vectors of the single layer. If, as in the case of wollastonite, the vectors defining the single layer are **b**, **c** (translation vectors) and $\mathbf{a}_0$ (not a translation vector), the vectors **A**, **B**, and **C** of the family structure are such that:

$$\mathbf{B} = \mathbf{b}/q,$$
$$\mathbf{C} = \mathbf{c}/t, \text{ and}$$
$$\mathbf{A} = p\mathbf{a}_0$$

with q, t, and p integer numbers.

The results of the formal derivation of Appendix 2.2 are reported in Fig. 2.13, which presents schematic drawings corresponding to both MDO structures, to the 'family structure' and their diffraction patterns: the family structure of wollastonite, which may be derived from either MDO polytypes by applying the superposition vector **b**/2, has space group symmetry $C12/m1$; $q = 2$, $t = 1$ [see (2.4) in the following], $p = 2$.

The number p of layers for each **A** translation (in general, for the translation in the direction of missing periodicity) in the 'family structure', may be derived also

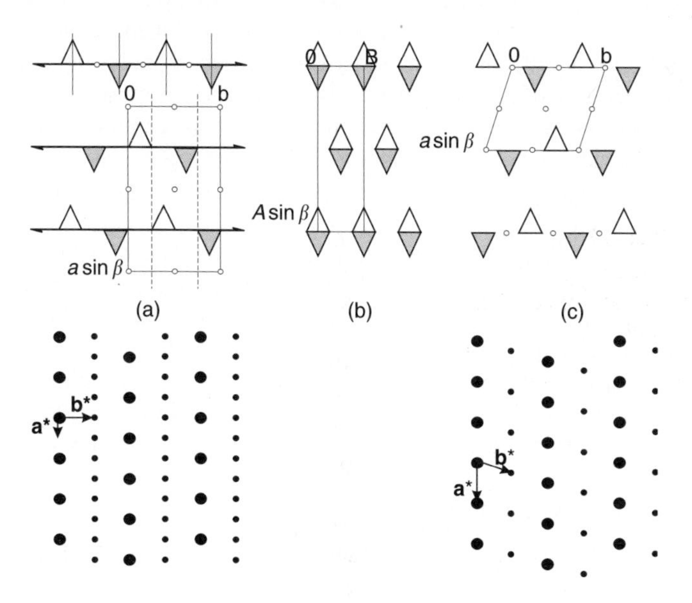

Fig. 2.13. (a) Schematic drawing of the MDO_2 structure of wollastonite (with indication of the symmetry elements of the structure and of the single layer) and corresponding diffraction pattern. (b) Family structure. (c) Schematic drawing of the MDO_1 structure of wollastonite (with indication of the symmetry elements of the structure) and corresponding diffraction pattern.

without any previous knowledge of the structure. According to Dornberger-Schiff and Fichtner (1972), p is the product of three factors: $p = p_1 \cdot p_2 \cdot p_3$; p_1 depends on the category of the OD structure ($p_1 = 1$ for categories I and II; $p_1 = 2$ for category III); p_2 depends on the isogonality relationships of the operations in the first and second lines of categories I and II, in the second and third lines for category III ($p_2 = 1$ if the operations in the two lines are isogonal; $p_2 = 2$ if the operations in the two lines are not isogonal); p_3 depends on the Bravais lattice of the 'family structure' ($p_3 = 1$ for a primitive lattice and for a lattice centred on the plane corresponding to the translation vectors of the layer; $p_3 = 2$ for the other centred lattices). In the wollastonite family, $p = p_1 \cdot p_2 \cdot p_3 = 2$, as $p_1 = 1$ (category I); $p_2 = 1$ (the operations in the first and second lines of (2.6) are isogonal); $p_3 = 2$ (the lattice of the 'family structure' is C centered, whereas the translations vectors of the OD layers are **b** and **c**).

2.3.7.1 *Derivation of the OD-groupoid family from the diffractional features*

Whereas the diffractional aspects of an OD compound may be derived from the knowledge of its OD-groupoid family, we are generally faced with the opposite problem, namely, we have to realize whether the examined compound has OD character and, if it has OD character, to derive the OD-groupoid family from the observation of the diffractional features of that compound. The OD character may be clearly displayed by the presence of one-dimensional streaking or diffuseness, or it may be hidden in more subtle features of the diffraction pattern: partial enhancement of symmetry for a particular set of reflections, non-space-group absences, polysynthetic twinning. The correct derivation of the OD-groupoid family is a necessary step in obtaining the structural arrangement of the compound, mainly when the crystals contain domains of various possible polytypes (generally the MDO polytypes), giving rise to a very complex diffraction pattern which may be properly interpreted once the OD aspects are fully appreciated.

In the following, we shall use once again the example of wollastonite. The OD-groupoid family which describes the symmetry properties common to all the members of the family has been derived in the preceding chapters just by looking at the structural arrangement in one member. It seems useful to consider how those symmetry properties could be obtained without any previous knowledge of structural arrangements, but only relying on general features of the diffraction pattern, following a procedure similar to that leading to the space group in the case of a 'normal' structure. The diffraction pattern of the monoclinic polytype shows a peculiar distribution of the reciprocal lattice points with:

(1) systematic absences limited to the reflections with $k = 2n$, which are absent for $h + 2k = 4n$;
(2) systematic absences valid for the whole pattern: reflections $0k0$ are absent for $k = 2n + 1$.

Rule (1) is not required by any monoclinic space group. However, it becomes an ordinary rule for reflections with even k values considered by themselves. These reflections, the 'family reflections', correspond to a reciprocal lattice with vectors $\mathbf{A}^*\mathbf{B}^*\mathbf{C}^*$ related to the vectors $\mathbf{a}^*\mathbf{b}^*\mathbf{c}^*$ of the monoclinic polytype as here indicated:

$$\begin{aligned} \mathbf{A}^* &= \mathbf{a}^* \\ \mathbf{B}^* &= 2\mathbf{b}^* \\ \mathbf{C}^* &= \mathbf{c}^* \end{aligned} \tag{2.4}$$

For these reflections, rule (1) becomes: HKL are absent for $H + K = 2n + 1$, which points to the following space groups for the family structure:

$$C\,1\,2/m\,1 \quad C\,1\,2\,1 \quad C\,1\,m\,1$$

From (2.4) we obtain that $\mathbf{b} = 2\mathbf{B}$ and $\mathbf{c} = \mathbf{C}$; therefore $q = 2$ and $t = 1$.

Systematic absences not limited to the family reflections can only be due to symmetry elements of the single layer (λ-operations). In the present case, the only absences registered for reflections with $k = 2n + 1$ are given in (2). Therefore, the possible symmetries of the single layer are:

$$P\,(1)\,2_1\,1 \quad P\,(1)\,2_1/m\,1 \tag{2.5}$$

The table given in Appendix 2.1, which presents all the OD-groupoid families of monoclinic and orthorhombic symmetries for OD structures built up with equivalent layers, is extremely useful at this stage. We may search in that table for OD-groupoid families presenting a first line of the type (2.5). We easily find two of them:

$$\begin{array}{ccc} P(1) & 2_1 & 1 \\ \{(1) & 2_{s-1} & 1\} \end{array} \qquad \begin{array}{ccc} P(1) & 2_1/m & 1 \\ \{(1) & 2_{s-1}/a_2 & 1\} \end{array} \tag{2.6}$$

The paper by Dornberger-Schiff and Fichtner (1972) indicates also how to proceed for obtaining the value of s (r and s in more general cases), using the information which may be obtained from the symmetry of the 'family structure'.

1. The operators which are present in the possible space groups of the 'family structure' are:

$$\begin{array}{ccccccccc} & & & & & & & 2 & \\ & 2 & & & m & & & 2_1 & \\ 1 & & 1 & 1 & & 1 & 1 & \text{——} & 1 \\ & 2_1 & & & a & & & m & \\ & & & & & & & a & \end{array}$$

2. Both λ and σ operators in the symbols (2.6) are considered and the translational components of any glide and screw are modified in agreement with the passing from the dimensions of the single layer to the dimensions of the 'family structure'. Here, the single layer has basis vectors $\mathbf{b}$, $\mathbf{c}$ (translation vectors) and $\mathbf{a}_0$ and the 'family structure' has translations $\mathbf{B} = \mathbf{b}/2, \mathbf{C} = \mathbf{c}$ ($q = 2, t = 1$), $\mathbf{A} = 2\mathbf{a}_0$ ($p = 2$). Therefore, we have to double the translational components which

refer to the **b** axis and to divide by two the translational components which refer to the **a** axis. We obtain:

$$\begin{matrix} 1 & 2_2 & 1 \\ 1 & 2_{2s} & 1 \end{matrix} \qquad \begin{matrix} 1 & 2_2/m & 1 \\ 1 & 2_{2s}/a & 1 \end{matrix}$$

3. Assuming $s = 1/2$, the operators reproduce those found in the space group of the 'family structure' and we obtain the two possible OD-groupoid families for the structures in the wollastonite family:

$$\begin{matrix} P(1) & 2_1 & 1 \\ \{(1) & 2_{1/2} & 1\} \end{matrix} \qquad \begin{matrix} P(1) & 2_1/m & 1 \\ \{(1) & 2_{1/2}/a_2 & 1\} \end{matrix}$$

Whereas in wollastonite the OD character, the symmetry of the OD layers and σ-operations relating subsequent layers are clearly understandable by looking at the known structures, in other cases the structural arrangement is unknown and the detailed derivation we have just described may give precious information on the symmetry properties, which, combined with the knowledge of the average structure (obtained on the basis of the family reflections) and correct crystal chemical reasoning, may be very helpful in revealing the actual structure of the compound under study.

2.4 Examples of OD families built up with equivalent layers

In the presentation of the fundamental concepts of OD theory, we have already discussed, together with 'schematic' examples, the OD character of few concrete structural arrangements: wollastonite, foshagite, decaborane. We shall now present and discuss additional examples of OD structures consisting of equivalent layers, selected, among the huge number of OD compounds, in some cases for the particular appeal of the structural arrangements (zeolite beta–tschernichite; balangeroite–gageite), but mainly as they are particularly appropriate to demonstrate the capability of an OD approach: (a) to rationalize the relationships among compounds (seidozerite–götzenite–rinkite); (b) to suggest the possible existence of new phases (new minerals in case of natural compounds) polytypically related to already known compounds; (c) to solve important problems where structural disorder prevented the understanding of the 'real' structures, as in the case of the phases of tobermorite group, or α-$PtCl_2$.

2.4.1 *Zeolite beta—tschernichite*

Zeolite beta is a large-pore high-silica zeolite, which presents a useful catalytic activity in a wide spectrum of hydrocarbon conversion processes. Its natural counterpart, **tschernichite**, has been found at Goble, OR (Boggs *et al.* 1993) and at Mt. Adamson, Antarctica (Galli *et al.* 1995). The various preparations of zeolite beta and specimens of tschernichite have different contents of 'zeolitic' cations and water molecules, as well as different Si/Al ratios, but present the same type of tetrahedral framework. Moreover, both the synthetic and the natural products display a high degree of structural disorder, which prevented for a long time their structure determination. At last, the framework structure of this zeolite was solved by Newsam *et al.* (1988) and Higgins *et al.* (1988) through a clever combination of various techniques, including X-ray powder diffraction, electron diffraction, and high-resolution transmission electron microscopy. On the basis of these results, the structure of zeolite beta (and tschernichite as well) may be described in terms of the OD layers represented in Fig. 2.6(a). The layers have basic vectors **a**, **b** (translation vectors of the layer, with $a = b = 12.5$ Å) and $\mathbf{c}_0$ ($c_0 = 6.6$ Å), and layer group symmetry $P\bar{4}m2$, and are built up by connection of (Si, Al) tetrahedra in a three-level scaffolding, represented in Fig. 2.6(a). These three-storied layers follow each to the other through the σ-operations listed in the second row of the OD-groupoid family symbol (already presented in section 2.3.3 and recalled here once more):

$$\begin{array}{cccccc} P & m & m & (\bar{4}) & 2 & 2 \\ & \{2 & 2_{2/3} & (4_4/n_{0,2/3}) & n_{-1/3,2} & n_{1/3,2}\} \end{array}$$

It is proper to observe that all the σ-operations listed in the symbol and represented in Fig. 2.6(b) bring the second layer in one position relative to the first layer. The other geometrically equivalent positions of the second layer may be obtained by the action of a set of σ-operations related to the previous ones through the λ-operations of the layer. Let us look at the operation $n_{0,2/3} = b_{2/3}$ (normal to **c** in Fig. 2.6(b)); due to the symmetry plane normal to **b** in the layer, both translations by **b**/3 or by $-\mathbf{b}/3$ (after the mirror operation) may be applied; the two operations are denoted $b_{2/3}$ and $b_{-2/3}$ respectively and both give rise to geometrically equivalent pairs of layers. The new layer now presents the ribbon of upward-pointing tetrahedra running along **a** and the corresponding σ-operators are now denoted $a_{2/3}$ and $a_{-2/3}$. Infinite possible sequences may exist, corresponding to the infinite sequences of alternating operators $b_{\pm 2/3}$ and $a_{\pm 2/3}$. Two of them correspond to the MDO structures, each characterized by a generating operation. One such operation is that indicated 4_4 in Figure 2.6(b). By applying it (rotating by 90° counter clockwise and translating the whole vector $\mathbf{c}_0$) the new layer is in such position that it is possible to re-apply the operation 4_4, etc. Through a constant application of the operation 4_4 we build up the MDO_1 structure (Fig. 2.14(a)): the 4_4 partial operation becomes a total 4_1 operation in a tetragonal structure with $a = 12.5$ Å and $c = 4c_0 = 4 \times 6.6$ Å; one diagonal λ-operator 2 in each layer becomes total twofold operator (at $z = 1/8, 3/8, \ldots$), whereas the σ-operators 2 become total twofold axes at $z = 1/4, 1/2, \ldots$; the resulting space group is $P4_122$. MDO_1 corresponds to the structure-type A of Higgins *et al.* (1988) and Newsam *et al.* (1988).

A second generating operation is the glide operation $n_{1/3,2}$ [reflection in a plane normal to $\mathbf{a} - \mathbf{b}$ and translation of $\mathbf{c}_0 + (\mathbf{a} + \mathbf{b})/6$]. That operation brings the layer in such position that the same operation may be applied once more and its continuous

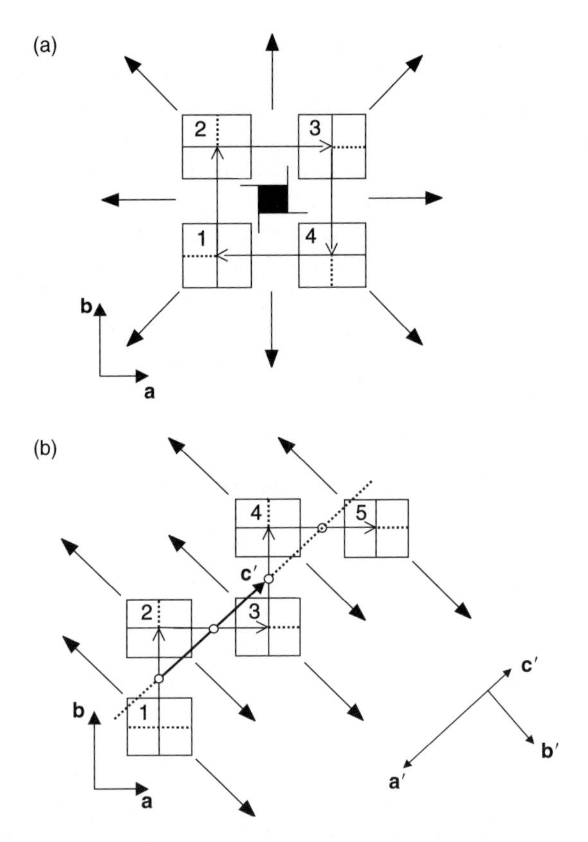

Fig. 2.14. Schematic reconstruction of the polytypes MDO_1 (space group $P4_122$) (a) and MDO_2 (space group $C2/c$) (b) of zeolite beta and tschernichite. The single layers have been represented by squares with crossing dotted and continuous lines, corresponding to the ribbons of down-pointing and up-pointing tetrahedra, respectively. The stacking sequence and the relative levels of the subsequent layers are indicated by the progressive numbers inside each square. The light arrows represent the path of the stacking. In (b), the direction of the vectors ($\mathbf{a}$, $\mathbf{b}$) of the single layer, as well as those of the vectors ($\mathbf{a}'$, $\mathbf{b}'$, $\mathbf{c}'$) of the MDO_2 polytype, are indicated. The inversion centers (small open circles) are placed at levels intermediate between those of adjacent layers.

application gives rise to the MDO_2 structure (Fig. 2.14(b)). To obtain the symmetry and metrics of it, we may firstly observe that the double application of the operation $n_{1/3,2}$ corresponds to a pure translation of $(\mathbf{a}+\mathbf{b})/3 + 2\mathbf{c}_0$. Assuming the reference frame with $\mathbf{a}' = -(\mathbf{a}+\mathbf{b})$, $\mathbf{b}' = \mathbf{a}-\mathbf{b}$, and $\mathbf{c}' = 2\mathbf{c}_0 + (\mathbf{a}+\mathbf{b})/3$, we observe that the $\mathbf{a}'$, $\mathbf{b}'$ plane is centred (C centring); the PO $n_{1/3,2}$ becomes a total glide operation c in a cell with the $\mathbf{c}'$ vector just indicated; moreover the twofold axes parallel to $(\mathbf{a}-\mathbf{b})$ of the single layers are total operators valid for the whole structure, which therefore has space group $C2/c$ ($a' = 17.9, b' = 17.9, c' = 14.6$ Å, $\beta = 114.6°$) and corresponds to the structure B of Higgins *et al.* (1988) and Newsam *et al.* (1988).

It is interesting to observe that in this particular OD family, not only the pairs but also the triples of layers are equivalent in all the members of the family. The peculiarity of the two MDO structures is that each of them presents only one kind of quadruple of layers (and one kind of quintuple, . . . n-tuple).

The operator 4_4 may act also in another position, related to the first one through the mirror plane normal to **a** in the single layer. In this case, the constant application of the clockwise rotation by 90°, followed by the translation $\mathbf{c}_0$, gives rise to the structure MDO'_1, with space group symmetry $P4_322$, enantiomorphous of MDO_1. Moreover, four distinct operators of type $n_{1/3,2}$ may be active, related each to the other by the mirror planes of the single layer: they give rise to four twin-related MDO_2 structures.

In conclusion, there will be the following MDO structures:

MDO_1 $P4_122$ $P4_322$ (enantiomorphous structures)

MDO_2 $C2/c$ (four possible twin-related orientations).

The OD approach allows us an easy interpretation of the peculiar diffraction patterns of specimens of zeolite beta or tschernichite. As a matter of fact, it was instrumental in the interpretation of the diffraction patterns of tschernichite crystals from Mt. Adamson and in the careful collection of data from both MDO structures, which consented a reliable refinement of the real structures of both polytypes (Alberti *et al.* 2002).

2.4.2 *The euristic power of the OD approach: the phases of the tobermorite group*

The OD character of wollastonite is clearly dependent on its peculiar crystal chemistry, in particular on the metrical relationships between the calcium polyhedral module, with the repeat of 3.65 Å, and the tetrahedral chains with their typical repeat of 7.3 Å. Figure 2.2 indicates that the tetrahedral chain may be connected to the calcium polyhedral ribbon in two distinct but equivalent positions, shifted by 3.65 Å in the **b** direction. Similar relationships exist in the wide family of **calcium silicate hydrates** (foshagite, **xonotlite, hillebrandite**, tobermorite, clinotobermorite, . . .) and Fig. 2.15 presents the connection schemes between silicon tetrahedral and calcium polyhedral modules in xonotlite and in the minerals of the tobermorite group.

Most of the listed phases, which in nature occur as alteration products of calcium carbonate rocks and as vesicle fillings in basaltic rocks, are also known to form when Portland cement reacts with water during the process of cement binding and are indicated as **CSH compounds** in cement chemistry. All of them display OD character. We shall discuss in some detail the case of the minerals of the tobermorite group, as it was just through the application of an OD approach that their real structures have been revealed. The procedures here presented may be usefully applied in other similar cases.

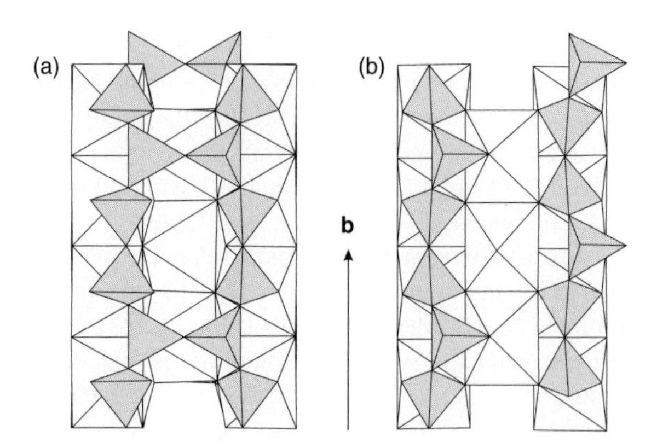

Fig. 2.15. The connection between the silicon tetrahedral (dark grey) and calcium polyhedral (light grey) modules (a) in xonotlite (b) and in the phases of the tobermorite group.

2.4.2.1 *Clinotobermorite*

The OD character of clinotobermorite is clearly displayed by its diffraction pattern (streaks, diffuse reflections, unusual absence rules), which may be referred to a unit cell with parameters $a = 11.27, b = 7.34, c = 22.64$ Å, $\beta = 97.28°$, thus showing the 7.3 Å translation, which is typical of CSH compounds. It was observed that whereas the reflections with $k = 2n$ were always sharp (they correspond to the 'family structure'), continuous streaks parallel to $\mathbf{c}^*$, or diffuse maxima on rows parallel to $\mathbf{c}^*$ were found for $k = 2n + 1$. As regards the systematic absences, the following conditions have been observed: (a) a C centring condition, valid for all the reflections, namely $h + k = 2n$; (b) a condition valid for reflections with $k = 2n$ (and therefore $h = 2n$), which are present only for $k + 2l = 4n$.

The second condition becomes an ordinary rule for reflections with $k = 2K$ (and thus $h = 2H$) considered for themselves. These reflections, the *family reflections*, correspond to a reciprocal lattice with vectors $\mathbf{A}^*$, $\mathbf{B}^*$, and $\mathbf{C}^*$ related to the vectors $\mathbf{a}^*$, $\mathbf{b}^*$, and $\mathbf{c}^*$ in this way: $\mathbf{A}^* = 2\mathbf{a}^*$, $\mathbf{B}^* = 2\mathbf{b}^*$, and $\mathbf{C}^* = \mathbf{c}^*$. For these reflections, the condition (b) becomes: reflections HKL are present for $K + L = 2n$, which corresponds to A centring of the lattice of the family structure.

The family structure, with space group symmetry $A2/m$, and $A = 5.638$, $B = 3.672$, $C = 22.636$ Å, $\beta = 97.28°$, was determined by Hoffmann and Armbruster (1997). Their results present the actual arrangement of the calcium 'polyhedral' sheets, which have the same sub-period, 3.67 Å, of the family structure, and point to the presence of tetrahedral chains of wollastonite type. An OD-groupoid family presenting λ and σ-POs in keeping with the space group $A12/m1$ of the family structure is:

$$\begin{array}{l} C\,1 \quad 2/m \quad (1) \\ \{1 \quad 2_{1/2}/c_2 \quad (1)\,\} \end{array} \qquad (2.7)$$

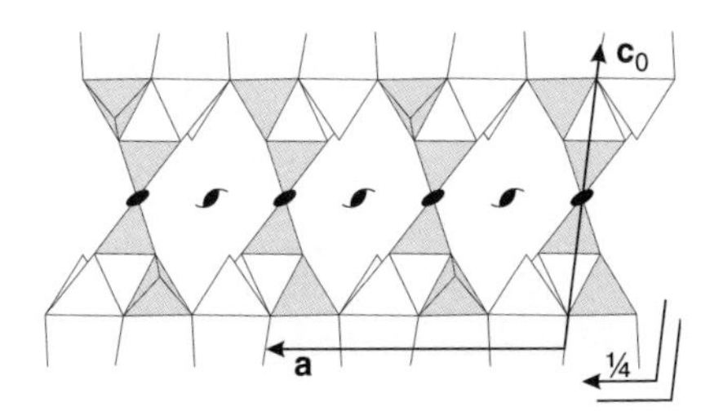

Fig. 2.16. The OD layer in clinotobermorite, as seen along **b**, with indication of the λ-POs. Double wollastonite-type chains, with $2/m$ symmetry, run parallel to **b**. Adjacent chains in the layer are shifted by **b**/2.

A reliable structural model for the layer, in keeping with the layer group symmetry in (2.7) and with the results of the study by Hoffmann and Armbruster (1997), could be guessed, as represented in Fig. 2.16. In this figure, the $[Si_3O_9]$ chains, running along **b**, are condensed to give double chains; however the same layer symmetry would be obtained if single chains occur, with facing chains displaced by **b**/2.

According to (2.7), layers with $C12/m(1)$ symmetry and basis vectors **a**, **b** (translation vectors of the layer, with $a = 11.27, b = 7.34$ Å) and $\mathbf{c}_0(c_0 = c/2 = 11.32$ Å; $\beta = 97.28°)$ may follow each other in the **c** direction, related by the operator $2_{-1/2}$ or by the operator $2_{1/2}$. Two MDO polytypes are possible in this family: MDO_1, which corresponds to the sequence in which the operators regularly alternate: $2_{-1/2}/2_{1/2}/2_{-1/2}/2_{1/2}/\ldots$ and has [– c_2 –] as generating operation; MDO_2, which corresponds to the sequence in which one operator is constantly active: $2_{-1/2}/2_{-1/2}/2_{-1/2}/\ldots$. (the sequence $2_{1/2}/2_{1/2}/2_{1/2}/\ldots$. corresponds to the twin structure) and has the translation $\mathbf{t} = \mathbf{c}_0 - \mathbf{b}/4$ as generating operation. The symmetry and metrics of the two MDO polytypes may be derived by looking at the corresponding sequences of λ and σ-POs (Fig. 2.17).

As regards MDO_1 we may observe that: (a) the partial glide c_2 between L_1 and L_2 layers has continuation between L_2 and L_3 layers, becoming a total operator c, in a structure with a parameter $c = 2c_0$; (b) due to the relative position of the successive layers, the operator [– 2 –] (λ operator of the single layers) becomes a total operator; (c) the translation operator $\mathbf{a}/2 + \mathbf{b}/2$ is valid for all the layers. Therefore the whole structure has symmetry $C2/c$, with cell parameters: $a = 11.27, b = 7.34, c = 22.64$ Å, $\beta = 97.2°$.

As regards MDO_2, we may observe that: (a) the partial glide c_2 between L_1 and L_2 layers has no continuation in the successive layers; (b) due to the relative position of the successive layers, the operator [– 2 –] is not valid for the whole structure; (c) both λ and σ inversion centres are total operators; (d) the translation operator $\mathbf{a}/2 + \mathbf{b}/2$ is valid for all the layers. Therefore the whole structure has $C\bar{1}$ symmetry, with cell parameters: $a = 11.27, b = 7.34, c = 11.47$ Å, $\alpha = 99.2°, \beta = 97.2°, \gamma = 90.0°$.

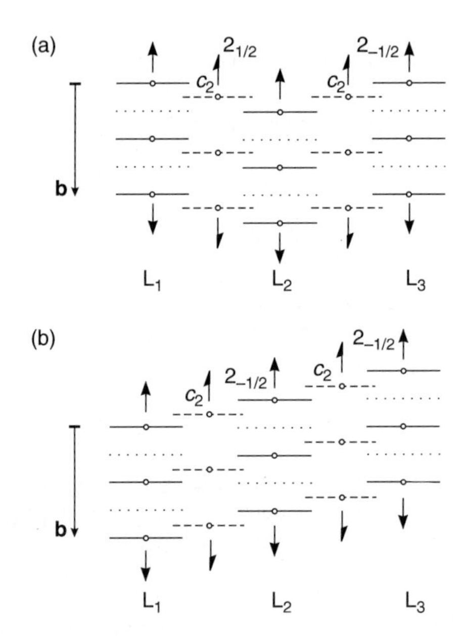

Fig. 2.17. The sequences of λ-POs (light marks) and σ-POs (heavy marks) in the monoclinic MDO_1 (a) and triclinic MDO_2 (b) structures of clinotobermorite, as seen along **a** (**b** vertical); modified after Merlino (1997*a*,*b*).

The relationship between the two MDO structures, and more generally among the various polytypes in the family, may be easily appreciated by observing that layers related by the operators $2_{-1/2}$ and $2_{1/2}$ are translationally equivalent and related by stacking vectors $\mathbf{t}_1 = \mathbf{c}_0 - \mathbf{b}/4$ and $\mathbf{t}_2 = \mathbf{c}_0 + \mathbf{b}/4$, respectively. In the monoclinic polytype, the two stacking vectors regularly alternate, whereas in the triclinic polytype the vector $\mathbf{t}_1$ is constantly applied (the constant application of vector $\mathbf{t}_2$ gives rise to the twin structure). The general discussion of the diffractional effects in the whole family (Merlino 1997a) is made easier by the translational equivalence of the layers.

The crystals of clinotobermorite are characterized by the presence of domains of the MDO_1 polytype, as well as of both twin individuals of the MDO_2 polytype. Diffraction data of both polytypes have been collected and the corresponding structures have been refined (Merlino *et al.* 1999, 2000*a*). In this way, the details of the real structure have been assessed, the chain condensation to build double wollastonite chains confirmed, and the positions of the 'zeolitic' calcium cations and water molecules placed in the channels of the structure defined, with resulting crystal chemical formula $Ca_5[Si_6O_{17}] \cdot 5H_2O$. The final refinements were carried out in the space groups Cc and $C1$ for MDO_1 and MDO_2: the descent in symmetry with respect to the OD models, is connected with the ordering of Ca cations and water molecules in the channels of the structure (Fig. 2.18).

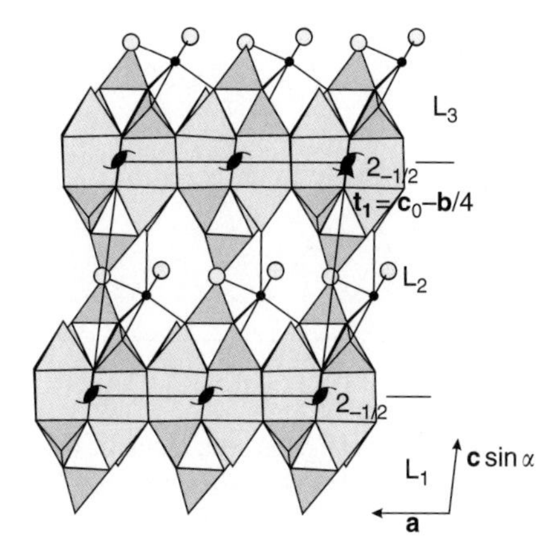

Fig. 2.18. The crystal structure of the triclinic MDO_2 polytype of clinotobermorite as seen along **b**. The 'framework' is built up by double chains of silicon tetrahedra (dark grey) and layers of sevenfold-coordinated calcium polyhedra (light grey). The 'zeolitic' calcium cation (small filled circle) is tightly bonded to oxygen atoms of the framework and to the 'zeolitic' water molecules (large grey circles). The sequence of the layers and of the σ-POs $2_{-1/2}$ (they are valid only for the framework), as well as the generating operation $\mathbf{t}_1 = \mathbf{c}_0 - \mathbf{b}/4$, are indicated.

2.4.2.2 *Tobermorite 11 Å*

The main outlines of its structure were sketched more than 40 years ago by Megaw and Kelsey (1956), who studied its 'subcell' structure (family structure). Their results, together with the C centring condition displayed by the whole diffraction pattern (McConnell 1954), the orthorhombic symmetry of the family structure and the close metrical relationships between clinotobermorite and tobermorite 11 Å point to an OD layer, described in Fig. 2.19, characterized once again by double wollastonite chains, with $C2mm$ symmetry, periodicities $a = 11.2$, $b = 7.3$ Å, and width $c_0 = 11.2$ Å.

For a layer with symmetry $C2mm$ and **c** as the direction of missing periodicity, OD theory suggests two possible OD-groupoid families (Appendix 2.1). One of them

$$\begin{array}{llll} C\,2 & m & (m) \\ \{2_{1/2} & n_{2,1/2} & (n_{1/2,1/2})\} \end{array} \quad (2.8)$$

is the correct one, as its λ and σ operations closely correspond to the operations of the space group $I2mm$ of the subcell structure.

The set of σ-operations in (2.8) indicates that pairs of successive layers are related by stacking vectors: $\mathbf{t}_1 = \mathbf{c}_0 + (\mathbf{a} + \mathbf{b})/4$ or $\mathbf{t}_2 = \mathbf{c}_0 + (\mathbf{a} - \mathbf{b})/4$. Once again, two MDO polytypes are possible: MDO_1, corresponding to the sequence $\mathbf{t}_1\mathbf{t}_2\mathbf{t}_1\mathbf{t}_2\ldots$, in which the stacking vectors regularly alternate and presenting $[-\, n_{2,1/2}\, -]$ as generating operation; MDO2, corresponding to the sequence $\mathbf{t}_1\mathbf{t}_1\mathbf{t}_1\mathbf{t}_1\ldots$. in which one stacking vector, the generating operation, is constantly active (the stacking sequence

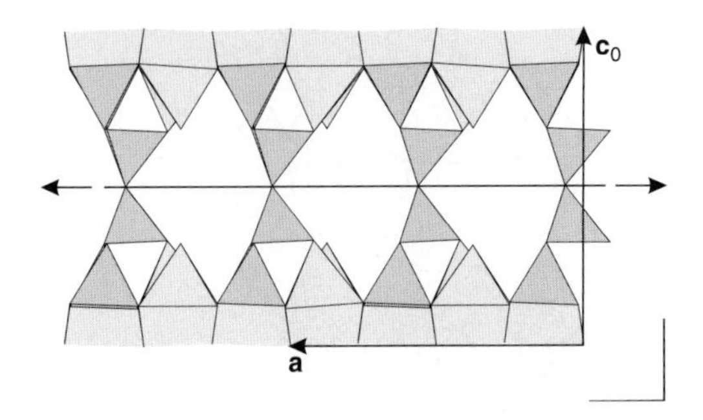

Fig. 2.19. The OD layer in tobermorite 11 Å, as seen along **b**, with indication of the λ-POs. Double wollastonite-type chains, with $2mm$ symmetry, run parallel to **b**. Adjacent chains in the layer are shifted by **b**/2.

$\mathbf{t}_2\mathbf{t}_2\mathbf{t}_2\mathbf{t}_2\ldots$ corresponds to the twin structure MDO_2'). Their unit cell dimensions and space group symmetries, which may be obtained following the same procedures as applied in the case of clinotobermorite, are:

MDO_1—$a = 11.2, b = 7.3, c = 44.8$ Å, space group $F2dd$;
MDO_2—$a = 11.2, b = 7.3, c = 22.4$ Å;

the corresponding cell is four-fold, being C centred and presenting additional lattice points at $\frac{1}{4}\frac{1}{4}\frac{1}{2}$ and $\frac{3}{4}\frac{3}{4}\frac{1}{2}$; the λ-PO $[- - m]$ is total operation, valid for the whole structure. A convenient cell is derived through the transformations $\mathbf{a}' = (\mathbf{a}+\mathbf{b})/2, \mathbf{b}' = -\mathbf{b}, \mathbf{c}' = -\mathbf{c}$, thus obtaining a B centred monoclinic cell, with parameters $a' = 6.7, b' = 7.3, c' = 22.4$ Å, $\gamma = 123°$, space group $B11m$.

Crystals from two different localities, with different calcium content, have been studied. In both cases, diffraction data of the two MDO polytypes have been collected on one crystal and the corresponding structures have been refined (Merlino *et al.* 1999, 2001): the chain condensation described in Fig. 2.19 has been fully confirmed and the channel content has been defined. The resulting crystal chemical formulas were $Ca_{4.5}[Si_6O_{16}(OH)] \cdot 5H_2O$ for one specimen (**normal tobermorite** from Baŝcenov, Urals, with 'zeolitic' calcium cations, as well as water molecules, in the channels) and $Ca_4[Si_6O_{15}(OH)_2] \cdot 5H_2O$ for the low-calcium specimen (**anomalous tobermorite** from Wessels mine, Kalahari, with no 'zeolitic' calcium cations in the channels). The structures of both MDO polytypes of this specimen are presented in Fig. 2.20.

2.4.2.3 *'Clinotobermorite 9 Å'*

Crystals of clinotobermorite heated at 300°C for a few hours, topotactically transform to a new phase in which the calcium layers parallel to (001), which characterize the whole group of tobermorites, are now closer to each other and connected through single chains of wollastonite type, derived from de-condensation of the double chains which are present in clinotobermorite. Four of the five water molecules are lost, the

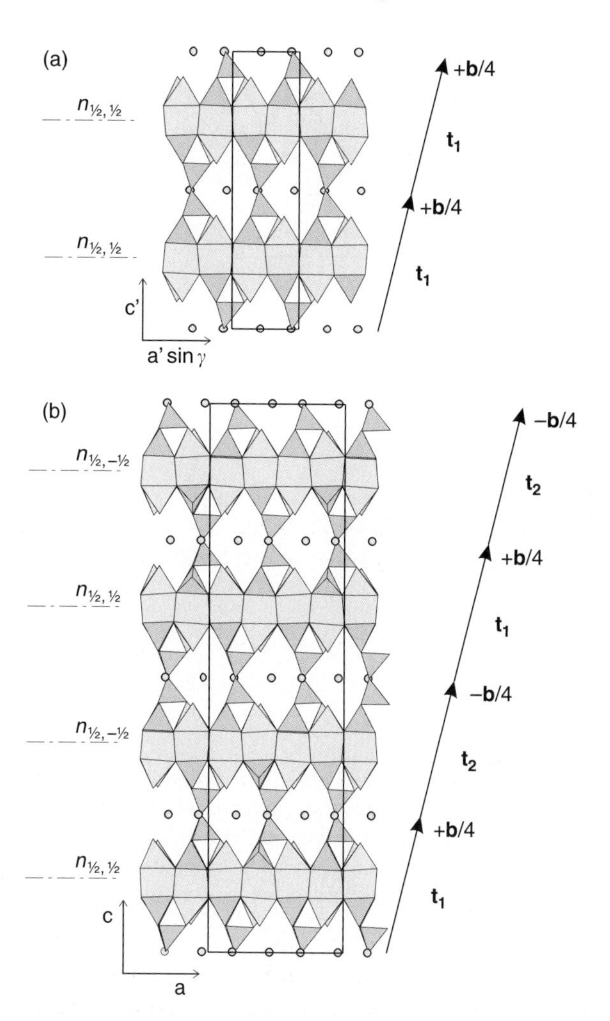

Fig. 2.20. The crystal structures of the monoclinic MDO_2 (a) and orthorhombic MDO_1 (b) polytypes of tobermorite 11 Å, as seen along **b**. Silicate double chains and layers of sevenfold coordinated calcium polyhedra are drawn in dark and light grey respectively. Grey circles indicate the 'zeolitic' water molecules. The sequences of σ-operators $[--n_{1/2,1/2}]$, $[--n_{1/2,-1/2}]$ and stacking vectors $\mathbf{t}_1$, $\mathbf{t}_2$, with indication of their components along **b**, are shown.

fifth being present as hydroxyls in SiOH groups. The new phase too has OD character and the OD layer resulting from the dehydration process is represented in Fig. 2.21: it has the same layer group symmetry $C2/m$ and the same translation vectors **a** and **b** as clinotobermorite, but a different $\mathbf{c}_0$ ($c_0 = 9.4$ Å; $\beta = 92.8°$). Also the OD-groupoid family and the space group of the family structure are the same as in clinotobermorite.

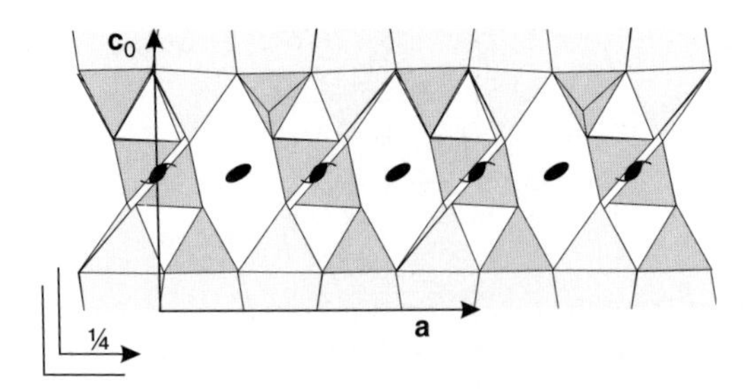

Fig. 2.21. The OD layer in 'clinotobermorite 9 Å', as seen along **b**, with indication of the λ-POs.

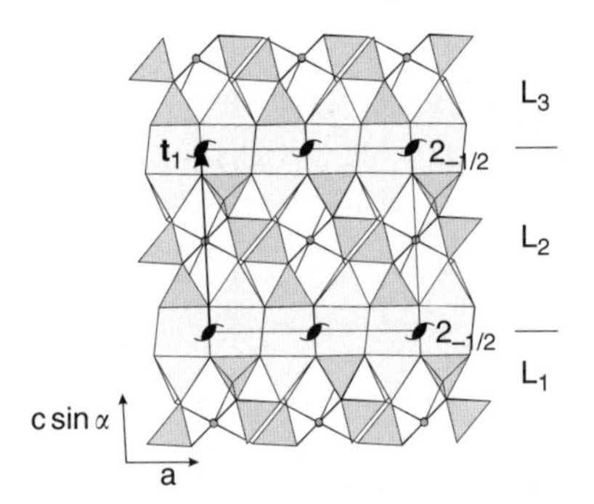

Fig. 2.22. The crystal structure of the triclinic MDO_2 polytype of 'clinotobermorite 9 Å', as seen along **b**. The 'framework' is built up by single chains of silicon tetrahedra (dark grey) and layers of sevenfold-coordinated calcium polyhedra (light grey). An additional calcium cation (small filled circle) is tightly bonded to oxygen atoms of the framework. The sequence of the layers and of the σ-POs $2_{-1/2}$, as well as the generating operation $\mathbf{t}_1 = \mathbf{c}_0 - \mathbf{b}/4$, are indicated.

As in that case, the OD layers follow each other in the **c** direction, related by the operators $2_{-1/2}$ or $2_{1/2}$ and the two MDO polytypes are: MDO_1 (generating operation [- c_2 -]), with space group *C2/c* and cell parameters: $a = 11.16, b = 7.30, c = 18.78$ Å, $\beta = 92.8°$; MDO_2 (generating operation $\mathbf{t}_1 = \mathbf{c}_0 - \mathbf{b}/4$), with space group $C\bar{1}$ and cell parameters: $a = 11.16, b = 7.30, c = 9.58$ Å, $\alpha = 101.3°, \beta = 92.8°, \gamma = 90.0°$. The crystal structure of this last polytype (crystal chemical formula $Ca_5[Si_3O_8(OH)]_2$) is shown in Fig. 2.22.

2.4.2.4 *Tobermorite 9 Å (riversideite)*

Also normal tobermorite 11 Å dehydrates at 300°C, and, unlike in the previous case, the dehydration product has been found also in nature as minor constituent of submicroscopic intergrowths with other forms of tobermorite at Crestmore, Riverside Co., CA, which explains the name riversideite given to this phase, as suggested also by McConnell (1954).

The diffraction patterns of tobermorite 9 Å point to an OD structure with an orthorhombic 'subcell' with $a = 5.58, b = 3.66, c = 18.70$ Å, as compared with the monoclinic subcell of 'clinotobermorite 9 Å', and a C centred 'true' cell with $a = 11.16, b = 7.32, c = 18.70$ Å (Taylor 1959).

To guess a reliable model for riversideite, we may start from the close metrical relationships of its cell parameters with those of 'clinotobermorite 9 Å' as well as from the similar chemical compositions of the corresponding parent compounds, normal tobermorite 11 Å and clinotobermorite. It appears highly probable that the dehydration process proceeds in similar way and gives rise to the same kind of connection between adjacent calcium polyhedral sheets, which means that the OD family of riversideite should present the same OD layer as found in 'clinotobermorite 9 Å' (Fig. 2.21). Once again, OD theory reveals its usefulness in suggesting appropriate structural models; in fact, besides the σ-operators indicated in (2.8) as compatible with the λ-operators of an OD layer with layer group symmetry $C\ 1\ 2/m\ (1)$, OD theory points to a second possible set of σ-operators compatible with that layer group symmetry, as indicated by Appendix 2.1, namely:

$$\begin{array}{cccc} C & 1 & 2/m & (1) \\ \{2_{1/2}/n_{1/2,2} & 1 & (2_2/n_{1/2,1/2})\} \end{array} \tag{2.9}$$

From (2.9), the space group of the subcell structure $P2_1/n\,2/m\,2_1/n$, more simply *Pnmn*, may be easily derived. Moreover, (2.9) indicates that in riversideite structural OD layers as found in 'clinotobermorite 9 Å' follow each other in the **c** direction related by $[- - n_{1/2,1/2}]$ or $[- - n_{1/2,-1/2}]$ operators, giving rise to infinite possible polytypes or disordered structures, characterised by distinct sequences of the indicated operators. As in the other families of the tobermorite group, two MDO polytypes exist in the structure-type of riversideite:

- MDO_1, corresponding to the sequence in which the operators $[- - n_{1/2,1/2}]$ and $[- - n_{1/2,-1/2}]$ regularly alternate. Its generating operation is $[n_{1/2,2} - -]$; its continuation to total operation gives rise to a glide $[d - -]$ in a structure with space group symmetry $Fd2d$, and unit cell parameters $a = 11.16, b = 7.32, c = 37.40$ Å (the glides $[- - d]$ correspond to the operations $[- - n_{1/2,1/2}]$ and $[- - n_{1/2,-1/2}]$, which have **reverse continuation**) (Fig. 2.23).
- MDO_2, corresponding to the sequence in which only one operator, for example $[- - n_{1/2,1/2}]$, is constantly active (the constant application of the other operator corresponds to the twin structure). Its generating operation is $[- - 2_2]$; its continuation to total operation gives rise to $[- - 2_1]$ in a structure with space group symmetry $C112_1/d$ (non-standard space group; the glide $[- - d]$ corresponds to the operation $[- - n_{1/2,1/2}]$, which has reverse continuation), and unit cell parameters $a = 11.16, b = 7.32, c = 18.70$ Å, $\gamma = 90°$. A standard space group is derived through the same transformations we have already applied in discussing the MDO_2 structure of tobermorite 11 Å (Section 2.4.2.2), namely $\mathbf{a}' = (\mathbf{a} + \mathbf{b})/2, \mathbf{b}' = -\mathbf{b}, \mathbf{c}' = -\mathbf{c}$, thus obtaining $a' = 6.7, b' = 7.32, c' = 18.70$ Å, $\gamma' = 123.5°$, space group $P112_1/a$.

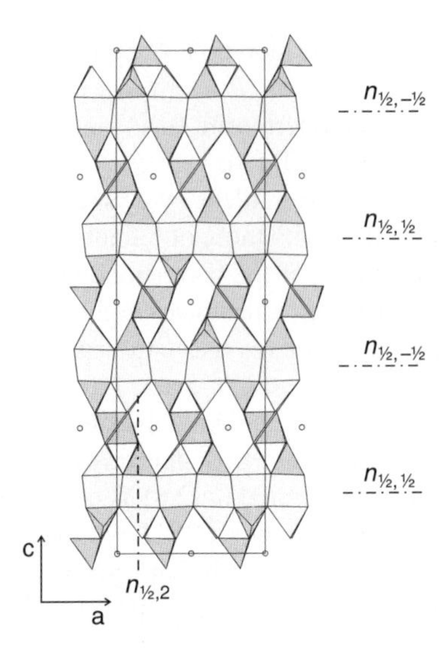

Fig. 2.23. The crystal structure of the orthorhombic MDO_1 polytype of tobermorite 9 Å (riversideite), as seen along **b**. The 'framework' is built up by single chains of silicon tetrahedra (dark grey) and layers of sevenfold-coordinated calcium polyhedra (light grey); additional calcium cations (small filled circles) are located in the cavities and linked to the oxygen atoms of the framework. The sequence of the alternating σ-POs $[- - n_{1/2,1/2}]$ and $[- - n_{1/2,-1/2}]$, as well as the generating operation $[n_{1/2,2} - -]$, are indicated.

Unlike what happens in the other phases of the tobermorite group, the diffraction patterns of tobermorite 9 Å indicate that the crystals are not built up mainly by domains of the two MDO polytypes; instead, they are characterized by highly disordered sequences of OD layers (Bonaccorsi *et al.* to be published).

2.4.3 *Seidozerite–götzenite–rinkite relationships*

The interconnection between two distinct OD families through a common OD layer, just discussed for the two families of 'clinotobermorite 9 Å' and riversideite, is displayed also by the minerals of the götzenite–seidozerite and rinkite groups which we shall now present.

2.4.3.1 *Seidozerite–götzenite OD relationships*

The minerals of the götzenite–seidozerite group [götzenite, $Na_2Ca_5Ti(Si_2O_7)_2(OH,F)_4$, $a = 9.667, b = 5.731, c = 7.334$ Å, $\alpha = 90°, \beta = 101.05°, \gamma = 101.31°$, $P\bar{1}$; seidozerite, $(Na,Ca)_4(Zr,Ti,Mn)_4(Si_2O_7)_2(O,F)_4$, $a = 5.53, b = 7.10, c = 18.30$ Å, $\beta = 102.7°$, $P2/c$] display triclinic or monoclinic symmetry and present crystal chemical formulas of the type $X_8(Si_2O_7)_2(O,OH,F)_4$, where X denotes

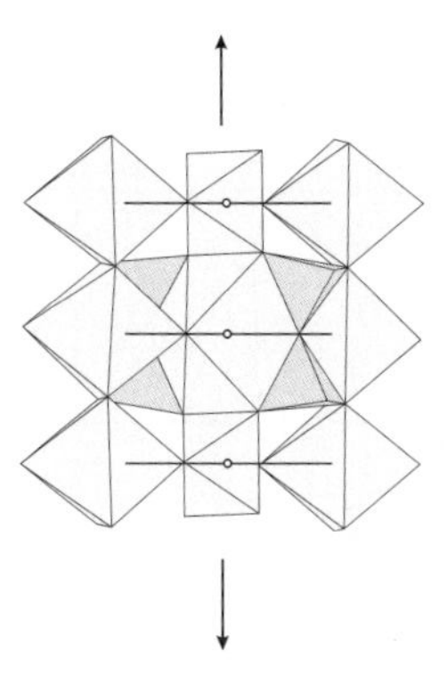

Fig. 2.24. The OD layer in the götzenite–seidozerite family, as seen along **a** (**b** vertical; with reference to the parameters of seidozerite), with indication of the λ-POs.

cations with various possible charges and radii from relatively large (Na, Ca) to medium (Ti, Zr, Mn, Fe), characterized by six to eight fold coordination.

The features common to the minerals of the group are: (a) the infinite 'polyhedral' layers, which host Ti, Na, Ca cations in götzenite and **hainite**, Ti, Na, Ca, Zr in **rosenbuschite** and Ti, Na, Mn in the monoclinic seidozerite; (b) the 'octahedral' ribbons, two columns large, which host Ca cations in götzenite and hainite, Ca and Zr cations in rosenbuschite, Na and Zr cations in seidozerite. The infinite layers and the ribbons are held together by disilicate groups Si_2O_7 grasped to both of them.

The three triclinic members of the group, namely götzenite, hainite and rosenbuschite, are isostructural; the ordering of the cations in the 'octahedral' ribbons and in the 'polyhedral' layers results in doubling the unit cell of rosenbuschite with respect to those of götzenite and hainite. The structure-type of the monoclinic member, seidozerite, is OD related to the structure-type of the triclinic members. In the OD family, the layer (Fig. 2.24) has translation vectors **a** and **b** ($a = 5.53$ Å, $b = 7.10$, assuming the values corresponding to the parameters of seidozerite) and third vector $\mathbf{c}_0$ ($c_0 = 9.15$ Å, $\beta = 102.7°$). The OD-groupoid family symbol is:

$$\begin{array}{cccc} P & 1 & 2/m & (1) \\ & \{1 & 2_{1/2}/c_2 & (1)\}. \end{array}$$

The layers succeed each to the other related by the operations [– $2_{1/2}$ –] or [– $2_{-1/2}$ –]. When the two operations regularly alternate, the structure-type of seidozerite is obtained. It corresponds to MDO_1 with generating operation [– c_2 –], which is continued to total glide c in a structure with space group $P2/c$, and $\mathbf{c} = 2\mathbf{c}_0$. When $2_{-1/2}$ is constantly operating, the structure-type of götzenite is obtained. It corresponds to MDO_2 with generating operation the translation $\mathbf{t} = \mathbf{c}_0 - \mathbf{b}/4$; the structure, with basis vectors $\mathbf{a}, \mathbf{b}, \mathbf{c} = \mathbf{c}_0 - \mathbf{b}/4$, is triclinic, space group $P\bar{1}$; the parameters closely correspond to those given above for götzenite, if we perform a cyclic axis transformation ($\mathbf{a} \to \mathbf{b} \to \mathbf{c} \to \mathbf{a}$) and take into account the larger

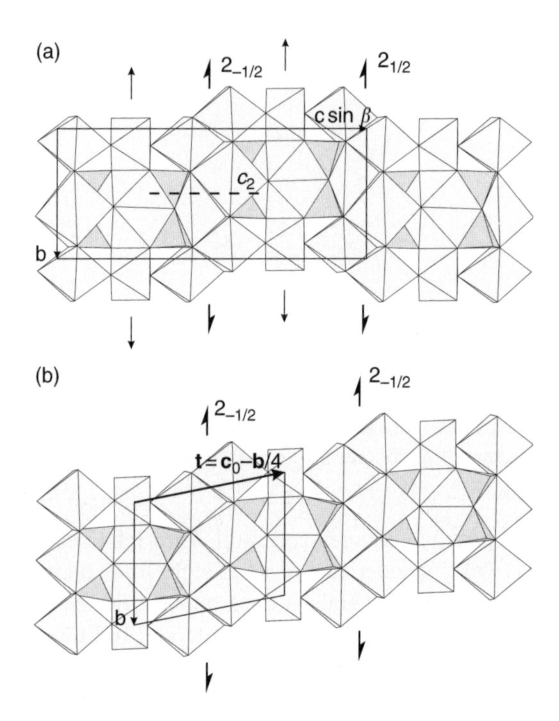

Fig. 2.25. The two MDO structures in the götzenite-seidozerite family, as seen along **a**. In MDO_1 (a) and MDO_2 (b) the sequences of the σ-POs, [– $2_{1/2}$ –] and, [– $2_{-1/2}$ –], as well as the generating operations glide [– c_2 –] and translation $\mathbf{t} = \mathbf{c}_0 - \mathbf{b}/4$, respectively, are indicated. In (a) also the position of the λ-operator [– 2 –] is shown.

average radii of the cations in götzenite; the operation $2_{1/2}$ would give rise, by constantly acting, to the twin structure MDO_2'. Actually, lamellar twinning on (100) is normally observed in götzenite. The two structure-types are represented in Fig. 2.25.

2.4.3.2 *OD character of rinkite*

The OD character of rinkite [$(Ca,Ce)_4Na_2Ca(Ti,Nb)(Si_2O_7)_2(O,OH,F)_4$, $a = 7.437$, $b = 5.664$, $c = 18.843$ Å, $\beta = 101.38°$, $P2_1/c$] is highlighted by the 'partial enhancement of symmetry' of its diffraction patterns, as reported by Gottardi (1966). In fact, reflections with even h indices display orthorhombic symmetry and point to a 'family structure' with $A = a/2$, $B = b$, $C = c \sin\beta$ with space group *Pmnn*. The true monoclinic symmetry of rinkite is revealed by the reflections with $h = 2n+1$. The single layer in the OD family of rinkite has nearly the same structure of the single layer already found in the götzenite–seidozerite family. The layer has translation vectors **a**, **b** ($a = 7.437$, $b = 5.664$ Å) and third basic vector $\mathbf{c}_0$ ($c_0 = C/2 = 9.236$ Å).

By looking at Appendix 2.1, we may find an OD-groupoid family symbol compatible with the $P2/m1(1)$ symmetry of the layer and with the *Pmnn* symmetry of the

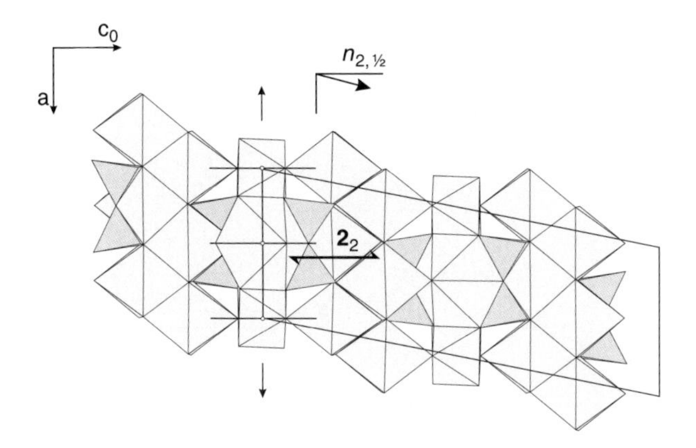

Fig. 2.26. The structure of the monoclinic MDO polytype in the rinkite family, with indication of the λ-POs, as well as of the σ-PO [– $n_{2,1/2}$ –], the generating operation of this polytype. Moreover it is also indicated the position of the σ-PO[– 2_2] which, by continuation to total operation, would give rise to the orthorhombic MDO polytype.

family structure:

$$\begin{array}{ccccc} P & 2/m & 1 & (1) & \\ \{ & 1 & 2_1/n_{2,1/2} & (2_2/n_{1/2,1})\} & \end{array}$$

The two MDO structures in this family have generating operations [– $n_{2,1/2}$ –] and [– – 2_2]. The first one continues to a total glide c in a structure with $\mathbf{c} = 2\mathbf{c}_0 + \mathbf{a}/2$, space group $P2_1/c$, namely the structure type of rinkite (Fig. 2.26). The second one continues to a total [– – 2_1] axis in a structure with $\mathbf{c} = 2\mathbf{c}_0$. In this structure the λ-PO [2 – –] and the σ-PO [– 2_1 –] are also total operations, giving rise to the space group symmetry $P22_12_1$. It would be interesting and hopefully rewarding to look for this polytype in rinkite specimens from various localities; it is also possible that it exists as small domains inside crystals of 'normal' rinkite.

2.4.4 *Real structures of gageite and balangeroite*

Gageite and balangeroite are fibrous silicates with crystal chemical formula $M_{42}O_6(OH)_{40}(Si_4O_{12})_4$, M indicating cations in octahedral coordination, with Mn and Mg dominant in gageite and balangeroite, respectively. The X-ray and electron diffraction patterns of crystal specimens of both compounds, revealing non-space-group absences, diffuseness in definite classes of reflections, and distinct polytypic forms, clearly point to their OD character. Moore (1969) was able to outline the sub-structure (family structure), with orthorhombic cell and parameters $A = 13.79$, $B = 13.68$, $C = 3.279$ Å. A reliable structural assessment was eventually found on the basis of the following points: (a) the structural work of Moore on gageite; (b) the indication of the trebling of the c parameter in the real structures, as

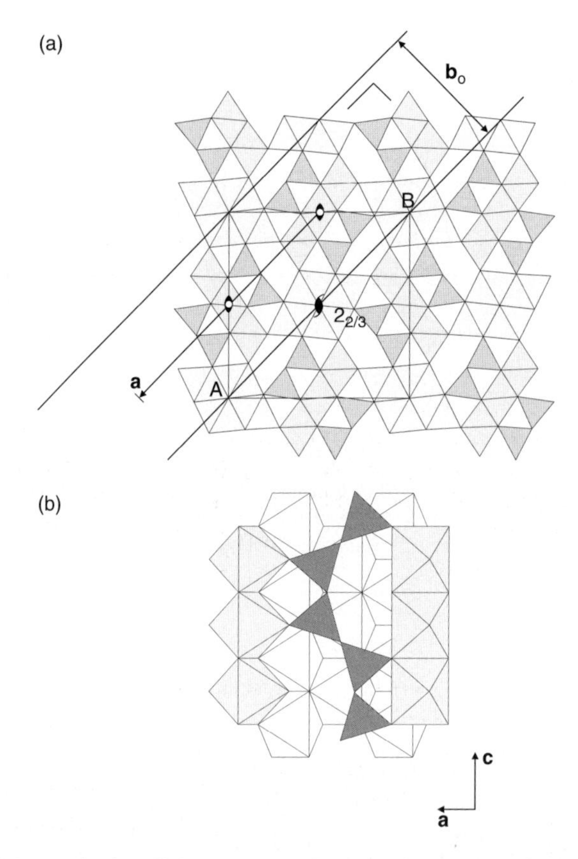

Fig. 2.27. (a) [001] projection of the structures of gageite–balangeroite family, with indication of the 'subcell', the OD layer and its λ-POs, and the σ-PO $2_{2/3}$ relating adjacent layers. (b) The connection of the tetrahedral chain (dark grey) with the octahedral framework, as seen along the [110] direction of the 'subcell'.

indicated by the [001] rotation photographs of balangeroite (Compagnoni *et al.* 1983); (c) the relationships between the subcell and the true cells as revealed by the electron diffraction patterns; (d) the chemical data; (e) the infrared spectrum of balangeroite, typical of chain silicates in the range 900–1100 cm^{-1}; and (f) the fibrous nature of both minerals (Ferraris *et al.* 1987). Such assessment, described in Fig. 2.27, is characterized by a framework built up by octahedral walls three-chains wide (3×1 walls), which share corners with (2×2) octahedral bundles, with four-repeat silicate chains connected to the octahedral scaffolding as shown in Fig. 2.27(b).

The real structures of gageite and balangeroite, the nature of their disorder, and the polytypic variants may be easily discussed on the basis of the OD character of the family. OD layers with translation vectors $\mathbf{a} = \mathbf{A} - \mathbf{B}$, $\mathbf{c} = 3\mathbf{C}$ and third basic vector $\mathbf{b_0} = (\mathbf{A} + \mathbf{B})/2$, layer group symmetry $P112/m$, follow each other as indicated by the OD-groupoid family symbol:

$$\begin{array}{cccc} P & 1 & (1) & 2/m \\ & \{1 & (1) & 2_{2/3}/n_{1,2}\,\}. \end{array}$$

Pairs of adjacent layers are related through the operations $[--2_{2/3}]$ or $[--2_{-2/3}]$, or, which is the same, by the stacking vectors $\mathbf{t_1} = -\mathbf{a}/2 + \mathbf{b_0} + \mathbf{c}/3$ or $\mathbf{t_2} = -\mathbf{a}/2 + \mathbf{b_0} - \mathbf{c}/3$. Two MDO structures exist in this family: $\mathrm{MDO_1}$ is obtained when the stacking vectors $\mathbf{t_1}$ and $\mathbf{t_2}$ regularly alternate; the generating operation is $[--n_{1,2}]$, which is continued to total operation n in a structure with space group $P112/n$ (the twofold axis of the single layer is valid for the whole structure), and basis vectors $\mathbf{a}_m = \mathbf{a}, \mathbf{b}_m = 2\mathbf{b_0}, \mathbf{c}_m = \mathbf{c}$; $\mathrm{MDO_2}$ is obtained when the translation $\mathbf{t_1}$, the generating operation of this polytype, is constantly applied, giving rise to a structure with space group $P\bar{1}$ (only the λ- and σ-inversion centres are now total operations), and basis vectors $\mathbf{a_t} = \mathbf{a}/2 + \mathbf{b_0} + \mathbf{c}/3$, $\mathbf{b_t} = -\mathbf{a}/2 + \mathbf{b_0} + \mathbf{c}/3$, $\mathbf{c_t} = \mathbf{c}$. The $\mathrm{MDO_1}$ polytype (Fig. 2.28(a)) has been found both in balangeroite and gageite, obviously together with disordered sequences of the stacking vectors, whereas the $\mathrm{MDO_2}$ polytype has been found only in specimens of gageite (Fig. 2.28(b)).

2.4.5 *The OD character of* WO_2Cl_2

The crystal structure of the compound has been determined by Jarchow *et al.* (1968) who pointed to the evidences of one-dimensional disorder: the reflections for $h = 2n$ were sharp, whereas the reflections for $h = 2n + 1$ showed diffuseness in $\mathbf{c}^*$ direction, presenting in some specimens continuous streaks. The distribution of the sharp reflections, *family reflections*, conforms to an orthorhombic symmetry and the corresponding cell has parameters $A = 3.84$, $B = 3.89$, $C = 13.90$ Å. Among the various specimens displaying different degrees of disorder, it was possible to select a crystal in which the non-family reflections were sufficiently sharp, so that a 'conventional' structure analysis was possible. The structural results so derived obviously refer to the particular polytype realized in the crystal under study, but consented a general OD description of the whole family by Jarchow *et al.* (1968) and subsequently by Backhaus (1979*b*).

The structure is built up by OD layers with translation vectors $\mathbf{a} = 2\mathbf{A}$, $\mathbf{b} = \mathbf{B}$, third vector $\mathbf{c_0} = \mathbf{C}/2$ and layer group symmetry $P2_1am$. In the layer, W atoms are coordinated by four oxygen atoms in the **a**,**b** plane, and by two chlorine atoms on both sides of the plane (Fig. 2.29). The OD-groupoid family symbol is:

$$\begin{array}{cccc} P & 2_1 & a & (m) \\ & \{2_{-1/2} & n_{2,-1/2} & (n_{1/2,1})\} \end{array}$$

Any layer is converted into the succeeding one by the operations $[2_{1/2}--]$ or $[2_{-1/2}--]$, or, which is the same, by the translations $\mathbf{t_1} = \mathbf{a}/4 + \mathbf{b}/2 + \mathbf{c_0}$ or $\mathbf{t_2} = -\mathbf{a}/4 + \mathbf{b}/2 + \mathbf{c_0}$. By constantly applying the translation $\mathbf{t_1}$, the $\mathrm{MDO_1}$ polytype is obtained (Fig. 2.30(a)), with space group symmetry $A1a1$ and cell parameters

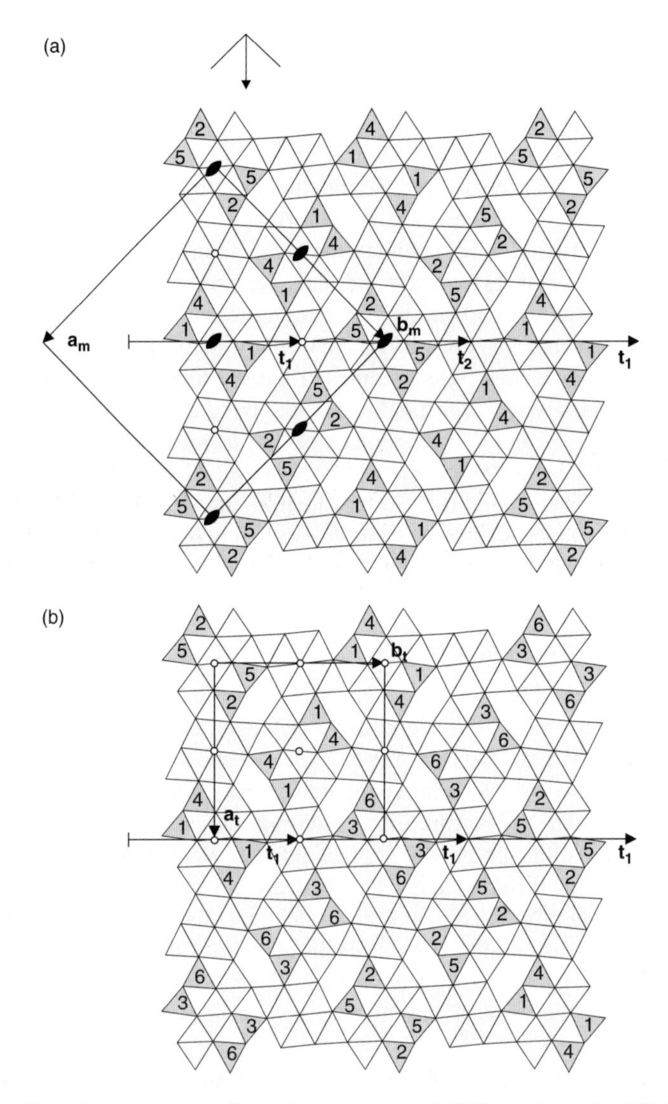

Fig. 2.28. Crystal structures of both polytypes MDO_1 (a) and MDO_2 (b) in gageite–balangeroite family, as seen along [001]. The relative positions of the various tetrahedral chains are indicated by giving the heights, in *c*/6 units, of the central oxygen atoms in both disilicate groups parallelel to [001] and connected to build up the tetrahedral four-repeat chain. (a) The $\mathbf{a}_m$ and $\mathbf{b}_m$ translation vectors of the MDO_1 structure are indicated, together with the symmetry elements: twofold axes, inversion centres and n glides at levels 0 and 3, in $c_m/6$ units. A non-standard origin has been chosen on the twofold axis to keep a closer relationship with Moore's cell. (b) The $\mathbf{a}_t$ and $\mathbf{b}_t$ translation vectors of the MDO_2 structure are indicated, both starting from the inversion centre at level 2, in $c_t/6$, assumed as origin, and ending on inversion centres at level 4. In both (a) and (b) the sequences of the stacking vectors giving rise to the two MDO structures are indicated.

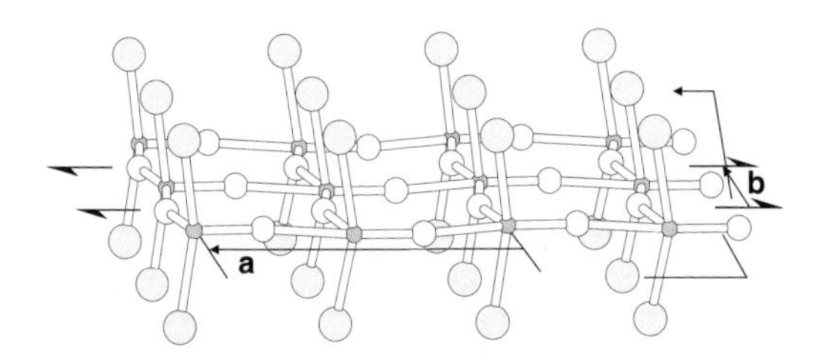

Fig. 2.29. The OD layer in WO_2Cl_2, as seen nearly along **b**; the drawing is slightly rotated about the horizontal and vertical axes for a better appreciation of the interatomic connections. The W, O, and Cl atoms are represented as small (dark grey), medium and large (light grey) circles. The translation vectors and the symmetry operators of the layer are indicated.

$a = 7.68, b = 3.89, c = 14.42$ Å, $\beta = 105.44°$, corresponding to the polytypic form studied by Jarchow *et al.* (1968). The constant application of the translation $\mathbf{t}_2$ gives the twin structure MDO'_1. The regular alternation of translations $\mathbf{t}_1$ and $\mathbf{t}_2$ gives rise to the MDO_2 polytype (Fig. 2.30(b)), with space group symmetry $P2_1am$ and cell parameters $a = 7.68, b = 3.89, c = 13.90$ Å.

2.4.6 *OD character of* $TeCl_4$

The structure of $TeCl_4$ has been determined by Buss and Krebs (1971) with polysynthetically (100) twinned crystals, in the space group $C2/c$, $a = 17.076, b = 10.404, c = 15.252$ Å, $\beta = 116.8°$ and the OD character of the compound has been discussed by Backhaus (1979*a*). The structure consists of Te_4Cl_{16} molecules (Fig. 2.31) arranged in layers with translation vectors **b**, **c** and symmetry $P(m)cm$, with third vector $\mathbf{a}_0$, normal to the layer plane [$a_0 = (a \sin\beta)/2 = 7.62$ Å].

The results of the structural study by Buss and Krebs (1971) indicate also the σ-POs converting a layer into the subsequent one and point to the OD-groupoid family symbol:

$$\begin{array}{ccc} P\ (m) & c & m \\ \{(n_{1,-1/2}) & n_{1/2,2} & n_{2,1}\} \end{array}$$

The *NFZ* relation indicates that two distinct positions are possible for each adjoining layer, corresponding to the operations $[n_{1,-1/2}--]$ and $[n_{1,1/2}--]$, or to the translations $\mathbf{t}_1 = \mathbf{a}_0 + \mathbf{b}/2 - \mathbf{c}/4$ and $\mathbf{t}_2 = \mathbf{a}_0 + \mathbf{b}/2 + \mathbf{c}/4$, respectively. The two possible MDO polytypes in this family correspond to the stacking sequences $\mathbf{t}_1\,\mathbf{t}_1\,\mathbf{t}_1\ldots$ and $\mathbf{t}_1\,\mathbf{t}_2\,\mathbf{t}_1\,\mathbf{t}_2\,\mathbf{t}_1\,\mathbf{t}_2\ldots$. The generating operation of the first polytype MDO_1 (Fig. 2.32(a)) is the translation $\mathbf{t}_1 = \mathbf{a}_0 + \mathbf{b}/2 - \mathbf{c}/4$, which, through its continuation, gives rise to a monoclinic structure with basis vectors $\mathbf{a} = 2\mathbf{a}_0 - \mathbf{c}/2$, **b**, **c**, and space group $C2/c$, corresponding to the structure of the specimen studied by Buss and Krebs (1971). The generating operation of the second polytype MDO_2 (Fig. 2.32(b)) is a glide normal to **c** with translational component $\mathbf{a}_0 + \mathbf{b}/2$; it may be continued to a normal glide $[--n]$ in a structure with $\mathbf{a} = 2\mathbf{a}_0$, **b**, **c**, and space group $Pmcn$. The polytype MDO_2

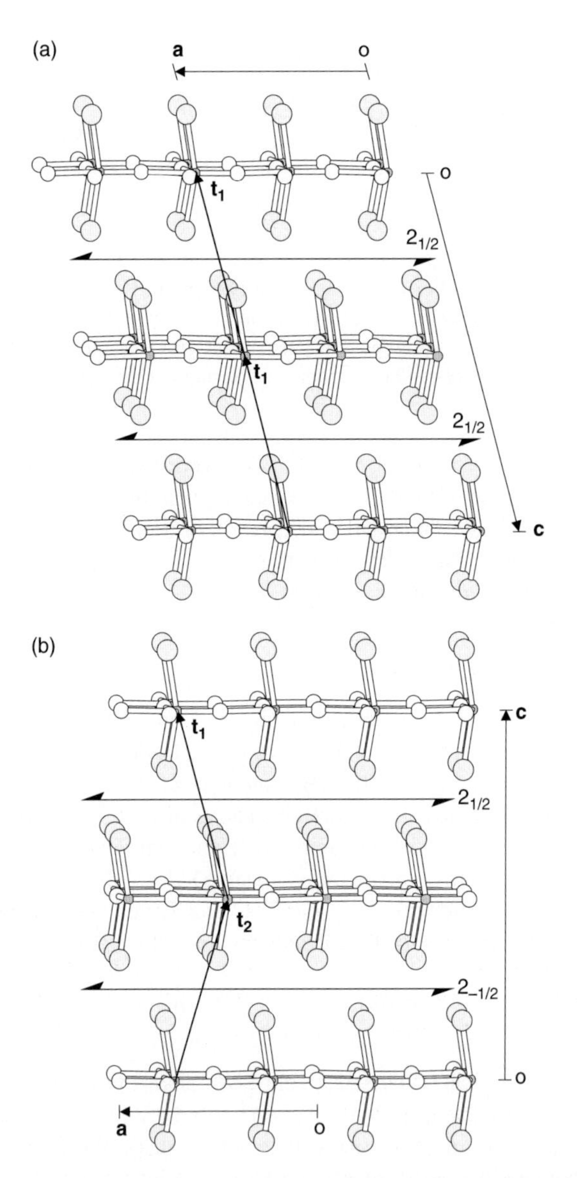

Fig. 2.30. The structures of the monoclinic MDO_1 (a) and orthorhombic MDO_2 (b) polytypes in WO_2Cl_2 family, as seen along **b**; the drawings are slightly rotated to appreciate the interatomic connections in the layers. In (a) the sequences of $2_{1/2}$ operations, as well as of the stacking vectors $\mathbf{t}_1$ are indicated. In (b) the regular alternation of $2_{1/2}$ and $2_{-1/2}$ operations, as well as of the stacking vectors $\mathbf{t}_1$ and $\mathbf{t}_2$ are indicated.

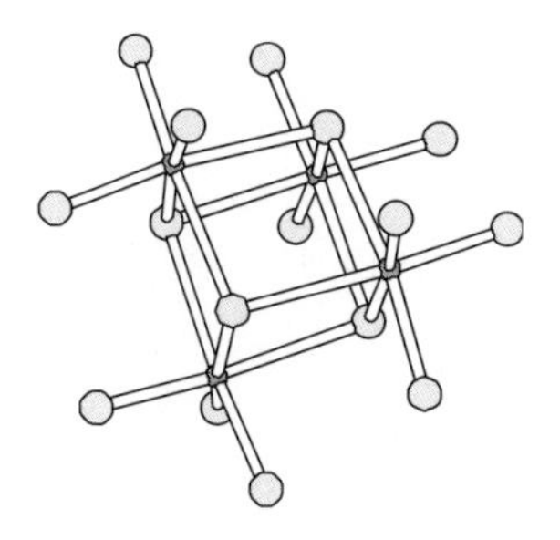

Fig. 2.31. The molecular structure of Te_4Cl_{16}, with Te and Cl atoms represented as small, dark-grey and large, light-grey circles, respectively. The single molecules have ideal $\bar{4}3m$ symmetry.

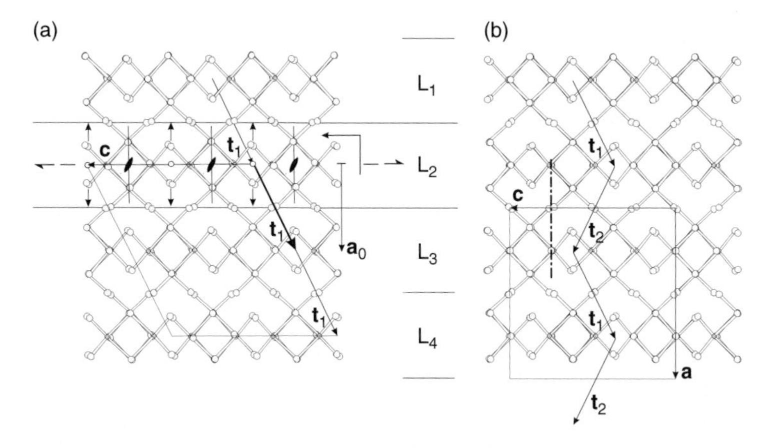

Fig. 2.32. The structures of the monoclinic (a) and orthorhombic (b) MDO polytypes in $TeCl_4$ family, as seen along **b**, with indication of the sequence of layers. In (a) the symmetry operations of the single OD layer are indicated, as well as the stacking of the layers according to the generating operation $\mathbf{t}_1$ (heavy marks). In (b) the stacking of the layers according to the sequence $\mathbf{t}_1\mathbf{t}_2\mathbf{t}_1\mathbf{t}_2$, as well as the generating operation $[--n]$ (heavy marks) are indicated.

has not yet found in specimens of $TeCl_4$. It may be realized as small, distinct domains inside the constantly twinned monoclinic crystals, where $\mathbf{t_1}\,\mathbf{t_1}\,\mathbf{t_1}\ldots$ (MDO_1) and $\mathbf{t_2}\,\mathbf{t_2}\mathbf{t_2}\ldots$ (MDO'_1) sequences coexist.

2.4.7 *Structural model for α-$PtCl_2$ on the basis of its OD character*

In their structural study of the high-temperature modification of $PtCl_2$, Krebs *et al.* (1988) observed a diffraction pattern pointing to a unit cell with $a = 13.258$, $b = 6.388$, $c = 6.802$ Å, $\beta = 107.75°$ characterized by the presence, besides sharp reflections for even k values, of diffuse reflections and streaks parallel to $\mathbf{a}^*$ for

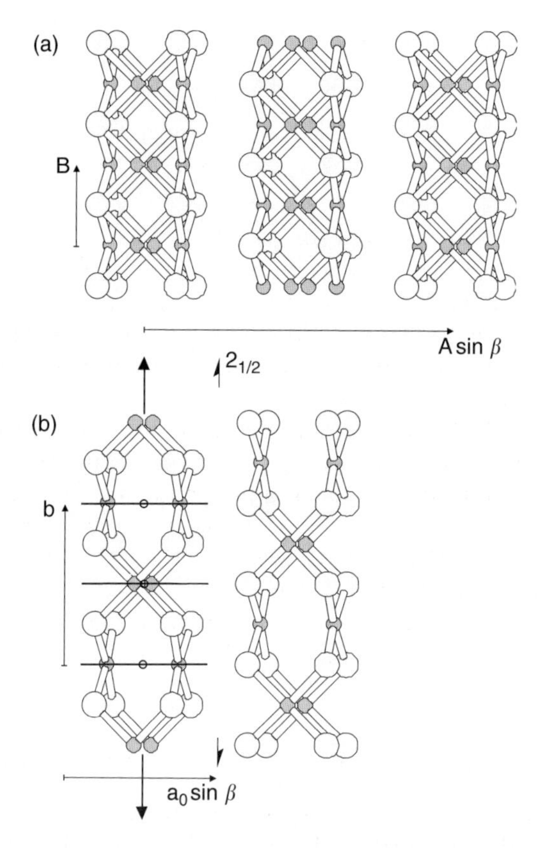

Fig. 2.33. The 'subcell' structure of α-$PtCl_2$ (a) as seen along **C**, with Pt atoms (half occupancy) represented as small dark-grey circles and Cl atoms represented as large light-grey circles. Two successive single layers (b), with $P2/m$ symmetry, stacked according to a [– $2_{1/2}$ –] screw rotation.

$k = 2n + 1$. These features point to an OD structure with OD layers stacking in the $\mathbf{a}^*$ direction. The sharp reflections define a 'subcell' with $\mathbf{A} = \mathbf{a}$, $\mathbf{B} = \mathbf{b}/2$, $\mathbf{C} = \mathbf{c}$, space group symmetry $C2/m$, a subcell corresponding to the family structure, which has been solved by Krebs *et al.* (1988) and is presented in Fig. 2.33(a): it is built up by tetragonal 'tubes', running along **B**, where the Pt atoms, in square planar coordination, have half occupancy.

The OD procedures are extremely helpful in guessing the real arrangement, corresponding to a definite ordering of the Pt atoms inside the tubes and definite stackings of the 'tube' sequences in the $\mathbf{a}^*$ direction. The translation vectors of the single layer are **b**, **c**, and the third basic vector is $\mathbf{a_0} = \mathbf{A}/2$ (the number p of OD layers for each **A** translation may be calculated following the procedure presented in Section 2.3.7). The operational elements of symmetry of the family structure are here recalled (left side) together with the OD groupoid family compatible with them (right side):

$$\begin{array}{ccccc} 2 & & & & \\ 2_1 & P & (1) & 2/m & 1 \\ 1\text{—}1 & & & & \\ m & & \{(1) & 2_{1/2}/a_2 & 1\} \\ a & & & & \end{array}$$

Layers with symmetry $P2/m$ follow each other related by the operations $2_{1/2}$ or $2_{-1/2}$ (or, which is the same, stacked according to $\mathbf{t_1} = \mathbf{a_0} + \mathbf{b}/4$ or $\mathbf{t_2} = \mathbf{a_0} - \mathbf{b}/4$ translations) and pairs of layers related by either operation are geometrically equivalent. Among the models suggested by Krebs *et al.* (1988), only one is in keeping with the $P2/m$ symmetry of the single layer and the vicinity condition for the layer stacking and is drawn in Fig. 2.33(b).

The two possible MDO structures in this family are: MDO_1, which is realized when the stacking vectors $\mathbf{t_1}$ and $\mathbf{t_2}$ regularly alternate; the generating operation is the glide normal to $\mathbf{b}$ with translational component $\mathbf{a_0}$, which is continued to a normal glide $[- a -]$ in a structure with $\mathbf{a} = 2\mathbf{a_0}$, $\mathbf{b}$, $\mathbf{c}$, and space group symmetry $P12/a1$ (Fig. 2.34(a)); MDO_2, which is realized when the stacking vector $\mathbf{t}_2$ is constantly applied (the constant application of $\mathbf{t}_1$ gives rise to the twin structure MDO_2'); the generating operation is obviously the translation $\mathbf{t}_2$; its continuation gives rise to a triclinic structure with $\mathbf{a} = \mathbf{a_0} - \mathbf{b}/4$, $\mathbf{b}$, $\mathbf{c}$, and space group symmetry $P\bar{1}$ (Fig. 2.34(b)).

2.4.8 *OD approach to the structure of tris(bicyclo[2.1.1]hexeno)benzene*

OD theory may be very useful in interpreting the results of the structural study by Birkedal *et al.* (2003), who described and studied three distinct diffraction patterns of tris(bicyclo[2.1.1]hexeno)benzene, corresponding to an ordered monoclinic form (M crystals), a growth twin of this form (M^{ht} crystals), and a disordered sequence of structural layers, with apparent hexagonal symmetry (H crystals, with $a = b = 9.01$, $c = 17.80$ Å). The streaking in the H crystals (diffuseness along $\mathbf{c}^*$), the non-space-group absences, the presence of both reflections common to the various diffraction patterns (subcell reflections), and characteristic reflections point to an OD structure, with $\mathbf{a},\mathbf{b}$ layers succeeding each to the other with ambiguities in the stacking.

The results of the study indicated that the single layer has $P\bar{6}2m$ symmetry, with $\mathbf{a}_1$ and $\mathbf{a}_2$ translation vectors ($a_1 = a_2 = 9.01$ Å), $\mathbf{c}_0$ third basic vector of the layer (not a translation vector; $\mathbf{c}_0 = \mathbf{c}_\text{H}/4 = 4.45$ Å). The OD theory indicates which σ-operations are compatible with the symmetry operations of the single layer. Among the few possibilities, the correct one, in keeping with the results of the structural study by Birkedal *et al.* (2003), is the following:

$$\begin{array}{cccccccc} P & 2 & 2 & 2 & (\bar{6}) & m & m & m \\ & \left\{ n_{s,2} \right. & n_{s',2} & n_{s'',2} & \begin{pmatrix} \bar{3} \\ 2_2 \\ 6_6 \end{pmatrix} & 2_s & 2_{s'} & \left. 2_{s''} \right\} \end{array}$$

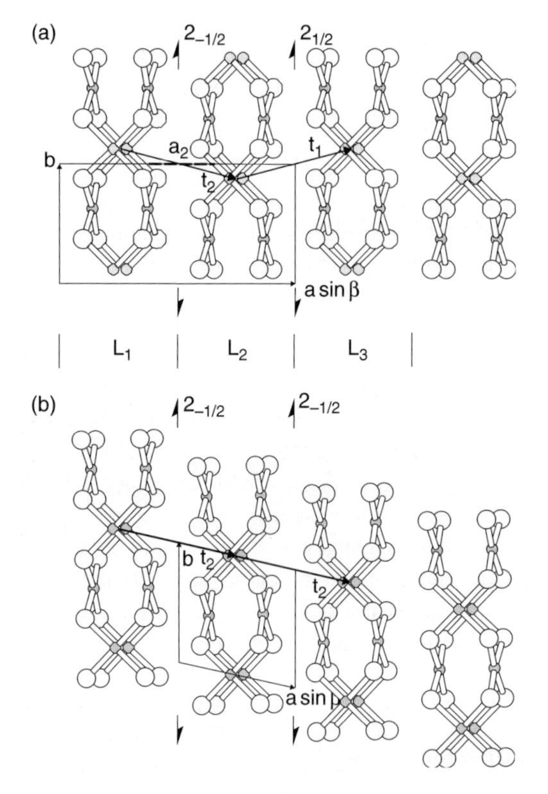

Fig. 2.34. The two main polytypes in α-$PtCl_2$: (a) the monoclinic polytype MDO_1, as seen along **c**, with indication of the alternation of the operations $2_{1/2}$ and $2_{-1/2}$, as well as of the two stacking vectors $\mathbf{t}_1$ and $\mathbf{t}_2$, and of the generating operation a_2; (b) the triclinic polytype MDO_2: the operation $2_{-1/2}$ and the vector $\mathbf{t}_2$ are constantly acting.

[with additional condition: $s+s'+s''=0$ (Dornberger-Schiff 1964)]. The parentheses in both lines indicate the direction of missing periodicity.

By examining the structural results [the single layer is presented in Fig. 2.35; the subsequent layer is placed at $(\mathbf{a}_1+\mathbf{a}_2)/3$, after rotation by 180°] we may find that $s=1/3$, $s'=-1/3$, $s''=0$. Therefore the OD-groupoid family symbol is:

$$\begin{matrix} P & 2 & 2 & 2 & (\bar{6}) & m & m & m \\ & \left\{ n_{1/3,2} \right. & n_{-1/3,2} & n_{0,2} & \begin{pmatrix} \bar{3} \\ 2_2 \\ 6_6 \end{pmatrix} & 2_{1/3} & 2_{-1/3} & \left. 2_0 \right\} \end{matrix} \tag{2.10}$$

The values of the parameters s, s', s'' may be derived also by considering the cell of the 'family structure', namely the sub-cell corresponding to the sharp and strong reflections. This subcell is hexagonal with vectors: $\mathbf{A}=\mathbf{a}/3$, $\mathbf{C}=2\mathbf{c}_0$. The λ- and

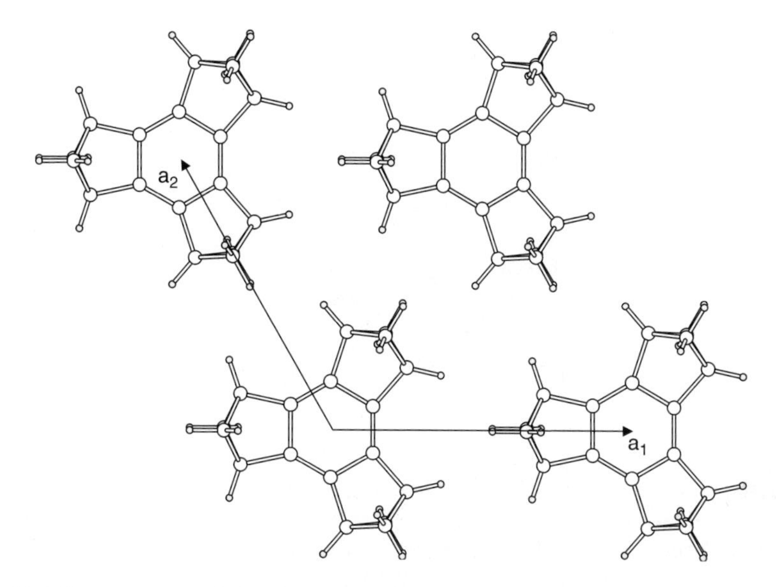

Fig. 2.35. The structure of the single layer of tris(bicyclo[2.1.1]hexeno)benzene: C and H atoms are drawn as large and small open circles, respectively.

σ-operators of the layer must be in keeping with the operators of the family structure, which has space group $P6_3/mmc$. To pass from the operators of the single layer to those of the family structure, the translational component referring to $\mathbf{a}_1$, $\mathbf{a}_2$, $\mathbf{a}_3$ and to $\mathbf{b}_1$, $\mathbf{b}_2$, $\mathbf{b}_3$ must be multiplied by 3 and those referring to $\mathbf{c}_0$ must be divided by 2. In this way, from (2.10) we obtain

$$\begin{matrix} P & 2 & 2 & 2 & (\bar{6}) & m & m & m \\ & \Bigg\{ n & n & c & \begin{pmatrix} \bar{3} \\ 2_1 \\ 6_1 \end{pmatrix} & 2_1 & 2_1 & 2 \Bigg\} \end{matrix} \qquad (2.11)$$

The operators in (2.11) closely correspond, also in their position and mutual orientation, to the operators in $P6_3/mmc$.

2.4.8.1 *MDO structures in tris(bicyclo[2.1.1]hexeno)benzene*

To find the MDO structures, we have to consider the σ–τ-operations: $n_{1/3,2}$ normal to $\mathbf{a}_1$, $n_{-1/3,2}$ normal to $\mathbf{a}_2$, $n_{0,2}$ normal to $\mathbf{a}_3$, and 2_2, 6_6 (clockwise rotation), 6_6 (counter-clockwise rotation) parallel to $\mathbf{c}_0$ (Fig. 2.36). Each of these operations—when constantly applied—gives rise to a MDO structure.

The operator $n_{-1/3,2}$, normal to $\mathbf{a}_2$, corresponds, by constant application, to a [– c –] glide in a monoclinic structure with $\mathbf{a}_m = -2\mathbf{a}_1 - \mathbf{a}_2$, $\mathbf{b}_m = \mathbf{a}_2$, $\mathbf{c}_m = 2\mathbf{c}_0 + (2\mathbf{a}_1 + \mathbf{a}_2)/3$. As along $\mathbf{b}_m = \mathbf{a}_2$ there is a twofold axis and the $\mathbf{a}_m$, $\mathbf{b}_m$ layer

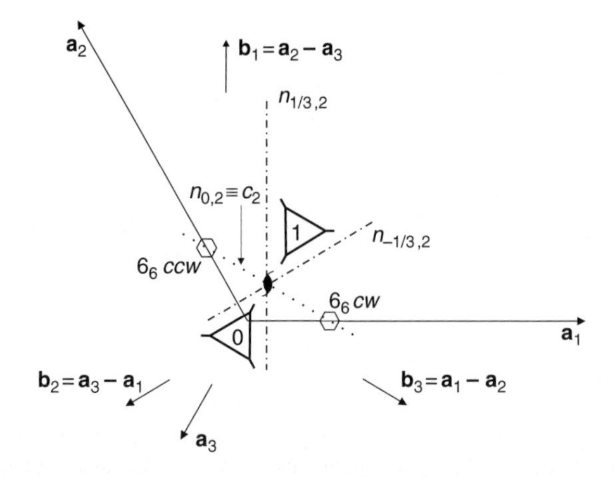

Fig. 2.36. Schematic representation of a pair of adjacent layers (layer 0 and layer 1), with indication of the $\mathbf{a}_1$, $\mathbf{a}_2$ translation vectors, the directions of $\mathbf{a}_3$, $\mathbf{b}_1$, $\mathbf{b}_2$, $\mathbf{b}_3$, and the $\sigma - \tau$-POs relating the layers. Only one molecule in each layer is drawn; it is schematically represented as a triangle with apical segments ($\bar{6}2m$ symmetry).

is centered, this first MDO structure has space group $C2/c$ and corresponds to that of the M polytype studied by Birkedal *et al.* (2003). If we constantly apply the operation $n_{1/3,2}$ (normal to $\mathbf{a}_1$), we obtain the same structure in twin orientation with respect to the preceding one.

Let us now consider the operator $n_{0,2}$ normal to $\mathbf{a}_3$: it corresponds, by continuous application, to a $[c - -]$ glide in an orthorhombic structure, with $\mathbf{a}_{\text{ort}} = \mathbf{a}_3$, $\mathbf{b}_{\text{ort}} = \mathbf{a}_2 - \mathbf{a}_1$, $\mathbf{c}_{\text{ort}} = 2\mathbf{c}_0 (a_{\text{ort}} = 9.01, b_{\text{ort}} = 15.61, c_{\text{ort}} = 8.90$ Å), space group *Ccmm*. The same structure may be obtained by constant application of the operation 2_2 parallel to $\mathbf{c}_0$.

Finally, by looking at Fig. 2.36, we may observe that the constant application of the operator 6_6 (in either position indicated in Fig. 2.36) gives rise to a hexagonal structure with $\mathbf{c}_{\text{hex}} = 6\mathbf{c}_0$ and space group symmetry $P6_122$ (or $P6_522$), with $a_{\text{hex}} = 9.01$, $c_{\text{hex}} = 26.80$ Å. It seems interesting to observe that so far only one MDO structure has been found, the monoclinic one. Possibly, under different conditions of crystallization, the orthorhombic and hexagonal forms may be obtained.

2.5 OD structures built up with $M > 1$ kinds of layers

The theory of OD structures consisting of equivalent layers has been developed in a systematic way and is presently substantially complete. The precise derivation of the 400 possible OD-groupoid families has been extremely important. They provide for each layer group symmetry the sets of σ-POs compatible with it, which allows us to derive the MDO structures, to sketch the corresponding diffraction patterns, and to

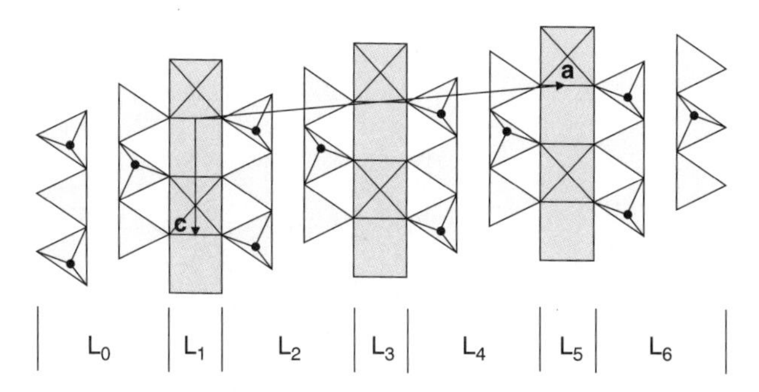

Fig. 2.37. The crystal structure of monoclinic stibivanite, as seen along **b**. The filled circles indicate the Sb atoms (this indication will be omitted in the Figs 2.38 and 2.41), which are at the apices of trigonal pyramids (light grey), pointing alternately up and down in the chains running along **c**. In the VO_5 tetragonal pyramids (dark grey) the apical oxygen atom point alternately up and down in the chains running along **c**. The sequence of layers is indicated.

decipher the generally complex patterns of OD crystals. Moreover, they sometimes provide, in association with proper crystal chemical reasoning, a formidable help in guessing reliable structural arrangements. On the other hand, the OD-groupoid family may be derived, in many cases, from the peculiarities of the diffraction patterns (streaks, diffuseness of reflections, non-space-group absences, ...), just as the possible space groups of 'fully ordered structures' may be derived from the 'normal' systematic absences.

However, the scope of OD theory goes beyond the families of structures consisting of equivalent layers and includes families of structures built up by two or more different kinds of layers. A good example is presented by the compound $\mathbf{Sb_2VO_5}$, synthesized and studied by Darriet *et al.* (1974), which has a natural counterpart in the mineral **stibivanite** (Kaiman *et al.* 1980). The structural studies by Darriet *et al.* (1976) and Szymański (1980) indicated that the compound is monoclinic, space group $C2/c$, $a = 17.989$, $b = 4.792$, $c = 5.500$ Å, $\beta = 95.15°$. The structure, represented in Fig. 2.37, may be described as consisting of two different kinds of layers, which are represented in Fig. 2.38: the first one is built up by chains of corner-sharing SbO_3 trigonal pyramids L_{2n}, the second one by chains of edge-sharing VO_5 tetragonal pyramids L_{2n+1}. The L_{2n} and L_{2n+1} layers, which we shall denote as SbO_3- and VO_5-layers, respectively, have the same **b** and **c** translations, but present different layer group symmetry, $P(1)2_1/c1$ and $P(2/m)2/c2_1/m$, respectively. Once again, we remark that the OD layers do not generally correspond to crystal-chemically significant modules; the merely geometrical nature of OD layers in the present case is clearly shown by the fact that the crystal-chemically significant modules are the ribbons built up by chains of VO_5 tetragonal pyramids with chains of SbO_3 pyramids grasped on both sides.

The fact that the symmetry of the L_{2n+1} layers is higher than that of the L_{2n} layers indicates the possibility of polytypic character. In the compound described

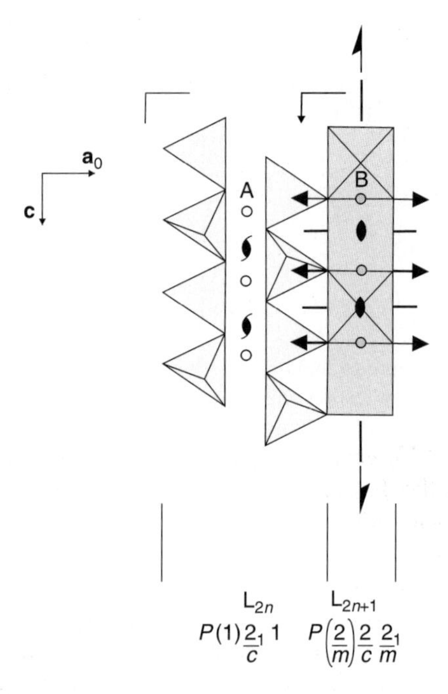

Fig. 2.38. The two distinct layers in stibivanite. The kind and position of the elements of symmetry, as well as the symbols of the two layer groups are indicated. The vector connecting the origins A and B in layers L_{2n} and L_{2n+1}, respectively, has components $r\mathbf{b}$ and $s\mathbf{c}$, with $r = 1/4$ and $s = -0.073$.

in Fig. 2.37, SbO_3-layers L_{2n} and L_{2n+2} are related by the twofold axes parallel to **b** and by the inversion centres which are λ–ρ-POs of the VO_5-layers L_{2n+1}: the pair $(L_{2n}; L_{2n+1})$ is converted by these operators into the pair $(L_{2n+2}; L_{2n+1})$. However, due to the symmetry of the VO_5-layer, there is an alternative way to connect SbO_3-layers on both sides of it: in the new arrangement the layers L_{2n} and L_{2n+2} are related by the mirror plane parallel to (100) and the screw axes parallel to **c**, which are the other λ–ρ-POs of the VO_5-layer. As in the preceding arrangement, the pair of layers $(L_{2n}; L_{2n+1})$ is converted by these operators into the pair $(L_{2n+2}; L_{2n+1})$. Therefore, the pairs of adjacent layers are geometrically equivalent in both arrangements, as are equivalent in any of the infinite possible structures corresponding to the various possible sequences of operators (twofold axes and inversion centres or mirror planes and screw axes) which may be active in the VO_5-layers. Consequently, it seems proper to consider these structures as belonging to one family of OD structures.

The geometric equivalence of all pairs of adjacent layers in the example we are discussing supports the expediency of extending the OD principles also to structures built up with $M > 1$ kinds of layers. This extension has been carried out by Dornberger-Schiff (1964) on the basis of the concepts introduced in dealing with OD structures consisting of equivalent layers: OD layers, λ partial operations, σ partial operations, categories of OD structures, *NFZ* relation, and MDO polytypes.

Obviously, this extension requires some revisions and generalizations. First, it seems useful to observe that the translation groups of the various OD layers building up the structure are identical (as it occurs in the case of OD structures consisting of equivalent layers and also in stibivanite) or present a common subgroup. Another concept which requires a more wide formulation is the 'vicinity condition'. As already mentioned in Section 2.3.2, we shall denote, as suggested by Grell and Dornberger-Schiff (1982), L for a non-specified OD layer, A any non-polar OD layer, namely a layer for which a λ–ρ-PO exists, and b or d for a polar OD layer, namely a layer for which no λ–ρ-PO exists. For example, the concrete layer sequence we have found in stibivanite corresponds to the general type $A_1^1\ A_2^2\ A_3^1\ A_4^2 \ldots$, where the superscripts are used to distinguish the kind of layer, whereas the subscripts indicate the position of the layer in the sequence.

The **vicinity condition**, stating the principle of OD structures, may be generalized in the following way to include OD structures containing $M > 1$ different kinds of OD layers: for any OD layer A all pairs of layers $(L_{n-1}; A_n)$ and $(A_m; L_{m+1})$ are equivalent (for every n and m value for which an OD layer of that kind exists); for any polar OD layer b all pairs $(L_{n-1}; b_n)$ and $(d_m; L_{m+1})$, as well as all pairs $(b_n; L_{n+1})$ and $(L_{m-1}; d_m)$, are equivalent (for every n for which an OD layer of kind b exists and any number m for which the corresponding OD layer of kind d exists).

2.5.1 *The four categories*

Four different **categories of OD structures with $M > 1$** have been established by Dornberger-Schiff (1964) on the basis of a set of theorems which are here shortly referred to.

1. No σ–τ-PO exists which links adjacent layers (in fact, the existence of a ${}_{p,p+1}\tau$-PO would indicate that the structure is built up by layers of one kind); therefore, adjacent layers are of a different kind or are related only by σ–ρ-POs.
2. If two OD layers p and p' are related through the operation ${}_{p,p'}\rho$ we define **ρ-plane** the plane which is transformed into itself by the operation ${}_{p,p'}\rho$. This plane occurs in position $(p + p')/2$ in the sequence of the layers and is therefore a λ–ρ-plane or a σ–ρ-plane according to the even or odd value of $p + p'$.
3. It may be shown that if an OD structure with $M > 1$ kinds of layer presents a ρ-plane, a second ρ-plane of different kind must be present and no more than two types of ρ-planes are possible.

The definitions and theorems just recalled allow a neat classification of OD-structures built up by $M > 1$ kinds of OD layers into four categories, as shown in Table 2.4.

Table 2.4 shows that an OD structure does not present ρ-planes (category II) or does display two distinct ρ-planes, which may be both σ–ρ-planes (category III), both λ–ρ-planes (category IV), or, finally, one λ–ρ- and one σ–ρ-plane (category I).

Table 2.4 The four categories of OD structures with $m > 1$ different kinds of OD layers.

Category	Number of distinct λ–ρ-planes	Number of distinct σ–ρ-planes	Periodicity of the layer sequence	Number of different types of pairs
I	1	1	$2M - 1$	M
II	0	0	M	M
III	0	2	$2M$	$M + 1$
IV	2	0	$2(M - 1)$	$M - 1$

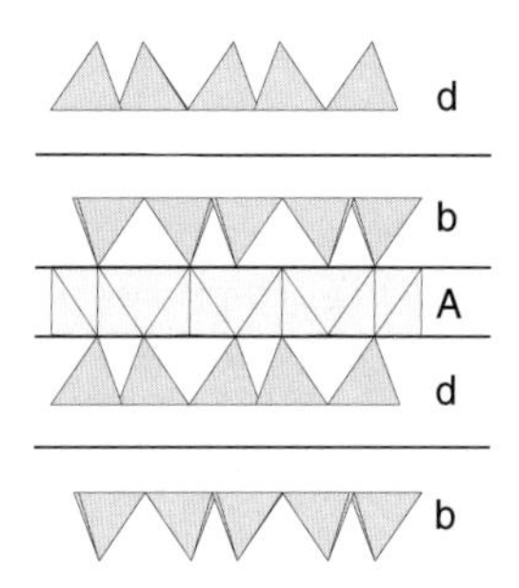

Fig. 2.39. The structural scheme for pyrophyllite and talc, with indication of the OD layers.

It is proper to observe that from Table 2.4, we may easily obtain, assuming $M = 1$, the three categories already known for OD structures consisting of equivalent layers; in fact for $M = 1$ category IV disappears (periodicity and number of different pairs are zero).

Category I. The OD structures in this category have one kind of non polar layers and $M - 1$ kinds of polar layers, as shown in the scheme:

$$\begin{array}{l} \mathrm{A}_1^1 \quad \mathrm{b}_2^1 \mid \mathrm{d}_3^1 \quad \mathrm{A}_4^1 \quad \mathrm{b}_5^1 \mid \mathrm{d}_6^1 \ldots \\ \lambda\text{–}\rho \qquad \sigma\text{–}\rho \\ \underbrace{\qquad\qquad\qquad\qquad}_{c_0} \end{array}$$

There are two kinds of ρ-planes, one of them is a λ-plane, and the other is a σ-plane. The origin is generally placed in the λ–ρ-plane and c_0, the distance between two nearest λ–ρ-planes, spans over $2(M - 1)$ polar and one non-polar layer, namely, over $2M - 1$ layers, as indicated in Table 2.4. Examples of OD structures in this category are: layered MX_2 compounds with M in octahedral coordination (CdI_2, ...), MX_2 compounds with M in trigonal prismatic coordination (MoS_2, ...), phases of the **talc-pyrophyllite group** (Fig. 2.39).

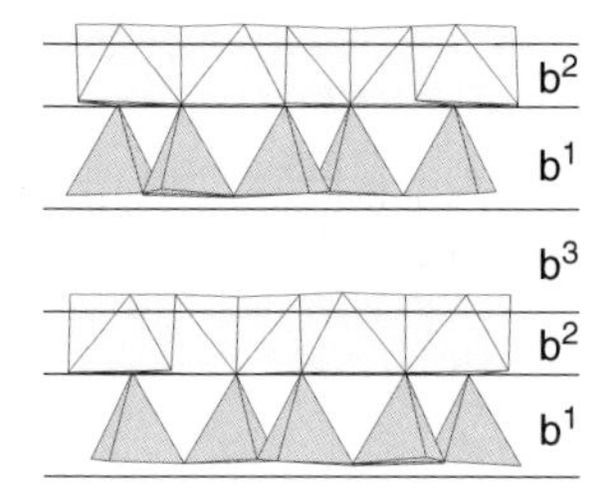

Fig. 2.40. The structural scheme for kaolinites, with indication of the OD layers.

Category II. All OD layers are polar as shown in the following scheme:

$$\mathrm{b}_1^1 \quad \mathrm{b}_2^2 \quad \mathrm{b}_3^1 \quad \mathrm{b}_4^2 \quad \ldots$$
$$\underbrace{\qquad\qquad}_{c_0}$$

The origin may be arbitrarily chosen; c_0, corresponding to the projection in the stacking direction of the vector connecting corresponding point in the nearest equivalent layers (b_1^1 and b_3^1 in the scheme), spans over M OD layers of different kind. Examples of OD structures in this category are given by the phases of the **serpentine–kaolinite** group (Fig. 2.40).

Category III. All OD layers are polar, with the sequences of M polar layers regularly inverting their polarity as shown in the scheme:

$$\mathrm{b}_1^1 \quad \underset{\sigma-\rho}{\mathrm{b}_2^2 \mid \mathrm{d}_3^2} \quad \underset{\sigma-\rho}{\mathrm{d}_4^1 \mid \mathrm{b}_5^1} \quad \mathrm{b}_6^2 \mid \mathrm{d}_7^2 \quad \ldots$$
$$\underbrace{\qquad\qquad\qquad\qquad}_{c_0}$$

There are two kinds of ρ-planes and both are σ–ρ-planes; the origin may be placed in one of them and c_0, the distance between the two nearest equivalent ρ-planes, spans over $2M$ OD layers, as indicated in Table 2.4. So far, no example of OD structures in this category has been found.

Category IV. The OD structures in this category are characterized by the presence of two kinds of non-polar layers. Additional $M - 2$ kinds of polar layers may be present (and their sequences regularly invert polarity), as in the left side of the scheme, or may not be present, as in the right side of the scheme:

$$\underset{\lambda-\rho}{\mathrm{A}_1^1} \quad \mathrm{b}_2^1 \quad \underset{\lambda-\rho}{\mathrm{A}_3^2} \quad \mathrm{d}_4^1 \quad \mathrm{A}_5^1 \quad \mathrm{b}_6^1 \quad \ldots \qquad \underset{\lambda-\rho}{\mathrm{A}_1^1} \quad \underset{\lambda-\rho}{\mathrm{A}_2^2} \quad \mathrm{A}_3^1 \quad \mathrm{A}_4^2 \quad \ldots$$
$$\underbrace{\qquad\qquad\qquad\qquad}_{c_0} \qquad\qquad \underbrace{\qquad\qquad}_{c_0}$$

There are two kinds of ρ-planes, corresponding to the λ–ρ-planes of the two kinds of non-polar layers. The origin may be placed in one of them; c_0, the distance between the two nearest equivalent ρ-planes, span over the two non-polar and $2(M - 2)$ polar layers. There are several examples of OD structures in this category, as those of stibivanite, we have just presented, $\mathbf{YCl(OH)_2}$, discussed by Dornberger-Schiff and Klevtsova (1967), the very complex structures of micas, chlorites, vermiculites. Other examples will be presented and discussed in the Section 2.6 or listed in Section 2.7.

2.5.2 *Symbols for the four categories*

OD-groupoid family symbols have been introduced also for the cases with $M > 1$. As for the OD structures with $M = 1$, these symbols fully describe the common symmetry properties of a whole OD family. Their first line presents the layer group symmetry of the M distinct kinds of layers, in the same order in which those layers follow each other in the stacking sequence.

The second line presents, for all the categories, the positional relations of subsequent layers of different kinds, which is done by giving in brackets the r and s components of the projection into the layer plane of the vector connecting the origins [chosen according to the International Tables for X-ray Crystallography, vol. I (1952)] of the two subsequent layers; the values r and s are referred to the basic vectors **a** and **b** of the common translation group (we obviously assume here that the stacking direction is $\mathbf{c}_0$). In categories I and III, where adjacent equivalent layers occur, the σ–ρ-POs relating them are indicated in curly brackets, in the same way as it occurs in OD-groupoid family symbols for OD structures with $M = 1$. We shall present, in the following, examples of symbols in the various categories, with reference, apart from category III, to concrete structural families.

Category I

Pyrophyllite (Ďurovič and Weiss 1983)

$$\begin{array}{ccc} P(\bar{3})12/m & & P(6)mm \\ & [-1/3, -1/3] & \left\{ 2_{-1/3}\, 2_{2/3}\, 2_{-1/3} \begin{pmatrix} \bar{3} \\ \bar{6} \\ n_{0,2/3} \end{pmatrix} 2_{1/3}\, 2\, 2_{-1/3} \right\} \end{array}$$

Talc (Ďurovič and Weiss 1983)

$$\begin{array}{ccc} H^{(*)}(\bar{3})12/m & & P(6)mm \\ & [-1/3, -1/3] & \left\{ 2_{-1/3}\, 2_{2/3}\, 2_{-1/3} \begin{pmatrix} \bar{3} \\ \bar{6} \\ n_{0,2/3} \end{pmatrix} 2_{1/3}\, 2\, 2_{-1/3} \right\} \end{array}$$

(*) H indicates that, with reference to the common basic vectors, the cell is triply primitive [International Tables for X-ray Crystallography (1952), Vol. I, p. 18].

Category II

Trioctahedral kaolinites (Dornberger-Schiff and Ďurovič 1975*a*,*b*)

$$\begin{array}{cccccc} P(6)mm & & H(3)1m & & H(6)mm & \\ & [1/3, 0] & & [1/3, 0] & & [1/3, 0] \end{array}$$

Dioctahedral kaolinites (Dornberger-Schiff and Ďurovič 1975*a*,*b*)

$$\begin{matrix} P(6)mm & & P(3)1m & & H(6)mm & \\ & [1/3, 0] & & [1/3, 0] & & [1/3, 0] \end{matrix}$$

Category III

No concrete example has been found so far

$$\begin{matrix} & \text{layer group of b}^1 & & \text{layer group of b}^2 & \\ \{\sigma\text{–}\rho\text{-POs}\} & & [r, s] & & \{\sigma\text{–}\rho\text{-POs}\} \end{matrix}$$

Category IV

Stibivanite (Merlino *et al.* 1989)

$$\begin{matrix} P(1)\ 2_1/c\ 1 & & P(2/m)\ 2/c\ 2_1/m \\ & [0.25, -0.0733..] & \end{matrix}$$

2.5.3 *NFZ relationships and MDO structures*

Assuming a definite position of a layer L_n the determination of the number Z of equivalent positions of an adjacent layer is made with the same *NFZ* relation we introduced and discussed in the case of $M = 1$. With $M > 1$ the relation may be expressed with the following concise statement. 'If two layers are non-equivalent, then $Z = N/F$. In the categories I and III, where adjacent equivalent layers are related by ρ-operations, $Z = N/F$ if at least one of these operations has a reverse continuation, otherwise $Z = 2N/F'$ (Ďurovič 1997*b*).

The systematic use of the *NFZ* relationship has been particularly fruitful in the analysis of the polytypic behaviour of the various classes of hydrous phyllosilicates (kaolinites, pyrophyllite and talc, chlorites, and micas). The *NFZ* relations, which are quite complex for those compounds, due the number of distinct OD layers and to their symmetry relationships, allow the calculation of all the stacking possibilities and eventually the derivation of all the MDO polytytpes. The presentation of the results of those investigations, which include not only the classification of the MDO polytypes and their characterization with descriptive symbols, but also the principles of their identification through diffraction patterns, cannot be carried on in this volume and we are forced to address the reader to the reviews by Ďurovič (1981, 1992), where the references to the original literature may be found.

We now go back to the OD family of **stibivanite**, applying to it the *NFZ* relation.

OD layer	Layer group	Subgroup of λ–τ-operations	N	F	Z
L_{2n+1}	$P2/m\ 2/c\ 2_1/m$	$P\ 2\ c\ m$	4 ↘		↗ 1
Symmetry of a layer pair →		$P\ 1\ c\ 1$		2	
L_{2n}	$P\ 1\ 2/c\ 1$	$P\ 1\ c\ 1$	2 ↗		↘ 2

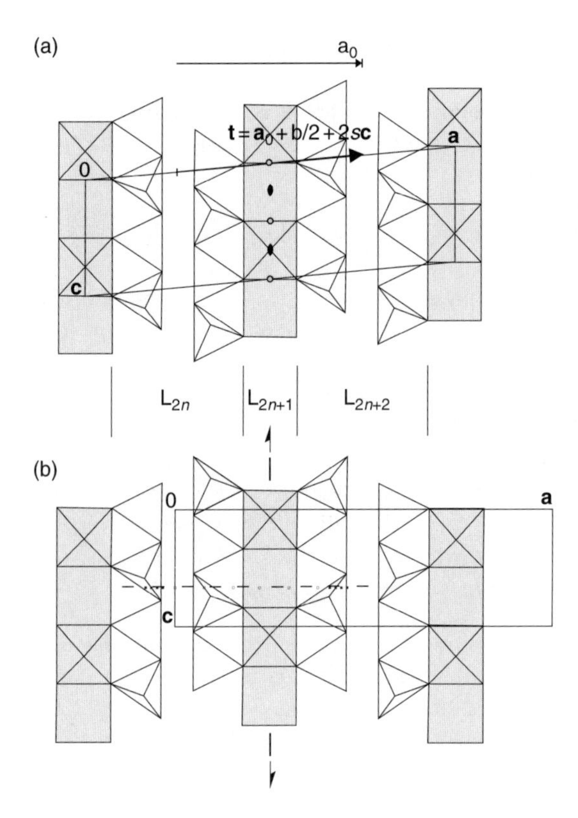

Fig. 2.41. The two possible types of (L_{2n}; L_{2n+1}; L_{2n+2}) triples in stibivanite and the resulting structural arrangements: (a) the monoclinic MDO polytype, (b) the orthorhombic polytype. In both cases the operations active in the L_{2n+1} layers and the generating operations, translation $\mathbf{t} = \mathbf{a_0} + \mathbf{b}/2 + 2s\mathbf{c}$ in (a) and glide normal to **c**, with translational component $\mathbf{a_0} + \mathbf{b}/2$, in (b), are indicated.

The scheme here illustrated indicates that assuming an arbitrary position of the SbO_3-layer L_{2n} the positions of the preceding and the subsequent layers L_{2n-1} and L_{2n+1} are uniquely determined ($Z = 1$). Consequently, only one kind of (L_{2n-1}; L_{2n}; L_{2n+1}) triples exists. On the contrary, there are two kinds of (L_{2n}; L_{2n+1}; L_{2n+2}) triples, corresponding to the two arrangements illustrated in Fig. 2.41. In fact, assuming an arbitrary position of the VO_5-layer L_{2n+1}, there are two positions ($Z = 2$) for each of the preceding and the subsequent layers L_{2n} and L_{2n+2} and pairs of the four resulting triples are geometrically equivalent.

The two structural arrangements described in Fig. 2.41 are the only ones presenting no more than two types of layer triples and correspond, according to the definition we have given in Section 2.3.6, to the MDO polytypes of this family. Each of them is defined by its own generating operation, namely 'the τ-operation with a translational component parallel to the stacking direction with magnitude equal to the distance between the two closest τ-equivalent layers' (Ďurovič 1997b).

The first one is the monoclinic form; it is obtained when twofold axes parallel to **b** and inversion centres are constantly operating in the VO_5-layers L_{2n+1}. The generating operation is the translation $\mathbf{t} = \mathbf{a_0} + \mathbf{b}/2 + 2s\mathbf{c}$, where $\mathbf{a_0}$, as indicated in Fig. 2.41(a), corresponds to the distance between the two nearest equivalent ρ-planes, and $s = -0.073$; the constant application of this operation generates a C-centred structure with $\mathbf{a} = 2\mathbf{a_0} + 4s\mathbf{c}$, **b, c** as basis vectors and $C2/c$ space group symmetry: the [– c –] glide is the common symmetry element of both OD layers, the [– 2 –] rotation axis of the VO_5-layer is valid for the whole structure.

The second one is obtained when the mirror planes parallel to (100) and the screw axes parallel to **c** are constantly operating in the VO_5-layers L_{2n+1}. The generating operation is a glide normal to **c** with translational component $\mathbf{a_0} + \mathbf{b}/2$. The constant application of this operation generates an orthorhombic structure with basis vectors $\mathbf{a} = 2\mathbf{a_0}$, **b**, and **c** and space group symmetry *Pmcn*: the [– – n] glide corresponds to the generating operation, the [– c –] glide is the common symmetry element of both OD layers, the [m – –] mirror of the VO_5-layers is valid for the whole structure. This MDO polytype corresponds to the form described by Merlino *et al.* (1989).

2.5.4 *Desymmetrization of OD structures*

The partial symmetry operations which build up the layer group of each OD layer are a central feature in OD theory. These operations are 'ideally' preserved in each polytype built up with equivalent layers or consisting of more kinds of OD layers. However, it is a common experience in the study of the structural arrangements of different polytypes of one family that those symmetry operations are actually only approximately preserved and light or severe distortions from ideal symmetry are observed, a phenomenon which has been called **desymmetrization** (Ďurovič 1979) and is dependent on the fact that in OD structures 'any OD layer is per definition situated in a disturbing environment, because its symmetry is different from that of the entire structure' (Ďurovič 1979). This disturbing effect, which weakens in disordered sequences due to the random distribution of the subsequent layers, is more drastic in MDO structures due to the regularity in the sequence of the layers, which may be expressed through the apparently paradoxical statement: 'the more disordered an OD structure is, the more symmetric it appears' (Ďurovič 1979).

The different desymmetrization which occurs in different polytypes of the same family is connected with the next-nearest-neighbour (*nnn*) interactions, as the vicinity condition requires identical nearest-neighbour interactions. The *nnn* interactions are obviously as stronger as shorter the corresponding interlayer distances are; therefore, a large desymmetrization is to be expected in those OD family consisting of $M > 1$ layers, in which one or more layers present a width corresponding to a single atomic plane.

This last feature—the introduction into the OD description of OD layers which appear artificial till the extreme choice of single atomic planes—is one of the main points raised in the debate about the relationships between OD structures and polytypes. These relationships will be discussed in the next chapter (Section 3.1.3); we now only recall that the OD approach is not concerned with the crystal-chemical aspects and looks at the symmetry of layers and symmetry relations between layers, and consequently, the choice of the OD layers (Grell 1984) is carried on in such a way that the vicinity condition is fulfilled and the results of OD theory may be applied. This choice may sometimes require an 'increase in the number of the layers in the structure on account of separation into several parts of those large layers which in unaltered form are present in all the polytypes' (Zvyagin 1987), as it occurs in the OD interpretation of kaolinite-type structures (Donberger-Schiff and Ďurovič 1975*a*,*b*), for which instead of the crystal-chemically significant two-storied layer OT, three distinct OD layers were introduced (Fig. 2.40).

2.6 Examples of OD structures consisting of two kinds of layers

We shall now present some examples of OD families built up by two different kinds of layers. Both layers are non polar, and therefore, all these families belong to category IV, which is the most frequently occurring among OD structures with $M > 1$.

2.6.1 *Shattuckite OD family*

The structure of shattuckite, $Cu_5Si_4O_{12}(OH)_2$, from Ajo (Arizona), was determined by Evans and Mrose (1977) as orthorhombic, space group *Pbca*, $a = 19.832$, $b = 9.885$, $c = 5.383$ Å, and consisting of tetrahedral chains similar to those in pyroxenes, running on both sides of brucite-like CuO_2 layers; the resulting complex TOT modules are linked by additional Cu atoms in square-planar coordination. Dódony *et al.* (1996), by selected area electron diffraction and high-resolution transmission electron microscopy, found that crystals from the same locality were commonly disordered along **a** and interpreted the defects as stacking faults that change the orientation of the octahedral layers. In ordered orthorhombic shattuckite, the orientations of the adjacent octahedral sheets can be represented as $+-+-+-\ldots$ (where $+$ and $-$ denote the sign of the tilting of the octahedra with respect to the **c** direction. Besides various isolated faults, they observed extended regions with either $++++++\ldots$ or $-------\ldots$ octahedral sequences, giving rise to a new monoclinic form, space group $P2_1/c$, $a = 9.92$, $b = 9.88$, $c = 5.38$ Å, $\beta = 92°$.

All these aspects may be conveniently discussed in terms of OD theory. In fact, the structure of shattuckite may be described as consisting of two distinct structural layers, both having translation vectors **b**, **c** ($b = 9.885$, $c = 5.383$ Å), regularly alternating each other in the $\mathbf{a}_0$ direction ($a_0 = 9.916$ Å): the first is the complex TOT layer with brucite-like CuO_2 sheets sandwiched by pyroxene-like chains on both sides (L_{2n} layers, with symmetry $P2_1/c$); in the second, ladder-like ribbons of copper atoms in square coordination run along **c** (L_{2n+1} layers with symmetry *Pbcm*) (Fig. 2.42).

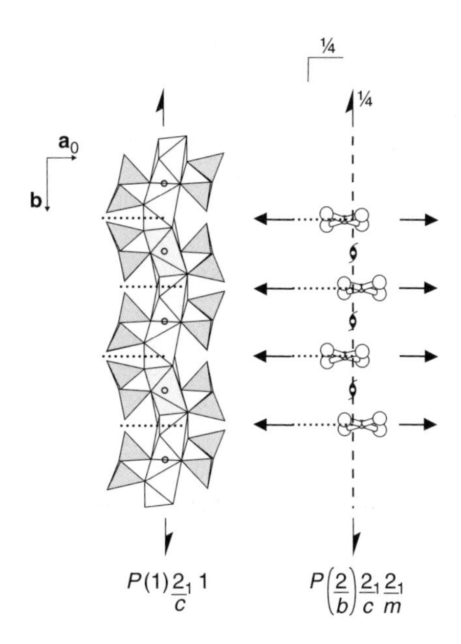

Fig. 2.42. The two distinct layers in shattuckite, as seen along **c** (**b** vertical). In the first one (left side), chains of silicon tetrahedra T (dark grey) are grasped on both sides of the layers of distorted copper octahedra O (light grey), forming the complex TOT layers. In the second one (right side), ladder-like strings of square-coordinated copper atoms run along **c**. The kind and position of the elements of symmetry, as well as the symbols of the two layer groups are indicated.

As in the case of stibivanite, the fact that the symmetry of the L_{2n+1} layers is higher than that of the L_{2n} layers points to the possibility of polytypic relationships. In the orthorhombic form, layers L_{2n} and L_{2n+2} are related through glide operators $[b - -]$ (and screw operators $[- - 2_1]$) which convert the layer L_{2n+1}, lying between them, into itself. Thus, the pair $(L_{2n}; L_{2n+1})$ is converted by these operations into the pair $(L_{2n+2}; L_{2n+1})$. However, because of the symmetry displayed by the L_{2n+1} layer, a different arrangement is also possible. In this arrangement, the layers L_{2n} and L_{2n+2} are related through the operators $[- \, 2_1 \, -]$ (and the inversion centres) which convert the layer L_{2n+1}, lying between them, into itself. As in the preceding case, the pair $(L_{2n}; L_{2n+1})$ is converted by these operations into the pair $(L_{2n+2}; L_{2n+1})$.

Infinite ordered polytypes and disordered structures are possible, corresponding to the various possible sequences of operators that may be active in the L_{2n+1} layers. As the present OD family belongs to category IV, the symmetry relations common to all the polytypes of the family are indicated by the OD-groupoid family symbol:

$$P\ (1)\ 2_1/c\ 1 \qquad\qquad P\ (2/b)\ 2_1/c\ 2_1/m$$
$$[0.0,\ -0.03]$$

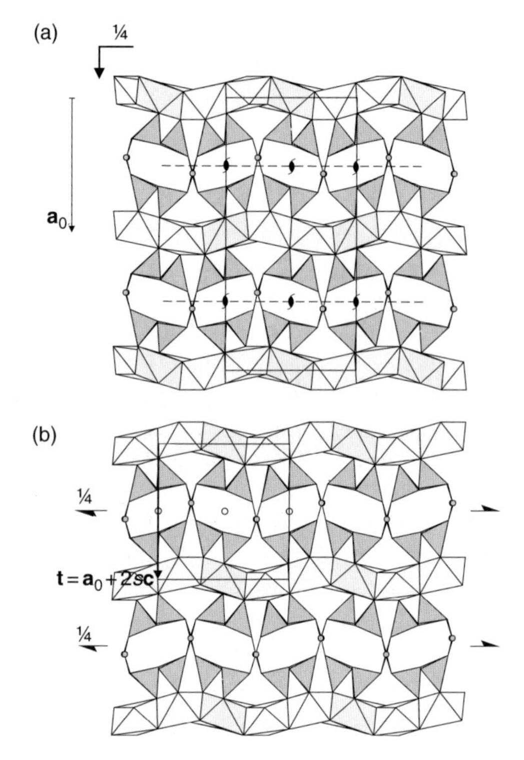

Fig. 2.43. The two MDO structures in shattuckite, as seen along **c** (**b** horizontal): (a) the orthorhombic polytype, (b) the monoclinic polytype. The complex TOT layers are connected through ladder-like strings of copper (grey circles) squares, running along **c**. In both (a) and (b) the operations active in the L_{2n+1} layers and the generating operations, namely glide normal to **c** with translational component $\mathbf{a_0}$ in (a), and translation $\mathbf{t} = \mathbf{a_0} + 2s\mathbf{c}$ in (b), are indicated.

An analysis similar to that presented for stibivanite indicates that only one kind of $(L_{2n-1}; L_{2n}; L_{2n+1})$ triples exists, whereas there are two kinds of $(L_{2n}; L_{2n+1}; L_{2n+2})$ triples. Consequently, as in stibivanite, two MDO polytypes are possible.

In the first of them (Fig. 2.43(a)) the operators $[b - -]$ and $[- - 2_1]$) are constantly active in the L_{2n+1} layers. The generating operation is a glide normal to **c** with translational component $\mathbf{a_0}$. The constant application of this operation generates an orthorhombic structure with basis vectors $\mathbf{a} = 2\mathbf{a_0}$, **b**, **c** and space group symmetry *Pbca*, corresponding to the form studied by Evans and Mrose (1977): the $[- - a]$ glide corresponds to the generating operation, the $[- c -]$ glide is the common symmetry element of both OD layers, and the $[b - -]$ glide of the L_{2n+1} layers is valid for the whole structure.

In the second MDO structure (Fig. 2.43(b)), the operators $[- 2_1 -]$ and inversion centres are constantly active in L_{2n+1} layer. The generating operation is the translation $\mathbf{t} = \mathbf{a_0} + 2s\mathbf{c}$. The continuation of this operation gives rise to a monoclinic polytype with basis vectors $\mathbf{a} = \mathbf{a_0} + 2s\mathbf{c}$, $\mathbf{b}$, $\mathbf{c}$, and space group symmetry $P2_1/c$: the [– c –] glide is the common symmetry element of both OD layers, the [– 2 –] rotation axis of the L_{2n+1} layer is valid for the whole structure, which corresponds to the monoclinic form described by Dodony *et al.* (1996).

2.6.2 *Planchéite OD family*

As in the case of shattuckite, the structure of planchéite $Cu_8Si_8O_{22}(OH)_4{\cdot}H_2O$ (Evans and Mrose 1977) is built up by two distinct structural layers, both having translation vectors $\mathbf{b}$, and $\mathbf{c}$ ($b = 19.043$, $c = 5.269$ Å; $a_0 = 10.065$ Å): one is the complex TOT layer with brucite-like CuO_2 sheets sandwiched by amphibole-like chains on both sides (L_{2n} layers, with symmetry $P2_1/c$); the other is constituted by ladder-like ribbons of copper atoms in square coordination running along $\mathbf{c}$ (L_{2n+1} layers with symmetry *Pncm*). Fig. 2.44 presents both layers and the distribution of symmetry operators in them.

The OD character of this family may be described following our discussion of the shattuckite family, substituting the glide [n – –] and the axes [– – 2] to the glide [b – –] and the screw axes [– – 2_1] as first pair of operators active in the L_{2n+1} layers. The second pair of operators active in that layers is given, exactly as in shattuckite, by the axes [– 2_1 –] and inversion centres. The symmetry relations common to all the polytypes of the family are indicated by the OD-groupoid family symbol:

$$\begin{array}{ccc} P\ (1)\ 2_1/c\ 1 & & P\ (2/n)\ 2_1/c\ 2/m \\ & [0.0,\ -0.04] & \end{array}$$

Also in the present case, infinite ordered polytypes as well as disordered structures are possible, corresponding to the various possible sequences of operators active in the L_{2n+1} layers; among them two MDO structures exist. In the first, the glide [n – –] and the axes [– – 2] are constantly active in the L_{2n+1} layers; the generating operation is a glide normal to $\mathbf{c}$ with translational component $\mathbf{a_0}$; the constant application of this operation generates an orthorhombic structure with basis vectors $\mathbf{a} = 2\mathbf{a_0}$, $\mathbf{b}$, and $\mathbf{c}$ and space group symmetry *Pnca*, corresponding to the form studied by Evans and Mrose (1977).

In the second MDO structure, the operators [– 2_1 –] and inversion centres, as in the monoclinic form of shattuckite, are constantly active in L_{2n+1} layer. The generating operation is the translation $\mathbf{t} = \mathbf{a_0} + 2s\mathbf{c}$. The continuation of this operation gives rise to a monoclinic polytype with basis vectors $\mathbf{a} = \mathbf{a_0} + 2s\mathbf{c}$, $\mathbf{b}$, $\mathbf{c}$ ($a = 10.074$, $b = 19.043$, $c = 5.269$ Å, $\beta = 92.4°$) and space group symmetry $P2_1/c$. In planchéite, as well as in the case of shattuckite, the a parameter of the monoclinic MDO structure is $a = a_0/\sin\beta$; the angle β may be easily obtained on the basis of the positional relationships of the two layers, given in the OD-groupoid family symbol and derived from the coordinates presented by Evans and Mrose (1977)].

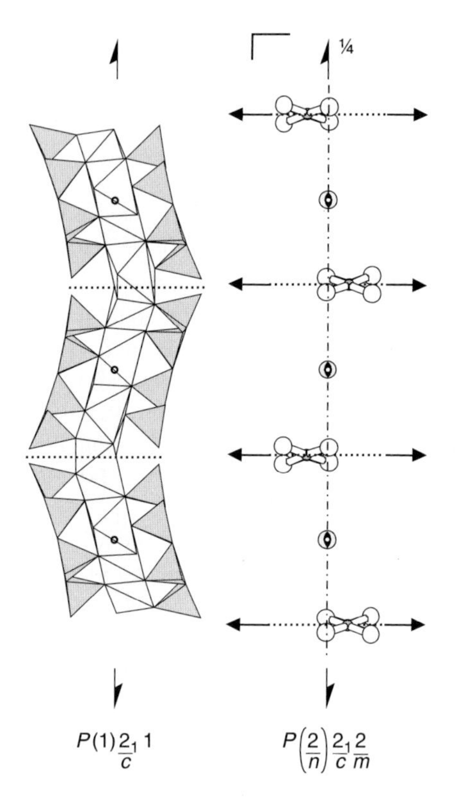

Fig. 2.44. The two distinct layers in planchéite, as seen along **c** (**b** vertical). In the first one (left side), double chains of silicon tetrahedra T (dark grey) are grasped on both sides of the layers of distorted copper octahedra O (light grey), forming complex TOT layers. In the second one (right side) row of water molecules (large grey circles) and ladder-like strings of square-coordinated copper atoms, both running along **c**, regularly alternate. The kind and position of the elements of symmetry, as well as the symbols of the two layer groups are indicated.

2.6.3 *Ericssonite–orthoericssonite; lamprophyllite-2M–lamprophyllite-2O*

Lamprophyllite $(Sr,Na)_2(Na_3Ti)Ti_2(Si_2O_7)_2(OH)_4$ and ericssonite $Ba_2Mn_4(Fe^{3+})_2(Si_2O_7)_2[O_2(OH)_2]$ display the same OD characters and are described by the same OD-groupoid family. The structure-type of these minerals, which, together with seidozerite and other natural phases, belong to the mero-plesiotypic series of bafertisite (Section 4.3.2.1 and Table 4.2), are characterized by HOH layers, composed of an octahedral O sheet sandwiched between two heterophyllosilicate H sheets (consisting of disilicate groups interlinked by Ti or Fe^{3+} square pyramids), with interlayer Sr or Ba cations. As in the previous examples, its OD family belongs to category IV, with two non-polar OD layers, corresponding to the O layer, with symmetry $P(1)2/m\ 1$, and the HAH (A corresponds to the interlayer cation sheet),

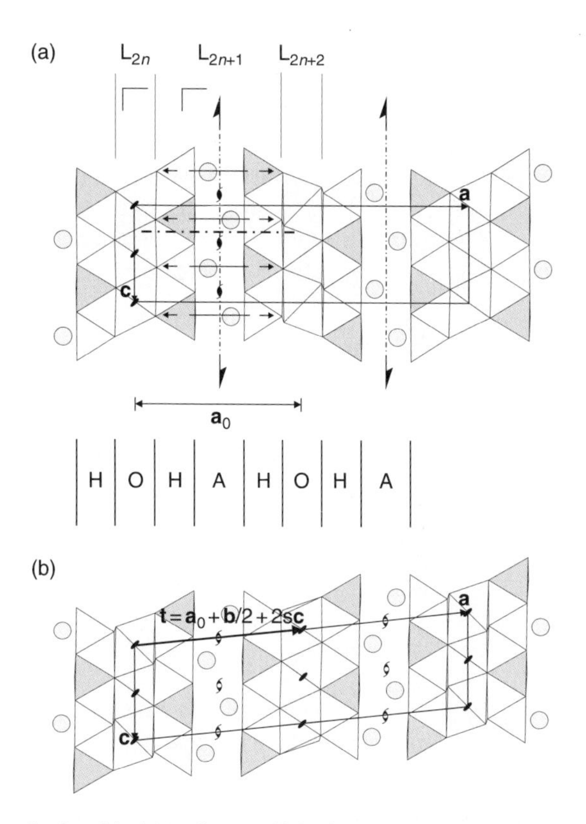

Fig. 2.45. The orthorhombic (a) and monoclinic (b) MDO polytypes for lamprophyllite and ericssonite, with indication of the sequence of the OD layers and of the structural modules H, O, A. In (a) the symmetry operations of the two OD layers are indicated. In both (a) and (b) the operations active in the L_{2n+1} layers and, in bold marks, the generating operations [glide normal to **c** with translation component $\mathbf{a_0} + \mathbf{b}/2$ in (a) and translation $\mathbf{t} = \mathbf{a_0} + \mathbf{b}/2 + 2s\mathbf{c}$ in (b)] are indicated. In the structural modules H the Si tetrahedra are drawn in dark grey, whereas the Ti (or Fe^{3+}) square pyramids are drawn in light grey.

with symmetry $P(2/n)\,2_1/m\,2_1/m$, and OD-groupoid family symbol:

$$\begin{array}{ccc} P(1)2/m1 & & P(2/n)\,2_1/m\,2_1/m. \\ & [0, s] & \end{array}$$

where s is -0.10 and -0.085 for lamprophyllite and ericssonite, respectively. Both layers have translation vectors **b**, **c** ($b \sim 7$, $c \sim 5.4$ Å; $a_0 \sim 10$ Å). In Fig. 2.45(a), the symmetry operations of both layers are reported. By applying the *NFZ* relation, we realize that there is only one kind of $(L_{2n-1}; L_{2n}; L_{2n+1})$ triples and two kinds of $(L_{2n}; L_{2n+1}; L_{2n+2})$ triples. Consequently, also in this case two MDO polytypes are possible.

In the first of them (Fig. 2.45(a)), the operators $[n --]$ (and $[-- 2_1]$) are constantly active in the HAH L_{2n+1} layers. The generating operation is a glide normal to **c** with translation component $\mathbf{a_0} + \mathbf{b}/2$: its continuation generates an orthorhombic structure with basis vectors $\mathbf{a} = 2\mathbf{a_0}$, **b**, **c** and space group symmetry *Pnmn*.

In the second MDO structure (Fig. 2.45(b)), the operators $[- 2_1 -]$ and inversion centres are constantly active in the L_{2n+1} layers. The generating operation is the translation $\mathbf{t} = \mathbf{a_0} + \mathbf{b}/2 + 2s\mathbf{c}$; its continuation gives rise to a monoclinic polytype with basis vectors $\mathbf{a} = 2\mathbf{a_0} + 4s\mathbf{c}$, **b**, **c**, and space group symmetry $C2/m$.

2.6.4 *Sursassite–pumpellyite–ardennite*

The close relationships among the structure-types of sursassite, $Mn_2Al_3[(OH)_3(SiO_4)(Si_2O_7)]$, pumpellyite, $Ca_2Al_3[(OH)_3(SiO_4)(Si_2O_7)]$, and ardennite, $Mn_4(Al,Mg)_6[(OH)_6(AsO_4)(SiO_4)_2(Si_3O_{10})]$, discussed in Section 1.8.3.1 and illustrated in Fig. 1.37, may be described in terms of OD theory (Merlino 1990*a*; Pasero and Reinecke 1991). They are built up by the two layers presented in Fig. 2.46, both having translation vectors **a**, **b** ($a \sim 8.8$, $b \sim 5.8$Å), regularly alternating each other in the $\mathbf{c}_0$ direction ($c_0 \sim 9.3$ Å). The OD-groupoid family symbol is:

$$\begin{array}{ccc} L_{2n} & & L_{2n+1} \\ P2/b2_1/m(2_1/m) & & C12/m(1) \\ & [0.068, 0] & \end{array}$$

At difference from the previous examples, in this case, as indicated by the *NFZ* scheme, both kinds of layers present alternate ways to connect adjacent layers.

OD layer	Layer group	Subgroup of λ–τ– operations	N	F	Z
L_{2n+1}	$C12/m1$	$C1m1$	4↘		↗2
	Symmetry of a layer pair →	$P1m1$		2	
L_{2n}	$P2/b2_1/m2/m$	$P\,b\,m\,2$	4↗		↘2

There are two kinds of $(L_{2n-1}; L_{2n}; L_{2n+1})$ triples, as well as two kinds of $(L_{2n}; L_{2n+1}; L_{2n+2})$ triples; consequently, four MDO structures are possible.

MDO_1 occurs when $[-2_1-]$ axes are constantly active both in L_{2n} and in L_{2n+1} layers: the generating operation is the translation $\mathbf{t} = \mathbf{c}_0 + 2(s - 1/4)\mathbf{a}$ ($s = 0.068$). The continuation of this operation gives rise to a monoclinic polytype with basis vectors **a**, **b**, $\mathbf{c} = \mathbf{c}_0 + 2(s - 1/4)\mathbf{a}$ and space group $P2_1/m$ (structure type of sursassite, Fig. 1.37(a)).

MDO_2 occurs when $[- 2_1 -]$ and $[- 2 -]$ axes are constantly active in L_{2n} and in L_{2n+1} layers, respectively: the generating operation is the translation $\mathbf{t} = \mathbf{c}_0 + 2s\mathbf{a} + \mathbf{b}/2$; its continuation gives rise to a monoclinic polytype with basis vectors **a**, **b**, $\mathbf{c} = 2\mathbf{c}_0 + 4s\mathbf{a}$, space group $A2/m$ (structure type of pumpellyite, Fig. 1.37(b)).

MDO_3 occurs when $[-- m]$ mirrors in L_{2n} and $[-2_1-]$ axes in L_{2n+1} layers are constantly active: the generating operation is a glide normal to **a** with translational

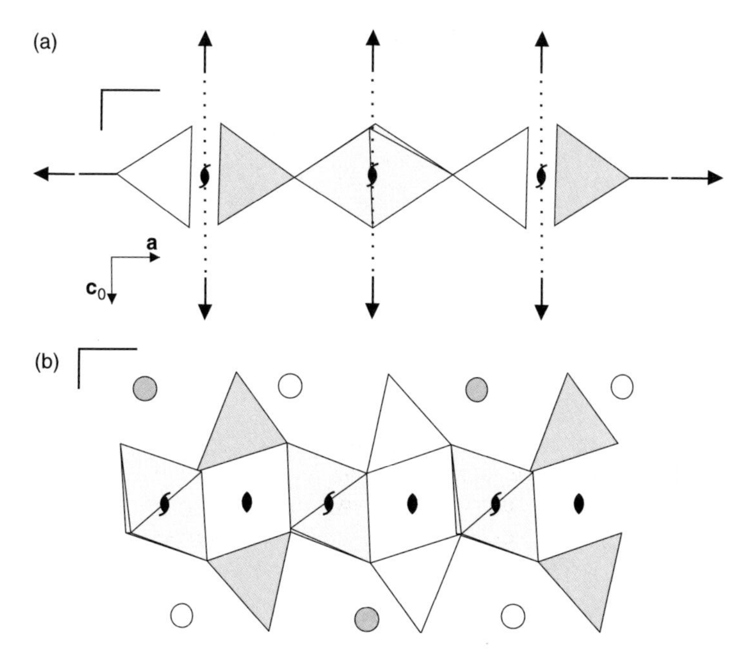

Fig. 2.46. The L_{2n} (a) and L_{2n+1} (b) OD layers in sursassite–pumpellyite–ardennite family, as seen along **b**. For each layer the kind and the position of the symmetry elements are represented. In both (a) and (b) white and dark grey tetrahedra correspond to silicon sites at $y = 3/4$ and $y = 1/4$, respectively. In (b) white and dark grey circles correspond to cation sites at $y = 3/4$ and $y = 1/4$, respectively.

component $\mathbf{b}/2+\mathbf{c}_0$; its continuation gives rise to an orthorhombic structure with basis vectors **a**, **b**, $\mathbf{c} = 2\mathbf{c}_0$ and space group *Pnmn* (structure type of ardennite, Fig. 1.37(c)).

MDO4 occurs when $[- - m]$ mirrors in L_{2n} and $[- 2 -]$ axes in L_{2n+1} layers are constantly active: the generating operation is a glide normal to **a** with translational component $\mathbf{c}_0$; its continuation gives rise to an orthorhombic structure with basis vectors **a**, **b**, $\mathbf{c} = 2\mathbf{c}_0$ and space group *Pcmn* (structure type of a new MDO modification, Fig. 1.37(d), not yet found so far).

2.6.5 *OD structures in spinelloids*

Polytypism in the family of spinelloids (the name indicates that the structure-type of spinel belongs to this family) has been discussed by various authors (Horiuchi *et al.* 1980; Hyde *et al.* 1982; Price *et al.* 1985) and spinelloid polytypes have been observed in several systems: $\mathbf{NiAl_2O_4}$–$\mathbf{Ni_2SiO_4}$ (Ma 1974; Ma *et al.* 1975; Ma and Tillmanns 1975; Ma and Sahl 1975; Horioka *et al.* 1981*a*,*b*), $\mathbf{MgGa_2O_4}$–$\mathbf{Mg_2GeO4}$ and $\mathbf{MgFe_2O_4}$–$\mathbf{Mg_2GeO_4}$ (Barbier 1989), $\mathbf{NiGa_2O_4}$–$\mathbf{Ni_2SiO_4}$ (Hammond and Barbier 1991), $\mathbf{Fe_3O_4}$–$\mathbf{Fe_2SiO_4}$ (Ross II *et al.* 1992), $\mathbf{Co_2SiO_4}$ (Akimoto and Sato 1968),

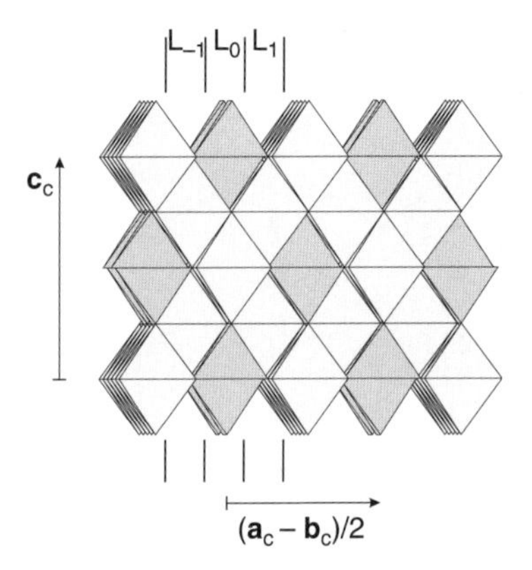

Fig. 2.47. The 'idealized' crystal structure of spinel as seen along [110] (with slight rotation about the vertical axis), with indication of the building layers.

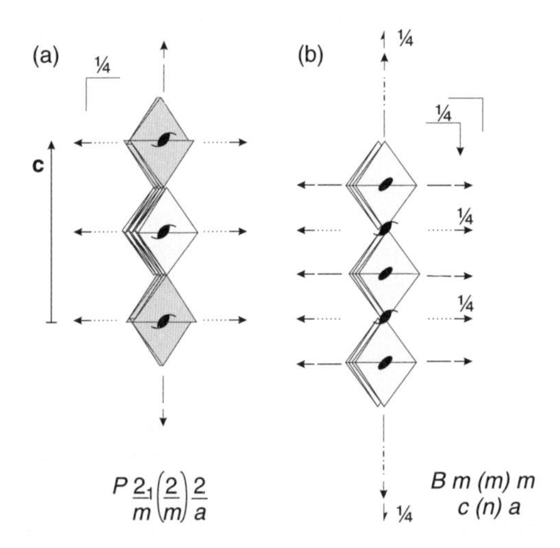

Fig. 2.48. The two OD layers building up the structures of spinelloids, as seen along **a** (with slight rotation about the vertical axis), with indication of the symmetry operations in each layer.

$\mathbf{Mg_2SiO_4}$ (Moore and Smith 1969, 1970; Ringwood and Mayor 1970), **manganostibite** (ideal composition Mn_7SbAsO_{12}; Moore 1970*b*). The OD approach may present a comprehensive and simple description of the structural features of this family.

The structure of spinel is represented in Fig. 2.47 as seen normal to the (110) plane. It may be described as a member of an OD family of structures built up with two distinct layers, L_0 and L_1.

A different choice of the layers, in which, while keeping the symmetry and metrics of the layers, the polyhedra around the cations are completed, is represented in Fig. 2.48. The common translation vectors are $\mathbf{a} = (\mathbf{a}_c + \mathbf{b}_c)/2$ and $\mathbf{c} = \mathbf{c}_c$,

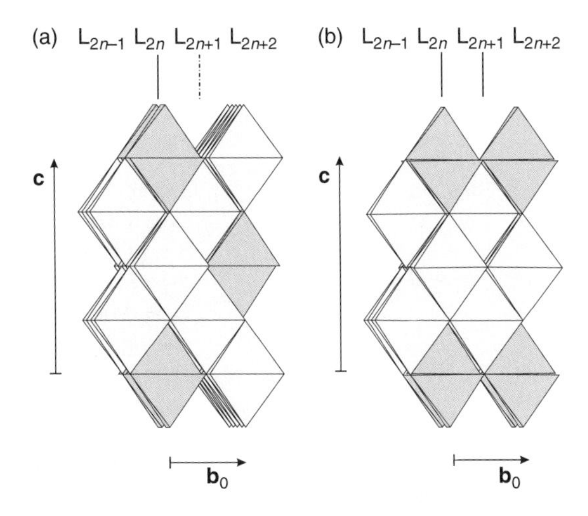

Fig. 2.49. A three-layer stacking $[L_{2n}; L_{2n+1}; L_{2n+2}]$, with symmetry operators [–*n*–] (a), or [–*m*–] (b) active in L_{2n+1} layer. A slight rotation about the vertical axis has been applied.

with $\mathbf{b}_0 = (\mathbf{a}_c - \mathbf{b}_c)/4$ (where the index c refers to the cubic cell of spinel and $a = 5.7, c = 8.1, b_0 = 2.85$ Å) and the OD-groupoid family symbol is:

$$\begin{array}{cc} L_{2n} & L_{2n+1} \\ P2_1/m(2/m)2/a & B2/m(2/m)2/m \\ \multicolumn{2}{c}{[1/4,\ 1/4]} \end{array}$$

Whereas the layers L_{2n-1} and L_{2n+1} on both sides of L_{2n} are always related by the symmetry plane [– *m* –], as well as by twofold axes parallel to **c**, by 2_1 axes parallel to **a** and by the inversion centres, there are two different ways to relate the layers L_{2n} and L_{2n+2} using the symmetry operators in L_{2n+1}: either through the glide [– *n* –], or the equivalent operators, as in Fig. 2.49(a), or through the mirror plan [– *m* –], or equivalent operators, as in Fig. 2.49(b).

Obviously, an infinite number of structures are possible in this family, corresponding to the regular alternation of layers L_0 and L_1 and differing for the sequence of operators *n* or *m* active in the layer L_1 (only the operator *m* being active in layer L_0). The structures will be disordered or ordered (polytypes), depending on the occurrence of random or ordered sequences of the operators *n* and *m* which are active in L_1. A similar analysis has been carried on by Hyde *et al.* (1982) who discussed the sequences of layers of type $[(L_{-1})/2\ L_0\ (L_1)/2]$, which may follow each other according to a stacking of spinel-type S, or of twin-type T, corresponding to *n* and *m*, respectively, in our analysis. Table 2.5 corresponds to the Table 1 presented by Hyde *et al.* (1982), reinterpreted in terms of the present analysis.

The structure-types I and III are represented in Figs 2.50 and 2.51, whereas the structure-type V is represented in Fig. 3.12; it is remarkable the high degree of **desymmetrization** which occurs in the OD layers of these structures, in accordance with our considerations in Section 2.5.4. The phases denoted *n* and *m* in Table 2.5, namely

Table 2.5 Sequences of layers, with indication of the active operators (n or m) in L_{2n+1} layers, in spinelloids. The structure types refer to the compounds in the $NiAl_2O_4$–Ni_2SiO_4 system

L_0	L_1	L_2	L_3	L_4	L_5	L_6	L_7	Symbol	Phase
m	$\boldsymbol{n}$	m	$\boldsymbol{n}$	m	$\boldsymbol{n}$	m	$\boldsymbol{n}$	$\boldsymbol{n}$	Spinel
m	$\boldsymbol{m}$	m	$\boldsymbol{n}$	m	$\boldsymbol{n}$	m	$\boldsymbol{m}$	$\boldsymbol{mn^2}$	Type V (Horioka *et al.*1981*a*; Ross II *et al.*1992)
m	$\boldsymbol{m}$	m	$\boldsymbol{m}$	m	$\boldsymbol{n}$	m	$\boldsymbol{n}$	$\boldsymbol{m^2n^2}$	Type I (Ma *et al.* 1975)
								$\boldsymbol{m^2nm^2n}$	Type II (Ma and Tillmanns 1974); manganostibite (Moore 1970*b*)
m	$\boldsymbol{n}$	m	$\boldsymbol{m}$	m	$\boldsymbol{n}$	m	$\boldsymbol{m}$	$\boldsymbol{mnmn}$	Type III (Ma and Sahl 1975); wadsleyite (Moore and Smith 1969, 1970; Ringwood and Mayor 1970)
								$(\boldsymbol{mn^2mn})^2$	Type IV (Horioka *et al.* 1981*b*)
m	$\boldsymbol{m}$	m	$\boldsymbol{m}$	m	$\boldsymbol{m}$	m	$\boldsymbol{m}$	$\boldsymbol{m}$	Structure ω

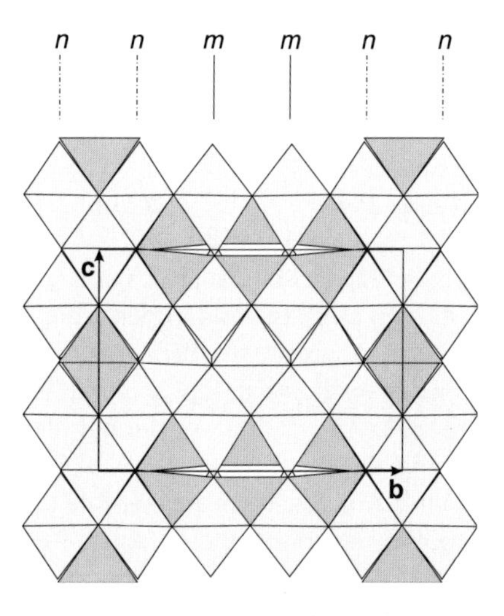

Fig. 2.50. The structure-type I in the system $NiAl_2O_4$–Ni_2SiO_4, as seen along **a**, with indication of the operators $[-\,n\,-]$ or $[-\,m\,-]$ active in the subsequent L_{2n+1} layers.

spinel and the unknown **phase** $\boldsymbol{\omega}$ (Fig. 2.52), correspond to the two MDO structures of this family.

The cell parameters and space group symmetries for any possible structure in the family may be easily derived from the OD-groupoid family symbol and the particular sequence of m and n operators in L_1 characteristic of that particular structure.

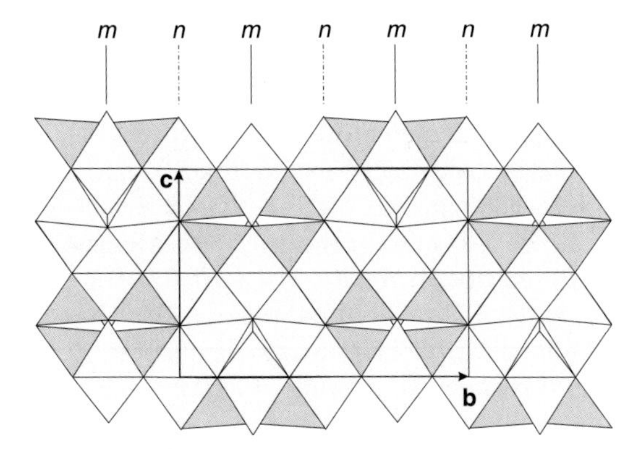

Fig. 2.51. The structure-type III in the system $NiAl_2O_4$–Ni_2SiO_4, as seen along **a**, with indication of the operators [– n –] or [– m –] active in the subsequent L_{2n+1} layers.

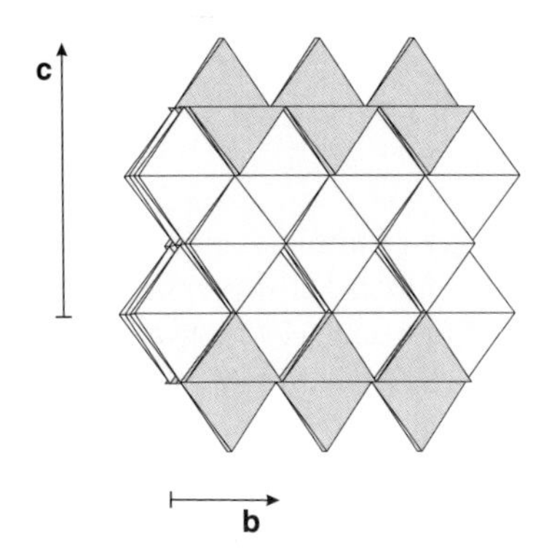

Fig. 2.52. The crystal structure of the hypothetical phase ω, as seen along **a** (with slight rotation about the vertical axis).

2.7 List of other OD structures

Examples of polytypic structures presented and discussed according to an OD approach have been listed by Dornberger Schiff (1964, 1979) and Ďurovič (1992). They include minerals, inorganic compounds, metal complexes, and organic compounds. We add here some further examples of OD structures consisting of equivalent layers or built up by two distinct OD layers.

OD structures consisting of equivalent layers — burpalite–låvenite, natural phases of the cuspidine group (Merlino *et al.* 1990*b*); actually OD relationships exist between the other pairs of structure-types in the cuspidine family, as described by Christiansen (2003); **fiedlerite**, $Pb_3Cl_4F(OH) \cdot H_2O$ (Merlino *et al.* 1994*a*); $\mathbf{Pb_2Fe^{3+}Cl_3(OH)_4 \cdot H_2O}$, a phase found in Baratti (Tuscany, Italy) formed by the action of sea water on ancient Etruscan metallurgical slags (Pasero *et al.* 1997); **hillebrandite**, $Ca_2SiO_3(OH)_2$ (Dai and Post 1995; Xu and Buseck 1996; Merlino 1997*a*); silver hydrogen sulphate, **Ag(O₃SOH)** (Belli Dell'Amico *et al.* 1998); **sapphirine**, **aenigmatite**, synthetic **SFCA** and other **aenigmatite-like compounds** (Dornberger-Schiff and Merlino 1974; Merlino 1990*a*; Merlino and Zvyagin 1998; Zvyagin and Merlino 2003); synthetic **calcium ferrites** $\mathbf{CaFe_3^{3+}AlO_7}$ and **SFCA-I** (Arakcheeva *et al.* 1991; Mumme *et al.* 1998; Zvyagin and Merlino 2003); **calcioaravaipaite**, $PbCa_2Al(F, OH)_9$ (Kampf *et al.* 2003); **brochantite**, $Cu_4SO_4(OH)_6$, (Cocco and Mazzi 1959; Merlino *et al.* 2003).

OD structures built up with two OD layers — lovdarite, $K_4Na_{12}(Be_8Si_{28}O_{72})\cdot 18H_2O$, a zeolite-like mineral (Merlino 1990*b*); synthetic porous **zincosilicate VPI-7**, $Na_4[Zn_2Si_7O_{18}]\cdot 5H_2O$ (Röhring *et al.* 1994); **zoisite –clinozoisite** (Merlino 1990*a*); **laurionite–paralaurionite**, PbCl(OH) (Merlino *et al.* 1993); **penkvilksite**, $Na_4Ti_2Si_8O_{22}\cdot 4H_2O$ (Merlino *et al.* 1994*b*); **surinamite**, $Mg_3Al_4Si_3BeO_{16}$, and the synthetic analogue $\mathbf{Mg_4Ga_4Ge_3O_{16}}$ (Moore and Araki 1983; Barbier 1998; Zvyagin and Merlino 2003); borates of the **kurchatovite group** (Yakubovich *et al.* 1976; Callegari *et al.* 2003; Belokoneva. 2003).

Finally, OD concepts have been fruitfully applied in discussing polytypy and twinning in complex sulphosalt structures, as **sinnerite** (Makovicky and Skinner 1975), **imhofite** (Balić-Žunić and Makovicky 1993), **owyheeite** (Makovicky *et al.* 1998), and **gillulyite** (Makovicky and Balić-Žunić 1999).

Appendix 2.1 Table of monoclinic and orthorhombic OD-groupoid families

$P\ 1\ 1\ (\bar{1})$ $\{1\ 1\ (2_2/n_{r,s})\}$	$P\ 1\ 1\ (m)$ $\{1\ 1\ (n_{r,s})\}$	$P\ 1\ 1\ (a)$ $\{1\ 1\ (n_{r-1,s})\}$	$P\ 1\ 1\ (2/m)$ $\{1\ 1\ (2_2/n_{r,s})\}$
$P\ 1\ 1\ (2/a)$ $\{1\ 1\ (2_2/n_{r-1,s})\}$	$P\ 1\ 1\ (2)$ $\{1\ 1\ (2_2)\}$ $[r/4\ s/4]$	$P\ 1\ 1\ (1)$ $\{1\ 1\ (n_{r,s})\}$ $\{1\ 1\ (n_{r',s'})\}$	$P\ 1\ 1\ (2)$ $\{1\ 1\ (n_{r,s})\}$ $\{1\ 1\ (n_{r',s'})\}$
$P\ 1\ 1\ \ (\bar{1})$ $\{1\ 2_s/c_2(1)\}$	$P\ 1\ 2\ (1)$ $\{1\ 2_s(1)\}$	$P\ 1\ 2_1\ \ (1)$ $\{1\ 2_{s-1}(1)\}$	$P\ 1\ 2/m\ (1)$ $\{1\ 2_s/c_2(1)\}$
$P\ 1\ 2_1/m\ \ (1)$ $\{1\ 2_{s-1}/c_2(1)\}$	$P\ 1\ 2/a\ (1)$ $\{1\ 2_s/c_2(1)\}$	$P\ 1\ 2_1/a\ \ (1)$ $\{1\ 2_{s-1}/c_2(1)\}$	$P\ 1\ m\ (1)$ $\{1\ c_2(1)\}$ $[s/4]$
$P\ 1\ a\ (1)$ $\{1\ c_2(1)\}$ $[s/4]$	$P\ 1\ 1\ (1)$ $\{1\ 2_s(1)\}$ $\{1\ 2_{s'}(1)\}$	$P\ 1\ m\ (1)$ $\{1\ 2_s(1)\}$ $\{1\ 2_{s'}(1)\}$	$P\ 1\ a\ (1)$ $\{1\ 2_s(1)\}$ $\{1\ 2_{s'}(1)\}$
$C\ 1\ 1\ \ (\bar{1})$ $\{1\ 2_s/c_2(1)\}$	$C\ 1\ 2\ (1)$ $\{1\ 2_s(1)\}$	$C\ 1\ 2/m\ (1)$ $\{1\ 2_s/c_2(1)\}$	$C\ 1\ m\ (1)$ $\{1\ c_2(1)\}$ $[s/4]$
$C\ 1\ 1\ (1)$ $\{1\ 2_s(1)\}$ $\{1\ 2_{s'}(1)\}$	$C\ 1\ m\ (1)$ $\{1\ 2_s(1)\}$ $\{1\ 2_{s'}\ (1)\}$		

$P\,1\quad 1\ (m)$ $\{n_{s,2}\ 2_s\ (1)\}$	$P\,1\quad 1\ (n)$ $\{n_{s,2}\ 2_{s-1}\ (1)\}$	$P\,1\quad 1\ (a)$ $\{n_{s,2}\ 2_s\ (1)\}$	$P\,1\quad 1\ (a)$ $\{2_{r-1}\ n_{2,r}\ (1)\}$
$P\,1\ \ 2\,(1)$ $\{2_r\ 1\ (2_2)\}$	$P\,1\ \ 2_1\,(1)$ $\{2_r\ 1\ (2_2)\}$	$P\,1\ \ 2\,(1)$ $\{n_{s,2}\ 1\ (n_{r,s})\}$	$P\,1\ \ 2_1\,(1)$ $\{n_{s-1,2}\ 1\ (n_{r,s})\}$
$P\,1\quad 1\quad (2/m)$ $\{2_r/n_{s,2}\ 2_s/n_{2,r}\ (1)\}$	$P\,1\quad 1\quad (2/a)$ $\{2_{r-1}/n_{s,2}\ 2_s/n_{2,r}\ (1)\}$	$P\,1\quad 1\quad (2/n)$ $\{2_{r-1}/n_{s,2}\ 2_{s-1}/n_{2,r}\ (1)\}$	
$P\,1\quad 2/m\,(1)$ $\{2_r/n_{s,2}\ 1\quad (2/n_{r,s})\}$	$P\,1\quad 2_1/m\,(1)$ $\{2_r/n_{s-1,2}\ 1\quad (2_2/n_{r,s})\}$	$P\,1\quad 2/a\,(1)$ $\{2_{r-1}/n_{s,2}\ 1\quad (2_2/n_{r,s})\}$	
$P\,1\quad 2_1/a\,(1)$ $\{2_{r-1}/n_{s-1,2}\ 1\quad (2_2/n_{r,s})\}$			
$P\,2\ \ 2\ (2)$ $\{2_r\ 2_s\ (2_2)\}$	$P\,2_1\ \ 2\ (2)$ $\{2_{r-1}\ 2_s\ (2_2)\}$	$P\,2_1\ \ 2_1\ (2)$ $\{2_{r-1}\ 2_{s-1}\ (2_2)\}$	$P\,2\ \ 2\ (2)$ $\{n_{s,2}\ n_{2,r}\ (n_{r,s})\}$
$P\,2_1\ \ 2\ (2)$ $\{n_{s,2}\ n_{2,r-1}\ (n_{r,s})\}$	$P\,2_1\ \ 2_1\ (2)$ $\{n_{s-1,2}\ n_{2,r-1}\ (n_{r,s})\}$	$P\,2\ \ m\ (m)$ $\{2_r\ n_{2,r}\ (n_{r,s})\}$	$P\,2_1\ \ a\ (m)$ $\{2_{r-1}\ n_{2,r}\ (n_{r,s})\}$
$P\,2_1\ \ m\ (a)$ $\{2_{r-1}\ n_{2,r}\ (n_{r-1,s})\}$	$P\,2\ \ a\ (a)$ $\{2_r\ n_{2,r-1}\ (n_{r-1,s})\}$	$P\,2\ \ m\ (b)$ $\{2_r\ n_{2,r}\ (n_{r,s-1})\}$	$P\,2_1\ \ a\ (b)$ $\{2_{r-1}\ n_{2,r-1}\ (n_{r,s-1})\}$
$P\,2\ \ a\ (n)$ $\{2_r\ n_{2,r-1}\ (n_{r-1,s-1})\}$	$P\,2_1\ \ m\ (n)$ $\{2_{r-1}\ n_{2,r}\ (n_{r-1,s-1})\}$	$P\,2\ \ m\ (m)$ $\{n_{s,2}\ 2_s\ (2_2)\}$	$P\,2_1\ \ a\ (m)$ $\{n_{s,2}\ 2_s\ (2_2)\}$
$P\,2_1\ \ m\ (a)$ $\{n_{s,2}\ 2_s\ (2_2)\}$	$P\,2\ \ a\ (a)$ $\{n_{s,2}\ 2_s\ (2_2)\}$	$P\,2\ \ m\ (b)$ $\{n_{s,2}\ 2_{s-1}\ (2_2)\}$	$P\,2_1\ \ a\ (b)$ $\{n_{s,2}\ 2_{s-1}\ (2_2)\}$
$P\,2\ \ a\ (n)$ $\{n_{s,2}\ 2_{s-1}\ (2_2)\}$	$P\,2_1\ \ m\ (n)$ $\{n_{s,2}\ 2_{s-1}\ (2_2)\}$	$P\,m\ \ m\ (m)$ $\{2_r/n_{s,2}\ 2_s/n_{2,r}\ (2_2/n_{r,s})\}$	

$P\,m\ \ a\ (a)$ $\{2_r/n_{s,2}\ 2_s/n_{2,r-1}\ (2_2/n_{r-1,s})\}$	$P\,b\ \ a\ (n)$ $\{2_r/n_{s-1,2}\ 2_s/n_{2,r-1}\ (2_2/n_{r-1,s-1})\}$	$P\,b\ \ m\ (m)$ $\{2_r/n_{s-1,2}\ 2_{s-1}/n_{2,r}\ (2_2/n_{r,s})\}$
$P\,m\ \ m\ (a)$ $\{2_{r-1}/n_{s-1,2}\ 2_s/n_{2,r}\ (2_2/n_{r-1,s})\}$	$P\,b\ \ m\ (n)$ $\{2_{r-1}/n_{s-1,2}\ 2_s/n_{2,r}\ (2_2/n_{r-1,s-1})\}$	$P\,b\ \ a\ (a)$ $\{2_r/n_{s-1,2}\ 2_{s-1}/n_{2,r-1}\ (2_2/n_{r-1,s})\}$
$P\,b\ \ a\ (m)$ $\{2_{r-1}/n_{s-1,2}\ 2_{s-1}/n_{2,r-1}\ (2_2/n_{r,s})\}$	$P\,m\ \ a\ (b)$ $\{2_{r-1}/n_{s,2}\ 2_{s-1}/n_{2,r}\ (2_2/n_{r,s-1})\}$	$P\,m\ \ m\ (n)$ $\{2_{r-1}/n_{s,2}\ 2_{s-1}/n_{2,r}\ (2_2/n_{r-1,s-1})\}$

$P\,1\ \ 1\ (2)$ $\{n_{s,2}\ n_{2,r}\ (1)\}$	$P\,1\ \ m\,(1)$ $\{n_{s,2}\ 1\ (2_2)\}$	$P\,1\ \ a\,(1)$ $\{n_{s,2}\ 1\ (2_2)\}$	$P\,m\ \ m\ (2)$ $\{n_{s,2}\ n_{2,r}\ (2_2)\}$
$P\,m\ \ a\ (2)$ $\{n_{s,2}\ n_{2,r-1}\ (2_2)\}$	$P\,b\ \ a\ (2)$ $\{n_{s-1,2}\ n_{2,r-1}\ (2_2)\}$	$P\,1\ \ 1\,(1)$ $\{1\ \ 1\ (n_{r,s})\}$ $\{2_{r'}\ 1\ (1)\}$	$P\,1\ \ 1\ (1)$ $\{2_r\ 1\ (1)\}$ $\{1\ \ 2_{s'}\ (1)\}$
$P\,1\ \ 1\ (2)$ $\{1\ \ 1\ (n_{r,s})\}$ $\{2_{r'}\ 2_{s'}\ (1)\}$	$P\,1\ \ 1\ (2)$ $\{2_r\ \ 2_s\ (1)\}$ $\{2_{r'}2_{s'}\,(1)\}$	$P\,1\ \ m\,(1)$ $\{2_r\ \ 1\ (n_{r,s})\}$ $\{2_{r'}\ 1\ (n_{r',s'})\}$	$P\,1\ \ a\,(1)$ $\{2_{r-1}\ \ 1\ (n_{r,s})\}$ $\{2_{r'-1}\ 1\ (n_{r',s'})\}$
$P\,1\ \ m\ (1)$ $\{1\ \ 2_s\ (1)\}$ $\{2_{r'}\ 1\ \ (n_{r',s'})\}$	$P\,1\ \ a\ (1)$ $\{1\ \ 2_s\ (1)\}$ $\{2_{r'-1}1\ \ (n_{r',s'})\}$	$P\,m\ \ m\ (2)$ $\{2_r\ \ 2_s\ (n_{r,s})\}$ $\{2_{r'}\ 2_{s'}\ (n_{r',s'})\}$	$P\,m\ \ a\ (2)$ $\{2_{r-1}\ \ 2_s\ (n_{r,s})\}$ $\{2_{r'-1}\ 2_{s'}\ (n_{r',s'})\}$
$P\,b\ \ a\ (2)$ $\{2_{r-1}\ 2_{s-1}\ (n_{r,s})\}$	$C\,1\ \ 1\ (m)$ $\{n_{s,2}\ 2_s\ (1)\}$	$C\,1\ \ 1\ (a)$ $\{n_{s,2}\ 2_s\ (1)\}$	$C\,1\ \ 1\ (d)$ $\{n_{s,2}\ 2_{s-1,2}\ (1)\}$
$C\,1\quad 1\quad (2/m)$ $\{2_r/n_{s,2}\ 2_s/n_{2,r}\ (1)\}$	$C\,1\quad 1\quad (2/a)$ $\{2_r/n_{s,2}\ 2_s/n_{2,r}\ (1)\}$	$C\,1\quad 1\quad (2/d)$ $\{2_{r-1}/n_{s,2}\ 2_{s-1}/n_{2,r}\ (1)\}$	$C\,1\ \ 2\,(1)$ $\{2_r\ 1\ (2_2)\}$
$C\,1\ \ 1\,(1)$ $\{n_{s,2}\ 1\ (n_{r,s})\}$	$C\,1\quad 2/m\,(1)$ $\{2_r/n_{s,2}\ 1\quad (2_2/n_{r,s})\}$	$C\,m\ \ 2\ (m)$ $\{n_{s,2}\ 2_s\ (n_{r,s})\}$	$C\,m\ \ 2\ (a)$ $\{n_{s,2}\ 2_s\ (n_{r-1,s})\}$
$C\,m\ \ 2\ (m)$ $\{2_r\ n_{2,r}\ (2_2)\}$	$C\,m\ \ 2\ (a)$ $\{2_r\ n_{2,r}\ (2_2)\}$	$C\,2\ \ 2\ (2)$ $\{2_r\ 2_s\ (2_2)\}$	$C\,2\ \ 2\ (2)$ $\{n_{s,2}\ n_{2,r}\ (n_{r,s})\}$

$C\,m \quad m \quad (m)$ $\{2_r/n_{s,2}\ 2_s/n_{2,r}\ (2_2/n_{r,s})\}$	$C\,m \quad m \quad (a)$ $\{2_r/n_{s,2}\ 2_s/n_{2,r}\ (2_2/n_{r-1,s})\}$	$C\,1 \quad 1 \quad (2)$ $\{n_{s,2}\ n_{2,r}\ (1)\}$	$C\,1 \quad m\,(1)$ $\{n_{s,2}\ 1\ (2_2)\}$
$C\,m \quad m \quad (2)$ $\{n_{s,2}\ n_{2,r}\ (2_2)\}$	$C\,1 \quad 1\,(1)$ $\{1 \quad 1\,(n_{r,s})\}$ $\{2_{r'}\,1\,(1)\}$	$C\,1 \quad 1 \quad (1)$ $\{2_r\ 1 \quad (1)\}$ $\{1\ 2_{s'}\,(1)\}$	$C\,1 \quad 1 \quad (2)$ $\{2_r\ 2_s\ (1)\}$ $\{2_{r'}2_{s'}\,(1)\}$
$C\,1 \quad 1 \quad (2)$ $\{1 \quad 1 \quad (n_{r,s})\}$ $\{2_{r'}2_{s'}\,(1)\}$	$C\,1 \quad m\,(1)$ $\{2_r \quad 1 \quad (n_{r,s})\}$ $\{2_{r'}\,1 \quad (n_{r',s'})\}$	$C\,1 \quad m \quad (1)$ $\{1 \quad 2_s\ (1)\}$ $\{2_{r'}\,1 \quad (n_{r',s'})\}$	$C\,m \quad m \quad (2)$ $\{2_r \quad 2_s\ (n_{r,s})\}$ $\{2_{r'}2_{s'}\,(n_{r',s'})\}$

Appendix 2.2 Diffraction effects in wollastonite

The Fourier transform of the whole structure may be obtained summing the contributions of the single layers

$$F(\mathbf{r}^*) = \Sigma_p\, F_p(\mathbf{r}^*).$$

The Fourier transform can be different from zero only in points ηkl of the reciprocal space, if the layers are periodic with translation vectors **b**, **c**. The succession of the layers may be periodic or aperiodic: therefore, the Fourier transform may be different from zero for discrete value of η or for any value of η. Consequently, we shall obtain either diffraction patterns with only sharp spots or diffraction patterns with diffuse streaks; moreover, as a disordered structure may contain ordered domains, we may observe sharp maxima placed on the diffuse streaks. For a correct interpretation of the diffraction pattern of an OD structure, a fundamental role is played by those reflections which are independent from the kind of sequence and are, therefore, indicated as 'family reflections'.

As we have previously shown, in wollastonite all the layers are translationally equivalent and the translation vector which connects succeeding layers is

$$\mathbf{t}_q = \mathbf{a}/2 + (\beta_q/4)\mathbf{b}$$

with $\beta_q = \pm 1$ and $\mathbf{a} = 2\mathbf{a}_0$ corresponding to the unit cell parameter of wollastonite $2M$.

The position of the layer L_p, relative to the layer L_0, may be given as:

$$\mathrm{L}_p = \mathrm{L}_{p-1} + (1/2, \beta_q/4, 0) = \mathrm{L}_0 + [p/2, (\sum_q \beta_q), 0]$$

where the terms in parentheses give the coefficients of **a**, **b**, and **c**, respectively, in the translation vector.

The sum $\sum_q \beta_q$ is even or odd according to whether p is even or odd; therefore it may be substituted with

$$\sum_q \beta_q = 2m_p + p \tag{A2.2.1}$$

where m_p are integer numbers, depending on the kind of succession

$$\mathrm{L}_p = \mathrm{L}_0 + (p/2, m_p/2 + p/4, 0).$$

Therefore, the Fourier transform of the layer L_p is related to the Fourier transform of L_0 by the relation:

$$F_p(\eta kl) = F_0(\eta kl)\exp\{2\pi\mathrm{i}[\eta p/2 + km_p/2 + kp/4]\}$$

and the Fourier transform of the structure is

$$F_p(\eta kl) = \sum_p F_p(\eta kl) = S(\eta kl)F_0(\eta kl) \qquad \text{(A2.2.2)}$$

with

$$\mathrm{S}(\eta kl) = \sum_p \exp\{2\pi\mathrm{i}[(\eta/2 + k/4)p + km_p/2]\}. \qquad \text{(A2.2.3)}$$

For $k = 2n$, the expression of $\mathrm{S}(\eta kl)$ simply becomes

$$\mathrm{S}(\eta kl) = \sum_p \exp\{2\pi\mathrm{i}[(2\eta + k)/4]p\}.$$

With a large number of layers, this expression vanishes except for integral values of $(2\eta + k)/4$, that is, for integral values h of η, for which

$$2h + k = 4n. \qquad \text{(A2.2.4)}$$

Therefore, the diffraction patters of any polytype or disordered structure in the family will present, as a common feature, sharp reflections for $k = 2n$, and these reflections will fulfil the rule (A2.2.4). Moreover, as the single layer displays monoclinic symmetry and $\mathrm{S}(hkl)$ has the same value for this set of reflections, the whole set will display monoclinic symmetry, in keeping with the observations of the various authors, as we have previously recalled (partial enhancement of symmetry). They are what we defined 'family reflections' and correspond to a reciprocal lattice with vectors $\mathbf{A}^*$ $\mathbf{B}^*$ $\mathbf{C}^*$ related to the vectors $\mathbf{a}^*$ $\mathbf{b}^*$ $\mathbf{c}^*$ as here indicated: $\mathbf{A}^* = \mathbf{a}^*$, $\mathbf{B}^* = 2\mathbf{b}^*$, and $\mathbf{C}^* = \mathbf{c}^*$. Consequently the relationships between the direct vectors are:

$$\mathbf{A} = \mathbf{a} = 2\mathbf{a}_0$$
$$\mathbf{B} = \mathbf{b}/2$$
$$\mathbf{C} = \mathbf{c}.$$

The diffraction pattern of the various polytypes (those diffraction patterns differ only in reflections for which $k = 2n + 1$) will be obtained from the expressions (A2.2.2) and (A2.2.3), on the basis of the actual sequence of layers. We shall here consider only the two MDO structures in the family.

For the polytype MDO_2, the layers regularly follow each other with alternation of $\beta_q = +1$ and $\beta_q = -1$. Therefore, after a sequence of $p = 2n$ layers, $\sum_q \beta_q = 0$, which requires $2m_p = -p$ in (A2.2.1). Substituting $m_p = -p/2$ in (A2.2.3) we obtain:

$$S(\eta kl) = \sum_p \exp[2\pi i(\eta/2)p].$$

With a large number $p(= 2n)$ of layers, this expression vanishes except for integral values h of η.

For the polytype MDO_1 the layers regularly follow each other with $\beta_q = -1$. With a sequence of p layers $\sum_q \beta_q = -p$, which requires $m_p = -p$ in (A2.2.1). The expression (A2.2.3) is now:

$$S(\eta kl) = \sum_p \exp[2\pi i(\eta/2 - k/4)p].$$

With a large number p of layers this expression vanishes except for half integer values h' of η, for which $2h' - k = 4n$. Therefore for $k = 4n + 1$ there are reflections at $h' = 2m + 1/2$; for $k = 4n + 3$ there are reflections at $h' = 2m - 1/2$.

It is easy to show that in the case of the polytype MDO'_1 (the twin structure), in which the layers regularly follow each other with $\beta_q = +1$, the condition becomes: $2h' + k = 4n$. Therefore for $k = 4n + 1$ there are reflections at $h' = 2m - 1/2$; for $k = 4n + 3$ there are reflections at $h' = 2m + 1/2$.

A beautiful comparison of the diffraction patterns of polytypes $1A(MDO_1)$ and $2M(MDO_2)$, twinned crystal ($MDO_1 + MDO'_1$) and crystal with evidence of disorder is presented by Müller (1976) for the isostructural mineral **pectolite**.

3

Polytypes and polytype categories

3.1 Introduction

According to a number of authors, **polytypes** belong to modular structures (Krypyakevich and Gladyshevskii 1972; Thompson 1978; Parthé *et al.* 1985; Angel 1986; Zvyagin 1987, 1993). Here, modules (layers, rarely rods or blocks) are of one or two (in rare cases even more) distinct kinds in strictly regular alternation (Ďurovič 1992). In ideal polytypes, this results in constant chemical composition of different polytypes composed of the same kinds of layers or, in more relaxed classifications, in constant element proportions. The well-defined geometric relations on the interfaces of two adjacent layers do not determine an unambiguously defined three-dimensional structure, that is, a unique stacking of moduli.

3.1.1 *Configurational (= heterochemical) polytypic series*

The strict definition of polytypy stresses the chemical (near-)identity of different polytypic modifications. Everyday usage exceeds these strict criteria to a different degree, from small compositional differences among different polytypes of 'one compound' to the polytype 'families' based on the same layer types and stacking principles but comprising a broad spectrum of chemical compositions based on elements with similar bonding properties. Thus, polytypes can be defined either on a crystal-chemical or on a configurational level. The former gives isochemical series of polytypes or series of crystal-chemically analogous compounds with certain, relatively small compositional differences, (which were called **polytypoids** in the fields of micas and pyriboles by Bailey (1984) and Guinier *et al.* (1984)). Polytypes with radically different chemical compositions but with the same layer configurations and layer-stacking principles form configurational (= heterochemical) polytypic series (Makovicky 1997*c*). A particular example of configurational polytypes and of the semantic problems involved is the **αTlI–FeB series** (Grin' 1992) encompassing 11 intermediate members (mostly alloys of Ni or Cu with other elements). On the one hand, it was described as ordered combinations of h and c stacking (i.e. as stacking variants or polytypes) ignoring chemical differences (Parthé 1976, 1990), and on the other as a unit cell intergrowth series (so-called 'linear inhomogeneous series') of αTlI (CrB), Pt_2U, and FeB structure types (Grin' 1992) (Fig. 3.1). A similar example of mineralogical and geological significance are the compositionally diverse spinelloids that are dealt with in several sections of this volume. Let us examine still another example of configurational polytypes, the series of hexagonal perovskites in more detail.

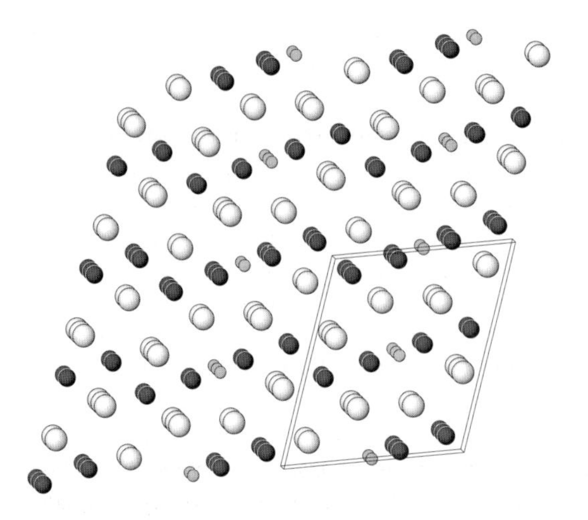

Fig. 3.1. The crystal structure of Y_4Co_4Ga (Grin' 1992). The prominent sequence of Co atoms (dark) alludes to the polytypic character of the structure whereas the distribution of Y and Ga (the latter interstitial to the polytypic stacking; small circles) is better described as an intergrowth of one period of the 'Ga-stuffed' Pt_2U-like structure (around $y = 1/2$) with three periods of α-TlI-like structure (centred on $y = 0$).

3.1.2 *Hexagonal perovskites and problems of polytype nomenclature*

Hexagonal perovskites display hexagonal close packing of AX_3 layers (A = large cation with 12-fold coordination) and small B cations in octahedral interstices as BX_6 polyhedra.

In pure hexagonal stacking, denoted as polytype 2H, BX_6 octahedra share faces (Fig. 4.2), unlike those in a pure cubic packing (3C), where they only share edges. A number of mixed stackings are known, such as 4H, 6H, 8H, 10H, 9R, 12R, 15R and 27R (Mitchell 2002) in which the two stacking principles combine in different proportions and sequences. These structures are generally known under the name 'hexagonal perovskite polytypes' and no questions are posed about their polytypic nature. A closer look reveals a number of problems and illustrates well the broad definition and understanding of polytypism in current practice.

There are no definition problems with polytypism of pure AX_3 layers without octahedral infilling: these are OD-polytypes. If, on introducing octahedrally coordinated B cations (as a single atom type), we define unit layers in the crystal-chemically most natural way, as octahedral layers, the ccp-built c and the hcp-obeying h layer pairs are non-equivalent. Thus, on this definition level, the studied polytypes are not OD-polytypes.

In order to convert them into OD-polytypes, we have to define them as polytypes composed of two types of unit layers: (1) the AX_3 layers and (2) the 'naked' B cation layers. Although this description is fully justified from the point of view of geometry and symmetry, it is rather far from giving a clear and useful geometric picture of the structure. We feel that the OD description loses much of its descriptional advantage/value, once the unit OD layers become separate layers of 'naked' cations and anions.

On co-ordination grounds, the family of hexagonal perovskite polytypes gives two extensive 'homologous' series (Mitchell 2002). The first one is based on a pair of face-sharing octahedra (an element of h stacking), separated in consecutive members of the series by zero, one, two or three octahedral layers thick sequences with c stacking (Fig. 3.2). The other one has a single octahedron thick layer that is bounded by c contacts; this layer interleaves with one, two or even four (above mentioned) h layer pairs separated by c contacts (Fig. 3.3). Yet another series has triple octahedral layers with mutual h contacts (face sharing of octahedra), separated by a 0, 1, . . . , octahedron thick layer situated between contacts.

Mitchell (2002) stresses that the pairs of octahedra related via h-type stacking contain highly charged cations that form metal–metal bonds (such as Ni, Ru, Nb, Ta) or tolerate close contacts like Ti or, alternatively, contain a combination of a highly charged and a low-charged cation such as W–Li. The c-based sequences contain preferably other cations than those above or diverse cations are ordered in

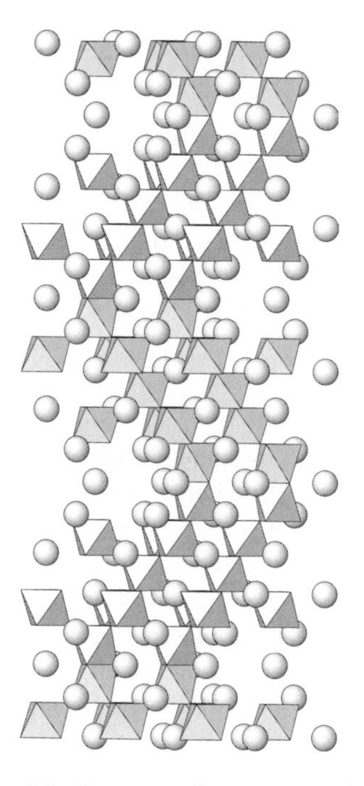

Fig. 3.2. The crystal structure of the 10H polytype of $Ba_{10}Ta_{7.04}Ti_{1.2}O_{30}$ (Shpanchenko *et al.* 1995).

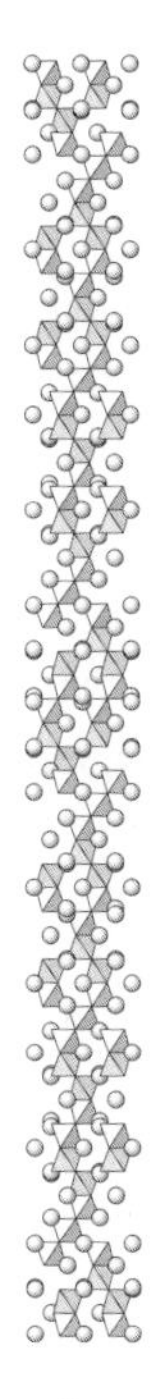

Fig. 3.3. The crystal structure of the 27R polytype of $BaCrO_3$ (Haradem *et al.* 1980).

them into distinct layers of the c pack—thus the layers become chemically distinct. Both the modular and occupational features of these compounds point towards true homologous series rather than to the standing polytype definitions. Their classification as 'polytypes' is essentially based on the simple stacking principles and (especially) on the geometric similarity of the configurations observed in the boundary— and in the accreting layer of these structures. Constancy of overall stoichiometry has been an added reason for classifying them as 'polytypes'. In agreement with Mitchell (2002), the compounds of the first mentioned series (Fig. 3.2) can be considered as example of unit cell twinning applied with different frequency to the structure of cubic perovskite. A similar situation occurs in the long-period (non-OD) micas.

Last, but not least, distinct members of these polytype series have distinct, often crystal-chemically strictly determined compositions and some compositions cannot occur as some of the polytypic sequences. For example, 2H polytype is known for $BaRuO_3$ (Hong and Sleight 1997), $BaTiO_3$ and a number of fluorides; 8H and 10H (Fig. 3.2) occur for the Ti–Ta mixture but they differ in the Ta : Ti ratio ($Ba_8Ta_4Ti_3$ (vacancy) O_{24} and $\mathbf{Ba_{10}Ta_{7.8}Ti_{1.2}O_{30}}$, respectively; Shpanchenko *et al.* 1995). The 9R polytype is $BaRuO_3$ (Donohue *et al.* 1965), 12R is $Ba_{12}Ir_{9.5}Nb_{2.3}O_{36}$ (Wilkens and Müller-Buschbaum 1991) whereas 27R is $\mathbf{BaCrO_3}$ (Haradem *et al.* 1980) (Fig. 3.3). What do they have in common is that all are formed with those B metals that allow close metal–metal contacts. Thus, these compounds do not satisfy the compositional part of the narrow polytype definition (Guinier *et al.* 1984), they are configurational polytypes in which configurations are preserved but cations and even anions may differ from a member to a member of the series.

Relativity of nomenclature and clash between practice and the strict commission definitions are not limited to this series. Especially the alloy literature abounds with such examples (Parthé 1976) and these interpretation problems should be kept in mind when defining any new series.

3.1.3 *Polytype categories*

Most of the polytype descriptions concentrate upon possible layer or block reflections, rotations and shifts. Combination of this approach with symmetry considerations was presented by Ito (1950). A nice example of these procedures is the detailed treatment of the **pinakiolite series** by Takéuchi (1997). Crystal-chemical point of view is less frequent. Calculations of relative stability of distinct polytypes (e.g. Angel *et al.* 1985) belong here. Analysis of layer rotations and shifts is in agreement with the frequent occurrence of excessively symmetrical configurations and/or subperiodicities (such as 1 : 2, 2 : 3, 3 : 5, etc.) on the layer surfaces. A combination of symmetry and geometry with a crystal-chemical approach will be used in this section.

In the broadest definition (Guinier *et al.* 1984), polytypes of one family differ in the stacking sequences of layers which have (nearly) identical structures and compositions. Significantly, differences in **stacking principles** are not mentioned, that is, not considered critical by this broad IUCr definition. This conforms with the viewpoints of Thompson (1981), Angel (1986), and others. In the definition forwarded by Ďurovič (1992), a polytypic family must not only satisfy the same structural principles but also the same symmetry principle—polytypes of one family must belong to the same OD-groupoid family (Dornberger-Schiff 1964; Fichtner 1979; Ďurovič 1992). This calls for a single type (Merlino 1990) of geometric configuration on each type of layer interface and identity of all layer pairs (of one or just a few kinds) in the structure (Dornberger-Schiff 1964). A strict definition appears to counter the general usage of the term and instead of trying to restrict the definition, we propose the following subdivision of polytypes.

3.1.3.1 *OD-polytypes*

OD-polytypes are a family of OD structures belonging to one groupoid family (Dornberger-Schiff 1966; Ďurovič 1981; Merlino 1990). These structures have one, two (or more) kinds of unit layers, and only one kind of layer pairs for each adjacent-layer combination present. First coordination spheres of atoms (sometimes those of vacant polyhedra as, e.g. in ccp and hcp metals) on the layer boundaries are preserved from one polytype to another. Polytypes become OD-polytypes only when the appropriate number and types of layers are defined. Exhaustive description of OD-polytypes was given in Chapter 2; here an example will be given in order to be compared with the next categories.

Oxyborates of the **pinakiolite–orthopinakiolite family** M_3BO_5 (M = Mg, di- and trivalent Mn and Fe) are typical polytypes (polytypoids) that can be described using one kind or two kinds of layers (Fig. 3.4). For only one kind of unit layers, the layer symmetry is $P(1)12/m$, with alternating local vertical reflection planes and twofold axes on layer interfaces (Fig. 3.4). They result in two kinds of layer pairs, represented by the 'S on S' and 'I on I' sequences. This family becomes a family of OD-polytypes when two kinds of alternating unit layers are selected, each one octahedron thick (Fig. 3.4). Their symmetries are $P(2/b)2_1/m2/m$ and $P(1)12/m$, respectively. Two equivalent positions of the latter layer after the former one are a cause of the OD character of the structure. Exceptionally, these polytypes received distinct mineral names, also due to their in part heterochemical character (Takéuchi 1997).

3.1.3.2 *Non-OD-polytypes*

Non-OD-polytypes cannot be described as a family of OD structures. Non-equivalent stacking principles of the selected kind of layers are active in different end-members or they combine in the observed sequences of layers (i.e. more than one kind of layer pairs are present for given type(s) of unit layers); the same holds true for possible interlayer symmetry operators. For example, these structures may have layers of one kind but with (at least) two distinct configurations on interlayer contacts; these distinct types of layer contacts can occur as pure sequences. Therefore, they do not represent the case of a single polar layer with reversals after every layer and with two kinds

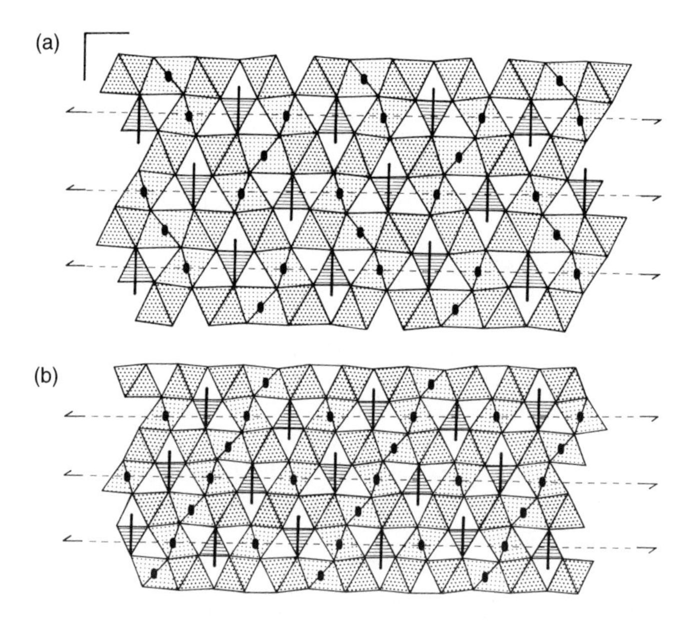

Fig. 3.4. Fredrickssonite (a) and pinakiolite, (b) two boundary structures of the $Me_3(BO_3)O_2$ oxyborate family (Me = Mg, Mn, Fe) (Moore and Araki 1974; Burns *et al.* 1994). OD-polytypes composed of two types of unit layers with layer symmetry, $P(1)12m$ and $P(2/b)2_1/m2/m$, respectively.

of OD interfaces, which occur in strictly regular alternation, these being one of the families of proper polytypes (Section 2.3.2).

Distinct types of stacking result in different first coordination spheres for boundary cations (or for boundary vacancies). Typical examples are the pairs **bayerite–gibbsite** for the case of $Al(OH)_3$ and **kaolinite/dickite** versus **nacrite**; a more complex example are borates **pringleite** and **ruitenbergite** (Grice *et al.* 1994) (Fig. 3.5), with two characteristic configurations ($\underline{X}$ and $\underline{Y}$) of the borate framework; they meet either in the X–Y configuration (pringleite) or in the X–X and Y–Y configurations (ruitenbergite).

Non-OD polytypy may be tied to a particular **choice of unit layers**. As already mentioned, non-OD-polytypes (e.g. the latter example) can presumably be converted into OD-polytypes by increasing the number of unit layers from one to two or more. While this procedure, expresses well the geometry of the structure (see Section 2.5.4), it may lose its crystallographic value when it becomes necessary to slice the structure into separate cationic and anionic layers, destroying the coordination polyhedra in the process. More so, when (some of) these layers are not 'selfsupporting' arrays (e.g. close-packed layers with ions at more or less van der Waals distances) but loose configurations. Thus, for example, although it is possible to describe MoS_2 with its trigonal prismatic coordination of Mo and SnS_2 with octahedral coordination of Sn as two OD-polytypes with two kinds of layers, such a description is more of a stacking

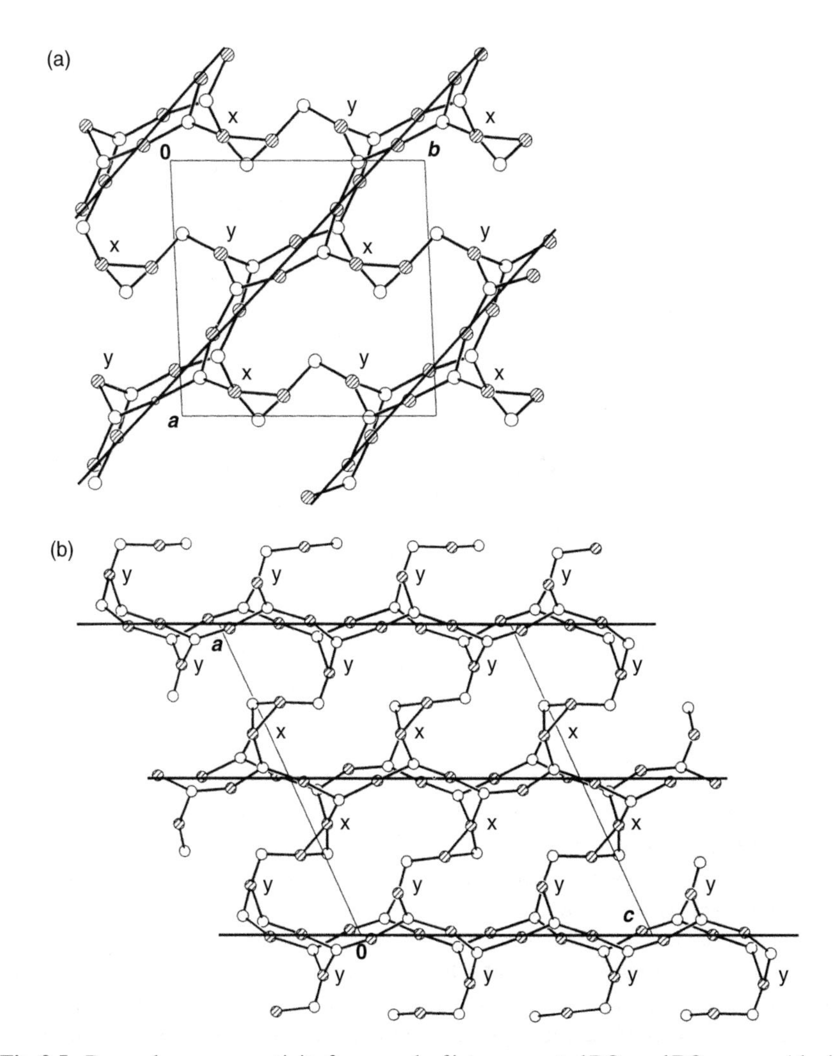

Fig. 3.5. Boron–boron connectivity framework of interconnected BO_4 and BO_3 groups (shaded and void circles, respectively) in the crystal structures of $Ca_9B_{26}O_{34}(OH)_{24}Cl_4 \cdot 13H_2O$ dimorphs: pringleite (a) (space group $P1$) and ruitenbergite (b) (space group $P2_1$). x and y denote characteristic configurations at the boundaries of unit layers (110) of pringleite/(100) of ruitenbergite.

formula than a penetrating structure analysis. A good example of this procedure is the family of LSeF polytypes (L = Y, Do, Er, ...) (Nguyen and Laruelle 1977, 1980). Dividing them into pure Y nets and pure 'Se and 2F' nets, with two possible positions of the latter after the former, describes the polytypy but hides the fact that the Y atoms change their coordinations (from CN = 7 to CN = 6 and 8) according to the position of the following anionic net.

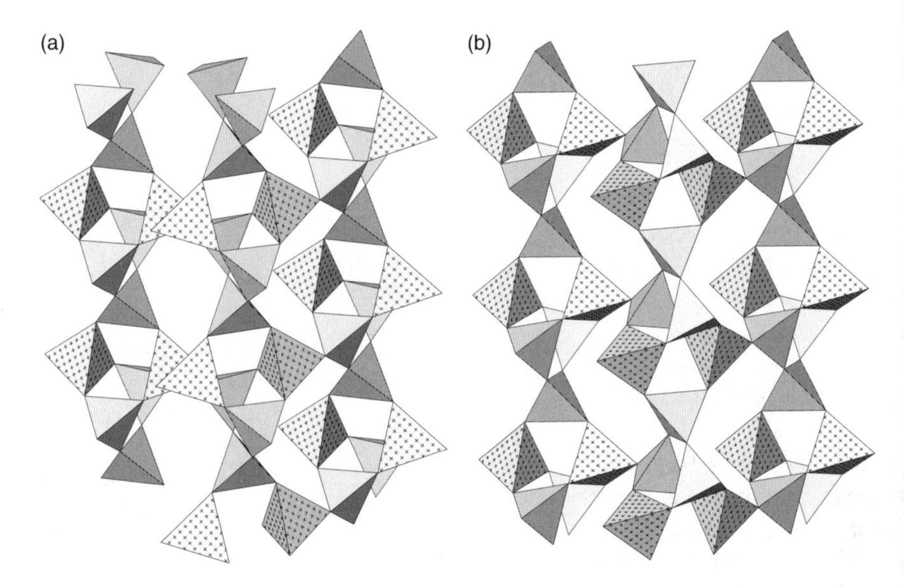

Fig. 3.6. Tetrahedral framework in the crystal structures of (a) natrolite $Na_4Al_4Si_6O_{20}{\cdot}4H_2O$ (antiparallel 'natrolite-like' chains) and (b) edingtonite $Ba_2Al_4Si_6O_{20}{\cdot}8H_2O$ (parallel chains).

Care should be excercised when, as in some **phyllosilicates**, several families of OD-polytypes occur for one ideal composition and layer type, related mutually as non-OD polytypes. Although it may be possible to convert the non-OD polytypy into an OD one by further subdivision of layers, the presence of several separate OD families on a certain level of slicing is an inherent property of these compounds and should not be obscured by the quest for universality.

The structures of the fibrous zeolites of **natrolite group** (Alberti and Gottardi 1975) consist of parallel silicate chains composed of repeating $(Si,Al)_5O_{10}$ units (Meier *et al.* 1996); those chains are loosely interconnected via free tetrahedral vertices and enclose channels with extra-framework cations and H_2O molecules. Considering any pair of chains, two modes of interconnection are possible (Fig. 3.6); in a given direction, the interconnecting tetrahedra can be surmounted by a single tetrahedron of a chain or, alternatively, by a pair of them, connected to other neighbouring chains. In **natrolite–mesolite–scolecite series** $Na_{16}Al_{16}Si_{24}O_{80} \cdot 16H_2O$–$Ca_8Al_{16}Si_{24}O_{80} \cdot 24H_2O$, any two adjacent chains have opposite orientation; in edingtonite $Ba_2[Al_4Si_6O_{20}] \cdot 8H_2O$ (Mazzi *et al.* 1984), they have parallel orientation. In thomsonite $Na_4Ca_8[Al_{20}Si_{20}O_{80}] \cdot 24H_2O$, these two principles alternate in a layer-like fashion. The two arrangements modify the channels, resulting in different coordination sites.

Therefore, the natrolite group are configurational non-OD-polytypes based on **rods** rather than layers, with altered coordinations for cations and H_2O molecules on rod interfaces.

3.1.3.3 *Proper and improper polytypes*

In agreement with the late Dr. Zvyagin, in the following we shall consider OD- and non-OD polytypes as **'proper polytypes'**, when they have configurationally unmodified layers, whereas the polytype-like cases (commonly described as polytypes in literature), in which configurational modifications occur, will be put in the **'improper polytype'** category.

The best example of improper polytypes is the pair **götzenite–rinkite** (Christiansen and Rønsbo 2000). **Götzenite** $Ca_2Na_2CaTiCa_2Si_2O_7(OH,F)_4$ is a heterophyllosilicate (for definition see Section 4.3.2) with Si_2O_7 groups that line octahedral layers being separated by diagonal walls of Ca octahedra; **rinkite** $(Ca,Ce)_2Na_2Ca(Ti,Nb)\,(Ca,REE)_2(Si_2O_7)_2(O,OH,F)_4$ has walls composed of seven-coordinated (Ca,Ce) polyhedra, due to mutual displacement of octahedral layers. In götzenite, all octahedral sheets are oriented in the same way, whereas in rinkite the adjacent sheets are oriented in the opposite way. Thus, these two structures should be two OD-polytypes with two types of layers, but the change in the coordination of Ca(Ce) polyhedra caused by the mutual shift in the position of T–O–T layers make them a good example for a pair of improper polytypes (Fig. 3.7(a, b)). Detailed treatment of the OD character of these structure types is in Section 2.4.3.

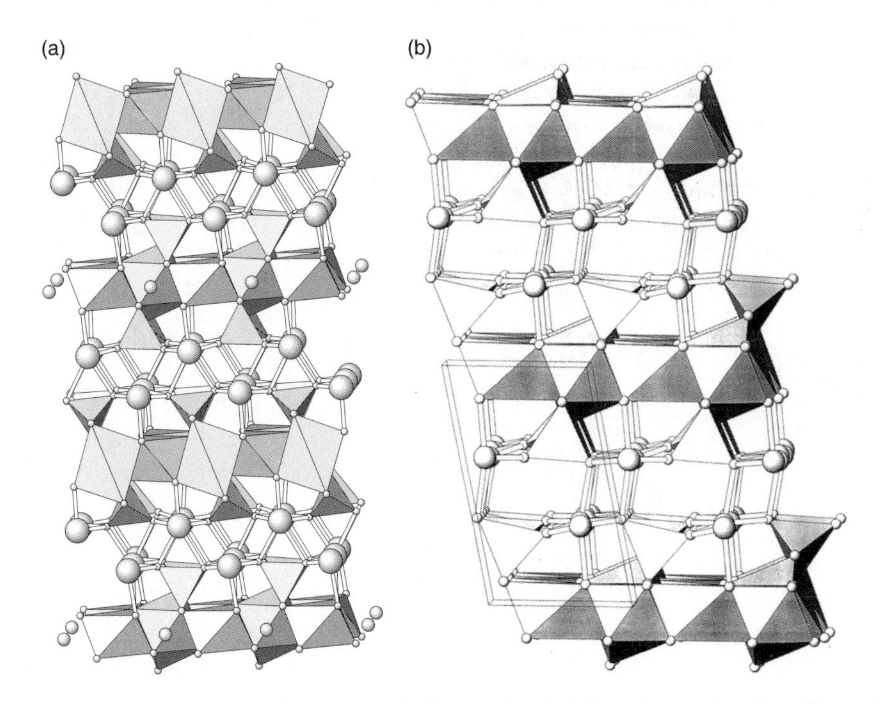

Fig. 3.7. Crystal structures of rinkite (a) (Galli and Alberti 1971) and götzenite (Cannillo *et al.* 1972), a pair of improper polytypes from the heterophyllosilicate family. Octahedral layers are decorated by Si_2O_7 groups and interconnected by coordination polyhedra of Ca/REE with respectively CN = 6 and CN = 7 (Christiansen and Rønsbo 2000).

Another example is the pair **ekanite–steacyite** described below, in which one member has boundary sites filled by large cations whereas in the other these sites are vacant (Szymanski *et al.* 1982).

3.1.3.4 *Pseudopolytypes*

Those cases of proper or improper polytypes, composed of crystal-chemically or only configurationally identical layers with different stacking modes, in which different 'polytypes' create different secondary and tertiary configurations in the process of stacking (e.g. different types of strongly bonded chains which transcend the boundaries between unit layers) are in a way at the opposite limit of polytype definition. For these, the term **pseudopolytype** is proposed here.

Typical example is the pair **stibnite** (Sb_2S_3)– **pääkkönenite** (Sb_2S_2As) (Bonazzi *et al.* 1995) (Fig. 3.8). Unit layers contain both Sb_4X_6 ribbons and the weak inter-ribbon interactions. Layer symmetry being in both cases $P(1)12_1/m$ (Bonazzi *et al.* 1995), interlayer operators are two-fold axes in Sb_2S_2As. This creates tightly bonded

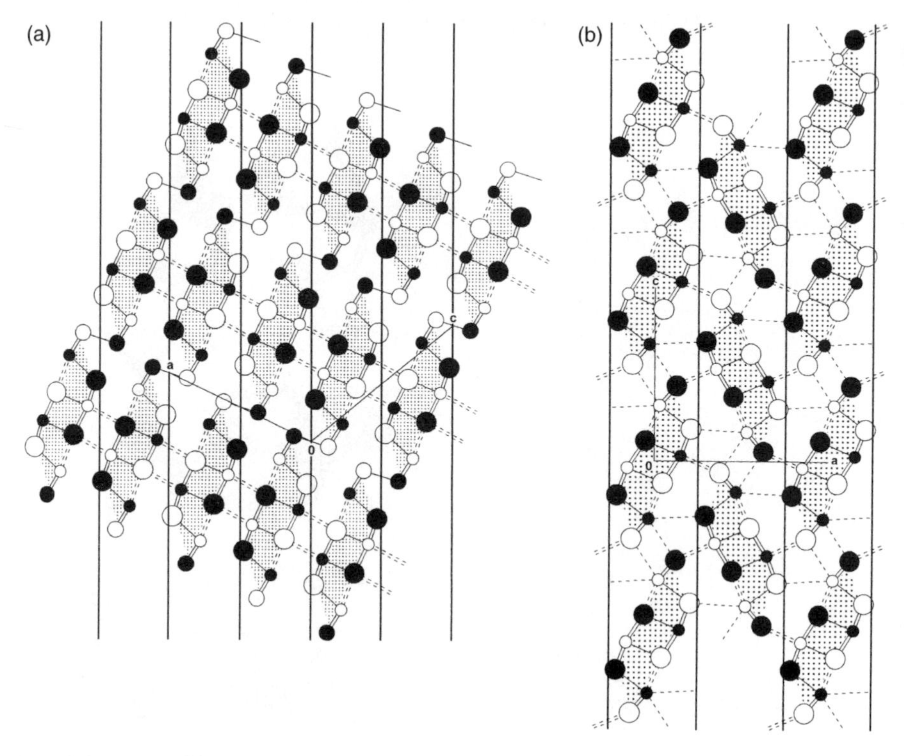

Fig. 3.8. The pääkkönenite (Sb_2S_2As) stibnite (Sb_2S_3)–pair (Bonazzi *et al.* 1995) of pseudopolytypes. Unit layers are indicated. Note the fundamentally different interconnection of unit ribbons (stippled) in the two structures.

layers $(\bar{2}01)$ via As–As pairs. Alternatively, glide planes between Sb_4S_6 ribbons acting as interlayer operators in Sb_2S_3 create the typical rod stacking in stibnite, without strongly bonded continuous layers.

Another typical example are **spinelloids**, in which single tetrahedra are altered into pairs, triples, etc. of (mostly) SiO_4 tetrahedra.

The pair **ekanite** $ThCa_2Si_8O_{20}$ (Szymanski *et al.* 1982)–**steacyite** $ThNa_xCa_{2-x}K_{1-x}Si_8O_{20}$ (Richard and Perrault 1972) is a good example of pseudopolytype relationship. Presence of additional potassium in steacyite and the difference in space groups ($I422$ and $P4/mcc$ in the above order) are the outward signs of structural differences. These were described by Perrault and Szymanski (1982) as the presence of a puckered sheet of SiO_4 tetrahedra with a composition Si_8O_{20} in ekanite (i.e. a phyllosilicate structure) versus discrete 'pseudocubic' Si_8O_{20} units in steacyite (i.e. a double-ring cyclosilicate structure) (Figs 3.9 and 3.10).

Both structures can be sliced into unit layers (001) $P422$ with boundaries drawn via the connecting oxygen atoms between two tetrahedra positioned above each other

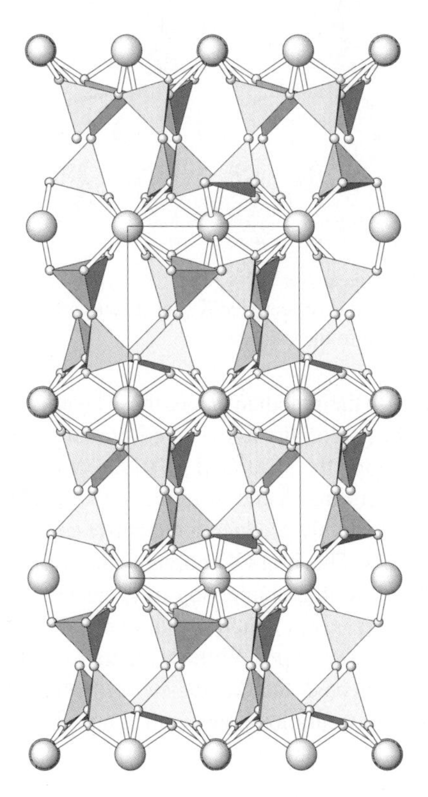

Fig. 3.9. Crystal structure of ekanite $ThCa_2Si_8O_{20}$ (Szymanski *et al.* 1982) with puckered Si_8O_{20} sheets of tetrahedra. Unit layers of the pseudopolytypic description have boundaries at $z = 1/4$ and 3/4. Details in text.

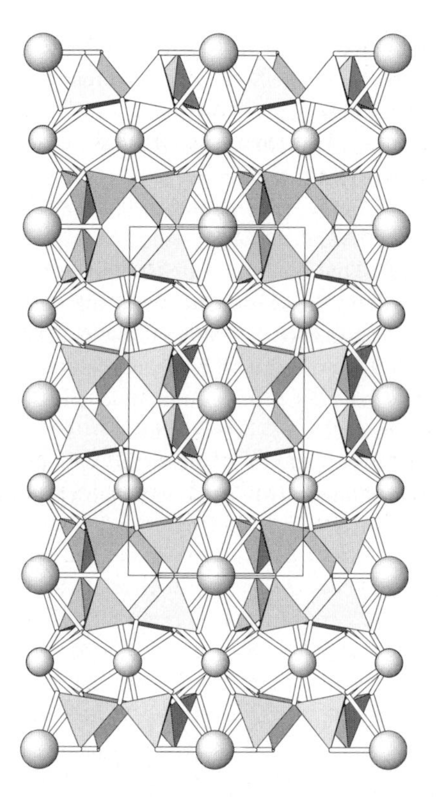

Fig. 3.10. Crystal structure of steacyite $ThNa_xCa_{2-x}K_{1-x}Si_8O_{20}$ (Richard and Perrault 1972) with Si_8O_{20} units separated by partly occupied K sites. Pseudopolytypic relationship to ekanite (Fig. 3.9) is described in detail in the text.

along the [001] direction. In their median planes these layers contain square antiprisms of Th and the Ca sites. The potentially K-occupied sites are on layer surfaces, between Si_4O_{10} groups adjacent along the [100] and [010] directions.

In steacyite, the unit layers are related by reflection on (001) boundaries, in ekanite by 2_1 glide-rotations, respectively producing the above-mentioned closed tetrahedral groups and puckered layers. This non-OD polytype description can be altered to OD description by separating the interconnecting oxygens at $z = 0.25$ and 0.75 as the second type of (loose!) unit layers which contain both the m plane and the 2_1 axes used alternatively in the two polytypes. Without regard to which mode of description, OD or non-OD, has been selected, the ekanite–steacyite pair has profoundly different strongly bonded tertiary configurations, that is, it is a pair of pseudopolytypes.

3.1.3.5 *Pyroxenes and amphiboles*

Pyroxenes and **amphiboles** are the most striking example of improper polytypes, which, however, do not result in **pseudopolytypy**. Numerous configurational and coordination adjustments, which take place when going from a monoclinic, $C2/c$ and $C2/m$ (for pyroxene and amphibole, respectively) member of these series to the two varieties of orthorhombic structures (*Pbca* and *Pnma* for orthopyroxene and orthoamphibole, respectively; *Pbcm* and *Pnmn* for their proto-varieties), together with chemical changes in a particular, 'adjustable' structure site (M2 in pyroxenes, M4 in amphiboles) cause that practically nobody treats these structures as (configurational) polytypes; a notable exception is Sedlacek *et al.* (1979).

The most natural approach is to consider them as polytypes composed of rods. These are fragments of the tetrahedral–octahedral–tetrahedral layer, broader in amphiboles than in pyroxenes. The central octahedral ribbons in the rods can have two orientations, either pointing 'up' or 'down' in the front face of the layer when looking along the $\boldsymbol{a}^*$ direction. Oxygens on the surface of octahedral ribbon determine the position of tetrahedra attached to them. Attempts to fix these families of structures to ccp or hcp of anions by rotating tetrahedra in alternative directions give the absolute 'end-members' but these are rather artificial, as any look at a real structure reveals. Rods are interconnected by a few ligands forming part of marginal tetrahedra and at the same time belonging to the octahedral ribbons of the adjacent rods.

Differences between 'polytypes' dwell in distinct orientation schemes of octahedral layer fragments in the structure (Fig. 3.11). Due to interconnections, coordinations of marginal cations of octahedral layers (M1 in pyroxenes, M4 in amphiboles) and rotations of (marginal) tetrahedra depend on the orientation of octahedral layers adjacent along [110]: they differ between the situation when both are up (down), and when they are opposite, up–down (Fig. 3.11). For the marginal cations, in most cases (but not always), it also means the presence of other cation species, Ca versus (Mg, Fe).

In a layer description of these structures, the interrupted octahedral (100) layers are 'polytype-active' whereas the one tetrahedron thick interlayers (100), in which all tetrahedra are concentrated, are to a large extent 'neutral'. Both the 'active' and 'passive' layers are substantially modified from a polytype to a polytype, that is, we deal with improper polytypes. These two structure families, pyroxenes and amphiboles, are also very good examples of the combination of polysomatic and polytypic principles.

3.1.3.6 *Endopolytypes*

In usual polytypes, it is expected that the alternative positions of the $(n+1)$th unit layer after the nth layer result in altered positions of both the cations and of their ligands (anions). The same holds for all the molecules in molecular crystals. However, there exist instances of OD or non-OD polytypy in which the anion arrays of the structure (effectively) do not change their configuration and position in space when the cation arrays assume the alternative positions of the $(n+1)$th unit layer. 'Effectively' means the slight degree of idealization of atomic positions as it is common in polytype families—abstracting from fine adjustments (desymmetrization) observed in every individual polytype. These cases will be called **endopolytypes** in order to stress the presence of polytypic phenomena *inside* a fixed structure framework.

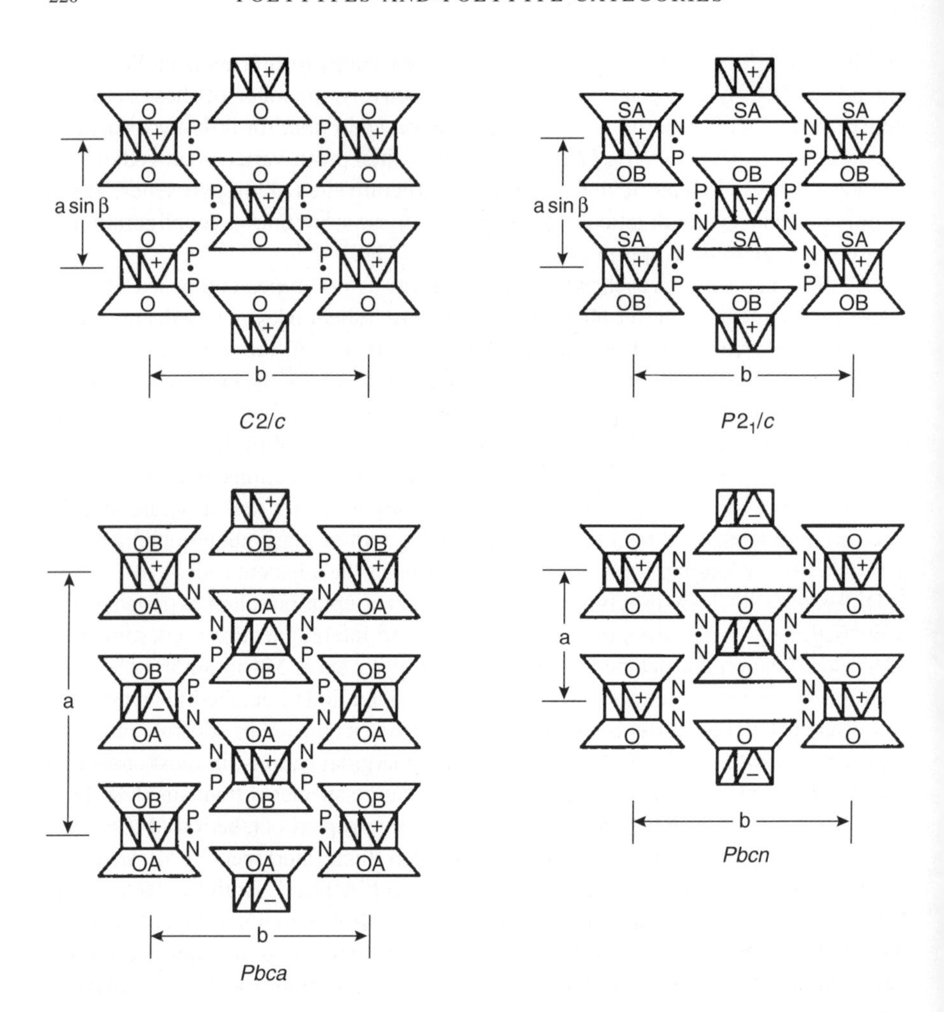

Fig. 3.11. Structural schemes for pyroxenes (biopyriboles $N = 1$). Each ‘I-beam’ consists of a double-octahedral column that is flanked by sites for potentially larger cations (indicated by dots) and is sandwiched between two pyroxene-like chains. Orientation of octahedra is indicated by the + and − signs, distortion sense and degree of silicate chains by the letters O, S and A, B, respectively. P and N denote a kind of chain-chain contacts, that is, of the anion configurations coordinating the larger cations. Redrawn with permission from Cameron and Papike (1981).

As the first example, let us quote **spinelloids**. In their crystal structure, anions form a cubic close packed array, with partly filled octahedral and tetrahedral voids. These coordination polyhedra share corners; the octahedra also edges (Fig. 3.12). If we select a unit layer (010) one tetrahedron (octahedron) thick, spinelloids are non-OD-polytypes and, because strongly bonded Si_2O_7 or even Si_3O_{10} groups form instead of isolated SiO_4 tetrahedra (i.e. tertiary configurations change) in the process

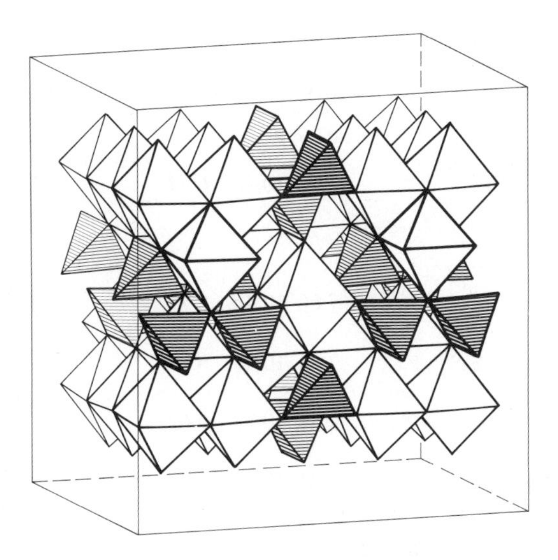

Fig. 3.12. The crystal structure of spinelloid Fe_2SiO_4(40)–Fe_3O_4(60) (Ross II *et al.* 1992); homologue N_1, N_2 = 1, 2 of a spinelloid series. Single and paired tetrahedra alternate in successive (010) slabs. Unit layers (010) are one tetrahedron thick; five such layers are shown.

of polytype building, they are pseudopolytypes (Makovicky 1997*c*). When isolated alternating layers (010), built of cations and anions in combination, are selected instead, the spinelloids become OD-polytypes (as discussed in detail in Section 2.6.5) but retain their pseudopolytype character. In these processes, the anion array remains unchanged, that is, we deal with an endopolytype.

The crystal structures of another endopolytype example, Cu_3SbS_3 and Cu_3BiS_3 (**skinnerite** (Makovicky and Balič Žunič 1995; Makovicky 1994; Pfitzner 1994) and **wittichenite** (Matzat 1972; Kocman and Nuffield 1973) are based on a unit cell twinned hcp array of sulfur. Twinning on $(11\bar{2}2)_{\text{hcp}}$, after every $3d_{11\bar{2}2}$, creates trigonal prismatic voids for Sb/Bi^{3+} whereas below the freezing temperature of these ionic conductors (Table 3.1) Cu atoms settle in trigonal coordinations in the walls of octahedral voids or in the walls that separate tetrahedral voids of the hcp.

Makovicky (1994) showed that the two above 2c-$P2_1/c$ (Fig. 3.13) and 1c-$P2_12_12_1$ structures, as well as a hypothetical 1c-$P2_1/n$ structure can be constructed by stacking wavy unit layers (001) of Cu atoms (Fig. 3.14) distributed in an unchanging S–(Sb,Bi) framework in two possible modes:

(1) either the nth and $(n + 1)$th layers are congruent and the stacking operation is a partial 2_1 axis parallel to [001], or
(2) these two layers are enantiomorphs and the operation is a centre of symmetry between the layers.

If the enantiomorphs are denoted as α and β, respectively, and the analogues is generated by a screw-axis are denoted by priming their symbols, the $P2_12_12_1$ structure

Table 3.1 Wittichenite homeotypes

Mineral	Formula	Lattice parameters				Space group	Reference
		a	*b*	*c*			
Wittichenite	Cu_3BiS_3	7.72	10.40	6.72		$P2_12_12_1$	Kocman and Nuffield (1973)
Intermediate wittichenite	Cu_3BiS_3	7.66	10.45	6.72[a]		Modulated	Makovicky (1994)
High wittichenite	Cu_3BiS_3[b]	7.66	10.49	6.71[b]		*Pnma*	Makovicky (1994)
Low skinnerite (γ)	Cu_3SbS_3—$P2_12_12_1$	7.88	10.22	6.62		$P2_12_12_1$	Pfitzner (1994)[c]
Intermediate skinnerite (β)	Cu_3SbS_3—$P2_1/c$	7.81	10.24	13.27	β 90.29°	$P2_1/c$	Makovicky and Balič Žunič (1995)[d]
High skinnerite (α)	Cu_3SbS_3—*Pnma*	7.81	10.25	6.59	–	*Pnma*	Pfitzner (1998)
Synth.	Cu_3SbSe_3[e]	7.99	10.61	6.84		*Pnma*	Pfitzner (1995)
Synth.	Li_3AsS_3[e]	8.05	9.82	6.63		$Pna2_1$	Seung *et al.* (1998)

[a] At 142°C.
[b] At 350°C.
[c] Also Whitfield (1980).
[d] Also Pfitzner (1994), who gives transformation temperatures as −9°C and +121°C, respectively.
[e] Cation arrangements differ from the Cu_3BiS_3–Cu_3SbS_3 scheme.

corresponds to the sequence

$$-\alpha-\alpha'-\alpha-\alpha' \quad \text{or} \quad -\beta-\beta'-\beta-\beta'-,$$

the $P2_1/c$ sequence

$$-\alpha-\alpha'-\beta'-\beta-\alpha-\alpha'-\beta'-\beta-,$$

and the $P2_1/n$ sequence

$$-\alpha-\beta-\alpha-\beta-\alpha-.$$

In all cases, the sulphur array of these configurational non-OD-polytypes remains unchanged, with the (ideal) space group *Pnma*. See also Section 3.1.3.8.

A case of OD-polytype with S array unchanged is **diaphorite**, $Pb_2Ag_3Sb_3S_8$, yet another endopolytype (Armbruster *et al.* 2002). In this structure, layers with

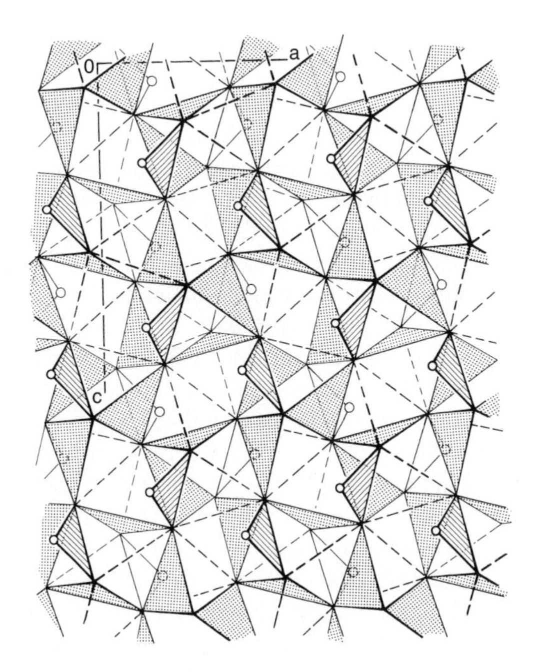

Fig. 3.13. The crystal structure of $P2_1/c$ skinnerite Cu_3SbS_3 (Makovicky and Balič Žunič 1994). Stippled triangles are the triangular CuS_3 coordinations, ruled are SbS_3 coordination pyramids. Compare with Fig. 3.14.

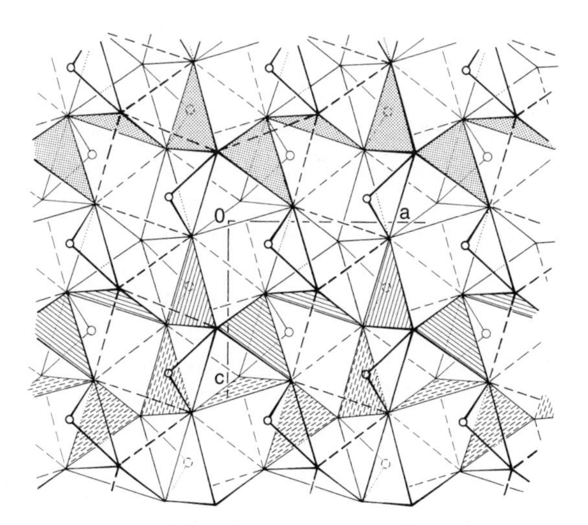

Fig. 3.14. Wavy unit layers (001) of Cu coordination triangles embedded in the hexagonal close packing of sulphur atoms, twinned on $(11\bar{2}2)_{hcp}$. Upper portion: an isolated unit layer, below: two layers related by 2_1 operator parallel to [001].

alternating Pb and Ag polyhedra and Sb coordination pyramids are separated by Ag and Sb containing layers with a periodicity halved in respect to the former. Thus, the Pb and Ag layers can, after each Ag and Sb layer, assume two positions, resulting in a cation disorder that is not followed by that in the sulphur array.

Finally, two sorosilicate groups are typical endopolytypes *on configurational level*. It is the **låvenite–wöhlerite group** with chemistry ranging from $Na_8Ca_8(P_2O_7)F_8$ to zirconosilicates wöhlerite and låvenite (the latter being $Na_4(Na,Ca)_4(Mn,Ca,Fe,Ti)_4(Zn,Nb)_4(Si_2O_7)_4O_4F_4$) and the **götzenite–seidozerite–rosenbuschite group** of zirconium-titanium silicates, the title compound being rosenbuschite $Zr_2Ca_2(Na,Ca)_4Ca_4Na_2ZrTi(Si_2O_7)_4F_4O_4$ (Christiansen *et al.* in press). Polytypic descriptions of these structures are presented by Betti (1998), Merlino *et al.* (1990*a*, *b*) and by Christiansen *et al.* (in press).

The låvenite–wöhlerite group contains a framework composed of walls of edge-sharing octahedra four columns wide whereas the götzenite–seidozerite–rosenbuschite group has infinite layers of edge-sharing octahedra in combination with ribbons two octahedra wide (see Section 2.4.3.1). Endopolytypy on configurational level (i.e. ignoring the distribution of the above quoted cations over the distinct octahedral sites of the walls and ribbons) consists of distinct disposition patterns of the rows of Si_2O_7 dimers along the triangular channels of these two unchanging frameworks. The z levels of these dimers lie at $c/8$ intervals, at up to four distinct levels (Fig. 3.15).

3.1.3.7 *Commensurate, semi-commensurate, and non-commensurate polytypes*

OD phenomena in the crystal structure of **wollastonite** $Ca_3Si_3O_9$ (Ohashi and Finger 1978) are based on the 2:3 correspondence between a period of two octahedra matching three tetrahedra of the typical 'wollastonite' silicate chain (Fig. 2.2). Similarly, in the **gageite** structure type (Section 2.4.4) the corresponding ratio is 3 : 4 (Fig. 2.27). Layer shifts that play substantial role in the polytypy of homooctahedral layer silicates are based on the 1 : 3 periodicity ratio of tetrahedral and octahedral layers. Polytypy in **kermesite** Sb_2S_2O (Kupčík 1967; Bonazzi *et al.* 1995) is based on a 2 : 1 match of oxide versus sulphide layers; 2 : 1 match of substitutional type occurs also in **imhofite** (Balič Žunič and Makovicky 1993) and the 2 : 1 ratio is also behind the OD character of **clinozoisite/zoisite** (Fig. 3.16). A more complicated match, 3.5 : 1 occurs on configurational level in astrophyllite (Zvyagin and Vrublevskaya 1976; Christiansen *et al.* 1999) (Fig. 1.21b). These non-(or semi-) commensurate ratios can become very complicated for non-commensurate misfit layer structures mentioned in section 1.7.

Contrary to these 'semi-commensurate' to non-commensurate cases are the 'commensurate' cases in which layers display 1 : 1 ratio of periodicities—for example, the structures of **pinakiolite family** (Fig. 3.4), the pair rosenbuschite–seidozerite (see Section 2.4.3.1, Fig. 2.25) or the sulphosalt **owyheeite** ($Ag_3Pb_{10}Sb_{11}S_{28}$; Makovicky, Olsen and Nielsen, unpublished).

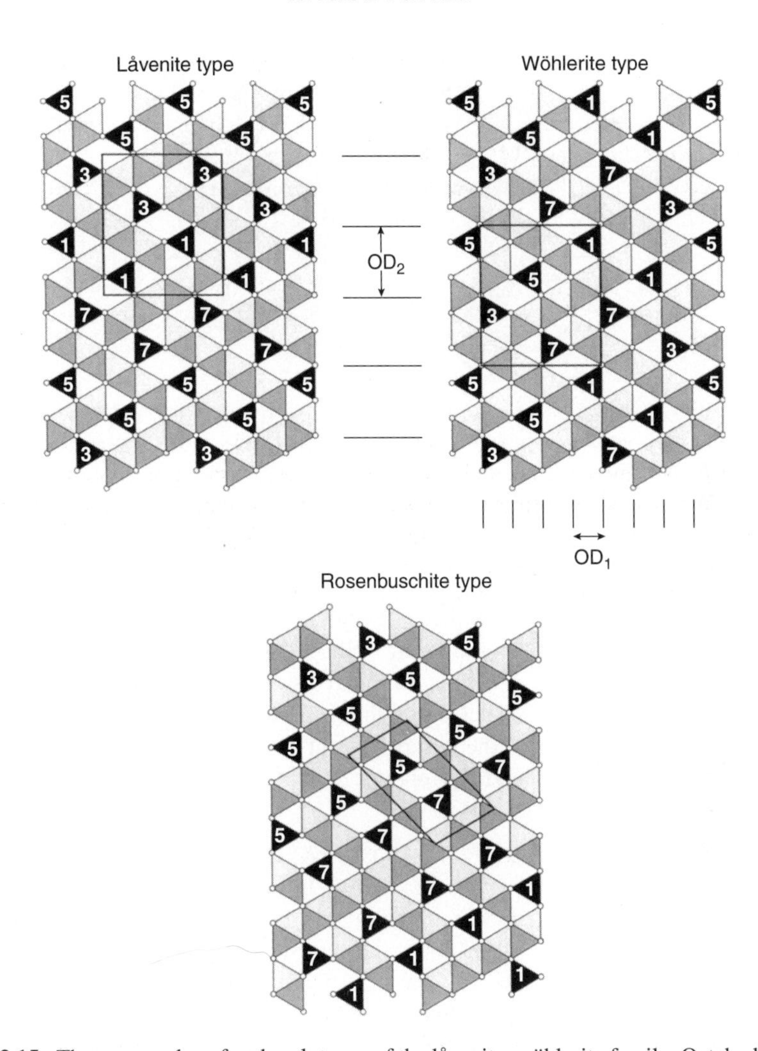

Fig. 3.15. Three examples of endopolytypes of the låvenite–wöhlerite family. Octahedral ribbons and walls enclose columns of Si_2O_7 groups. In the two octahedra high period these groups are positioned at four possible levels 1/4 period apart. Two types of OD layers indicated were used by Christiansen *et al.* (in press) for the description of configurational endopolytypy in the låvenite–wöhlerite group.

3.1.3.8 *Layered racemate intergrowths*

In relatively rare instances, unit layers obtained by slicing of the left-handed (S) and right-handed (D) enantiomorphs of the same structure type can form regular SDSD intergrowths. These cases are at the limits of the polytype definition of Guinier *et al.* (1984). If 'the same kind of layers' in the definition includes both enantiomorphs, the SSS, DDD and SDSD sequences are non-OD **racemate polytypes**. If

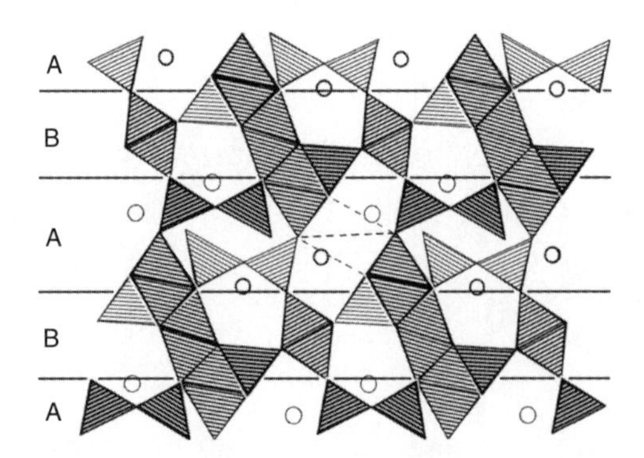

Fig. 3.16. Crystal structure of clinozoisite $Ca_2Al_3[SiO_4][Si_2O_7]O(OH)$ (Dollase 1968). OD structure with two types A and B of unit layers (Merlino 1990a). Every second octahedral column in B layers is flanked by additional octahedra; single SiO_4 tetrahedra and Si_2O_7 dimers are in the B and A layers, respectively. Circles: Ca atoms.

this definition excludes enantiomorphs, they are **layered racemate intergrowths**. The SDSD sequences are not (automatically) polytypes composed of two types of unit layers. The best example is **moganite**, SiO_2 (Miehe and Graetsch 1992), a regular SDSD intergrowth of unit $(11\bar{2}1)$ layers of left and right quartz. Random S-D sequences form chalcedony (ibid.). Further examples are skinnerite-wittichenite (Section 3.1.3.7), the hexaborates of the tunellite $SrB_6O_9(OH)_2.3H_2O$ family (Belokoneva 2003) and the organic compound hexahelicene (Green and Knossow 1981).

4

Application of modularity to structure description and modelling

4.1 Introduction

In the general survey of modularity reported in Chapter 1, the identification between **accretional homologous series** and **polysomatic series** has been discussed. In particular, it has been stated that whereas the homology treatment privileges the accretional aspect of a basic structural motif, the polysomatic description slices the structures in chemically and structurally different modules. These modules, normally not more than two, are combined in different proportions to generate chemically and structurally different compounds. One of the modules used by a polysomatic description of a structure corresponds to the accreting interface of a homologous description. In Chapter 1, comparative examples of the two viewpoints have been given and, merits and demerits have been discussed. A polysomatic description may offer advantages in cases like: (i) each building module does exist as independent structure (end-member), (ii) only loose bonds do exist across the building modules, the strongest bonds being inside the modules, (iii) one or more modules are constant in a series and are intercalated by other modules (**merotype series**; see Section 1.8). In general, it can be affirmed that a polysomatic description has advantages when a group of structures shows aspects that are amenable to intergrowth mechanisms between different structures. The examples reported in this chapter intend to show the polysomatic, merotype, and plesiotype (see Section 1.8) aspects of some series of compounds which are important either methodologically or for their technological and mineralogical relevance. More generally, the advantages of considering an intrinsic modularity in describing and modelling crystal structures are illustrated.

The number of crystal structures described as consisting of two or more chemically and crystallographically distinct modules M, M′, M″, . . . is increasing quite fast, particularly in some fields of technological interest, like **superconductors**. When the same (normally two) modules combine in different way to generate a group of stoichiometrically different compounds, the members of the group can be expressed by a general formula $M_m M'_{m'}$ and are said to form a **polysomatic series**, that is, as mentioned above, a specific case of homologous series where, in principle, each module may exist as an end member. The members of a polysomatic series are clearly collinear in composition. Excluding the incommensurate crystal structures, the cell parameters of the members are linearly correlated too, because the parallel lattice periodicities of two adjacent modules must be commensurate in order to match the interfaces. Often, the building modules are planar layers (**layered structures**) and the members of the series share (multiples of) the lattice periodicities that occur in the layers. In this case, the third lattice periodicity must be a linear combination of the layer thickness t_M and $t_{M'}$. If the building modules are rods or blocks, only one or none of the lattice parameters is fixed, unless these modules are arranged in layers. In the latter case, rods and blocks may be emphasized but just for some descriptive reason, a layer being the effective building module.

When known, the combinatorial mechanisms that generate a polysomatic series can be exploited both to model unknown structures of compounds which, on chemical and lattice dimension grounds, are recognized to belong to a given series, and to synthesize new tailored members as well. Sometimes, the combination of modules is ruled by some more or less subtle symmetry constrains (cf. Zvyagin 1993, 1997) similar to those widely described for polytypism (see Chapter 2). However, in the polysomatic and merotype series, the most important aspect ruling the fitting of modules at their interfaces is a dimensional one. Thus, it can be affirmed that polysomatism is a kind of polysynthetic **syntaxy at cell level**; syntaxy reduces to epitaxy if only plane modules are involved.

The examples of modular structures given in this chapter are presented in groups that are characterized by at least a recurring module. Most modules occurring in these structures are based on more or less compact packing of anions, in particular O^{2-}. In this case, the matching (syntaxy/epitaxy) of different modules is assured by the periodicities related to the anion packing. This aspect is particularly evident in simple structures (e.g. perovskite, spinel, NaCl, corundum). Note that several structures described in this chapter contain octahedral layers with a typical **close-packing** configuration. Of course, intercalation of completely different modules between close-packing based modules (e.g. in the organic–inorganic perovskites of Section 4.2.3.3) requires an appropriate choice of the former ones. If the linkage between modules is due to weak and/or flexible bonds, like coordination bonds of low charge cations and hydrogen bonds, the matching problem is greatly simplified.

4.2 The perovskite module

The **perovskite-type structure** offers to materials science some of the most basic and important building modules. In fact, many members of the perovskite family and their structures play a fundamental role in several technological applications because of their electrical, magnetic, optical, catalytic, and other properties.

Perovskite s.s. corresponds to the mineral **$CaTiO_3$**. Ideally, its crystal structure is cubic ($Pm\bar{3}m$, $a_c \sim 3.8$ Å) and is usually represented (Fig. 4.1) by emphasizing the octahedra surrounding the small cation Ti^{4+}. This cation is situated at the corners of the unit cell; the centre of the cell is instead occupied by the large cation Ca^{2+} in dodecahedral (cubo-octahedral) coordination. The oxygen ions O^{2-} are at centre of each cell edge. It may be useful to consider the large cation as part (1/4) of a cubic-close-packing (ccp) array which also includes the oxygen atoms; in this ccp array, the

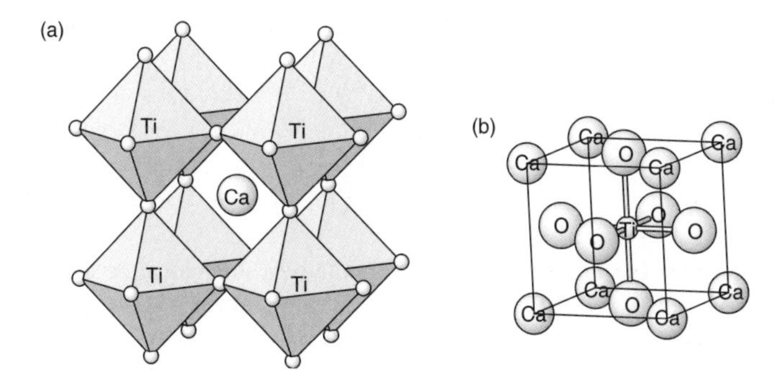

Fig. 4.1. Structure of perovskite s.s. ($CaTiO_3$) represented as (a) packing of corner-sharing Ti-octahedra, with the large cation Ca^{2+} in dodecahedral coordination at the centre of the cell, and (b) as cubic close packing of $Ca^{2+} + O^{2-}$ ions, with Ti^{4+} in octahedral coordination at the centre of the cell.

small Ti occupies 1/4 of the octahedral sites (precisely, the one at the centre of the cell). Actually perovskite s.s., $CaTiO_3$, is distorted down to orthorhombic symmetry where commonly it is described either in the space group *Pnma* or *Pbnm* (N. 62; cf. Section 5.1.2.3 and Fig. 5.1); Ca assumes a $8 + 4$ coordination. With reference to the ideal cubic cell and a general formula ABX_3, the bond length of A–X (dodecahedral site) and B–X (octahedral site) is, in the order, $a_c/2$ and $a_c/(2)^{1/2}$.

The representation of the crystal structure as a three-dimensional network of corner-sharing octahedra is particularly useful to describe the structural distortions resulting from rotation or tilting of the octahedra. Instead, the representation based on the stacking of close packing hexagonal layers is useful to understand the hexagonal polytypes and their derivative structures (Section 4.2.1). A detailed analysis and description of derivative structures is found in Darriet and Subramanian (1995).

The perovskite-type structure is the archetype for a large number of synthetic and natural compounds with stoichiometry ABX_3 and their derivative structures. These are formally obtained from the archetype through mechanisms like anion or cation deficiency and splitting into independent subsets of crystallographic position sets that are equivalent in the ideal cubic structure (cf. Darriet and Subramanian 1995; Mitchell 2002). In the ideal formula ABO_3 the cations A and B are not bound to have charge 2+ and 4+, respectively, as in perovskite s.s.; what must be kept constant and equal to 6+ is the sum of the positive charges. For example, $\mathbf{La^{3+}Cr^{3+}O_3}$ and $\mathbf{Na^{1+}W^{5+}O_3}$ are known.

The most apparent feature of the crystal structures that can be derived from an ideal perovskite structure is often a tilting of the octahedra. Consequently, in these structures the dodecahedral cage is distorted, as often is the octahedron too. The departure from the ideal perovskite structure usually depends from the size of the A and B cations relative to the anion X. Rhombohedral, tetragonal, and orthorhombic symmetries are typical of distorted perovskite structures. Complex perovskite-like structures characterized by occupancy and ordering of different cations on either

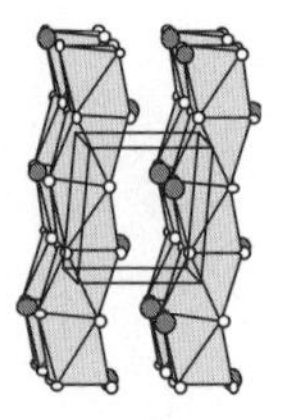

Fig. 4.2. Structure of the polytype $2H$ of a perovskite-type structure as occurring in $BaNiO_3$ ($P6_3/mmc$; $a = 5.629$, $c = 4.811$ Å; Takeda *et al.* 1976). Blank and grey circles represent oxygen and Ba atoms, respectively; Ni, not shown, is at the centre of each face sharing octahedron.

the A or B sites (**double perovskites**) or on both sites (**quadruple perovskites**) are known. In this case, cation ordering on independent sites is possible and the symmetry can be as low as monoclinic and triclinic. Either cation- or anion-deficient non-stoichiometric perovskites are also known.

4.2.1 *Perovskite polytypes*

In the ideally cubic perovskite-type structure, the AX_3 layer is stacked along the [111] cube diagonal according to a sequence **AB** (bold **A** and **B** to represent close packed layers). Other stacking sequences of the layers **A** and **B** are possible and form various kinds of polytypes among which the hexagonal polytype $2H$ is of interest for our purposes. In this polytype (Fig. 4.2), the stacking sequence is **ABC** along the [001] direction of a hexagonal cell whose parameters are simply related to those of the cubic polytype: $a_{\text{hex}} = a_c(2)^{1/2}$ and $c_{\text{hex}} = a_c(3)^{1/2}$. Apart from the symmetry, the main difference between the cubic and hexagonal polytypes is that whereas each coordination octahedron shares only corners in the former, in the latter it shares two faces thus forming columns along $[001]_{\text{hex}}$. Due to the face sharing, in a hexagonal polytype, the B octahedral cations are closer to each other than in the cubic polytype, thus decreasing the stability of the structure. Usually the **hexagonal perovskites** are stabilized by the formation of metal–metal bonds between the B atoms. A mixed stacking of corner-sharing cubic modules (c) and face-sharing hexagonal modules (h) originates complex polytypes (cf. Mitchell 2002).

4.2.2 *Slicing perovskites*

Modules (fragments) of perovskite-type (shortly perovskite) crystal structure alternated with other structural modules occur in several crystalline materials that are of paramount interest for science and technology. These types of structures are known as **hybrid** or **intergrowth perovskites** and some have been known for more than 100 years (Topsöe 1884) even if most of the types have been discovered in recent years, particularly in connection with the continuing search for superconducting materials.

An updated survey of perovskite structures, including series described in the following, can be found in Mitchell (2002). In this chapter, a perovskite layer (slab)

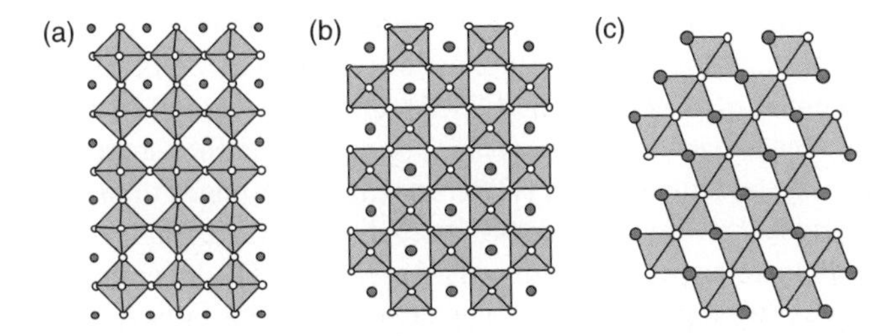

Fig. 4.3. Projection of the perovskite crystal structure in the plane (001) (a), (110) (b), and (111) (c) to show the different stacking of the octahedral sheets. Blank and grey circles represent X anions and dodecahedral cations, respectively.

means a planar fragment of a perovskite-type structure irrespective of its chemical nature.

Fragments of perovskite structure are present in composite structures with different dimensionality. Three-dimensional (3D), if the sharing of octahedral corners extends in three non-coplanar directions; two-dimensional (2D), if the sharing of octahedral corners extends only in two directions such that one-octahedron thick layers are formed; mono-dimensional (1D), if the sharing of octahedral corners develops along one direction only such that rows of octahedra result; zero-dimensional (0D), if only isolated octahedra occur. It is intended that the 0D and 1D cases must preserve the distribution of isolated octahedra (0D) and rows of octahedra (1D) that is typical of the perovskite structure.

The 3D and 2D fragments can be cut in different ways from a perovskite structure (Fig. 4.3); layers perpendicular to [100] or [110] and symmetry related directions, that is, perpendicular to the sets of directions $\langle 100 \rangle$ (the most common ones) or $\langle 110 \rangle$, are well documented, but also compounds with layers perpendicular to $\langle 111 \rangle$ are known (cf. Mitzi 2001; Section 4.2.3.3). Being different the periodicities in the (100), (110), and (111) planes, the nature of the interlayer (complex) ion proves to be selective in orienting the perovskite module as shown below.

4.2.3 *High temperature superconductors*

Superconductive materials are characterized by the absence of electrical resistivity below a critical temperature (T_c). The era of the so called high T_c superconductors began with the discovery (Bednorz and Müller 1986) of the superconductivity at 36 K in $(La,Ba)_2CuO_4$ followed by $YBa_2Cu_3O_7$ (Wu *et al.* 1987), which is a superconductor at the temperature of liquid nitrogen, that is, at a temperature not so expensive to be maintained. Since these discoveries, various research groups began to hunt **high-temperature superconductors** (HTSC) and nowadays the crystal structures of some hundreds of compounds of this class are known (cf. Wong-Ng 1997).

The crystal structures of a large amount of the HTSC, particularly the cuprate family, can be described as built by layer modules with perovskite features plus other layers which may range from sodium chloride like structures to sheets of either cations or anions and even organic molecules (cf. Raveau *et al.* 1991; Shekhtman 1993; Vainshtein *et al.* 1994; Mitchell 2002). Often the perovskite module consists only of CuO_2 sheets that correspond to plane sections of corner-sharing octahedra. In few cases, the crystal structure of a HTSC exactly corresponds to that of perovskite, like $Ba_{1-x}K_xBiO_{3-y}$ (Hinks *et al.* 1988). Sometimes, octahedral layers of perovskite occur, either alone or alternating with square pyramids. The latter feature occurs, for example, in magnetic superconductors that are obtained by replacing some Cu by Fe (cf. Mochiku *et al.* 2002). Aleksandrov and Beznosikov (1997) have analysed the structures of perovskite-based compounds by including vacancies as vertices of the BX_6 octahedra. They considered these structures as intergrowth systems of stacks consisting either of *n* layers, which are connected via the vertices of the octahedra, or of combinations of layers, made of octahedra, pyramids, and squares, with block layers of various types. Based on this classification, various possible superconductor structures were predicted.

4.2.3.1 *Tallium cuprates*

The family of compounds with general formula $\mathbf{Tl_mBa_2Ca_{n-1}Cu_nO_{2n+m+2}}$ ($m = 1$, $n = 1$–5; $m = 2$, $n = 1$–4), where Tl, Ba, and Ca can be replaced by other cations, offers a didactic example to describe the type of multi-module modularity that may occur in HTSC (cf. Zvyagin and Romanov 1991; Vainshtein *et al.* 1994; Ferraris 2002). In the two series presented here ($m = 1$ and 2), four types of (001) plane modules (sheets) are necessary to build each structure; these sheets are conventionally indicated as A = TlO, B = BaO, C = CuO_2, and D = Ca. The CuO_2 sheet formally corresponds to the section of a perovskite-type structure through the octahedrally coordinated Cu atom. In some cases, the stacking of the four sheets is such that octahedral and pyramidal perovskite modules occur (Figs 4.4 and 4.5).

Conventionally, the family of superconductors we are dealing with are labelled by a set of four digits corresponding, in the order, to the stoichiometric coefficients of Tl, Ba, Cu, and Ca. For $m = 1$, the Bravais lattice is primitive P and the number of layers occurring in one unit cell corresponds to the sum of the four digits; for $m = 2$, the lattice is body centred I and the number of layers occurring in one unit cell corresponds to two times the sum of the four digits. The metric (but not always the symmetry) of the lattices is tetragonal and the dimensions of the periodicity in the (001) layer plane ($a \sim 3.85$ Å) corresponds to two times the length of the Cu–O bond. From n to $n + 1$ members, the c parameter increases by about $m3.1$ Å. In conclusion, the knowledge of m, n and the physical role of the different layers allow fixing the structure of a member of the series.

Ivanova and Frank-Kamenetskaya (2001) have theoretically analysed the profiles of X-ray diffraction patterns expected from chemically non-homogeneous single crystals, which belong to series like those described in this section, when random stacking defects of the building layers occur, a phenomenon not rare in a modular structure. Leonyuk *et al.* (1999b), following their polysomatic analysis of complex cuprates

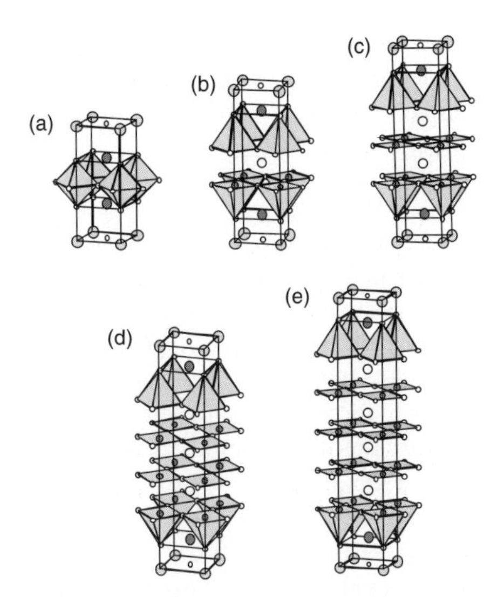

Fig. 4.4. Modular representations of the crystal structures of the tetragonal $P/4mmm$ superconductors $TlBa_2Ca_{n-1}Cu_nO_{2n+3}$ with $a = 3.85$ Å. (a): $n = 1$, $c = 9.54$ Å (Matheis and Snyder 1990); (b): $n = 2$, $c = 12.72$ Å (Kolesnikov *et al.* 1989); (c): $n = 3$, $c = 15.92$ Å (Morosin *et al.* 1988); (d): $n = 4$, $c = 19.00$ Å (Ogborne and Weller 1994); (e): $n = 5$, $c = 12.72$ Å (Weller *et al.* 1997). (001) slabs of perovskite structure are represented by grey-shaded squares, pyramids, and octahedra. Small blank, large blank, small dark-grey, large dark-grey, and light-grey circles represent O, Ca, Cu, Ba, and Tl atoms.

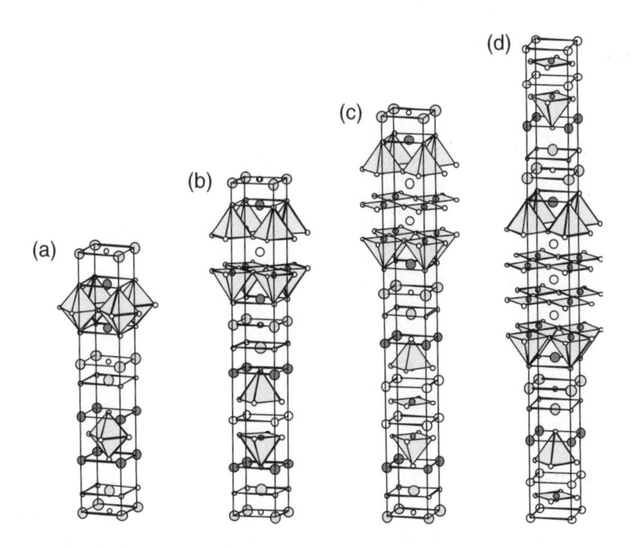

Fig. 4.5. Modular representations of the structures of the tetragonal $I/4mmm$ superconductors $Tl_2Ba_2Ca_{n-1}Cu_nO_{2n+4}$ with $a = 3.85$ Å. (a): $n = 1$, $c = 23.13$ Å (Shimakawa *et al.* 1988); (b): $n = 2$, $c = 29.22$ Å (Subramanian *et al.* 1988); (c): $n = 3$, $c = 35.66$ Å (Cox *et al.* 1988); (d): $n = 4$, $c = 42.05$ Å (Ogborne and Weller 1992). Symbols as in Fig. 4.4.

(Leonyuk *et al.* 1998, 1999a), suspected the occurrence of polysomatic defects, which alter the stoichiometry, as a possible source of anomalies noted in the superconducting behaviour of single crystals belonging to compounds of the series described here.

4.2.3.2 *Superconducting oxycarbonates*

Superconducting cuprates related to those of Figs 4.4 and 4.5 have been obtained by insertion of carbonate groups in the structure. Ignoring disorder of the latter groups, the crystal structures of some **oxycarbonates** belonging to the A_mB_n polysomatic series with composition $\mathbf{(Sr_2CuO_2CO_3)_m(X_pSr_2CuO_5)_n}$ $(m > n)$ are shown in Fig. 4.6. Each A module consists of a Sr_2CO_3 layer and a CuO_2 sheet. These modules are connected in groups of m to form a slab that alternates with n B modules, each one containing an Sr_2CuO_4 perovskite-like layer plus an X_pO sheet. The members (Huvé *et al.* 1993; Pelloquin *et al.* 1993; Nakata *et al.* 1995) shown in Fig. 4.6 correspond to A_1B_0 ($I\bar{4}$, $a = 7.81$, $c = 14.99$ Å), A_1B_1 [$X_p = $ (Tl, Pb), $P4/mmm$, $a = 3.82$, $c = 16.52$ Å], and A_2B_1 ($X_p = Bi_2O$, $Fmmm$, $a = 5.47$, $b = 5.48$, $c = 54.26$ Å). The symmetry of the A_2B_1 member is orthorhombic but its metric is very close to

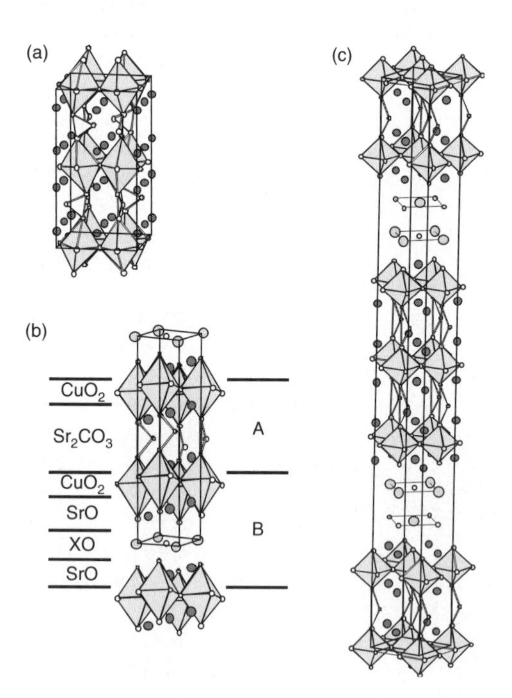

Fig. 4.6. Crystal structure of the members A_1B_0 (a), A_1B_1 (b), and A_2B_1 (c) of the oxycarbonate superconductors series $(Sr_2CuO_2CO_3)_m(X_pSr_2CuO_6)_n = A_mB_n$. Octahedral perovskite modules are shaded; triangles represent (disordered) CO_3 planar groups. Blank, dark-grey, and light-grey circles represent oxygen, Sr and X atoms (see text, also for reference). In (b), the composition of the layers and the modules A and B are shown.

tetragonal. The end-member A_1B_0 lacks the B module and consists of one-octahedron thick perovskite layer alternated with CO groups only.

Actually, each A module can also be interpreted as built by an Sr_2CuO_4 perovskite-like layer plus a CO sheet. Thus, the most apparent feature of the A_1B_0, A_1B_1, and A_2B_1 structures is a module consisting, in the order, of perovskite and CO layers in the ratio $1:1, 2:1, 3:2$ and intercalated by nX_pO layers. The a parameter of the three members is related to the Cu–O bond length of 1.9 Å; precisely, the length of a is $4 \times 1.9, 2 \times 1.9$, and $\sim 2(2)^{1/2} \times 1.9$ Å for A_1B_0, A_1B_1, and A_2B_1 respectively. Being about 7.5 Å the thickness of a layer consisting of one perovskite plus one CO sheet and about 2.0 Å that of an X_pO layer, the expression $c = 2^{(m-n)}[(m+n)7.5 + p(2.0)]$ Å gives the approximate periodicity along [001].

4.2.3.3 *Organic–inorganic hybrid perovskites*

Mitzi and co-workers (see Mitzi 1999, 2001; Mitzi *et al.* 2001a,b), and other groups as well, have synthesized and characterized families of **organic–inorganic perovskites** belonging to the wider family of organic–inorganic hybrid materials. The crystal structure of these hybrids consists of layer perovskite modules, which act as complex anion, and interlayers of organic molecules, which act as cation. The combination of the structural diversity, plastic mechanical properties and efficient luminiscence of the organic part with the electrical and mechanical properties of the inorganic part allows to engineer materials with interesting magnetic, electrical, and optical characteristics. The examination of a series of these materials, each one characterized by a fixed layer of perovskite and a variable organic interlayer, enables a correlation between the structure and its properties and an easier creation of tunable functional materials. Note that the tailoring of layer organic–inorganic perovskite materials can be obtained by playing on various parameters of the basic ABX_3 structure, as follows:

- Nature of the A (dodecahedral; usually a monovalent organic cation) and B (octahedral; usually a bivalent metal) cations.
- Nature of the X anion (usually a halide).
- Orientation and thickness of the perovskite layer.

Hybrids between [100]-, [110]-, and [111]-oriented layers of perovskite and organic anions are known (Figs 4.7–4.9). In 3D hybrids, the size of the organic A cation is limited by the dodecahedral coordination. Typical A cations of 3D hybrids are methyl-ammonium, $(CH_3NH_3)^+$, and formamidinium $(NH_2CH{=}NH_2)^+$. In the mono-layer hybrids, the organic A cation enters only as interlayer and the constraint of the ionic radius drops. In the multi-layer hybrids, the cations A (usual of two different chemical species) occupy both the dodecahedral site within the perovskite layer and the interlayer. Bonding at the organic–inorganic interface is often provided by hydrogen bonding. Of course, the interlayer organic part must also satisfy some size requirements, like matching the ideally quadratic two-dimensional array of X anions. The periodicity of this array depends on the radius of the X and B ions. The following

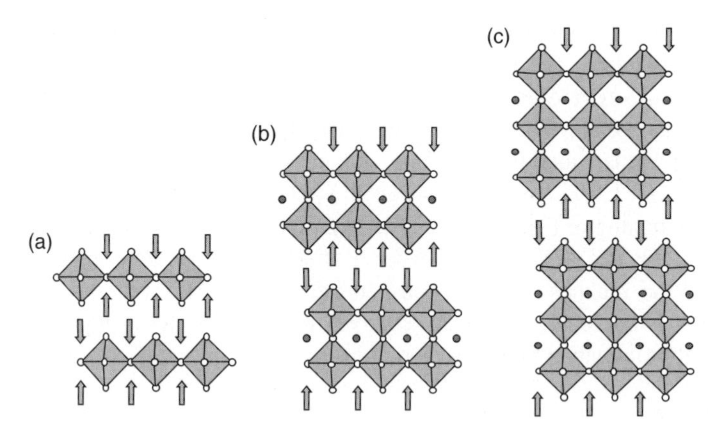

Fig. 4.7. Projection along [010] of the crystal structure of organic-inorganic perovskites based on n-octahedra thick (100) layers of perovskite alternated with organic cations RNH_3; the latter are schematically indicated by arrows. For the series $(RNH_3)_2A_{n-1}B_nX_{3n+1}$, (a), (b), and (c) represent the members with $n = 1, 2, 3$, in the order. Blank and grey circles represent X and A ions, respectively. (Modified after Mitzi 2001.)

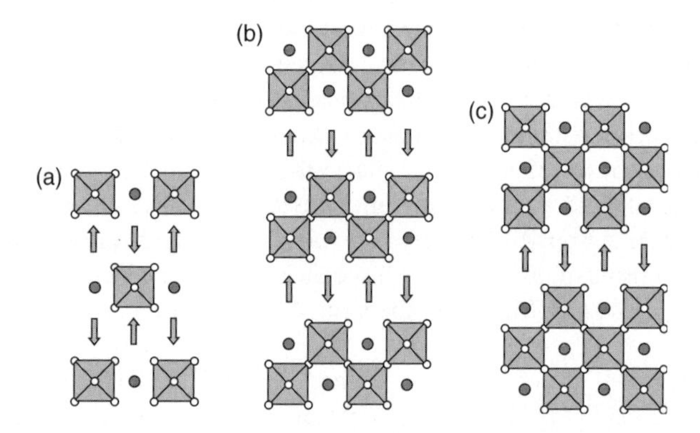

Fig. 4.8. Projection along [001] of the crystal structure of organic–inorganic perovskites based on n-octahedra thick (110) layers of perovskite alternated with organic cations A′; the latter are schematically indicated by arrows. For the series $A'_2A_mB_mX_{3m+1}$, (a), (b), and (c) represent the members with $n = 1, 2, 3$, in the order. Blank and grey circles represent X and A ions, respectively. (Modified after Mitzi 2001.)

examples show the type of series that have been so far obtained in the field of layer organic–inorganic perovskites. Further examples can be found in the reference cited.

Series of organic–inorganic layered perovskites can be built in at least three ways: (i) keeping the thickness of the perovskite layer fixed and changing the organic interlayer, (ii) keeping the organic interlayer fixed and changing the thickness of the perovskite layer, (iii) varying the thickness of both layers, if the organic layer can be incremented by polymerization. As mentioned above, the perovskite layer can have

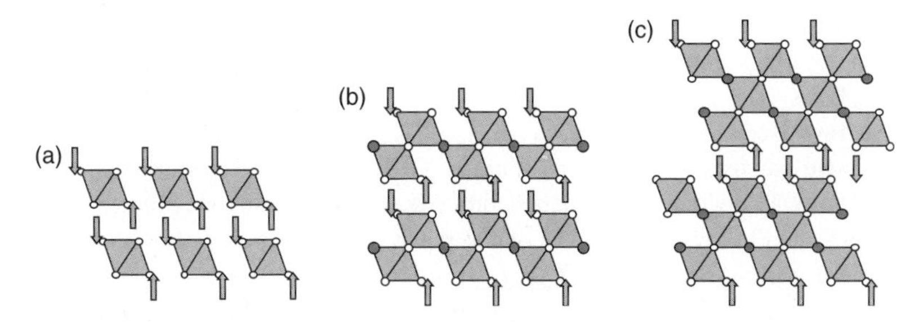

Fig. 4.9. Projection along [111] of the crystal structure of organic–inorganic perovskites based on n-octahedra thick (111) layers of perovskite alternated with organic cations A′; the latter are schematically indicated by arrows. For the series $A'_2A_{q-1}B_qX_{3q+3}$, (a), (b), and (c) represent the members with $n = 1, 2, 3$, in the order. Blank and grey circles represent X and A ions, respectively. (Modified after Mitzi 2001.)

three different orientations which are stabilized by appropriate dimensions of the organic layers. In case (i), for each chemical composition, thickness, and orientation of this module, a **merotype series** might be obtained by inserting different organic modules. The cases (ii) and (iii) represent **polysomatic series**.

A large family of **hybrid compounds** based on (100) perovskite layers (Fig. 4.7) represented by the general formula $\mathbf{(RNH_3)_2A_{n-1}B_nX_{3n+1}}$ is known for different R, A, B, and X. The running index n indicates the number of octahedral sheets which form the perovskite module. For $n = 1$, the perovskite layer is only one-octahedron thick and the A cation, which would occupy the dodecahedral cage of the corresponding 3D perovskite structure, is obviously absent; $(RNH_3)_2$ is the interlayer organic part. For R = C_4H_9 (butyl), A = $(CH_3NH_3)^+$, B = Sn^{2+} and X = I^-, **polysomatic series** with $1 \leq n \leq 5$ are known (Mitzi *et al.* 1994). Similar polysomatic series have been reported for Sn^{2+} replaced by Pb^{2+}, I^- by Br^-, and butyl by nonyl or phenyl (Calabrese *et al.* 1991). The properties of the $\mathbf{(C_4H_9NH_3)_2(CH_3NH_3)_{n-1}Sn_nI_{3n+1}}$ **polysomatic series** (Mitzi 2001) well illustrate the engineering of materials which can be realized as a function of n. The $n = 1$ compound is a fairly large band gap **semiconductor**. By increasing n, the resistivity of the materials rapidly decreases, with a transition to metallic behaviour for $n \geq 3$. The $n = \infty$ material, $\mathbf{(CH_3NH_3)SnI}$, is a tridimensional perovskite with the properties of a low-carrier density p-type metal. Both the Sn and Pb halide organic–inorganic perovskites exhibit **non-linear optical properties** and **electroluminiscence**.

The general formula $\mathbf{(RNH_3)_2PbCl_4}$ represents a **merotype series** where the constant module is represented by a mono-octahedron thick module of perovskite, $(PbCl_4)^{2-}$, and the variable organic interlayer by different dye molecules. Interlayers with R = 2-phenylethyl, 2-naphthylmethyl, or 2-anthrylmethyl are known (Era *et al.* 1997; Braun *et al.* 1999). All members of the series are **photolomuniscent**, but the characteristics of the emission spectrum strongly depends on the interlayer composition. A group of materials for thin-film **transistors** with general formula

(m-Fluorophenethylammonium)$_2$SnI$_4$ forms a **merotype series** (m = 2, 3, 4) based on the perovskite monolayers $(SnI_4)^{2-}$ (Mitzi *et al.* 2001b).

$\mathbf{A'_2A_mB_mX_{3m+1}}$ is the general formula for a family (Fig. 4.8) of layered **hybrid perovskites** based on (110) layers. In case of $\mathbf{[NH_2C(I){=}NH_2]_2(CH_3NH_3)_nSn_nI_{3n+2}}$ **polysomatic series**, methylammonium, (CH_3NH_3), and iodoformamidinium, $[NH_2C(I){=}NH_2]$, occupy the dodecahedral site and the interlayer, respectively (Mitzi *et al.* 1995). (111) layers (Fig. 4.9) of the perovskite structure occur in compounds with general formula $\mathbf{A'_2A_{q-1}B_qX_{3q+3}}$.

Note that, due to the different boundaries of the layers, the stoichiometry of the perovskite part is different in different cuts, that is, $A_{n-1}B_nX_{3n+1}$, $A_mB_mX_{3m+3}$, and $A_{q-1}B_qX_{3q+3}$ for (100), (110), and (111) layers, respectively. Except for the (110) cut, the dodecahedral cation A is absent in the one-octahedron thick layer compounds. This layer consists of isolated and corner sharing octahedra in the (111) and (100) cuts, respectively; in the order, the two cases are also known as 0D and 1D perovskite structures (Section 4.2.2).

4.2.4 *Ruddlesden–Popper and related series*

The $\mathbf{A_{n+1}B_nX_{3n+1}}$ and $\mathbf{A'A_{n-1}B_nX_{3n+1}}$ **series** share the same type of (100) perovskite layer; they are, respectively, known as **Ruddlesden–Popper** and **Dion–Jacobson series**. The compounds of the latter series with $A' = Bi_2O_2$ are known as **Aurivillius phases**. These layered perovskites are characterized by anion excess relative to ideal perovskite structures. The $\mathbf{A_nB_nX_{3n+2}}$ and $\mathbf{A_{n+1}B_nX_{3n+3}}$ **series**, which contain (110) and (111) perovskite layers, respectively, show anion excess too. The latter compounds can also be regarded as hexagonal perovskite-like members (Section 4.2.6) of the $\mathbf{(BaX)_2A_{n+1}B_nO_{3n+3}}$ **polysomatic series** (cf. Wilkens and Müller-Buschbaum 1991) where perovskite layers with thickness increasing versus n and distorted rock-salt AX layers are the building blocks.

The layered two-dimensional perovskites of the mentioned anion excess series show technologically important **photocatalytic**, **ionic conduction**, dielectric, magnetic, and **luminiscence** properties.

4.2.4.1 *Ruddlesden–Popper series*

The $\mathbf{A_{n+1}B_nX_{3n+1}}$ **polysomatic series** became known as the **Ruddlesden–Popper series** (Beznosikov and Aleksandrov 2000) subsequent to the earlier work of Ruddlesden and Popper (1958) on this type of compounds. The series is also known as $(AX + nABX_3)$ to emphasize the combination of perovskite slabs consisting of n (100) perovskite layers (ABX_3 = P) intercalated with one sodium chloride-like slab (AX = N) (Fig. 4.10). Shortly, the series is also described as NP_n thus emphasizing its polysomatic nature. Given the cell parameter a_p of the basic perovskite structure (its exact value depends on the nature of A, B, and X), two consecutive P n-layers sandwiching an N layer must be offset by $a_p/(2)^{1/2}$ in order to match the **sodium–chloride module**. The thickness of the latter module is approximately $a_p/(2)^{1/2}$. Consequently, the knowledge of n allows to model the crystal structure; in particular,

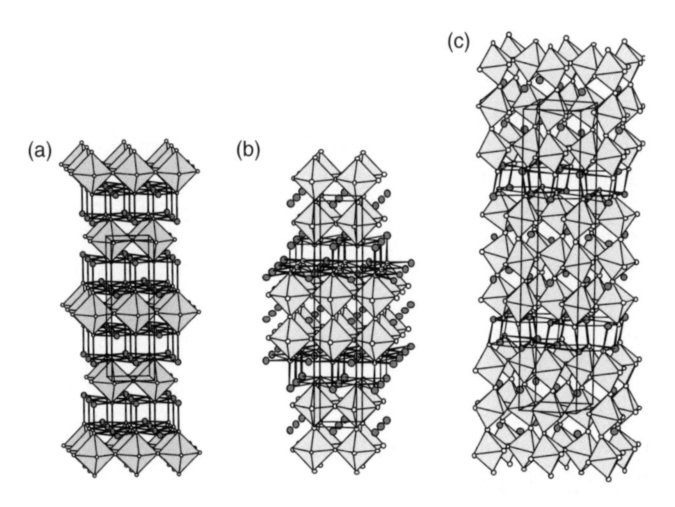

Fig. 4.10. Crystal structures of Sr_2TiO_4 (a), $Sr_3Ti_2O_7$ (b), and $Ca_4Ti_3O_{10}$ (c) representing, in the order, the members with $n = 1$, 2, and 3 of the Ruddlesden–Popper $A_{n+1}B_nX_{3n+1}$ series. The perovskite and sodium chloride modules are represented by shaded and not shaded polyhedra, respectively. The vertical axis is c in (a) and (b), and b in (c). (See text for reference.)

the value of the c parameter is $[2na_p + a_p/(2)^{1/2)}]$. So far, only members with $n < 6$ are known (e.g. the **$Sr_{n+1}Ti_nO_{3n+1}$ series**; Haeni *et al.* 2001). Some crystal structures have been experimentally determined, for others only theoretical models are known (Beznosikov and Aleksandrov 2000). The structures of **Sr_2TiO_4** ($n = 1$; $I4/mmmm$, $a = 3.90$, $c = 20.38$ Å; Ruddlesden and Popper 1957), **$Sr_3Ti_2O_7$** ($n = 2$; $I4/mmmm$, $a = 3.90$, $c = 20.38$ Å; Ruddlesden and Popper 1958) and of **$Ca_4Ti_3O_{10}$** ($n = 3$; $Pcab$, $a = 5.408$, $b = 27.143$, $c = 5.434$ Å; Elcombe *et al.* 1991) are shown in Fig. 4.10.

The structure of the member with $n = 1$ contains monolayers of perovskite and is also known as **K_2NiF_4** structure type. The most famous member with $n = 1$ is **$(La,Ba)_2CuO_4$** for which Bednorz and Müller (1986) discovered **high-T_c superconductivity** with $T_c \sim 30$ K. Members of the **$Ba_{n+1}(Pb_{1-x}Bi_x)nO_{3n+1}$ polysomatic series** are among materials which are intensively studied (cf. Antipov *et al.* 2002) to search new superconductors. The member with $n = 2$ and composition **$Sr_{1.8}La_{1.2}Mn_2O_7$** shows the phenomenon of **colossal magnetoresistance** (CMR; Moritomo *et al.* 1996). Transmission electron microscopy studies (Seshadri *et al.* 1997; Sloan *et al.* 1998; Bendersky *et al.* 2001) show complex and disordered intergrowths of different **polysomes** including some with $n > 5$, that is, of members which are not known as stable phases. The occurrence of polysomes that are known only as defects within a stable matrix is a well-documented phenomenon; cf., in particular the **biopyriboles** (Veblen and Buseck 1979; Veblen 1991) and the **inophites** (Section 4.4.5.2).

4.2.4.2 *(110) perovskite layers*

Niobates and **titanates** belonging to the series based on [110]-oriented perovskite modules, have been recently reviewed by Lichtenberg *et al.* (2001). As mentioned above, the series has general formula $A_nB_nO_{3n+2}$. Some members (e.g. $\mathbf{CaNbO_{3.50}}$ and $\mathbf{LaTiO_{3.50}}$) are among the highest-T_c ferroelectrics. If n is even, the coefficients can be divided by 2, thus the series looks like $\mathbf{A_{n'}B_{n'}O_{2n'+2}}$. Ishizawa *et al.* (1980 and references therein) have studied members with $n = 2$, A = (Ca, Sr), and B = (La, Ta), some of which are **piezoelectric** and **ferroelectric**. $\mathbf{Sr_2Mg_2F_4}$ belongs to the same series with F as anion and is a material with possible optical applications (Ishizawa *et al.* 2001). In these compounds, the appearance of the perovskite modules is the same as in Fig. 4.8.

4.2.5 *Anion deficient derivatives of cubic perovskites*

Some series of compounds alternate (100) perovskite layers with other types of inorganic layers (cf. Mitchell 2002). In the $\mathbf{A_nB_nO_{3n-1}}$ **series**, perovskite blocks $(ABO_3)_n$ consisting of n octahedral sheets alternate with one block of **brownmillerite-type structure** $A_2B_2O_5$. In its turn, the **brownmillerite** crystal structure ($\mathbf{Ca_2FeAlO_5}$; $n = 2$ member of the series; *Ibm*2, $a = 5.559$, $b = 14.507$, $c = 5.342$ Å; Jupe *et al.* 2001) consists of alternating octahedral and tetrahedral layers (Fig. 4.11). The B cations are represented by Fe and Al and statistically occupy both the tetrahedral and octahedral sites. Brownmillerite is a major component of **Portland cement**. Among minerals, also **srebrodolskite**, $\mathbf{Ca_2Fe_2O_5}$, adopts the brownmillerite structure. Members with $n = 3$ (e.g. $\mathbf{Ca_3Fe_2TiO_8}$; Grenier *et al.* 1976) and 4 (e.g. $\mathbf{Ca_4Fe_2Ti_2O_{11}}$; Gonzales-Calbet and Vallet-Regí 1987) are known.

In the $\mathbf{La_nNi_nO_{3n-1}}$ **polysomatic series**, which is known for the members $n = 2$ ($\mathbf{La_2Ni_2O_5}$), $n = 4$ ($\mathbf{La_4Ni_4O_{11}}$), and $n = \infty$ ($\mathbf{LaNiO_3}$), the tetrahedral layer of brownmillerite is substituted by a layer of NiO_4 squares (Gonzalez-Calbet *et al.* 1989).

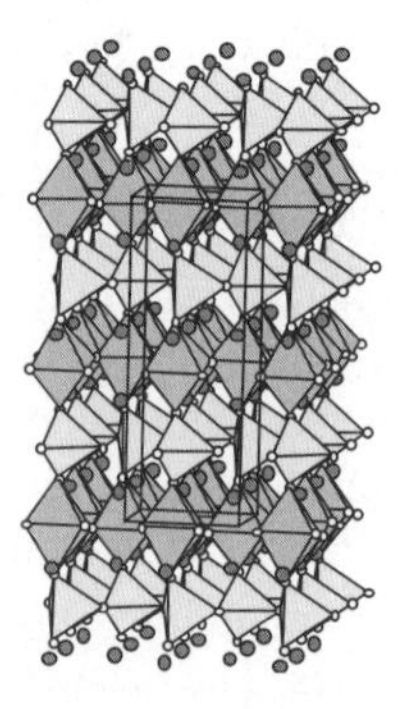

Fig. 4.11. Perspective view of the browmillerite crystal structure. Fe and Al are statistically distributed both in the octahedral (grey) and tetrahedral (light grey) sites. The grey circles represent Ca. The direction [010] is vertical. (See text for reference.)

4.2.6 *Derivatives of hexagonal perovskites*

As mentioned above (Section 4.2.1), the hexagonal polytypes of perovskite intrinsically show a low stability because of the presence of columns of face-sharing octahedra (Fig. 4.2). A periodic intercalation of different modules stabilizes a large variety of modular structures (cf. Mitchell 2002).

4.2.6.1 *Palmierite derivatives*

A group of hexagonal perovskite derivatives is based on alternating layers of hexagonal perovskite and **palmierite-type structure** [**palmierite** s.s. is the mineral $\mathbf{(K,Na)_2Pb(SO_4)_2}$]. Several compounds with $A_3B_2O_8$ stoichiometry adopt palmierite-type structure. It is based on the structure of the $9R$ perovskite polytype with partially occupied tetrahedral and octahedral sites. Figure 4.12 shows the structure of $\mathbf{Ba_3MoTiO_8}$ ($R\bar{3}m$, $a = 9.57$, $c = 21.29$ Å; Mössner and Kemmler-Sack 1985) as representative of the palmierite-type structure. Being based on the same sheets of closest-packing oxygen atoms, layers of the hexagonal perovskite- and palmierite-type structures can easily match and several mixed layer compounds of this kind are known (cf. Darriet and Subramanian 1995). In the structure of $\mathbf{Ba_8V_7O_{22}}$

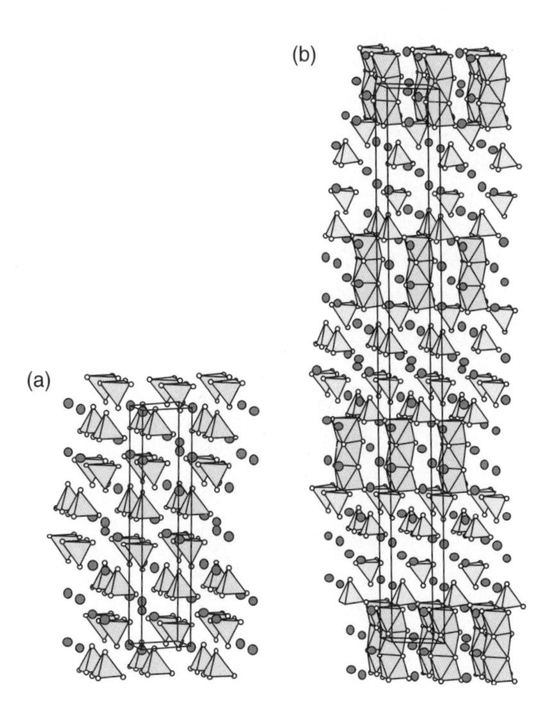

Fig. 4.12. Perspective view of the crystal structure of (a) Ba_3MoTiO_8 (palmierite-type) and (b) $Ba_8V_7O_{22}$; palmierite-type (block of tetrahedra) and $2H$-perovskite-type (block of face-sharing octahedra) alternate. (See text for reference.)

(Fig. 4.12; $R\bar{3}m$, $a = 5.784$, $c = 57.074$ Å; Liu and Greedan 1994), palmierite-type and $2H$-perovskite modules alternate. This compound is unusual in that V is found in three oxidation states, with V^{5+} occurring in the tetrahedral sites, V^{3+} in the octahedral sites, and V^{4+} distributed between both sites.

4.2.6.2 *Mixed hexagonal layers*

As example of a series of mixed hexagonal layer compounds, the case of the compounds with general formula $\mathbf{A_{3m+3n}A'_nB_{3m+n}X_{9m+6n}}$ is illustrated (Perez-Mato *et al.* 1999). In the crystal structures of these compounds, two types of tripled AX_3 layers, one with stoichiometry A_3X_9 and a second one with stoichiometry A_3X_6 because of X vacancies, are hexagonally stacked. Besides the octahedral sites which normally occur within the building modules (B site), the stacking of the A_3X_6 modules leads to the creation of trigonal prismatic sites for a cation A′; thus the stoichiometry of the module becomes $A_3A'X_6$. In the general formula, m and n indicate the number of $(ABX_3)_3$ and $A_3A'X_6$ in the structure, respectively. The main feature of the resulting structures are chains of octahedral and face-sharing trigonal prisms; the chains are separated by the A cations.

The structures of the two end-members of the series correspond to that of the **2*H* perovskite polytype** ($n = 0$ and $m = 1$; Fig. 4.2) and of $\mathbf{Sr_4PtO_6}$ ($n = 1$ and $m = 0$; $A = A' = $ Sr; $R\bar{3}m$, $a = 9.748$, $c = 11.879$ Å; Wong-Ng *et al.* 1999; Fig. 4.13). Polymers of face sharing octahedra occur in the structures of members with $m > 0$ (cf. Mitchell 2002). Some members of the series show interesting **magnetic properties** (cf. Darriet *et al.* 1997); often, **modulated structures** occur (cf. Battle *et al.* 1998; Perez-Mato *et al.* 1999).

4.2.6.3 *Magnetoplumbite derivatives*

The hexagonal structure-type of the mineral **magnetoplumbite**, $\mathbf{Pb(Fe^{3+}, Mn^{3+})_{12}O_{19}}$ (cf. Holtstam 2003), is adopted by several **ferrites** with general formula

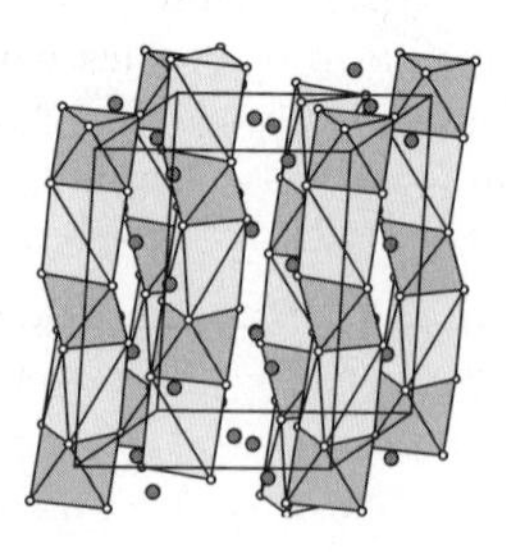

Fig. 4.13. Perspective view of the crystal structure of $\mathbf{Sr_3SrPtO_6}$, which is a member with $n = 1$ and $m = 0$ of the series $A_{3m+3n}A'_nB_{3m+n}X_{9m+6n}$. Light-grey trigonal prisms and grey octahedra contain A′ (Sr) and B (Pt) cations, respectively. The chains are separated by Sr cations (grey circles) in octahedral coordination corresponding to the A sites. (See text for reference.)

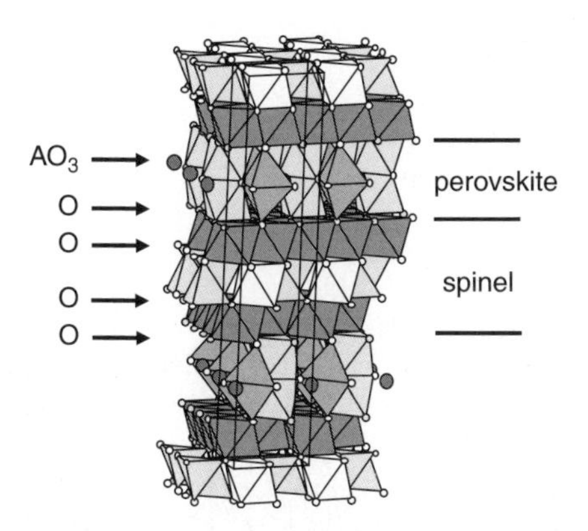

Fig. 4.14. Perspective view of the **magnetoplumbite-type structure** of $BaFe_{12}O_{12}$ showing alternating perovskite and spinel layers. Arrows indicate the AO_3 and O sheets mentioned in the text. The grey circles represent the large A (Ba) cations. (See text for reference.)

$AB_{12}O_{19}$. In Fig. 4.14, the type is represented by the structure of $\mathbf{BaFe_{12}O_{19}}$ ($P6_3/mmc$, a = 5.865, c = 23.099 Å; Collomb *et al.* 1986). The structure is essentially close-packed and consists of four oxygen sheets labelled O in Fig. 4.14, plus a sheet with stoichiometry AO_3 because 25 percent of the oxygen atoms are replaced by a large A cation. The AO_3 sheet, together with the two adjacent O sheets, forms a hexagonal perovskite layer where trigonal bypiramidal sites, instead of the usual prismatic ones, occur. The trigonal sites are occupied by the B (Fe) cation. The four O sheets form a **spinel layer** (Section 4.3.1) where both tetrahedral and octahedral sites are occupied by the B (Fe) cation. As a whole, the structure consists of alternating perovskite (with dimers of face-sharing octahedra) and spinel layers.

The crystal structure of **$\boldsymbol{\beta}$-Na-alumina, $\mathbf{NaAl_{11}O_{17}}$** (known as the mineral **diaoyudaoite**; Zhu *et al.* 1992), is a derivative of that of $\boldsymbol{\beta}$-$\mathbf{Al_2O_3}$ (**$\boldsymbol{\beta}$-alumina**) and can be described as a magnetoplumbite-type structure where the oxygen atoms of the fifth oxygen layer are 7 per cent substituted by A sites which, however, are only 50 per cent occupied. Thus, an AO sheet is realized instead of the AO_3 sheet described for magnetoplumbite. In other words, the β-Na–alumina structure (Collin *et al.* 1986) consists of a spinel layer plus an oxygen deficient perovskite layer where the A sites are inside infinite channels which allow them a high mobility. This structure is adopted by a series of **ionic conductors**, with the general formula $A_2O{\cdot}nBO_3$ (n = 5–11), which are known as **NASICON-type** (*Na* *S*uper*I*onic *CON*ductor) materials.

Complex intercalations between spinel and deficient perovskite layers, have been reported in some cases like the minerals **lindqvistite** [$\mathbf{Pb_2(Mg,Fe^{2+})Fe^{3+}_{16}O_{27}}$; Holtstam and Norrestam 1993] and **plumboferrite** [$\mathbf{Pb_2(Fe_{10.67}, (Mn,Fe)_{0.33})O_{18.33}}$; Holstam *et al.* 1995], $\mathbf{Ba_2Mg_6Al_{28}O_{50}}$ (Iyi *et al.* 1998), $\mathbf{Ca_2Mg_2Al_{28}O_{46}}$, $\mathbf{Ca_2Mg_2Al_{16}O_{27}}$ (Iyi *et al.* 1995), and $\mathbf{Sr_2MgAl_{22}O_{36}}$ (Iyi and Göbbels 1996).

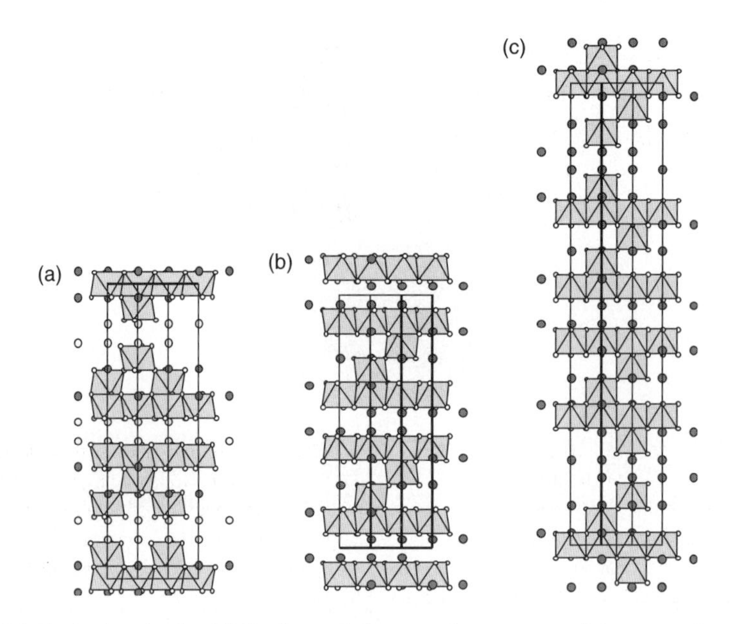

Fig. 4.15. Projection in the (100) plane of the crystal structures of three members of the series $A_nB_{n-1}O_{3n}$: (a) $La_4Ti_3O_{12}$ ($n = 4$), (b) $BaLa_4Ti_4O_{15}$ ($n = 5$) and (c) $Ba_2La_4Ti_5O_{18}$ ($n = 6$). The perovskite layer of the n-th member consists of $n - 1$ octahedral sheets. (See text for reference.)

4.2.6.4 *A cation deficient series*

Complex oxides with cation-deficient hexagonal perovskite structure and general formula $\mathbf{A_nB_{n-1}O_{3n}}$ ($n \geq 3$) have been synthesized for A = Ca, Sr, Ba, La, and B, that is, metals with oxidation number from two to six. The known crystal structures are trigonal and can be derived from a hexagonal perovskite-type by the introduction of B cation vacancies. Figure 4.15 shows the crystal structure of the three members of the series, where various B vacant layers of hexagonal perovskite-type structure are separated by a completely B vacant octahedral sheet of the same type. Each layer of the n member consists of $n - 1$ octahedral sheets. The described members (Harre *et al.* 1998; Teneze *et al.* 2000) correspond to $n = 4$ ($\mathbf{La_4Ti_3O_{12}}$; $R\bar{3}$, $a = 5.551$, $c = 26.178$ Å), $n = 5$ ($\mathbf{BaLa_4Ti_4O_{15}}$; $P\bar{3}c1$, $a = 5.567$, $c = 22.460$ Å) and $n = 6$ ($\mathbf{Ba_2La_4Ti_5O_{18}}$; $R\bar{3}$, $a = 5.581$, $c = 41.056$ Å).

4.3 Structures based on spinel modules

A large number of complex oxides TM_2O_4 (plus some non-oxygenated compounds) are based on the ideally cubic ($Fd\bar{3}m$) **spinel** ($\mathbf{MgAl_2O_4}$) structure-type where the O^{2-} ions form a ccp array. The spinel cell is eight times larger than that of the basic ccp cell and contains 32 oxygens (Fig. 4.16). The T and M atoms occupy 1/8 and 1/2 of

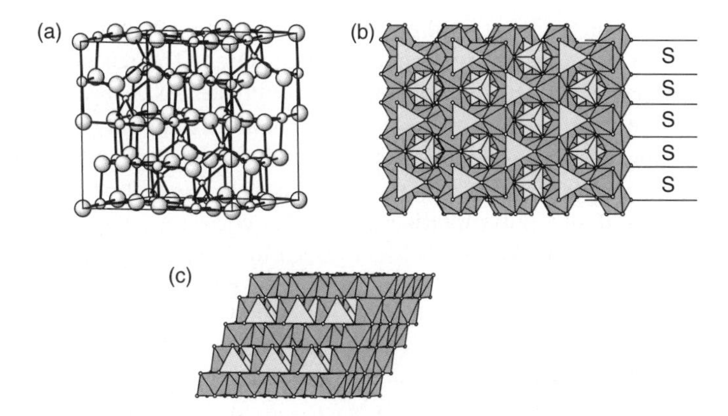

Fig. 4.16. Spinel-type structure shown (a) as perspective view of oxygen closest-packing (large spheres) with M (intermediate spheres) and T (small spheres) cations occupying 1/2 and 1/8 of the octahedral and tetrahedral sites, respectively. In (b), the packing of octahedra (grey) and tetrahedra (light grey) is projected into the $(1\bar{1}0)$ plane and the [011] direction is vertical; in (c), the packing of octahedra and tetrahedra is projected into the $(1\bar{1}1)$ plane and the [111] direction is vertical. The thickness of the [011] layers is shown in (b).

the tetrahedral and octahedral sites, respectively. In the **inverse spinel structure**, half of the M atoms occupy tetrahedral sites and the other half is statistically distributed among the octahedral sites. T and M normally, but not necessarily, represent T^{2+} and M^{3+} cations. To balance the eight negative charges in the formula unit, the constrain [$2\times$ (charge of M) + (charge of T) = 8+] holds.

Technologically important materials, like the magnetic **ferrites** ($T^{2+}Fe_2O_4$ with T = Fe, Ni, Cu, Mg), adopt the spinel and inverse spinel structures. $\mathbf{Fe^{2+}Fe_2^{3+}O_4}$ occurs in nature as **magnetite** and shows an inverse spinel structure, being half Fe^{3+} in tetrahedral sites. Even if not widespread as perovskite, modules with the spinel structure occur in several crystal structures and some have already been mentioned in the section dedicated to the perovskite modules (Section 4.2.6.3). Both (011) and (111) modules of spinel (Fig. 4.16) are known to occur in crystal structures. Note that the closest-packing sheets of oxygen ions are perpendicular to [111].

4.3.1 (111) *spinel layers*

Occurrence of (111) spinel layers has already been described in the magnetoplumbite- (Fig. 4.14) and β-Na-alumina-type structures (Section 4.2.6.3).

4.3.1.1 *Högbomite, nigerite, and taaffeite group*

The recognition of, often unsuspected, (approximate) closest-packing anions, even in complex structures (cf. Section 4.5.9), has in some cases suggested a modular description based on polytypic stackings of oxygen monolayers (sheets). However, at least in complex structures, such simple modular description cannot univocally constrain the crystal structure and the chemical composition. The recent review of the **högbomite**, **nigerite**, and **taaffeite structures** (Armbruster 2002), offers an emblematic case where a shift of the description viewpoint allows not only to synthetically express structure and composition, but also to predict unknown structures (Table 4.1). In this case, the author abandoned a previous polytypic description in favour of an interpretation based on a **polysomatic series** S_mN_n consisting of (111), **spinel** (S), and (001) **nolanite** (N) **layers**, as first recognized by Grey & Gatehouse (1979). A polytypic description of the group members was based first (McKie 1963) on the observation that, for each known member, $c = n4.6$ Å, and later (Peacor 1967) on the stacking of compact 2.6 Å-thick oxygen layers.

The hexagonal unit cell of the group members (Fig. 4.17) has $a = 5.72$ Å. Both the S, $T_2M_4O_8$, and N, TM_4O_8 or $TM_4O_7(OH)$, modules have the same 4.6 Å thickness. In fact, both modules are based on an oxygen closest-packing layer that has cubic (c) environment in spinel and hexagonal environment (h) in nolanite. T and M represent tetrahedrally and octahedrally coordinated cations, respectively. The composition of högbomite and nigerite group minerals thus depends on the composition of the two modules and the ratio m/n between the number of spinel and nolanite modules assembling the structure. Ti^{4+} and Sn^{4+} are the octahedral cation of the nolanite module in the högbomite and nigerite group, respectively.

All known minerals of the **högbomite** and **nigerite group** are composed of spinel modules belonging to the aluminum spinel subgroup. According to the tetrahedral content of the spinel module, a prefix is added to the group name. This prefix is 'zinco', 'ferro', or 'magnesio' according the **gahnite** component, $ZnAl_2O_4$, the **hercynite** component, $FeAl_2O_4$, or the spinel component, $MgAl_2O_4$, dominates the spinel module, in the order. To characterize the various **polysomes** found for each subgroup (**zincohögbomite**, **ferrohögbomite**, **magnesiohögbomite**), a hyphenated

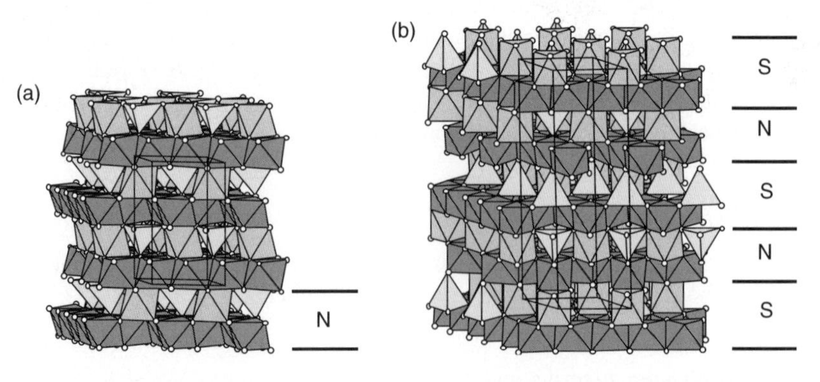

Fig. 4.17. Perspective view of the crystal structures of (a) nolanite $[(V,Fe,Ti,Al)_{10}O_{14}(OH)_2$; $P6_3mc$; $a = 5.897$, $c = 9.256$ Å; Gatehouse *et al.* 1983] and (b) ferrohögbomite-2N2S $[(Fe^{2+}_3ZnMgAl)(Al_{14}Fe^{3+}Ti^{4+})O_{30}(OH)_2$; $P6_3mc$; $a = 5.712$, $c = 18.317$ Å; Hejny *et al.* 2002]. The relevant labels indicate the spinel (S; Fig. 4.16) and nolanite (N) modules.

Table 4.1 Modular composition of the högbomite and nigerite-group polysomes and related structures. Names of minerals between quotation marks are not approved species. When not indicated, references to structural data can be found in Armbruster (2002)

Polytype Peacor (1967)	Space group	c_h(Å)	Sequence of cubic 'c' and hexagonal 'h' close packed oxygen layers	Sequence of spinel (S) and nolanite (N) modules	Mineral name (quotation marks for species not yet approved)
6*C*	$Fd3m$	13.8	chch	3S	
4*H*	$P6_3mc$	9.2	chch	NN	Nolanite
6*T*	$P\bar{3}m1$	13.8	2 × (c + ch)	NNS	Ferronigerite-2N1S
					'Zinconigerite-2N1S'
					Magnesionigerite-2N1S
8*H*	$P6_3mc$	18.4	2 × (cc + ch)	NSNS	Magnesiohögbomite-2N2S
					Ferrohögbomite-2N2S*
					Zincohögbomite-2N2S
10*T*	$P\bar{3}m1$	23.0	2 × (ccc + ch)	NSSNS	Magnesiohögbomite-2N3S
12*H*	$P6_3mc$	27.6	2 × (cccc + ch)	NSSNSS	Structure modelled
14*T*	$P\bar{3}m1$	32.2	2 × (ccccc + ch)	NSSSNSS	Structure modelled
16*H*	$P6_3mc$	36.8	2 × (cccccc + ch)	NSSSNSSS	Zincohögbomite-2N6S
18*R*	$R\bar{3}m$	41.4	3 × (cc + chhc)	3 × (NNS)	Known for taaffeite group
24*R*	$R\bar{3}m$	55.2	3 × (cccc + hcch)	3 × (NNSS)	Zincohögbomite-2N6S
					'Zinconigerite-6N6S'
					Magnesionigerite-6N6S
					Magnesiohögbomite-6N6S
30*R*	$R\bar{3}m$	69.0	3 × (cccccc + hcch)	3 × (NNSSS)	Predicted
36*R*	$R\bar{3}m$	82.8	3 × (cccccccc + hcch)	3 × (NNSSSS)	'Ferrohögbomite-6N12S'
					Predicted

*Hejny *et al*. 2002.

suffix composed of the total number of nolanite (N) and spinel (S) modules is attached. Table 4.1 shows how the polytype nomenclature proposed by Peacor (1967) is transformed to the new polysome nomenclature.

As mentioned above, the polytype symbol gives complete information neither on the structure (i.e. exact stacking of the modules) nor on the chemical composition. This type of information is instead fully contained in the sequences of symbols both in terms of c and h layers and of S and N modules. A chemical analysis alone gives information only on the ratio m/n, a quantity which is in common to different polysomes (Table 4.1).

The **taaffeite-group** minerals are composed of spinel and modified modules where Be occupies a tetrahedral site close to the hydrogen position in the nolanite module (Armbruster 2002).

4.3.2 (011) *spinel layers*

(011) Spinel layers (Fig. 4.16) are also present in spinelloids (cf. Section 3.1.3, 2.6.5).

4.3.2.1 *The sapphirine polysomatic series*

The members of the **sapphirine series** (Merlino and Pasero 1997; Merlino and Zvyagin 1998; Zvyagin and Merlino 2003) have been interpreted as polysomatic intergrowths of (010) **pyroxene**-like P modules with (011) **spinel**-like S modules in several distinct proportions (Fig. 4.18). The P modules are one-chain-width broad (010) slices of pyroxene structure represented in Fig. 4.18 by $\mathbf{MgGeO_3}$ ($C2/c$, $a = 9.640$, $b = 8.978$, $c = 5.173$ Å, $\beta = 101.14°$; Yamanaka *et al.* 1985), whereas the S modules are one-octahedron/tetrahedron broad (011) slices of spinel structure (Fig. 4.16). The ccp anion framework of both slice types is coherent across the slab boundaries.

The introduction of the P_mS_n polysomatic series has been largely influenced by HRTEM observation of intergrowth **polysomes** in **sapphirine** (Christy and Putnis 1988; Merlino and Pasero 1997) and **rhönite** (Bonaccorsi *et al.* 1990) matrices. Besides the end-members P and S, in their review of the P_mS_n polysomatic series, Merlino and Pasero (1997) and Zvyagin and Merlino (2003) report the members illustrated below (Fig. 4.18). The occurrence of polytypes in the P_mS_n polysomatic series is discussed in Sections 1.5.2 and 1.7.

PS polysome This polysome corresponds to the crystal structure of the **sapphirine–aenigmatite group** members whose general formula can be written as $A_2M_6T_6O_{20}$; A, M, and T indicate cations in eight-, six-, and four-fold coordination, respectively. Merlino and Pasero (1997) list 16 minerals and synthetic phases with the sapphirine-type structure. Some of the synthetic compounds are important industrial phases of the **system** $\mathbf{CaO–Fe_2O_3–Al_2O_3–SiO_2}$ (cf. Bonaccorsi *et al.* 1990). As shown by the **sapphirine-2*M*** structure [$\mathbf{(Mg_2Al_2)(Al_2Si)O_{10}}$; $P2_1/n$, $a = 9.814$, $b = 14.438$, $c = 9.957$ Å; $\beta = 110.39°$; Higgins and Ribbe 1979; cell parameters transformed

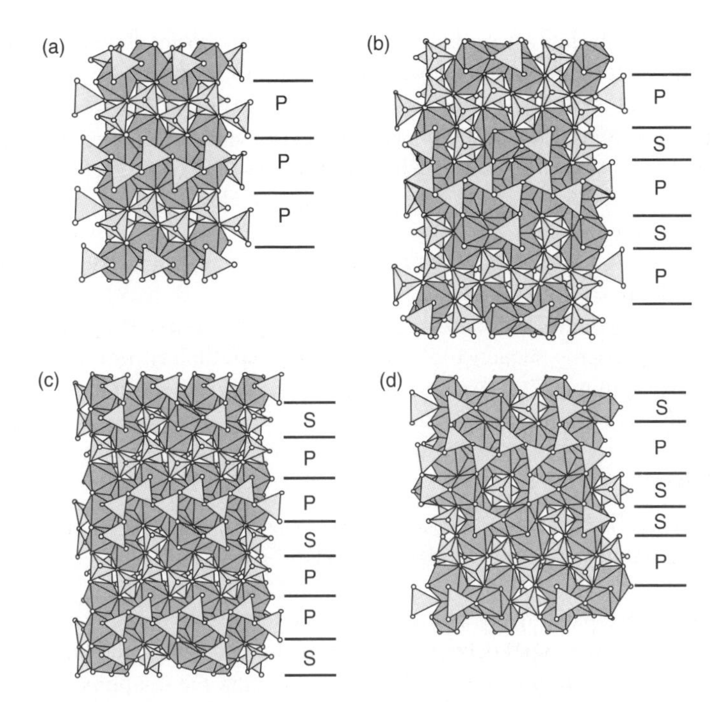

Fig. 4.18. Projection of the crystal structures of members belonging to the P_mS_n polysomatic series: (a) pyroxene projected along [111] to show the (010) P modules which occur in the sapphirine series, (b) sapphirine-2H, (c) surinamite, (d) $CaFe_3AlO_7$. The (011) S spinel layer that occurs in the sapphirine series is shown in Fig. 4.16. (See text for reference.)

according to Zvyagin and Merlino (2003) by $101/0\bar{1}0/00\bar{1}$], in the PS polysome the modules P and S alternate along [001], as first recognized independently by Christy and Putnis (1988) and Barbier and Hyde (1988).

P_2S polysome This polysome, with sequence PPS, occurs in **surinamite-1*M*** (**$Mg_3Al_4Si_3BeO_{16}$**; $P2/n$, $a = 9.631$, $b = 11.384$, $c = 9.916$ Å; $\beta = 109.3°$; Moore and Araki 1983; a and b interchanged). According to Zvyagin and Merlino (2003), the same sequence occurs in: (i) **$Mg_4Ga_4Ge_3O_{16}$** ($A2/a$, $a = 10.073$, $b = 23.733$, $c = 10.324$ Å; $\beta = 109.29°$; Barbier 1998; a and c interchanged), (ii) the hypothetical polytypes **surinamite-2*M*** ($A2/a$, doubled a), (iii) the **sapphirine polymorphs 1*A*** ($P\bar{1}$, $a = 9.35$, $b = 7.14$, $c = 9.35$ Å, $\alpha = 109.0$, $\beta = 109.5$, $\gamma = 83.80°$), and 2M ($P2_1/m$, $a = 9.35$, $b = 13.5$, $c = 9.35$ Å, $\beta = 109.5°$).

PS_2 polysome This polysome occurs, with sequence SSP, in the synthetic **ferrite $CaFe_3AlO_7$** ($P\bar{1}$, $a = 10.406$, $b = 11.805$, $c = 10.593$ Å, $\alpha = 94.17$, $\beta = 110.07$, $\gamma = 111.30°$; Arakcheeva *et al.* 1991; Hamilton *et al.* 1989; cell parameters of the latter paper transformed by $111/00\bar{1}/\bar{1}00$), and **SFCA-I** (**$Ca_{3.18}Fe^{2+}_{0.82}Fe^{3+}_{14.66}Al_{1.34}O_{28}$**; $P\bar{1}$, $a = 10.431$, $b = 11.839$, $c = 10.610$ Å,

$\alpha = 94.14$, $\beta = 110.27$, $\gamma = 111.35°$; Mumme *et al.* 1998; *b* and *c* exchanged). On the basis of the cell parameter values ($a = 9.87$, $b = 21.44$, $c = 10.76$ Å, $\beta = 106.6°$) and chemical composition, Bonaccorsi *et al.* (1990) and Merlino and Pasero (1997) attributed to the synthetic phase A of Hamilton *et al.* (1989) the same polysomatic sequence SSP occurring in $CaFe_3AlO_7$.

P_2S_3 polysome On the basis of the cell parameter values ($a = 10.32$, $b = 17.95$, $c = 10.42$ Å, $\beta = 109.4°$) and chemical composition, Bonaccorsi *et al.* (1990) and Merlino and Pasero (1997) attributed to the synthetic phase B of Hamilton *et al.* (1989) the polysomatic sequence SSPSP. This hypothesis has been experimentally confirmed by Mumme (2003) who has refined the crystal structure of a phase B (renamed **SFCA-II**) with composition $Ca_{5.1}Al_{9.3}Fe^{3+}_{18.7}Fe^{2+}_{0.9}O_{48}$ and triclinic cell $a = 10.338$, $b = 10.482$, $c = 17.939$ Å, $\alpha = 90.38$, $\beta = 89.77$, $\gamma = 109.40°$.

4.4 Slicing tetrahedral and octahedral layers

Octahedral O close-packing layers with fully (brucite-type) or partially (e.g. gibbsite-, spinel-, and corundum-type) occupied octahedra are recurrent in modular structures, as it has been shown in previous sections. In particular, the coupling of tetrahedral silicate T layers with dioctahedral (**gibbsite**) and trioctahedral (**brucite**) O layers constitutes the building blocks of TO (or 1 : 1) and TOT (or 2 : 1) layer silicates (Fig. 4.19). For example, as mentioned in Section 1.8.2.5, the 2 : 1 layer silicates can be grouped in an A_nB_m **merotype series** where the TOT module (A) is the fixed building module and B is the interlayer variable module. **Talc** is representative of $n = 1$ and $m = 0$; **micas** (B = alkaline or earth alkaline cation; Fig. 4.19), **chlorites** (B = octahedral sheet), and **'illites'** (B = H_2O) are well known members with $n = 1$ and

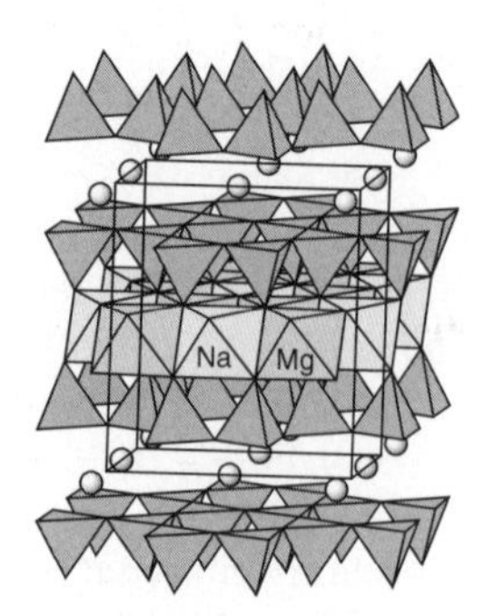

Fig. 4.19. Perspective view of the structure of the new trioctahedral mica **shirokshinite** [$\mathbf{K(NaMg_2)Si_4O_{10}F_2}$; *C*2/*m*, polytype 1*M*; $a = 5.269$, $b = 9.092$, $c = 10.198$ Å, $\beta = 100.12°$; Ivaldi *et al.* 2003] to show the (001) TOT layer. Na- and Mg-bearing octahedra are differently shaded.

$m = 1$. Because of the (pseudo)hexagonal symmetry of the oxygen sheets, polytypism is widespread in the layer silicates [cf. Nespolo and Ďurovič (2002) for micas].

Slices of mica (and talc) cut perpendicularly to the TOT occur in several silicate structures, for example in biopyriboles which represent the first polysomatic series introduced by Thompson (1978) (cf. Section 1.2.1.2). In the following, examples of other series based on TOT and/or TO modules are illustrated. Among these examples is the wide family of **heterophyllosilicates** that includes mainly titanosilicates, a kind of compounds that is increasingly attracting the attention of the materials scientists because of their microporous properties (cf. Rocha and Anderson 2000).

4.4.1 *Nafertisite and the heterophyllosilicates*

By analogy with **phyllosilicates**, a group of **titanium silicates** whose structures are based on TOT-like layers has been named **heterophyllosilicates** (Ferraris *et al.* 1997). In these structures, a row of Ti-polyhedra periodically substitutes a row of disilicate tetrahedra in the T tetrahedral sheet that is typical of the layer silicates (Fig. 4.20). HOH layers are thus obtained where H stands for *hetero* to indicate the presence of the rows of five- or six-coordinated Ti in a sheet corresponding to the T sheet of the layer silicates. The lengths of octahedral and tetrahedral $O \cdots O$ edges are very close, thus, the insertion of the octahedra in a T sheet does not produce strain.

As summarized by Ferraris (1997), depending on the periodicity of the Ti substitution, three types of HOH layers (Fig. 4.20) are known so far (in the following formulae I and Y represent interlayer cations and octahedral cations, respectively):

1. **Bafertisite-like $(HOH)_B$ layer**—A bafertisite-like module B = $I_2Y_4[Ti_2O_4Si_4O_{14}](O,OH)_2$ is one-to-one intercalated with a one-chain-wide mica-like module M = $IY_3[Si_4O_{10}](O,OH)_2$.
2. **Astrophyllite-like $(HOH)_A$ layer**—Relative to the $(HOH)_B$ layer, a second one-chain-wide mica-like module M is present between two bafertisite-like modules.
3. **Nafertisite-like $(HOH)_N$ layer**—Relative to the $(HOH)_B$ layer, a second and a third one-chain-wide mica-like module M are present between two bafertisite-like modules [alternatively, a second M module is intercalated in the $(HOH)_A$ layer].

4.4.1.1 *Modelling the structure of nafertisite*

A structural model for **nafertisite** (nfr) (Figs 4.20 and 4.21) $\{(Na,K,\square)_4(Fe^{2+}, Fe^{3+}, \square)_{10}[Ti_2O_3Si_{12}O_{34}](O,OH)_6$; $A2/m$, $a = 5.353$, $b = 16.176$, $c = 21.95$ Å, $\beta = 94.6°$; Ferraris *et al.* 1996} was obtained by comparison with the crystal structures of **bafertisite** {bft; $Ba_2(Fe,Mn)_4[Ti_2O_4Si_4O_{14}](O,OH)_2$; $P2_1/m$, $a = 5.36$, $b = 6.80$, $c = 10.98$ Å, $\beta = 94°$; Guan *et al.* 1963; Pen and Shen 1963; Rastsvetaeva *et al.* 1991} and **astrophyllite** {ast; $(K,Na)_3(Fe,Mn)_7[Ti_2O_3Si_8O_{24}](O,OH)_4$; $P\bar{1}$, $a = 5.36$, $b = 11.63$, $c = 11.76$ Å, $\alpha = 112.1$, $\beta = 103.1$, $\gamma = 94.6°$; Woodrow 1967; Shi *et al.* 1998}. The following points were noted:

1. The difference in composition between astrophyllite and nafertisite is about $(I,\square)(Y,\square)_3[Si_4O_{10}](OH,O)_2$; that corresponds to half the difference between

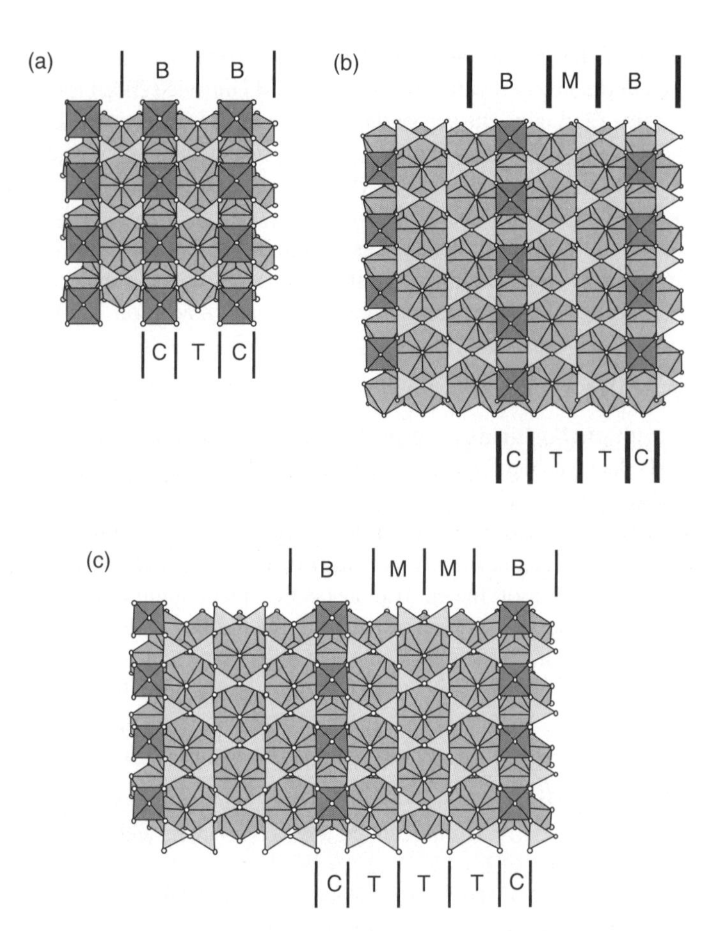

Fig. 4.20. Orthogonal projection of the bafertisite-like (a), astrophyllite-like (b), and nafertisite-like (c) HOH heterophyllosilicate layers. The modules B, M, C, and T used for different polysomatic descriptions are indicated (see text). Ti-octahedra belonging to the H hetero sheet are dark-shaded; octahedra of the O sheet are grey-shaded. (See text for reference.)

the compositions of nafertisite and bafertiste and is comparable to the composition of **mica**. Ignoring details, it turns out that, as mentioned above, an astrophyllite layer differs from a bafertisite layer only in having an additional mica-like module (M) between two (010) slabs built up by Ti- and Si-polyhedra (bafertisite-like module, B).

2. Bafertisite, astrophyllite, and nafertisite have a common value of $a \sim 5.4$ Å, which matches the a value of mica.
3. $(b_{\text{ast}} - b_{\text{bft}}) \sim 1/2(b_{\text{nfr}} - b_{\text{bft}}) \sim 4.7$ Å corresponds to $b/2$ in mica.
4. $(d_{002})_{\text{nfr}} = 10.94$ Å matches the thickness of one structural HOH layer in bafertisite and astrophyllite.

4.4.1.2 *The heterophyllosilicate polysomatic series*

Following Ferraris *et al.* (1996 1997), bafertisite, astrophyllite, and nafertisite (Figs 4.20 and 4.21) are members of a B_mM_n **polysomatic series**, which is based on B (bafertisite-like) and M (mica-like) modules and has a general formula $I_{2+n}Y_{4+3n}[Ti_2(O')_{2+p}Si_{4+4n}O_{14+10n}](O'')_{2+2n}$. In the formula, atoms belonging, even in part, to the H sheet are shown in square brackets. I are large (alkali) interlayer cations and Y are octahedral cations; O′ (bonded to Ti) and O″ (belonging to the octahedral sheet only) can be oxygen, OH, F, or H_2O; the $14 + 10n$ oxygens are bonded to Si. The value of p depends on the environment of Ti. This cation is five- or six-co-ordinated according to polyhedra that share one corner with the octahedral sheet and four corners with four Si-tetrahedra. In case of octahedral co-ordination, the sixth corner of the Ti-polyhedron can be (Figs 4.21 and 4.22) (i) unshared ($p = 2$), (ii) shared with a second Ti-octahedron ($p = 1$), or (iii) with an interlayer anion ($p = 0$); $p = 0$ occurs also in the cases of co-ordination number five and of an edge shared between two Ti-octahedra.

The heterophyllosilicates have also been described by using differently defined modules (Christiansen *et al.* 1999), a possibility which is not rare in modular crystallography (Merlino 1997*b*). Christiansen *et al.* (1999) consider the layer HHO instead of HOH and slice it into two rod-shaped modules (Figs 4.20 and 4.21): a T module with mica composition and a C module which consists of two corner sharing hetero-octahedra and one octahedron belonging to the O sheet.

4.4.2 *Branching the heterophyllosilicates*

Each of the three types of HOH hetero-layer occurs in isomorphous series originated by complex substitutions both in the H and O sheets. Polytypes are known for several

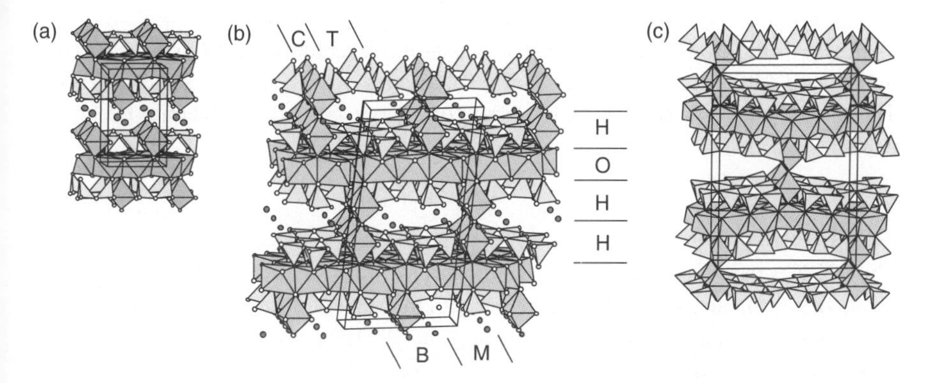

Fig. 4.21. Perspective view of the bafertisite (a), astrophyllite (b), and nafertisite (c) structures. In (b) the H and O sheets and the modules B, M, C, and T used for different polysomatic descriptions are indicated (see text). Note that in (b) and (c) the hetero Ti-octahedron bridges two HOH layers. (See text for reference.)

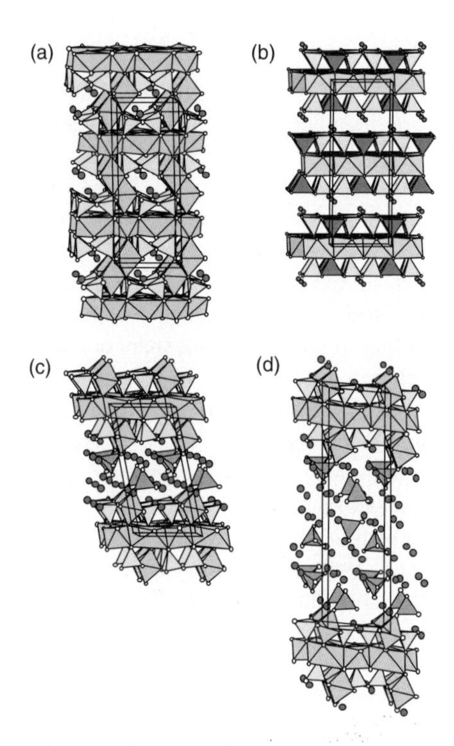

Fig. 4.22. Perspective view of the crystal structures of members of the bafertisite plesio-merotypes series to show different interlayer content (merotype) and modifications of the HOH layer (plesiotype). In the seidozerite structure (a), the hetero-octahedra share an edge, (b) in the lamprophyllite structure the hetero-polyhedron is a tetragonal pyramid, (c) in lomonosovite, cations, and phosphate tetrahedra occur in the interlayer, (d) in polyphite the interlayer consists of two nacaphite layers. (See text for reference.)

members of the isomorphous series (Zvyagin and Vrublevskaya 1976; Dornberger-Schiff *et al.* 1985; Christiansen *et al.* 1999; Piilonen 2001; Krivovichev *et al.* 2003). The $(HOH)_B$ layer is the building module of a large titanosilicate family described below.

4.4.2.1 *The bafertisite series*

The $(HOH)_B$ bafertisite-like layer is the most versatile of the three heterophyllosilicate layers being able to sandwich a large variety of interlayer fragments. As shown below, the variety of crystal structures containing the $(HOH)_B$ layer is by far larger than that reported for structures containing the TOT layer. However, the latter appears in important rock-forming minerals, while the titanosilicates are found only in rare hyperalkaline rocks (Khomyakov 1995).

Two main groups of $(HOH)_B$-bearing compounds are known: the götzenite group, which have Ca instead of Ti in the hetero-octahedra and is described elsewhere for its polytypism (Sections 2.1.3.3 and 2.4.3.1; cf. Christiansen and Rønsbo 2000), and the complex **seidozerite** or **bafertisite series** (Ferraris *et al.* 1997) where $(HOH)_B$ alternates with a large variety of interlayer contents. The latter titanosilicates are related by **merotypy** to the B_1M_0 member of the heterophyllosilicate polysomatic series. For B_1M_0 the general formula given above becomes $A_2\{Y_4[Z_2(O')_{2+p}Si_4O_{14}](O'')_2\}W$ ($A + W = I$ in the general formula). In this formula:

1. $[Z_2(O')_{2+p}Si_4O_{14}]^{\nu-}$ is a complex anion representing the heterophyllosilicate H sheet.
2. $\{Y_4[Z_2(O')_{2+p}Si_4O_{14}](O'')_2\}^{\mu-}$ is a complex anion representing the $(HOH)_B$ layer; it is within braces to leave in evidence the interlayer content A and W (Table 4.2).
3. A = large interlayer cations; W = (complex) interlayer content (cations, anions, water) different from A; Y = cations forming the O sheet; Z = cations occupying the hetero polyhedron (usually an octahedron) belonging to the H sheet.
4. p same meaning as in the general formula.

Fixing as c the cell parameter outside the $(HOH)_B$ layer, all the titanosilicates listed in Table 4.2 are characterized by similar values of $a \sim 5.5$ Å and $b \sim 7$ Å (or multiples) occurring in the layer. As in the layer silicates, two adjacent $(HOH)_B$ modules define an interlayer space that may contain either a single cation or an entire mineral-forming module, like **nacaphite** (**$Na_2Ca[PO_4]F$**; $C1$, $a = 10.654$, $b = 24.443$, $c = 7.102$ Å, $\alpha = 89.99$, $\beta = 90.01$, $\gamma = 90.01°$; Sokolova *et al.* 1989), which occurs in **quadruphite**, **polyphite** (Khomyakov *et al.* 1992) and **sobolevite** (Sokolova *et al.* 1988). Following Egorov-Tismenko (1998), Christiansen *et al.* (1999) recognize sobolevite and quadruphite as OD polytypes. Besides nacaphite, other stoichiometric compounds act as interlayer content: **barite, $BaSO_4$**, in **innelite** (Chernov *et al.* 1971), **Na_3PO_4** in **lomonosovite** (Belov *et al.* 1978), **vuonemmite** (Ercit *et al.* 1998), and **bornemanite** (Ferraris *et al.* 2001*b*; Section 4.4.2.2); **β-Na_3VO_4** in **$Na_8\{(NaTi)_2[Ti_2O_2Si_4O_{14}]O_2\}(VO_4)_2$** (Massa *et al.* 2000). Members of the group reported in Table 4.2 were originally called the seidozerite-derivatives by Egorov-Tismenko and Sokolova (1990).

The interlayer content determines the value of the c parameter which, of course, increases with the complexity of the sandwiched module (Table 4.2); on its own, the thickness of the $(HOH)_B$ layer is about 10 Å. Where the interlayer is simple, the c parameter is close to $n \times 10$ Å, with n indicating the number of independent $(HOH)_B$ layers. In Table 4.2, the compounds are listed in order of the increasing parameter $t = d(001)/n$ and, implicitly, in order of increasing complexity of the chemical formula. In **bornemanite** and M72 (Section 4.4.2.2), modules of seidozerite alternate with modules of lomonosovite and murmanite, respectively; thus, the value of t does no longer directly indicate the complexity of a single interlayer content.

The inclusion of compounds with unknown structure in the bafertisite series by Ferraris (1997) was done on the basis on their cell parameters and chemical data, that is, by an approach consistent with the structural application of modularity concepts (Merlino 1997*b*). In the case of **lamprophyllite**-group phases, **delindeite**,

Table 4.2 Members of the bafertisite mero-plesiotype series in order of increasing thickness $t = d(001)/n$, with n corresponding to the number of $(HOH)_B$ layers in the cell. The content of the heteropolyhedral H sheet is shown in square brackets and that of the $(HOH)_B$ layer is within braces; thus the composition of the interlayer appears outside the braces. When not given, a key reference can be found in the text or Ferraris *et al.* (2001*b*)

Mineral	Formula	t(Å)
Seidozerite	$Na_2\{(Na,Mn,Ti)_4[(Zr,Ti)_2O_2Si_4O_{14}]F_2\}$	8.93
Lamprophyllite	$(Sr,K,Ba)_2\{(Na(Na,Fe)_2Ti)[Ti_2O_2Si_4O_{14}](O,OH)_2\}$	9.68
Barytolamprophyllite	$(Ba,Na)_2\{(Na,Ti,Fe,Ba)_4[Ti_2O_2Si_4O_{14}](OH,F)_2\}$	9.80
Nabalamprophyllite[a]	$Ba(Na,Ba)\{Na_3Ti[Ti_2O_2Si_4O_{14}](OH,F)_2\}$	9.80
Orthoericssonite	$(Ba,Sr)_2\{(Mn,Fe)_4[Fe_2O_2Si_4O_{14}](OH)_2\}$	10.11
Ericssonite[b]	$Ba_2\{Mn_4[Fe_2O_2Si_4O_{14}](OH)_2\}$	10.16
Shkatulkalite[b]	$\{(Na,Mn,Ca,\square)_4[(Nb,Ti)_2(O,OH)_2Si_4O_{14}](OH,F)_2\}\cdot 2H_2O$	10.34
Surkhobite[c]	$(Ca,Ba,Na,K)_2\{(Fe,Mn)_4[Ti_2O_2Si_4O_{14}](F,OH,O)_3\}$	10.36
Jinshajiangite[b]	$(Na,Ca)(Ba,K)\{(Fe,Mn)_4[(Ti,Nb)_2O_3Si_4O_{14}](F,O)_2\}$	10.37
Perraultite	$(Na,Ca)(Ba,K)\{(Mn,Fe)_4[(Ti,Nb)_2O_3Si_4O_{14}](OH,F)_2\}$	10.38
Delindeite	$Ba_2\{(Na,\square)_3Ti\ [Ti_2(O,OH)_4Si_4O_{14}](\ H_2O,OH)_2\}$	10.73
Bafertisite	$Ba_2\{(Fe,Mn)_4[Ti_2O_2(O,OH)_2Si_4O_{14}](O,OH)_2\}$	10.85
Hejtmanite	$Ba_2\{(Mn,Fe)_4[Ti_2(O,OH)_4O_{14}](OH,F)_2\}$	10.90
Murmanite[d]	$(Na,\ \square)_2\{(Na,Ti,Ca,\square)_4[(Ti,Nb)_2(O,OH)_3Si_4O_{14}](OH)_2\}\cdot 2H_2O$	11.60
Epistolite[d]	$(Na,\ \square)_2\{(Na,Ti,Ca)_4[(Ti,Nb)_2(O,H_2O)_3Si_4O_{14}](O,F)_2\}\cdot 2(H_2O,\square)$	11.71
Bornemanite	$Na_3Ba\{(Na,Ti,Mn)_4[(Ti,Nb)_2O_2Si_4O_{14}](F,OH)_2\}PO_4$	11.99
M72[e]	$BaNa\{(Na,Ti)_4[(Ti,Nb)_2(OH,O)_3Si_4O_{14}](OH,F)_2\}\cdot 3H_2O$	12.70
Vuonnemite	$Na_8\{(Na,Ti)_4[Nb_2O_2Si_4O_{14}](O,OH,F)_2\}(PO_4)_2$	14.40
Lomonosovite	$Na_8\{(Na,Ti)_4[Ti_2O_2Si_4O_{14}](O,F)_2\}(PO_4)_2$	14.48
Synthetic[f]	$Na_8\{(NaTi)_2[Ti_2O_2Si_4O_{14}]O_2\}(VO_4)_2$	14.50
Innelite	$(Ba,K)_2Ba2\{(Na,Ca)_3Ti[Ti_2O_2Si_4O_{14}]O_2\}(SO_4)_2$	14.65
Yoshimuraite	$Ba_4\{Mn_4[Ti_2O_2Si_4O_{14}](OH)_2\}(PO_4)_2$	14.74
Bussenite[g]	$Ba_4Na_2\{(Na,Fe,Mn)_2[Ti_2O_2Si_4(OH)_2]\}(CO_3)_2F_2\cdot 2H_2O$	15.85
Quadruphite	$Na_{13}Ca\{(NaMgTi_2)[Ti_2O_2Si_4O_{14}]O_2\}(PO_4)_4F_2$	20.25
Sobolevite	$Na_{12}CaMg\{(NaMgTi_2)[Ti_2O_2Si_4O_{14}]O_2\}(PO_4)_4F_2$	20.28
Polyphite	$Na_{14}(Ca,Mn,Mg)_5\{(Ti,Mn,Mg)_4[Ti_2O_2Si_4O_{14}]F_2\}(PO_4)_6F_4$	26.49
M73[b]	$BaNa\{(Na,Mn,Ti,\square)_4[(Ti,Nb)_2O_2Si_4O_{14}](O,OH,F)_2\}\cdot 5H_2O$	h
M55[b]	$(Ba,K,\square)_2Na_4Ba\{(Ca,Nb,Mn,Fe,\square)_4[Ti_2O_2Si_4O_{14}]O_2\}(PO_4)_2$	h

Note: a Rastsvetaeva and Chukanov (1999); Chukanov *et al.* (2003).
b Structure unknown.
c Rozenberg *et al.* (2003).
d Németh *et al.* (2003).
e Németh *et al.* (2002); now approved as **bykovaite** (IMA Code 2003-044).
f Synthetic compound (Massa *et al.* 2000).
g M74 in Khomyakov (1995) and Ferraris *et al.* (1997).
h Cell unknown.

perraultite, **bornemanite**, **bussenite**, **M72** and **epistolite**, for which the structure was later determined, the prediction has been confirmed.

In terms of merotypy and plesiotypy (Section 4.4.2.1), the titanosilicates of Table 4.2 form a **mero-plesiotype series** (Ferraris *et al.* 2001*a*). This series is merotype because one module, namely the layer $(HOH)_B$, is constantly present (but see below) in all members whereas a second module, namely the interlayer part

(A + W), is peculiar for each member. At the same time, the series has a plesiotype character because the $(HOH)_B$ layer modifies not only its chemical composition but, more important, often also changes to some extent its topology, including the coordination number of the Z cations, which can be either five or six, and that of the Y cations.

4.4.2.2 *Seidozerite as building module*

An ideal formula for **bornemanite** (Men'shikov *et al.* 1975) can be written as $BaNa_3\{(Na,Ti)_4[(Ti,Nb)_2O_2Si_4O_{14}](F,OH)_2\}PO_4$. Electron diffraction data obtained from two different orientations of (001) bornemanite laths allowed Ferraris *et al.* (2001*b*) to index the powder diffraction pattern of this titanosilicate and to obtain the following crystal data: space group $I11b$; lattice parameters $a = 5.498$, $b = 7.120$, $c = 47.95(4)$ Å, $\gamma = 88.4°$.

The values of the cell parameters and the presence of the complex anion $[(Ti,Nb)_2O_2Si_4O_{14}]$ in the chemical formula support a strong analogy between bornemanite and members of the bafertisite **mero-plesiotype series**, as inferred by Ferraris *et al.* (1997) and Ferraris (1997). A comparison of the cell parameters shows that $c/2$ (23.97 Å) of bornemanite corresponds to the sum in thickness of one **lomonosovite-like module** (14.5 Å) and one **seidozerite-like module** (8.9 Å). Disregarding isomorphic substitutions (like Ba for Na and Nb for Ti), it turns out half the sum of the crystal-chemical formulae of **lomonosovite** and **seidozerite** (Table 4.2), $[Na_8\{(Na_2Ti_2)[Ti_2O_2Si_4O_{14}](O,F)_2\}(PO_4)_2 + Na_2\{(Na,Mn,Ti)_4[(Na,Ti,Zr)_2O_2Si_4O_{14}]F_2\}]/2$, well corresponds to the simplified crystal-chemical formula of bornemanite given above. On the basis of these indications, it has been possible to build a structure model for bornemanite based on alternating seidozerite-like and lomonosovite-like modules (Fig. 4.23). In practice, the structure of bornemanite can be described as a [001] stack of $(HOH)_B$ heterophyllosilicate layers in which the bornemanite and seidozerite interlayer contents alternate.

The structure of **M72** (Table 4.2) $[BaNa\{(Na,Ti)_4[(Ti,Nb)_2(OH,O)_3Si_4O_{14}](OH,F)_2\}\cdot 3H_2O$; $I11b$, $a = 5.55$, $b = 7.179$, $c = 50.94(1)$ Å, $\gamma = 91.10°]$ has been modelled (Németh *et al.* 2002) noting that, a part isomorphous substitutions, its chemical composition differs from that of bornemanite only for the interlayer content ($BaNa \cdot 3H_2O$, instead of $BaNa_3PO_4$), in agreement with its longer c parameter that allows H_2O to form hydrogen bonding. In other words, the murmanite-like interlayer content of M72 corresponds to the lomonosovite-like interlayer content of bornemanite.

The crystal structures of bornemanite and M72 can be thought as built by the same seidozerite-like module intercalated by different interlayer contents and considered the only two members so far known of a new branching of the heterophyllosilicate series: precisely, a merotype series based on the seidozerite structure [i.e. two coupled $(HOH)_B$ layers sandwiching Na only] as fixed module.

4.4.2.3 *The astrophyllite* $(HOH)_A$ *layer as building module*

Till recently the $(HOH)_A$ astrophyllite-like module was known only in members of an isomorphous series which includes seven species (**niobokupletskite**, **kupletskite**,

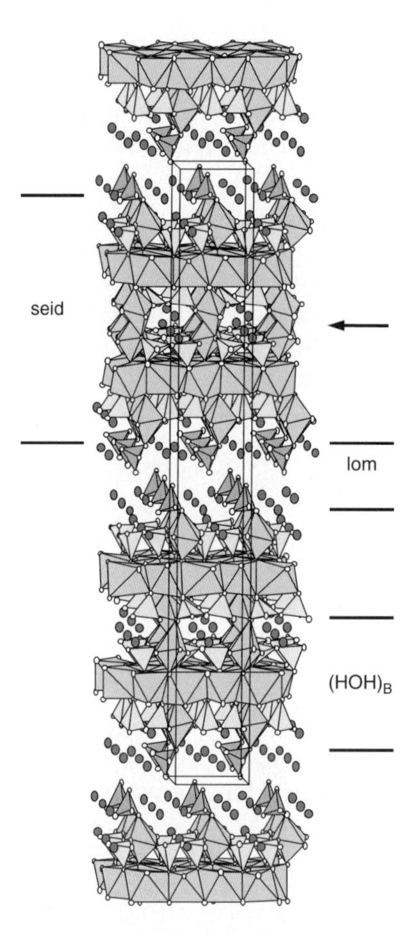

Fig. 4.23. Perspective view of the bornemanite crystal structure. The interlayer content between two $(HOH)_B$ heterophyllosilicate layers alternatively corresponds to that of seidozerite (Na only; indicated with an arrow) and of lomonosovite (lom; cf. Fig. 4.22). The entire module of seidozerite is indicated as seid. The structure of M72 (see text) differs only in the content between two seidozerite modules. (See text for reference.)

cesium kupletskite, **astrophyllite**, **magnesium astrophyllite**, **niobophyllite**, and **zircophyllite**; cf. Piilonen *et al.* 2003*a*,*b*) and their polytypes as mentioned above (cf. Piilonen *et al.* 2001).

Two basic astrophyllite-type structures are known (Fig. 4.21(b)).

- **Magnesium astrophyllite**—$(K_2Na)Na(Fe,Mn)_4Mg_2[Ti_2(OH)_2Si_8O_{24}](OH,F)_4$; $C2$, $a = 5.322$, $b = 23.129$, $c = 10.370$ Å, $\beta = 99.55°$; the coordination number of Ti is five, that is, $p = 0$ in the general formula given above (Shi *et al.* 1998).
- **Astrophyllite**—$(K_2Na)Fe_7[Ti_2(O,OH)_3Si_8O_{24}](OH,F)_4$; $A\bar{1}$, $a = 5.365$, $b = 11.88$, $c = 21.03$ Å, $\alpha = 84.87$, $\beta = 92.25$, $\gamma = 103.01°$; two hetero

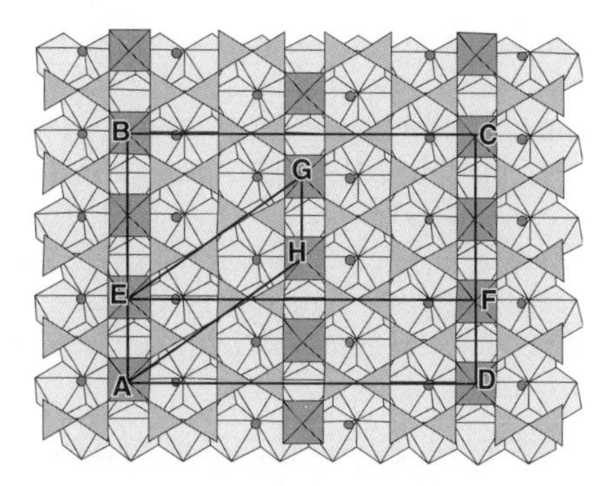

Fig. 4.24. Depending on the ordering of cations (in particular in the grey-shaded Ti-octahedra), three different two-dimensional *ab* cells are known for the $(HOH)_A$ layer: a basic AEGH cell, a double AEFD cell, and a sextuple ABCD cell.

Ti-octahedra share a corner across the interlayer, that is, $p = 1$ in the general formula (Yamnova *et al.* 2000).

Recently, Men'shikov *et al.* (2002) have characterized **eveslogite**, a new titanosilicate from Khibina massif, Russia $\{(Ca,K,Na,Sr,Ba)_{48}[(Ti,Nb,Fe,Mn)_{12}(OH)_{12}Si_{48}O_{144}](F,OH,Cl)_{14}$; $P2/m$, $a = 14.069$, $b = 24.937$, $c = 44.31$ Å, $\gamma = 95.02(4)°\}$. Eveslogite shows an *ab* base six times larger than that of astrophyllite and three times larger than that of magnesium astrophyllite (Fig. 4.24); its *c* parameter corresponds to the thickness of four astrophyllite-like $(HOH)_A$ layers. That supports a fitting of the eveslogite chemical data to a crystal chemical formula corresponding to six times that of a $(HOH)_A$ layer, that is, to an ideal formula $A_{18}Y_{42}[X_{18}O'_{12+6p}Si_{48}O_{144}]O''_{24}$ including some vacancies (for symbols, see Section 4.4.1.2). On this basis, Ferraris *et al.* (2002) proposed a model of the eveslogite structure and interpreted its chemical and crystallographic data by comparison with those of astrophyllites.

4.4.3 *The palysepiole polysomatic series*

The characterization of the new mineral **kalifersite** $\{(K,Na)_5(Fe^{3+})_7[Si_{20}O_{50}](OH)_6 \cdot 12H_2O$; $P\bar{1}$, $a = 14.86$, $b = 20.54$, $c = 5.29$ Å, $\alpha = 95.6$, $\beta = 92.3$, $\gamma = 94.4°\}$ was achieved by Ferraris *et al.* (1998) after discovering a modular relationship relating it with **sepiolite** {also known as 'attapulgite', ideally $Mg_8[Si_{12}O_{30}](OH)_4 \cdot 12H_2O$; *Pncn*, $a = 13.40$, $b = 26.80$, $c = 5.28$Å; Brauner and Preisinger 1956)} and **palygorskite** {ideally $Mg_5[Si_8O_{20}](OH)_2 \cdot 8H_2O$; two polytypes are known (Artioli and Galli 1994): $C2/m$, $a = 13.27$, $b = 17.868$, $c = 5.279$ Å, $\beta = 107.38°$; *Pbmn*, $a = 12.763$, $b = 17.842$, $c = 5.241$ Å}. In particular, it was noted:

(i) Kalifersite, sepiolite, and palygorskite have close values of their *a* and *c* parameters. The [001] direction corresponds to the fibre axis of these silicates and its periodicity to that of a pyroxene chain.

(ii) The b value of kalifersite is intermediate between that of palygorskite and sepiolite.

(iii) The crystal structures of sepiolite and palygorskite (Fig. 4.25) are based on a framework of interconnected [001] TOT ribbons which correspond to cuts with different width of the phyllosilicate 2 : 1 layer (cf. Fig. 4.19). These ribbons are chess-board arranged and delimit [001] channels. In the [010] direction, the $(TOT)_S$ ribbon of sepiolite is one chain wider than that, $(TOT)_P$, of palygorskite. This feature requires for sepiolite a b value about 9 Å longer than that of palygorskite, that is, about 4.5 Å per T chain.

Taking into account the above chemical and crystallographic aspects, a structure model for kalifersite based on a 1 : 1 chess-board arrangement of $(TOT)_P$ and $(TOT)_S$ [001] ribbons and filling of the channels with alkalis and water molecules was obtained.

Palygorskite (P), and sepiolite (S) are the end members of the **palysepiole** (**paly**gorskite + **sepiol**ite) polysomatic series P_pS_s defined by Ferraris *et al.* (1998); kalifersite is the P_1S_1 member. Interestingly, Martin-Vivaldi and Linares-Gonzales (1962) have interpreted diffraction patterns which are intermediate between those of palygorskite and sepiolite on the basis of a random intergrowth of $(TOT)_S$ and $(TOT)_P$ ribbons.

Falcondoite (a = 13.5, b = 29.9, c = 5.24 Å; Springer 1976) and **loughlinite** (Fahey *et al.* 1960) differs from sepiolite just for the presence of Ni and Na,

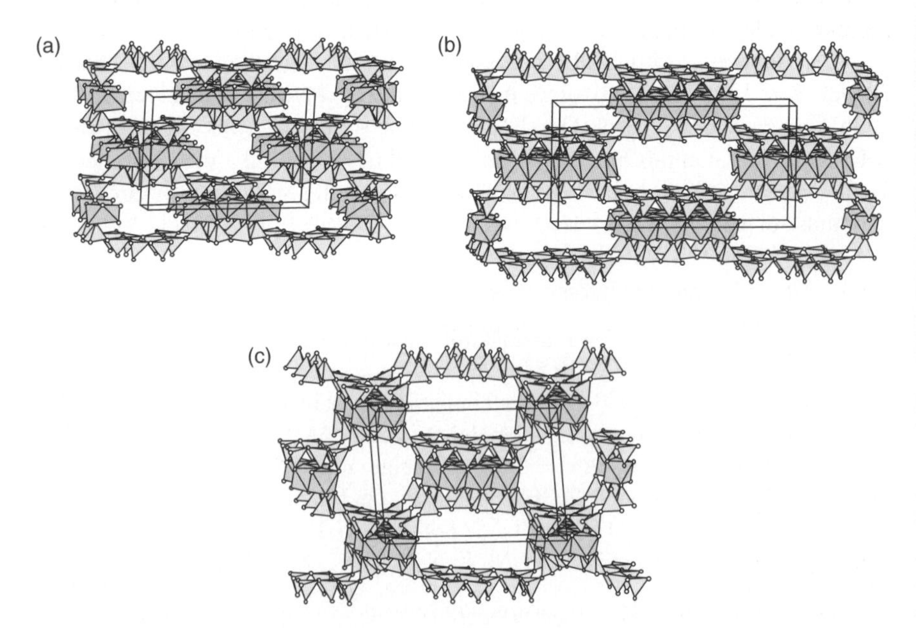

Fig. 4.25. Perspective view along [001] of the palygorskite (a), sepiolite (b), and kalifersite (c) structures. The content of the [001] channels is not shown. (See text for reference.)

respectively. **Yofortierite** (Perrault *et al.* 1975) and **tuperssuatsiaite** ($a = 13.92$, $b = 17.73$, $c = 5.30$ Å, $\beta = 104.78°$; von Knorring *et al.* 1992) are, in the order, the Mn and Fe equivalent of palygorskite. On the basis of the chemical composition and (if available) lattice dimensions, Ferraris *et al.* (1998) made hypotheses of isostructurality with sepiolite for falcondite and loughlinite, and with palygorskite for yofortierite and tuperssuatsiaite. The hypothesis has been subsequently confirmed for tuperssuatsiaite by Cámara *et al.* (2002), that is, the only compound of the group for which the crystal structure has been determined.

With their variably filled channels, the members of the palysepiole series show typical features of zeolite-like microporous materials, a property exploited both by nature (palygorskite and sepiolite are important clay minerals) and technology. In fact, a wide literature (e.g. Rytwo *et al.* 2002, and references therein) reports the capability of these materials to absorb various molecules, in particular to prepare pigments. Recently, Chiari *et al.* (2003) have shown that **Maya Blue**, a synthetic pigment produced by ancient Mayas, is a combination of palygorskite (and sepiolite) and the organic dye indigo (see also Hubbard *et al.* 2003).

4.4.3.1 *Merotypes and plesiotypes of palysepioles*

Raite {$Na_3Mn_3Ti_{0.25}[Si_8O_{20}](OH)_2{\cdot}10H_2O$; $C2/m$, $a = 15.1$, $b = 17.6$, $c = 5.290$ Å, $\beta = 100.5°$; Pushcharovsky *et al.* 1999} consists of a palygorskite-like framework, but the channel content differs substantially from that of palygorskite both in chemistry and structure. The two minerals can, therefore, be considered in **merotype** relationship.

In **intersilite** {$(Na,K)Mn(Ti,Nb)Na_5(O,OH)(OH)_2[Si_{10}O_{23}(O,OH)_2]{\cdot}4H_2O$; $I2/m$, $a = 13.033$, $b = 18.717$, $c = 12.264$ Å, $\beta = 99.62°$; Yamnova *et al.* 1996} two adjacent sepiolite-like ribbons partially overlap along [010] because of tetrahedral inversions within the same ribbon. The overlap reduces the length of b to 18.7 Å, as compared with 26.8 Å in sepiolite. Because of the substantial modification of the sepiolite framework, intersilite can be considered in **plesiotype** relationship with the palysepiole polysomatic series.

4.4.3.2 *An alternative modular description for palysepioles—Relation with biopyriboles*

The palysepiole series P_pS_s and, in particular, kalifersite can be compared with **mica** and, more generally, with **biopyriboles**. Zoltai (1981) named DiC and TriC the TOT layer modules corresponding, respectively, to one-chain wide (010) slabs of dioctahedral (Di) and trioctahedral (Tri) micas; the height t of these slabs is about 10 Å. C stands for interlayer cation and is omitted for pyroxenes and amphiboles with empty A site. The same author uses the symbol 1/2TriC to describe intermediate tri- and di-octahedral layers, as those that, reduced to ribbons, occur in palygorskite and sepiolite. The vertical shifts (along the [100] direction in palysepioles) between basic modules is nt ($n = 0$, 1/2, 3/4). The value $n = 0$ occurs in the major layer silicates, like micas; in **pyroxenes** and **amphiboles** $n = 1/2$, while $n = 3/4$ describes the shift between ribbons in palygorskite, sepiolite, and kalifersite. According to Zoltai's symbolism, palygorskite, sepiolite, and kalifersite represent the 1/2TriC–0–1/2TriC–3/4, 1/2TriC–0–1/2TriC–0–1/2TriC–3/4, and DiC–0–DiC–3/4–DiC–0–DiC–0–DiC–3/4 members of the biopyribole series.

4.4.4 *A polysomatic series related to biopyriboles*

The crystal structures (Evans and Mrose 1977) (cf. Section 2.6.1 for polytypism) of **shattuckite** [$\mathbf{Cu_5(Si_2O_6)_2(OH)_2}$; *Pbca*, $a = 19.832$, $b = 9.885$, $c = 5.382$ Å] and **planchéite** [$Cu_5(Si_4O_{11})_2(OH)_4 \cdot xH_2O$; *Pnca*, $a = 20.129$, $b = 19.043$, $c = 5.269$ Å] are to some extent related to those of pyroxenes and amphiboles for the presence of the typical single and double tetrahedral chains. However, both minerals contain a continuous trioctahedral sheet where Cu is octahedrally coordinated and which is sandwiched by the mentioned silicate chains (Fig. 4.26).

Merlino and Pasero (1997) have stressed the analogy of shattuckite and planchéite with **biopyriboles** and described their structures as members of a **polysomatic series** based on a **pyroxene-like module** of shattuckite [S = $Cu_5(Si_2O_6)_2(OH)_2$] and a **mica-like module** [M = $Cu_3(Si_4O_{10})_2(OH)_2 \cdot xH_2O$] (Fig. 4.26). The same authors extended the analogy with biopyriboles beyond the members S (shattuckite) and MS (planchéite) to a hypothetical member M_2S [$Cu_{11}Si_{12}O_{32}(OH)_6 \cdot 2xH_2O$; *Pbca*, $a = 19.97$, $b = 29.11$, $c = 5.32$ Å] corresponding to the pyroxene **jimthompsonite**, that is, the member with triple chains.

4.4.5 *The serpentine minerals*

The serpentine minerals are 1 : 1 layer silicates based on polar TO layers. The misfit between the dimensions of the tetrahedral (T) and three-octahedral (O) sheets, which in the TOT layer silicates is overcome mainly via isomorphous substitutions,

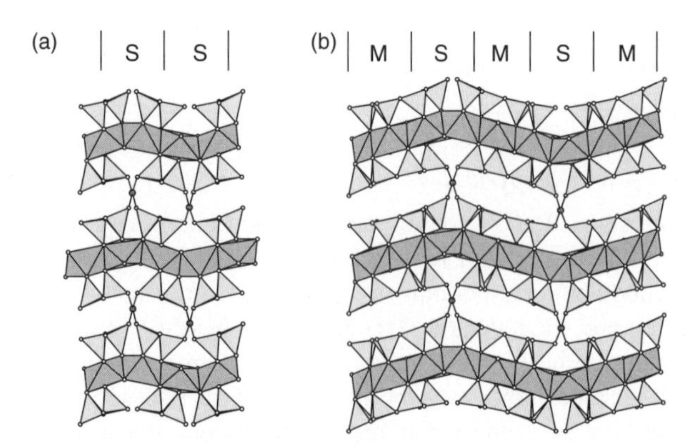

Fig. 4.26. Projection along [001] of the crystal structures of shattuckite (a) and planchéite (b). The building modules S (pyroxene-like) and M (mica-like) are shown. The grey larger circles represent four-coordinated Cu. (See text for reference.)

is dramatically enhanced by the different strain that occurs on the opposite surfaces of the polar TO layer. The classical magnesian members of the group are the species **antigorite**, **chrysotile** (actually *para*- and *ortho*-chrysotile), **lizardite** and the variety **polygonal serpentine**; to these species and varieties, **carlosturanite** is added mainly for structural reasons. Disregarding minor chemical differences, the four species and polygonal serpentine are characterized by the adoption of different mechanisms to overcome the misfit between the T and O sheets. Apart from the role of hydrogen bonding and very limited isomorphous substitutions in stabilizing the flat TO layer of lizardite (cf. Mellini and Viti 1994), these mechanisms include rolling of the layer to form cylindrical or spiral tubes (chrysotile and polygonal serpentine), waving (antigorite) and breaking (carlosturanite) of the T sheet.

The efforts to elucidate the crystal structures of these phases are as long as the existence of the structural crystallography and are still in action. However, the possibility of studying nanostructures (electron microscopy) and obtaining single-crystal X-ray diffraction data from submillimetric samples are dramatically accelerating the problem solution. HRTEM studies on polygonal serpentine (Baronnet *et al.* 1994*b*) and antigorite (Uehara 1998; Dódony *et al.* 2002; Grobéty 2003) and single-crystal X-ray work by Capitani and Mellini (2003) on antigorite are the most recent milestones on route.

Chrysotile (and related polygonal serpentine) and lizardite have the same ideal chemical composition $\mathbf{Mg_3Si_3O_5(OH)_4}$. They represent a special kind of polymorphism where the difference in structure is due to the secondary structure of chrysotile originated by the mentioned rolling of the TO layers. The composition of antigorite and carlosturanite is discussed below in connection with their polysomatic nature.

Chrysotile is the main **asbestos** mineral, a natural fibrous material widely used for its mechanical and phono/thermo isolating properties. Now the use of asbestos is prohibited in most countries because the inhalation of its micrometric dust is a risk for the human health.

4.4.5.1 *Antigorite polysomatic series: a hybrid between TO and TOT modules*

The long hunting for a sound crystal-structure model of **antigorite** is summarized by Capitani and Mellini (2003). The polysomatic nature of antigorite structure, based on a wave-like modulation of the T sheet along [100], and a general formula explaining a higher content of Si relative to the basic formula, $Mg_3Si_3O_5(OH)_4$, of the serpentine minerals is established since some years. Instead, the controversy on the exact structure at the tetrahedral sheet inversion (Fig. 4.27) is at moment soundly settled only for the 'odd' members, after Capitani and Mellini (2003) have determined the X-ray diffraction crystal structure of a polysome (see below).

The general formula of the antigorite **polysomatic series** is $\mathbf{M_{3m-3}T_{2m}O_{5m}}$ $\mathbf{(OH)_{4m-6}}$, where M and T correspond mainly to Mg and Si, respectively; m represents the number of independent tetrahedra along the modulation wave and the order of the polysome. The ratio Si/Mg $= 2m/(3m-3)$ is higher than the value of 2/3

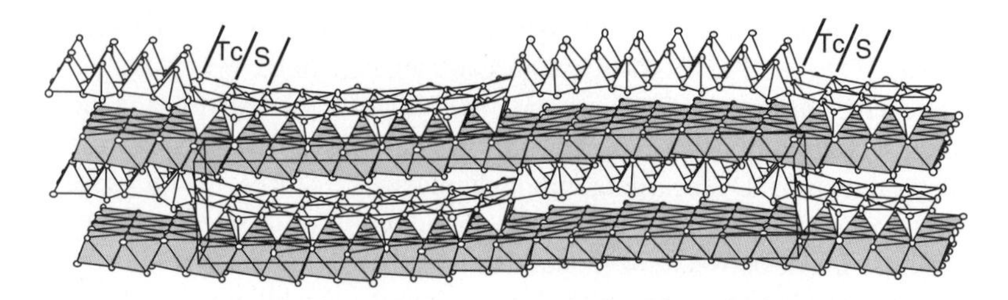

Fig. 4.27. Perspective view along [010] of the crystal structure of the antigorite polysome $m = 17$. Talc (T_c) and serpentine (S) modules are indicated. (See text for reference.)

which occurs in lizardite, as mentioned above. The occurrence of different polysomes in metamorphic rocks has been related to the degree of metamorphism (Mellini *et al.* 1987). The commonest polysome has $m = 17$, cell parameters $a = 43.505$, $b = 9.251$, $c = 7.263$ Å, $\beta = 91.32°$, and space group *Pm*. The periodicity of a wave is about $m2.6$ Å; the lattice is either *P* or *C* monoclinic for *m* odd or even, respectively; in the latter case, the periodicity along [100] spans two waves.

In their study of the polysome $m = 17$, Capitani and Mellini (2003) have confirmed the wave modulation of the antigorite structure (Fig. 4.27). In particular, they have shown: (i) the different length of the half waves (eight and nine independent tetrahedra for $m = 17$), (ii) the presence of six- (at $x = 0$) and eight-plus four-member (at $x = 1/2$) rings of tetrahedra alternatively at the inversion point of the tetrahedral sheets, (iii) the occurrence of a TOT **talc-like module** at the inversion corresponding to the six-member ring. Thus, the polysomatic nature of (at least the 'odd') antigorite can be expressed as indicated by Ferraris *et al.* (1986), that is, by the series formula $S^+_n S^-_{n'} Tc$. In this formula, S^+ and S^- represent serpentine (lizardite) TO modules with opposite polarity, and Tc represents a talc TOT module. All modules are about 2.7 Å wide along [100]; $n + n' + 1$ corresponds to *m* in the polysomatic formula given above. For *m* even, the TEM results mentioned above are contradictory on the nature of the rings that occur at the reversals and a sound crystal-structure determination is still lacking. It should, however, be noted that the presence in these members of six-member rings only seems ruled out if the ratio Si/Mg $= 2m/(3m - 3)$ must be maintained, according to the known chemical analyses. In fact, each six-member ring reversal introduces a Tc module that increases the above ratio.

4.4.5.2 *The carlosturanite polysomatic series*

Carlosturanite is a water and magnesium-rich, silicon-poor serpentine-like asbestiform mineral. It has ideal composition $Mg_{21}[Si_{12}O_{28}(OH)_4](OH)_{30} \cdot H_2O$ and shows rotational disorder about the [010] fibre axis. The modelling of its structure (Mellini *et al.* 1985) represented one of the first applications of polysomatism after the introduction of the biopyribole polysomatic series by Thompson (1978). X-ray (rotation about the fibre axis and powder pattern) and electron diffraction patterns supplied: $a = 36.7$, $b = 9.4$, $c = 7.3$ Å, $\beta = 101°$, space group *Cm*; a marked superlattice

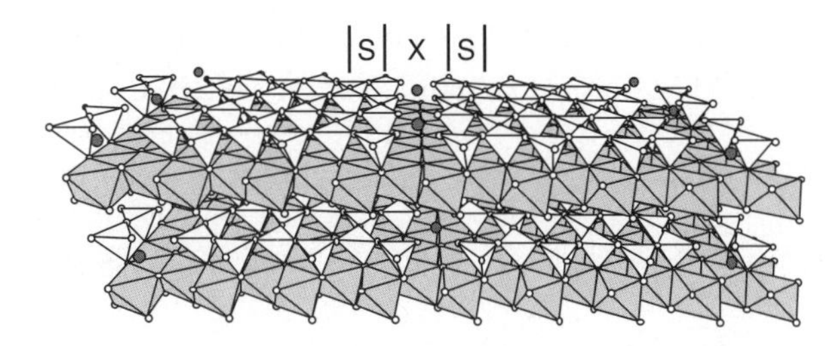

Fig. 4.28. Perspective view along [010] of carlosturanite structure. The hydrosilicate, X, and serpentine, S, modules are indicated (see text). Grey larger circles represent water molecules.

with $a' = a/7$ occurs. HRTEM images show an evident modification of the structure with periodicity $a/2$ and chain-multiplicity faults along [100].

Comparison with lizardite shows that **carlosturanite** has the same cell except a, that is 7 times the a of lizardite. This, together with the silicon-poor chemical composition, suggests that the structure of carlosturanite can be derived by inserting a silicon depleted module in a sevenfold lizardite cell. A model suitable to explain the observed data was obtained by preserving the octahedral sheet of lizardite and introducing a [010] row of tetrahedral vacancies at $x = 0$ and 1/2 (Fig. 4.28). Precisely, starting from a sevenfold lizardite cell with $a = 7\text{x}5.2$ Å and content $7Mg_6[Si_4O_{10}](OH)_8$, two $[Si_2O_7]^{6-}$ ditetrahedral groups are replaced by two $[(OH)_6H_2O]^{6-}$ groups (same arrangement), where H_2O and OH carry the bridging and non-bridging oxygen atoms, respectively. A structure model with composition $Mg_{42}[Si_{24}O_{56}(OH)_8](OH)_{60}(H_2O)_2$ is thus obtained which fits the unit cell content of carlosturanite obtained from the chemical analysis.

HRTEM images of carlosturanite (Mellini *et al.* 1985; Belluso and Ferraris 1991; Baronnet *et al.* 1994*a*; Ferraris 1997) show chain-multiplicity faults along [100]. Following the scheme of interpretation proposed for biopyriboles, the faults are explained as members of a polysomatic series (**inophite polysomatic series**) composed of two (100) modules: $S = Mg_3Si_2O_5(OH)_4$ and $X = Mg_6Si_2O_3(OH)_{14} \cdot H_2O$. The S module corresponds to the same (100) slab of serpentine (lizardite) structure with thickness $a_s/2$ ($a_s = 5.2$ Å $= a$ of serpentine) described for antigorite; the X slab has thickness a_s and corresponds to a hypothetical hydrosilicate. A generic S_nX member of the inophite series has: P lattice, $a = [(n+2)/2]a_s$ and $Z = 1$ for n even; C lattice, $a = (n+2)a_s$ and $Z = 2$ for n odd. Carlosturanite represents the S_5X member of the inophite series. As mentioned above, several polysomes with n in the range 1–23 have been observed as chain-width defects in a matrix of carlosturanite. This is in strict analogy with the results reported for the biopyribole polysomatic series.

4.4.6 *Series based on more than two modules*

Complex series involving three and more building modules are known as already shown for superconductors (Section 4.3).

4.4.6.1 *From palysepioles and zeolites to the lintisite group*

The crystal structure (Fig. 4.29) of **silinaite** (**$NaLiSi_2O_5 \cdot 2H_2O$**; $C2/c$, $a = 14.383$, $b = 8.334$, $c = 5.061$ Å, $\beta = 96.6°$; Grice 1991) shows a chess-board arrangement of channels, which are filled by Na and H_2O, and ribbons comparable with that of palysepioles (Fig. 4.25). Whereas in palygorskite and sepiolite, the inversion of the tetrahedral sheet is every four and six tetrahedra, respectively, in silinaite the same

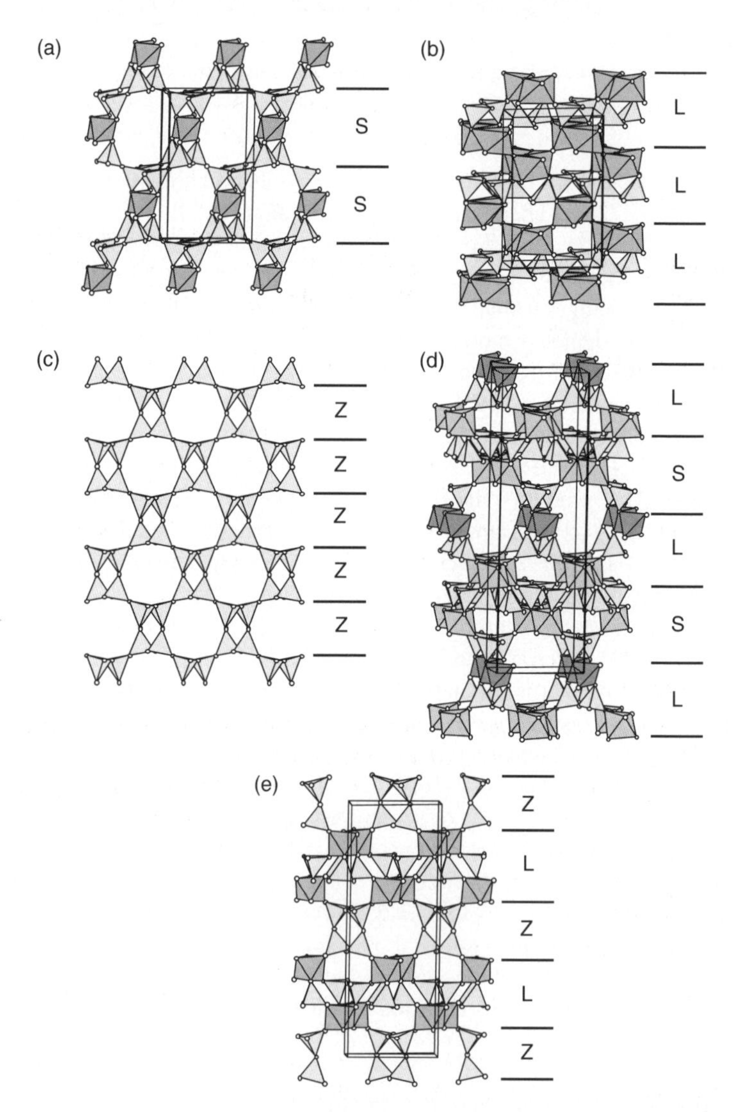

Fig. 4.29. Perspective view (about along [001]) of the crystal structures of (a) silinaite (S), (b) lorenzenite (L), (c) the zeolite like compound $(Na,K)Si_3AlO_8 \cdot 2H_2O$ (Z), (d) lintisite, and (e) vinogradovite. (See text for reference.)

inversion is every two tetrahedra. Besides, the O part of the silinaite TOT ribbon consists of Li-tetrahedra instead of octahedra as in palyspioles and, in general, in the phyllosilicates.

Merlino and Pasero (1997) used modules of silinaite (S) as one of three modules needed to describe the modularity in the **lintisite group** (Fig. 4.29). The other two modules are slabs of the structures of **lorenzenite** [L; $\mathbf{Na_4Ti_4(Si_2O_6)_2O_6}$; *Pnca*, $a = 14.487$, $b = 8.713$, $c = 5.233$ Å; Sundberg *et al.* 1987] and of the **zeolite-like phase** $\mathbf{(Na, K)Si_3AlO_8 \cdot 2H_2O}$ (Z). This phase has never been found, but is listed by Smith (1977) in his exhaustive enumeration of framework silicate structures. Besides the three end members, in this three-module group of compounds, Merlino and Pasero (1997) enumerated the members LS [**lintisite**, $\mathbf{Na_3LiTi_2(Si_2O_6)_2O_2 \cdot 2H_2O}$; $C2/c$, $a = 28.583$, $b = 8.600$, $c = 5.219$ Å, $\beta = 96.6°$; Merlino *et al.* 1990*a*] and LZ [**vinogradovite**, $Na_4LiTi_4(Si_2O_6)_2[(Si,Al)_4O_{10}]O_4 \cdot (H_2O, Na, K)_3$; $C2/c$, $a = 24.50$, $b = 8.662$, $c = 5.211$ Å, $\beta = 100.15°$; Rastsvetaeva *et al.* 1968]. **Kukisvumite** [$\mathbf{Na_6ZnTi_4(Si_2O_6)_4O_4 \cdot 4H_2O}$; *Pcnn*, $a = 29.029$, $b = 8.595$, $c = 5.209$ Å; Merlino *et al.* 2000*b*] and its Mn equivalent (**IMA No. 2002–029**; Grice and Ferraris 2003) are closely related to lintisite. In its turn, as supposed by Merlino and Pasero (1997), **paravinogradovite** [$\sim Na_2[Ti_3^{4+}Fe^{3+}\{Si_2O_6\}_2\{Si_3AlO_{10}\}(OH)_4 \cdot H_2O$; $P1$, $a = 5.2533$, $b = 8.7411$, $c = 12.9480$ Å, $\alpha = 70.47$, $\beta = 78.47$, $\gamma = 89.93°$; Khomyakov *et al.* 2003] is closely related to vinagradovite.

Even if the three modules S, L, and Z form their own compounds, that is, an important condition for a polysomatic description is satisfied, the lintisite group shows merotype more than polysomatic characteristics. In fact, never the three modules occur together and lintisite appears as the intersection between the SL and ZL merotype series (Section 1.8.1.2).

4.4.6.2 *Modelling tungusite and merotype series related to reyerite*

On the basis of chemical, X-ray (powder) and electron diffraction data, Ferraris *et al.* (1995) modelled the crystal structure of **tungusite** {$[Ca_{14}(OH)_8](Si_8O_{20})(Si_8O_{20})_2[(Fe^{2+})_9(OH)_{14}]$; $P\bar{1}$, $a = 9.714$, $b = 9.721$, $c = 22.09$ Å, $\alpha = 90.13$, $\beta = 98.3$, $\gamma = 120.0°$} by comparison with the crystal structures (Fig. 4.30) of **reyerite** [$Ca_{14}(Na,K)_2Si_{22}Al_2O_{58}(OH)_8 \cdot 6H_2O$; $P\bar{3}$, $a = 9.765$, $c = 19.067$ Å; Merlino 1988*a*] and **gyrolite** [$Ca_{16}NaSi_{23}AlO_{60}(OH)_8 \cdot 14H_2O$; $P\bar{1}$, $a = b = 9.74$, $c = 22.40$ Å, $\alpha = 95.7$, $\beta = 91.5$, $\gamma = 120.0°$; Merlino 1988*b*]. In particular, the following aspects were considered.

- The difference $(Ca,Na)O_2 \cdot 8H_2O$ between the chemical compositions of reyerite and gyrolite is accounted for by the presence in gyrolite of a partially filled (001) octahedral X sheet which is sandwiched between two centrosymmetrically related (001) S_1OS_2 layers (Merlino 1988*a*,*b*). The O sheet consists of edge-sharing Ca-octahedra; the S_1 and S_2 sheets are built up by six-member rings of tetrahedra pointing upwards and downwards in the ratio 1 : 1 and 3 : 1 for S_1 and S_2, respectively.

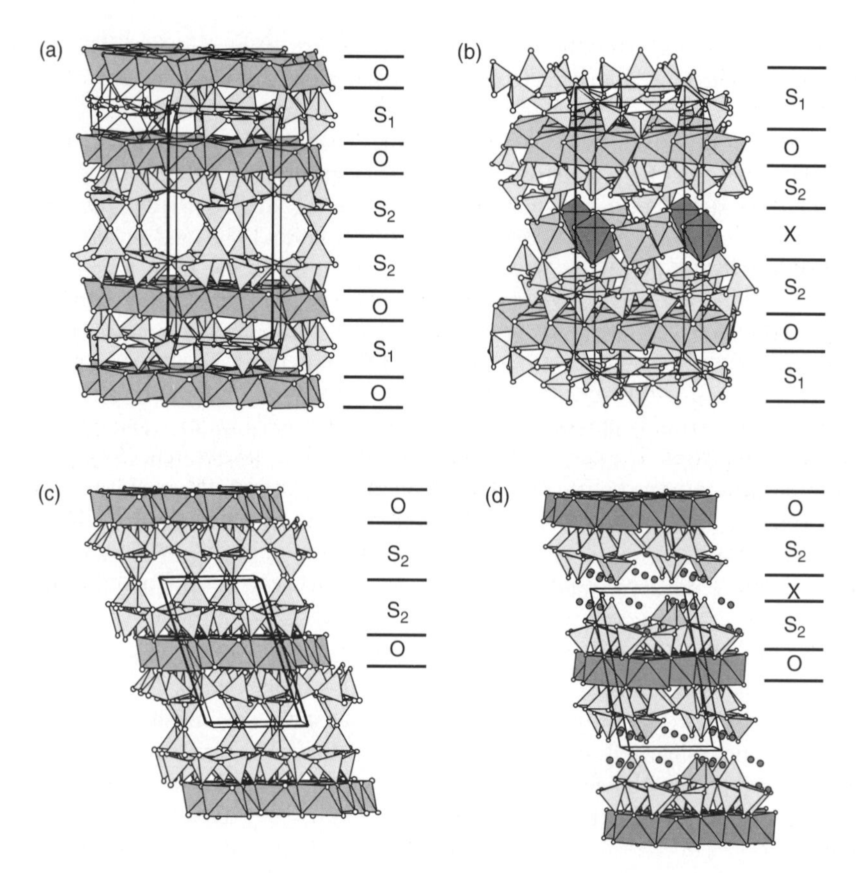

Fig. 4.30. Perspective view of the (a) reyerite, (b) gyrolite (and tungusite), (c) fedorite, and (d) IMA No. 2001-59 structures. The building modules S_1, S_2, O, and X are shown. The darker tetrahedra in IMA No. 2001-59 are occupied by boron. (See text for reference.)

- The ideal composition of tungusite differs from that of gyrolite by the presence of six further divalent cations, which complete the (001) X sheet whose dimensions ($a \sim b \sim 9.72$ Å, $\gamma = 120°$, thickness about 2.8 Å) correspond to those of a 3 × 3 trioctahedral sheet.

As shown by Merlino (1988*b*) also **fedorite** [$K_2(Ca_5Na_2)Si_{16}O_{38}(OH,F)_2 \cdot H_2O$; $P\bar{1}$, $a = b = 9.67$, $c = 12.67$ Å, $\alpha = 102.2$, $\beta = 71.2$, $\gamma = 120.0°$] and the **phase K** [$\mathbf{K_7Si_{16}O_{38}(OH)_2}$; $P\bar{1}$, $a = b = 9.70$, $c = 12.25$ Å, $\alpha = 108.5$, $\beta = 78.0$, $\gamma = 120.0°$] are based on O and S_2 modules (Fig. 4. 30). Also the **new mineral IMA No. 2002–026** [$(Na,Ca)_6(Ca,Na)_3Si_{16}O_{38}(F,OH)_2 \cdot 3H_2O$; $P\bar{1}$, $a = 9.613$, $b = 12.115$, $c = 9.589$ Å, $\alpha = 92.95$, $\beta = 119.81$, $\gamma = 96.62°$; Grice and Ferraris 2003] has a fedorite-like structure. According to Merlino (1988*b*), the introduction of an X module between two S_2 modules in the fedorite structure generates the **phase Z, $\mathbf{K_9Si_{16}O_{40}(OH)_2 \cdot (14 + x)H_2O}$**, for which $a = b = 9.65$ and $d_{001} = 15.3$ Å are known. The structure proposed for the phase Z has recently been found for the **new mineral IMA No. 2001–059** [$(Na,Ca)_{11}Ca_4(Si,S,B)_{14}B_2O_{40}F_2{\cdot}4H_2O$; $P\bar{1}$, a = 9.544, b = 14.027, c = 9.535 Å, α = 71.057, β = 119.79, γ = 105.85°; Grice and Ferraris 2002]; the X module of this structure is highly disodered. Besides, the crystal structures of **truscottite** [$(Na,K)_2Ca_{14}Si_{22}O_{58}(OH)_8{\cdot}6H_2O$; $P\bar{3}$, a = 9.767, c = 19.06 Å; Merlino 1988*b*] and **minehillite** [$(K,Na)_2Ca_{28}Zn_5Al_4Si_{40}O_{112}(OH)_{16}$; $P\bar{3}c1$, a = 9.777, c = 33.293 Å; Dai *et al.* 1995] can be strictly compared with that of reyerite which differs mainly in the topology of the double S_2 layer. All the phases here discussed contain the module OS_2 that stands alone in the crystal structure of fedorite (Fig. 4. 30). Thus, these phases can be described as members of a **merotype series** (Section 1.8.1.2) where the fixed module OS_2 may stand alone or be intercalated by either one or two different modules. The case of one module, is represented by both the intercalation of X in the phase Z and IMA No. 2001–059, and of S_1 in reyerite and isostructural phases. The case of two intercalated module, is represented by gyrolite and isostructural phases. Note that the structure of reyerite is obtained by inserting an S_1 module between two modules of fedorite; the structure of IMA No. 2001-059 is obtained from that of fedorite by insertion of a disordered X module between two S_2 modules; finally, the insertion of an X module between two S_2 modules in the reyerite structure leads to gyrolite.

The members of the reyerite merotype series, may show a quite complex polytypism (Zvyagin 1997; Merlino 1988*b*).

4.4.6.3 *A new family of layered bismuth oxyhalides*

For the first time a series of layered **bismuth oxyhalides** containing **fluorite-like** Bi_2O_2 modules alternating with **CsCl-like** and (001) **perovskite-like modules** of different thickness, has been recently synthesized by Charkin *et al.* (2003*a*). In particular, the crystal structures of $Bi_4NdTi_{1.6}Nb_{0.4}Cs_{0.6}O_{11}Br_2$ ($P4/mmm$, a = 3.857, c = 25.280 Å) and $Bi_4Nd_2Ti_{2.4}Fe_{0.6}Cs_{0.6}O_{14}Cl_2$ ($P4/mmm$, a = 3.842, c = 26.622 Å) have been established by using X-ray powder diffraction data. The perovskite layer is two- and three-octahedra thick in the Br and Cl compounds, respectively. In a large number of other bismuth oxyhalides studied by Charkin *et al.* (2003*b*) the fluorite- and perovskite-like layers occur alone.

4.5 Miscellaneous modular structures

Some structure building modules are now well known (e.g. mica, perovskite, spinel) and, as shown in the foregoing sections, they are widely used not only in the structure descriptions, but also mainly as a guide in searching for new tailored materials and in modelling unknown crystal structures. In recent years, an increasing attention is dedicated to the modular aspects of the crystalline structures and the number of crystal structure descriptions done on modular basis is sharply increasing, even if sometimes the character of series is not explicitly emphasized. In parallel, new building modules are put in evidence. In the following pages, some novel modules not yet introduced, or just mentioned before, are illustrated.

4.5.1 *Fluorite module*

The **fluorite** (CaF_2; $Fm\bar{3}m$, $a = 5.462$ Å; cf. Ferraris 2002) structure corresponds to a ccp array of Ca^{2+} cations with all the tetrahedral sites occupied by F^- anions; that is, the anion F^- occupies the centre of the eight cubes with edge $a/2$.

Wada and co-workers have synthesized and characterized several series of compounds which contain single and multiple **fluorite-type layers**, sometimes alternated with **perovskite-type modules** (cf. Wada *et al.* 1992). For example, double fluorite-type [$(Ho,Ce)O_2$] layers alternate with $(Pb,Cu)Sr_2(Ho,Ce)Cu_2O_{7+z}$ layers in the member with $n = 3$ [$(Pb,Cu)Sr_2(Ho,Ce)_3Cu_2O_{11+z}$; $P4/mmm$, $a = 3.826$, $c = 17.203$ Å] of the **series** $\mathbf{(Pb,Tl,Cu)(Sr,Ba,Ln)_2(Ln,Ce,Ca)_nCu_2O_{5+2n+z}}$ investigated for potential superconductivity (Wada *et al.* 1991; Ln for lantanon).

4.5.2 *Rutile module*

Rutile (TiO_2; $P4_2/mmm$, $a = 4.594$, $c = 2.958$ Å; cf. Ferraris 2002) is the titania phase, which is stable at room conditions. [001] rows of edge-sharing Ti-octahedra characterize its crystal structure; these rows are connected via shared corners to form (110) layers of octahedra (Fig. 4.31*c*). A (110) rutile-like layer (indices referred to the rutile cell) occurs in the **perrierite-** and **chevkinite-type structures** (synthetic perrierite, $\mathbf{Mg_2La_4Ti_3Si_4O_{22}}$; $P2_1/a$, $a = 13.818$, $b = 5.677$, $c = 11.787$ Å, $\beta = 113.85°$—**synthetic Co-chevkinite**, $\mathbf{Co_2Nd_4Ti_3Si_4O_{22}}$; $P2_1/a$, $a = 13.328$, $b = 5.727$, $c = 10.971$ Å, $\beta = 100.82°$; Calvo and Faggiani 1974). As first recognized by Bonatti and Gottardi (1966), the two structure types are very close but differ in the relative shifts of chains, that is, they are polytypes. These structures, where the mentioned rutile-like layer alternates with a mixed tetrahedral/octahedral

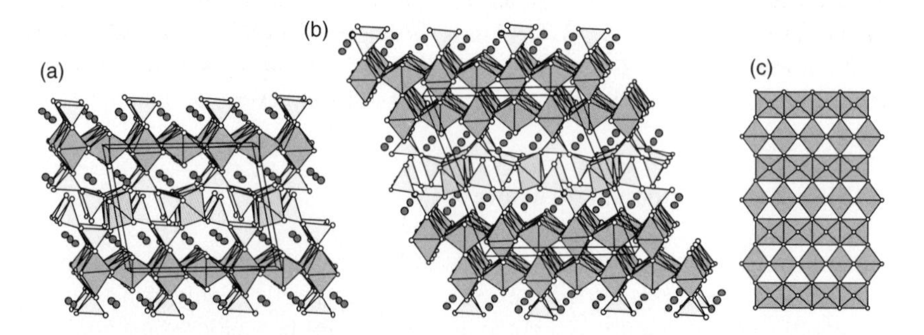

Fig. 4.31. Crystal structures of the members with $m = 1$ (a) and $m = 2$ (b) of the polysomatic series $La_4Ti(Si_2O_7)_2(TiO_2)_{4m}$. Ti occurs both in the dark-grey rutile-like layer (TiO_2) shown in (c) and in the light-grey octahedral silicate layer [$La_4Ti(Si_2O_7)_2$]. The grey circles represent La. (See text for reference.)

layer (Fig. 4.31), are of technological interest because of their strong **anisotropy** (cf. Wang *et al.* 1995); several analogue synthetic compounds have been prepared beginning with those reported by Ito (1967). The same structure types also occur in some minerals recently described: **rengeite** (Miyajima *et al.* 2001), **matsubaraite** (Miyajima *et al.* 2002), **polyakovite-(Ce)** (Popov *et al.* 2001), and a Fe-rich **chevkinite-(Ce)** (Yang *et al.* 2002).

The chevkinite/perrierite-type structures are actually the members with $m = 1$ of an **oxosilicate polysomatic series** of which two members are known in the **series** $\mathbf{La_4Ti(Si_2O_7)_2(TiO_2)_{4m}}$ (Wang *et al.* 1995). Figure 4.31 shows the crystal structures of the members with $m = 1$ ($\mathbf{La_4Ti_5Si_4O_{22}}$; $C2/m$, $a = 13.621$, $b = 5.673$, $c = 11.143$ Å, $\beta = 100.59°$; Wang and Hwu 1995) and $m = 2$ ($\boldsymbol{\alpha}$-$\mathbf{La_4Ti_9Si_4O_{30}}$; $C2/m$; $a = 13.545$, $b = 5.751$, $c = 15.189$ Å, $\beta = 110.92°$; Wang *et al.* 1995). In the member with $m = 2$ a double rutile-like layer occurs; consequently the c parameter increases by about 4 Å relative to that of the member with $m = 1$. Wang *et al.* (1995) also describes a β polymorph of the member with $m = 2$. Actually the two polymorphs are polytypes, as are perrierite and chevkinite, that is, the two members with $m = 1$ of the series. Presumably, most of the compounds belonging to the series can form polytypes. Members with Si partially substituted by Ge (Taviot-Guého *et al.* 1999) and P (Wang and Hwu 1995) are known.

4.5.3 *Corundum module*

The **corundum** structure (α-Al_2O_3; $R\bar{3}c$, $a = 4.751$, $c = 12.97$ Å; cf. Ferraris 2002) may be derived from a hcp array of oxygen atoms where 2/3 of the octahedral sites are occupied by Al atoms. To compare this structure with the basic hcp cell it must be kept in mind that the corundum cell is nine times larger and contains 18 oxygen atoms; in particular, along [001] there are six oxygen sheets instead of two as in the basic hcp.

$[Ti_2O_3]^{2+}$ corundum-type modules (Fig. 4.32) alternating with $\mathbf{LiNbO_3}$-type modules have been recently reported for technological materials like the H-phase of the compound $\mathbf{Li_2Ti_3O_7}$ ($R\bar{3}c$, $a = 5.074$, $c = 69.940$ Å; Bordet *et al.* 2000). A polymorph of this compound with a **ramsdellite** (MnO_2) structure type is known. Farber *et al.* (2002) by transmission electron microscopy (TEM) have shown the existence

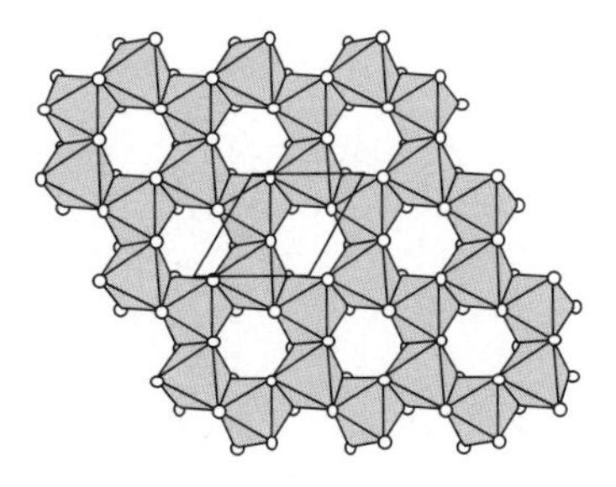

Fig. 4.32. The (001) layer of the corundum crystal structure. The layer is one octahedron thick.

of a **polysomatic series** $\mathbf{Li_{1-x}Nb_{1-x-3y}Ti_{x+4y}O_3}$ where a variable number of $LiNbO_3$-type layers alternate with a single layer corundum-type structure.

The crystal structure of **harrisonite** [$Ca(Fe^{2+},Mg)_6(SiO_4)_2(PO_4)_2$; $R\bar{3}m$, a = 6.248, c = 26.802 Å] has been described by Grice and Roberts (1993) as comprising three types of (001) layers: a corundum-type layer with composition $(FeO_6)^{10-}$, a layer of isolated $(SiO_4)^{4-}$ tetrahedra, and a $[Ca(PO_4)_2]^{4-}$ slab of $(CaO_6)^{10-}$ polyhedral trigonal antiprisms with six apical $(PO_4)^{3-}$ tetrahedra.

4.5.4 *Topaz module*

Topaz, $\mathbf{Al_2[SiO_4](OH,F)_2}$, is an orthorhombic nesosilicate (*Pmnb*; a = 8.394, b = 8.800, c = 4.650 Å; cf. Ferraris 2002) with a crystal structure based on [001] chains where pairs of Al-octahedra alternate with tetrahedra; further tetrahedra link the chains along [100].

Independently, Izokh *et al.* (1998) and Yakubovich *et al.* (2000) have recognized modules of topaz in the crystal structure of $\mathbf{CuAl_2(Si_2O_7)(F,OH)_2}$. In this structure, the topaz-like [001] crankshaft chains of edge-sharing (AlO_4F_2) octahedra and (SiO_4) tetrahedra is broken to insert an extra tetrahedron and Cu atoms. As a result, the c parameter (14.075 Å) of the copper silicate becomes about 1.5 times larger than the topaz a parameter. The two compounds instead show very close values of the other two lattice parameters, which are not modified by the insertions. Yakubovich *et al.* (2000) have given a list of hypothetical compounds generated by the described insertion mechanism.

4.5.5 *Gehlenite module*

Dódony and Buseck (2001) have interpreted the structure of $\mathbf{Ca_5Al_6MgSiO_{17}}$ **(Hanic phase**; *Pmmm*, a = 27.638, b = 10.799, c = 5.123 Å), a phase occurring in cements and ceramic materials, as the member GX of a **polysomatic series** G_nX_m based on modules of **gehlenite** (G) ($\mathbf{Ca_2Al_2SiO_7}$; $P\bar{4}2_1m$, a = 7.690, c = 5.067 Å) and $\mathbf{Ca_3Al_4MgSiO_{10}}$ (X) ($a \sim 16.6$, $b \sim 10.8$, $c \sim 5.12$ Å; Hanic *et al.* 1980). The same authors, by high resolution HRTEM images, have observed other polysomes of the series as defects in a matrix of the Hanic phase.

4.5.6 *Epidote modules*

According to Bonazzi *et al.* (2003), the new mineral **gatelite-(Ce)**, {$(CaREE_3)[Al_2(Al,Mg)(Mg,Fe,Al)][(Si_2O_7)[SiO_4](O,F)(OH,O)_2$; $P2_1/a$, a = 17.770, b = 5.651, c = 17.458 Å, β = 116.18°} can be described as a regular alternance along [001] of modules of **epidote-type structure** {E = $A_2M_3[Si_2O_7][SiO_4]O(OH)$; $P2_1/m$, $a \sim 8.7$, $b \sim 5.7$, $c \sim 10.1$ Å, $\beta \sim 115°$} and **törnebohmite-type structure** {T = $(REE)_2Al[SiO_4]_2(OH)$; $P2_1/c$, a = 7.383, b = 5.673, c = 16.937 Å, β = 112.04°}. Relative to their unit cells, the epidote modules are 10.37 Å thick (001) slabs, and the törnebohmite modules are 7.08 Å thick ($\bar{1}02$) slabs. Recently, the

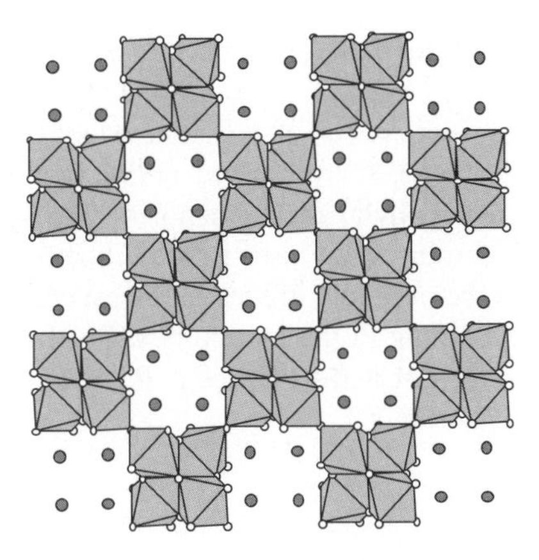

Fig. 4.33. Projection in the (100) plane of a pyrochlore slab; its thickness corresponds to that of two octahedra.

same type of modularity has been recognized in the new mineral **IMA No. 2002-025** [$Ce_3CaMg_2Al_2Si_5O_{19}(OH)_2F$; $P2_1/m$, $a = 8.939$, $b = 5.706$, $c = 15.855$ Å, $\beta = 94.58°$; Grice and Ferraris (2003)]. The close relation between **gatelite-(Ce)** and **IMA No. 2002-025** is better recognized if the cell of the latter mineral is transformed to $a = 17.878$, $b = 5.706$, $c = 17.569$ Å, $\beta = 115.90°$ by the matrix $\bar{2}00/010/101$.

An epidote building block is present in the compounds with general formula $A_2B(XO_4)_2(OH, H_2O)$ (**brackebushite group**; A = Ba, Ca, Pb, Sr, ...; B = Al, Cu, Fe, Mn, Zn, ...; X = As, P, S, V, ...). For example, see **bearthite** [$\mathbf{Ca_2Al(PO_4)_2(OH)}$; $P2_1/m$, $a = 7.231$, $b = 5.734$, $c = 8.263$ Å, $\beta = 112.57°$; Chopin *et al.* 1993].

4.5.7 *Pyrochlore modules*

The occurrence of pyrochlore modules (Fig. 4.33) has already been described in Section 1.8.2.3. In the **fersmanite** [$Ca_4(Na,Ca)_4(Ti,Nb)_4(Si_2O_7)O_8F_3$; $C2/c$, $a = 10.183$, $b = 10.183$, $c = 20.396$ Å, $\beta = 97.19°$] crystal structure, Sokolova *et al.* (2002) have emphasized the presence of Ti_4O_{18} tetramers which polymerize by corner-sharing of the Ti-octahedra to form (001) layers. The same layers tri-dimensionally polymerize to form a network in the **pyrochlore** crystal structure [$(Ca,Na)Nb_2O_6(OH,F)$; $Fd\bar{3}m$, $a = 10.39$ Å].

4.5.8 *Nacaphite modules*

Modules with the **nacaphite** structure [$\mathbf{Na_2Ca(PO_4)F}$; $C1$, $a = 10.654$, $b = 24.443$, $c = 7.102$ Å, $\alpha = 89.99$, $\beta = 94.58$, $\gamma = 90.01°$; metrically pseudo-orthorhombic and pseudo-hexagonal with $a' = c$ and $c' \sim 40$ Å] occur with the same chemical composition in members of the **mero-plesiotype bafertisite series** (Section 4.4.2.1) and, maintaining the topology only, in **sulphohalite** [$\mathbf{Na_6(SO_4)_2(F,Cl)_2}$; $Fm\bar{3}m$,

$a = 10.065$ Å], **schairerite** [$\mathbf{Na_{21}(SO_4)_7F_6Cl}$; $P3_1m$, $a = 12.197$, $c = 19.259$ Å], and **kogarkoite** [$\mathbf{Na_3(SO_4)F}$; $P2_1/m$, $a = 18.073$, $b = 6.949$, $c = 11.440$ Å, $\beta = 107.7°$] (Egorov-Tismenko *et al.* 1984). In their study of a **nacaphite dimorph** ($R3m$, $a = 7.072$, $c = 40.56$ Å), Sokolova *et al.* (1999) pointed out that whereas the module of nacaphite in heterophyllosilicate members corresponds to a block $7.1 \times 5.3 \times 6.0$ Å, oriented according to the pseudo-orthorhombic cell, the same type of module occurring in the other reported structures corresponds to a block $7.1 \times 7.1 \times 2.7$ Å and is oriented according to the pseudo-hexagonal cell. The latter block is also found in **arctite** [$\mathbf{(Na_5Ca)Ca_6Ba(PO_4)_6F_3}$; $R\bar{3}m$, $a = 7.094$, $c = 41.32$ Å; Sokolova *et al.* 1984].

4.5.9 *Apatites*

Compounds with general formula $A_5(BO_4)_3X$ (A = Na, Ca, Ba, Pb, Sr, Cd, Mn, K, Sn, REE, . . .; B = P, S, V, As, Ge, Si, . . .; X = F, Cl, Br, OH, . . .) and apatite-type structure are industrially important materials with applications in **catalysis**, **environmental remediation**, **bone replacement**, and **ceramic membranes** (White and ZhiLi 2003). These authors note that only about 50 per cent of a set of 74 chemically distinct apatites conform to the symmetry of the proper **apatite** [$\mathbf{Ca_5(PO_4)_3F}$, $P6_3/m$, $a = 9.363$, $c = 6.878$ Å], others show $P6_3$, $P\bar{3}$, and lower symmetry with increasing chemical complexity. White and ZhiLi (2003) construct the structures of the set by filling prismatic and tetrahedral interstices of an ideal hexagonal net and show schematically that the apatite is one end-member of an **apatite–nasonite polysomatic series**. By alternation of **nasonite** [$\mathbf{Pb_6Ca_4(Si_2O_7)_3Cl_2}$; $P6_3/m$(?), $a = 10.080$, $c = 13.270$ Å] and apatite modules along [001], the structure of **ganomalite** [$\mathbf{Pb_9Ca_5Mn(Si_2O_7)_3(Si_2O_4)_3}$, $P\bar{6}$, $a = 9.850$, $c = 10.130$ Å] is obtained (Fig. 4.34). The ratio of the *c* parameters of apatite, ganomalite, and nasonite is about 1 : 1.33 : 2.

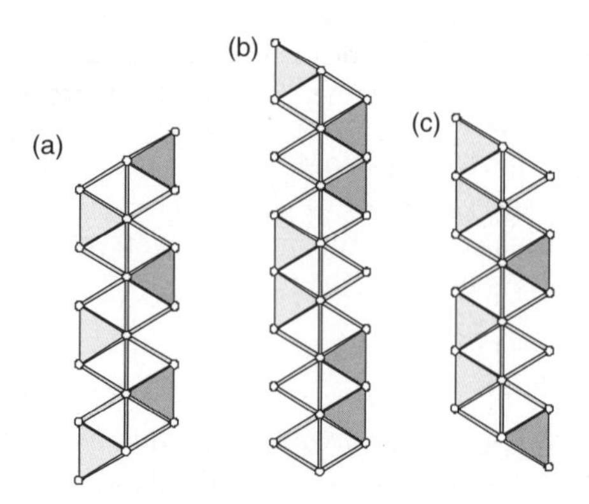

Fig. 4.34. Modules of apatite (a) alternate along [001] with modules of nasonite (b) in the structure of ganomalite (c). Only the tetrahedral sites are shown [see text; modified after White and ZhiLi (2003)]

4.5.10 *Fluorocarbonates*

The general formula $\mathbf{A}_m(\mathbf{REE})_n(\mathbf{CO}_3)_{n+m}\mathbf{F}_n$ represents the observed polysomes of two groups of **fluorocarbonates** with A = Ca, Ba. Each polysome, in its turn, can belong to a family of polytypes. These REE fluorocarbonates have generated considerable interest because of their economic, being the most abundant REE ore minerals, and crystallographic aspects.

4.5.10.1 *The huanghoite–zhonghuacerite polysomatic series*

With A = Ba and REE = Ce, the series $A_m(REE)_n(CO_3)_{n+m}F_n$ represents the **huanghoite–zhonghuacerite polysomatic series**. Yang *et al.* (1996) proposed a modular interpretation of the structure of **cebaite-(Ce)** [$\mathbf{Ba_3Ce_2(CO_3)_5F_2}$; $C2/m$, $a = 21.42$, $b = 5.078$, $c = 13.30$ Å, $\beta = 94.8°$] as the member HZ of a polysomatic series with end-members **huanghoite-(Ce)** [H = $\mathbf{BaCe(CO_3)_2F}$; $R\bar{3}m$, $a = 5.072$, $c = 38.46$ Å] and **zhonghuacerite-(Ce)** [$Z = \mathbf{Ba_2Ce(CO_3)_3F}$; $P2_1/m$, $a = 13.365$, $b = 5.097$, $c = 6.638$ Å, $\beta = 106.45°$]. They noted that the chemical formula of cebaite-(Ce) can be considered to be the sum of the formulae of H and Z, and all three compounds have almost the same b parameter. Polysomatic defects have been observed by electron microscopy in a matrix of cebaite-(Ce).

4.5.10.2 *The bastnäsite–synchisite polysomatic series*

With A = Ca the series $A_m(REE)_n(CO_3)_{n+m}F_n$ represents the **bastnäsite–synchisite polysomatic series**. The **bastnäsite**, $\mathbf{(REE)(CO_3)F}$ [for REE = Ce: $P\bar{6}2c$, $a = 7.118$, $c = 9.762$ Å; Wang *et al.* 1994], and **synchisite** groups, $\mathbf{Ca(REE)(CO_3)_2F}$ [for REE = Ce: $C2/c$, $a = 12.329$, $b = 7.110$, $c = 18.741$ Å, $\beta = 102.68°$; Ni *et al.* 1993], include mineral species which are differentiated by the dominating REE, substitution of F by OH, and polytypism. The bastnäsite and synchisite structures occur as building modules in other mineral species [**parisite**, $\mathbf{Ca(REE)_2(CO_3)_3F_2}$; Cc, $a = 12.305$, $b = 7.105$, $c = 28.250$ Å, $\beta = 98.26°$; Ni *et al.* 2000—**röntgenite**, $\mathbf{Ca_2(REE)_3(CO_3)_5F_3}$; $R3$, $a = 7.13$, $c = 69.4$; Donnay and Donnay 1953] or as defects (polysomes) revealed by HRTEM (Wu *et al.* 1998; Meng *et al.* 2001*a*; Meng *et al.* 2001*b*; Meng *et al.* 2002).

At least four modular description of the structures with general formula $\mathbf{Ca}_m(\mathbf{REE})_n(\mathbf{CO}_3)_{n+m}\mathbf{F}_n$ are known:

1. Donnay and Donnay (1953) used four chemically and structurally different types of hexagonal layers with composition CeF, Ca, and CO_3 (two types) to describe the (pseudo)hexagonal structures known at that time.
2. Van Landuyt and Amelinckx (1975) named d, f, e, and g the Donnay and Donnay's (1953) layers and used their different stacking to describe the polytypism observed for the members of a **polysomatic series** B_nS_m based on modules of

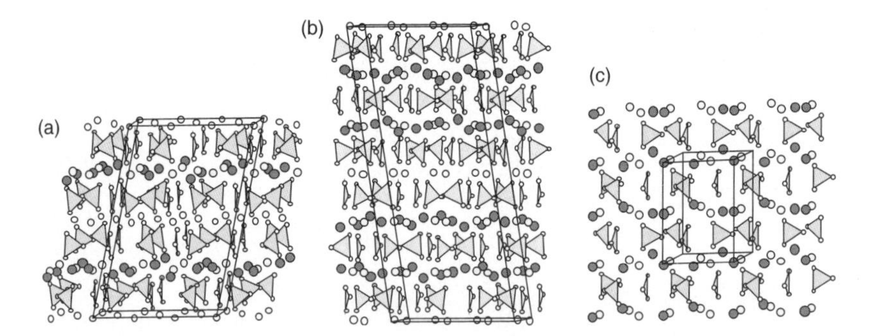

Fig. 4.35. Structures of synchisite-(Ce) (a), parisite-(Ce) (b), and bastnäsite-(Ce) (c). Triangles represent (CO_3) groups with their oxygens shown as small blank circles. Grey, large blank, and intermediate blank circles represent F, Ce, and Ca, respectively. A Ca sheet connects pairs of bastnäsite mono- and bi-layers in synchisite and parisite, respectively. (See text for reference.)

the two end-members bastnäsite (B = de; $n = \infty$, $m = 0$) and synchisite (S = dgfg; $m = \infty$, $n = 0$).

3. In front of the increasing number of observed polysomes and polytypes [more than 50, according to Table 3 in Yang *et al.* (1998)], Yang *et al.* (1998) proposed a notation based on two basic layers of CeO_6F_3 and CaO_8 polyhedra. Each layer exists in four different states: n, u, and their mirror-related $\bar{n}$ and $\bar{u}$, for the CeO_6F_3 layer; b, d, p, and q, for the CaO_8 layer.
4. Recently, Ni *et al.* (2000), after solving the structure of parisite-(Ce), have shown that the structures of the $\mathbf{Ca}_m\mathbf{(REE)}_n\mathbf{(CO_3)}_{n+m}\mathbf{F}_n$ polysomes can be described as composed of bastnäsite modules, $[(CeF)_n]^{2n+}[(CO_3)_{n+1}]^{2(n+1)-}$ (thickness increasing with n), which are connected by one Ca-layer only, independently of their thickness.

Following Ni *et al.* (2000), the linkage of two bastnäsite slabs (Fig. 4.35) by a Ca layer requires an offset between the slabs; this offset is the origin of the observed polytypism. The offset vector is **a**/3 along [100] of the (pseudo)hexagonal (001) two-dimensional lattice; analogously to micas, the angle between the offset vectors belonging to two adjacent slabs is either 120° or 240°. At variance with micas, a 60° angle cannot be realized because of hindrance between heavy atoms and carbonate groups, which would be aligned perpendicularly to the slabs. Bastnäsite does not show polytypism because the Ca layer is missing in its structure. The Ca-richest member of the series is the end-member synchysite ($n = 1$). In fact, as observed by Yang *et al.* (1994), a Ca layer cannot connect with itself otherwise a slab of **vaterite** structure, $\mathbf{CaCO_3}$, would be realized; however, this phase is not stable at ambient conditions. Bastnäsite slabs of different thickness can occur in the same structure; for example, röntgenite contains slabs $(CeF)(CO_3)_2$ and $(CeF)_2(CO_3)_3$ in a ratio 1 : 1.

In comparison with the formalism represented by the formula A_nS_m, which gives information only on the ratio n/m, the polysomatic picture proposed by Ni *et al.* (2000) allows to predict that for each polysome the number of bastnäsite slabs is equal to the number of Ca atoms in the unit formula. Thus, in principle, given the chemical composition the number of different slabs and their thickness can be determined. Anyway, due to polytypism, only a crystal structure determination can show the exact sequence of layers.

In conclusion, only an elaborate formalism, for example, that of Yang *et al.* (1998), can properly describe such sequences. In the order, for bastnäsite-(Ce), parisite-(Ce), röntgenite-(Ce), and synchysite-(Ce), the symbols according to the different formalism mentioned above are: B_1S_0 (*de*; *bd*), B_1S_1(*dedgfg*; *nbd$\bar{n}$pq*), B_1S_2 (*dedgfgdgfg*; *nbuqp*), B_0S_1 (*dgfg*; *nbuq*) [in parentheses the sequence according Donnay and Donnay (1953) and Yang *et al.* (1998) are shown, in the order]. The symbols that are necessary for a full description of the layer sequence, are reminiscents of those introduced for layer silicates (cf. Dornberger-Schiff *et al.* 1982; Ross *et al.* 1966; Takeda and Sadanaga 1969; Zvyagin 1985).

4.5.11 *The schafarzikite polysomatic series*

After an earlier modular description by Mellini and Merlino (1979), Ferraris *et al.* (1986) presented the following modular description of schafarzikite and related structures. The crystal structures of **schafarzikite** (**$Fe^{2+}_4Sb_8O_{16}$**; $P4_2/mbc$, $a = 8.568$, $c = 957$ Å; Fischer and Pertlik 1975), **versiliaite** (**$Fe^{2+}_4Fe^{3+}_8Sb_{12}O_{32}S_2$**; *Pban*, $a = 8.499$, $b = 8.326$, $c = 957$ Å; Mellini and Merlino 1979), and **apuanite** (**$Fe^{2+}_4Fe^{3+}_{16}Sb_{16}O_{48}S_4$**; *Pban*, $a = 8.499$, $b = 8.326$, $c = 957$ Å; Mellini and Merlino 1979) are based on two building modules (Fig. 4.36). One module, Sc, corresponds to a $c/2$ thick (001) slab of schafarzikite; the second one, U, corresponds to a slab of a hypothetical structure with composition **$Fe^{3+}_8Sb_4O_{16}S_2$**. The thickness of the U module is two times that of Sc, that is about 6 Å. Fe^{3+} is 50 per cent tetrahedral in versiliaite and apuanite and all tetrahedral in U.

To the Sc_mU_n polysomatic series, besides the polysomes Sc_2 (schafarzikite), Sc_2U (versiliaite), Sc_2U_2 (apuanite), and U (hypothetical end-member), belongs also an

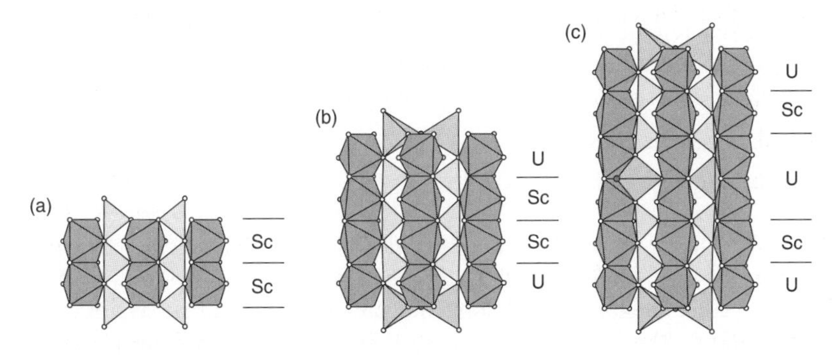

Fig. 4.36. Projection along [010] of the schafarzikite (a), versiliaite (b), and apuanite (c) structures. The building modules Sc and U are indicated (see text). (See text for reference.)

18 Å orthorhombic microstructure, Sc_4U, identified by TEM observation (Mellini *et al.* 1981). The same authors explained the non-stoichiometry usually shown by the three mentioned mineral species, as originated by polysomatic defects and intergrowths, which they abundantly observed by electron microscopy.

4.5.12 *Layered double hydroxides*

A book (Rives 2001) has been recently dedicated to the **layered double hydroxides**, a wide **merotype series** of natural and synthetic modular structures based on a fixed brucite-type octahedral layer (Section 4.4; see also Section 1.8.2.5), where a partial Mg^{2+}/Al^{3+} substitution has taken place (thus the name 'layered double hydroxide'), and a variable interlayer including either organic or inorganic complex anions. The mineral **hydrotalcite** can be considered the prototype of the series. Technological applications include catalysis and anion exchange; the restricted interlayer space also represents a sort of nanoreactor to perform chemical reactions in a constrained region.

4.5.13 *A mero-plesiotype series of microporous silicates*

It the article reporting the modelling of the crystal structure of the microporous titanosilicate **seidite-(Ce)**, $Na_4(Ce,Sr)_2\{Ti(OH)_2(Si_8O_{18})\}(O,OH,F)_4{\cdot}5H_2O$ (Ferraris *et al.* 2003), the occurrence of a layer of eight-membered channels delimited by corner-sharing SiO_4-tetrahedra is discussed by comparison with the following compounds that bear the same layer (Hesse *et al.* 1992; Rocha & Anderson 2000): **rhodesite**, $K_2Ca_4[Si_8O_{18}(OH)]_2{\cdot}12H_2O$, **macdonaldite**, $BaCa_4[Si_8O_{18}(OH)]_2{\cdot}10H_2O$, **delhayelite**, $K_7Na_3Ca_5[Si_7AlO_{19}]_2F_4Cl_2$, **hydrodelayelite**, $K_2Ca_4[Si_7AlO_{17}(OH)_2]_2{\cdot}6H_2O$, **montregianite**, $K_2Na_4Y_2[Si_8O_{19}]_2{\cdot}10H_2O$, $\mathbf{K_2Na_4Ce_2[Si_8O_{19}]_2{\cdot}10H_2O}$ and its Eu and Tb equivalents. In all these compounds, the silicate double layer, which shows pores up to about 3 × 5 Å wide, alternates with a layer of cations: chains of edge-sharing CaO_6 octahedra in rhodesite and macdonaldite, mixed $(Y,Na)O_6$ octahedra in montregianite, a continuous layer of octahedra in delhayelite, and isolated TiO_6 octahedra in seidite-(Ce). Thus, the rhodesite group represents a **mero-plesiotype series** (Section 4.4.2.1) because, apart from slight modifications (plesiotype aspect), the same double silicate layer occurs in all members, alternating with a variable layer module (merotype aspect).

4.6 Some conclusions

The type of building modules has been the guideline to group together the examples of polysomatic series illustrated in this chapter. That immediately shows the frequent presence in different structures of some modules, like those of **perovskite-**, **mica-**, and **spinel-type**. The modules of these recurrent structure-types are found also in different orientations. Examples of perovskite and spinel-type modules according to three different orientations have been discussed; the mica-type module occurs both as TOT layers and rods with different widths.

4.6.1 *Role of close packing structures*

If only examples of synthetic materials would be known, one could suppose that the high recurrence of a module-type (e.g. perovskite-type in superconducting materials) is an artefact due to the interesting physical properties of the structure-type that stimulate research in the field. However, not to mention the large variety of natural **perovskites** (cf. Mitchell 2002), variously sliced **mica-like modules** are present in minerals, that is, natural materials, namely **layer silicates** (from **talc** to **chlorites** and mixed-layer **clay minerals**), **biopyriboles**, **inophites**, **heterophyllosilicates**, **palysepioles**, and several related structures. It seems therefore reasonable to connect the variety and frequency of a module with its structural stability.

Other characteristics for sure favour the diffusion of a structure-type as structure building module. High flexibility to host isomorphous substitution of both cations and anions and the presence of some 'universal' reticular periodicity are among these characteristics. Isomorphism allows the formation of a large number of chemically different compounds belonging to the same structure-type; 'universal' periodicities, for example, those connected with the anion–anion distances, favour the matching of modules on their interfaces. An (approximate) close packing array of at least an atomic constituent (usually the anion) assures both aspects. In fact, close packing of atoms is found in structures with any kind of chemical bond: from ionic in halides, to covalent in diamond, and metallic in intermetallic compounds (cf. Ferraris 2002). Symmetry aspects, too, cannot be ignored, because a high symmetry favours the matching of modules according to different orientations. Once more, the close packing, which is based on a two-dimensional hexagonal periodicity, offers matching benefits connected with high symmetry.

4.6.2 *End-member modules*

When the polysomatic description of a structure is possible according to different choices of modules, preference for those modules that are able to form their own compounds (i.e. end-members of a polysomatic series) looks sensible, under several aspects. A modular description based on sheet of atoms (as in some **superconductors**) looks too much like an artefact, mainly because of the lack of a chemical meaning from a stoichiometric viewpoint. First, the modules of end-members tend to occur in different series and useful cross links can be established, as shown in several examples discussed in this book. Second, but for sure highly important when studying stacking defects, which often are related to solid state reactions (cf. Buseck 1992), these 'natural' modules have a high capability of polymerization, as proved by their infinitely extended linkage in forming their own compounds. It is, therefore, highly probable that these types of modules occur as width-defects in microstructures. Thus, their individuation can offer a key to detect and understand stoichiometry anomalies and related occasional variable physical properties (cf. Section 4.2.3.1 for superconductivity).

5

Modularity at crystal scale — Twinning

Originally, twins were studied from a morphological viewpoint as a special case of **oriented association** of individual single crystals belonging to the same phase, that is, a group of crystals (hereafter, individuals) that can be brought to the same orientation by a translation, rotation or reflection. Individuals related by a translation form a **parallel association**, a case that does not lead to a situation of particular crystallographic interest in the present context. Individuals related either by a reflection (mirror plane m or centre of symmetry $\bar{1}$), or a rotation form a twin (Friedel 1926). An element of symmetry crystallographically relating differently oriented crystals cannot belong to the individual. In fact, an element of symmetry belonging to the individual cannot orient it in a different way. The element of symmetry that originates a twin is called a **twinning element of symmetry** and the connected operation is a **twinning operation of symmetry**. Besides morphological consequences, the twinning operation brings a total or partial superposition of the nodes belonging to the Bravais lattices of different individuals with effects on the diffraction pattern that are discussed in the following.

The term 'twin' has been introduced and is widely used in literature also to indicate twinning-like **structure-building operations** at unit cell level (**cell-twinning**) that do not overlap the Bravais lattices or morphologically associate the individuals. Often these structure-building operations represent a mechanism generating a modular crystal structure and have been discussed and used in other sections of this book. If necessary, to avoid confusion, in this chapter, the term **classical twin(ning)** is used in opposition to cell-twin(ning); otherwise, the latter type of phenomenon is never meant, if only the term twin(ning) is used.

Whereas all other aspects of the modular materials discussed in this book occur at cell level (including cell-twinning), the classical twinning involves the orientation of modules, which correspond to the entire crystal structure of individuals, whose dimensions span from few cells [like micro-**polysynthetic twinning** and **domain twins** (cf. Hahn *et al.* 1999) originated by phase transitions] to macroscopic crystals. Undoubtedly, however, the classical twinning covers an aspect of structural modularity, because its presence means that the entire structure of a material is fragmented in oriented modules. Without realizing the presence of differently oriented fragments, it is normally impossible to properly solve the crystal structure of the individual.

5.1 Classification of twins

The introduction of automatic methods for the treatment of twinning in software devoted to the solution and refinement of the crystal structures (e.g. Herbst-Irmer and

Sheldrick 1998) has produced a revival of interest on twins. Thus, a topic which, for a long time, had been cultivated only by people interested in morphology, like mineralogists and crystal growers, suddenly has entered the world of the structural crystallographers, often just embedded in crystallographic computing packages. This may pose problems to some categories of users (cf. Nespolo and Ferraris 2003) in properly grasping crystallographic aspects, which may be weak or even absent in their basic preparation and even in books dealing with structural crystallography. A non-canonical, or just approximate, use of nomenclature and its related physical meaning is a main source of confusion in the structural treatment of twinning (see examples below). Attention shall be paid in the following pages to fix the proper nomenclature and procedural aspects that are involved with twinning problems.

The investigation on twins started long time ago and was brought to its modern form by the French school (Bravais 1851; Mallard 1885; Friedel 1926) and later papers (cf. Buerger 1945; Donnay and Donnay 1974; Santoro 1974). Buerger (1945) stressed that the geometrical conditions for twinning established by the French school are necessary, but not sufficient for the appearance of the phenomenon in a crystal structure. He introduced and illustrated the constraints that the atomic structure poses to the appearance of twinning. These type of conditions shall no longer be discussed here, but shortly can be summarized as the necessity that the crystal structure, at least locally at the interface between individuals (**twin wall**, **twin boundary**), matches the supplementary symmetry introduced by a twin operation. Updated references on this topic, also related to material properties and phase transitions, can be found, for example, in Salje (2002) and Heinemann *et al.* (2003).

The **Mallard's law** states that the twin element (i.e. the geometrical element relative to which the twin operation is defined) is restricted to a direct lattice element. Therefore, the twin laws, that is, the geometrical relations between pairs of individuals in a twin, are defined in terms of lattice nodes (**twin centres**), lattice rows (**twin axes**), and lattice planes (**twin planes**) in the direct space. In the order for centre, axis, and plane, the **twin law** is expressed by the symbol $\bar{1}$ (twin centre), [*uvw*], and (*hkl*). In case all symmetry equivalent (*hkl*) planes or [*uvw*] directions are meant, the symbols $\{hkl\}$ and $\langle uvw \rangle$ are used, in that order. Matrices transforming the cell parameters (and atomic coordinates) of one individual into those of a related one are routinely used. These matrices are usually needed in calculations dealing with twinning. For example, the twinning laws $\bar{1}$, (010) and [001] twofold axis are expressed by the following matrices, in the order: $\bar{1}00/0\bar{1}0/00\bar{1}/$; $\bar{1}00/010/00\bar{1}/$; $\bar{1}00/0\bar{1}0/001/$.

Whereas in morphological crystallography, twinning is always described in the direct space (real space), to the structural crystallographer, the twinning usually appears in the reciprocal space (diffraction space) via a diffraction pattern. Thus, the structural investigation tends to favour a description of the twin laws through twin elements defined in the reciprocal space. However, because twinning is physically governed by the direct lattice and by the structural relations between pairs of individuals, the final expressions of the twin laws should always be given in terms of the direct lattice. Note that, normally, the symmetry of a twin (**twin point group**) is that of the crystal point group augmented by the symmetry of the twin operation. However, a symmetry element that is oblique relative to the twinning element of symmetry, is absent in the twin (Friedel 1926; Buerger 1954). Classical examples are {111} **spinel twins** (Section 5.1.2.4), which show trigonal symmetry, and {101} **elbow twins** (for example in **rutile**); the latter are due to the presence of a pseudocubic supercell and display orthorhombic symmetry.

5.1.1 *Twinning related to lattice (pseudo)symmetry*

5.1.1.1 *Twinning by merohedry*

Twinning by **merohedry** is the oriented association of two or more individuals that are related by a **twin operation** belonging to the point group of the lattice, but not to the point group of the individual. The lattice common to the twinned individuals (i.e. to the twin) is called the **twin lattice**; in case of merohedry (but see below) coincides with the lattice of each individual. Merohedry is classified on the basis of the ratio between the order of the lattice point group (**holohedry**) and the order of the structure point group. This ratio is also known as **subgroup index** s **of the merohedry** or **twin order**, and is shown here in parentheses for the possible merohedries: **hemihedry** (2), **tetartohedry** (4), and **ogdohedry** (8). The possible crystallographic holohedries are those corresponding to the point groups of the Bravais lattices: $m\bar{3}m$ (cubic), $6/mmm$ (hexagonal), $\bar{3}m$ (rhombohedral), $4/mmm$ (tetragonal), mmm (orthorhombic), $2/m$ (monoclinic), and $\bar{1}$ (triclinic). For example, the holohedry of the tetragonal lattice (and of the tetragonal system) is $4/mmm$; relative to this holohedry (order 16), the symmetry of the structures with point group $4/m$ (order 8) or $\bar{4}$ (order 4) represents a hemihedry or a tetartohedry, respectively.

In connection with the deconvolution of the overlapped diffracted intensities in the process of solving a crystal structure (see below), Catti and Ferraris (1976) classified twins by merohedry into twins of **class I** (the twin operation belongs to the Laue group of the individual) and **class II** (the twin operation does not belong to the Laue group of the individual). In class I (e.g. Laue group $4/m$, crystal point group $\bar{4}$, and twin operation m), the twin operation always superimposes equivalent lattice nodes, that is, equivalent diffracted intensities produced by each individual. In class I, all twins are by hemihedry and necessarily contain only two individuals. In this case, the centre of symmetry $\bar{1}$ may always be assumed as twin law, even if, usually, planes and axes are preferred. In class II twins, instead, the superimposed lattice nodes (and the corresponding diffracted intensities) are not equivalent in the Laue group. All kinds of merohedries can occur for class II twins and, for a twin of order s, $s - 1$ twin operations are possible. A twin composed of s individuals is called **complete twin** (Curien and Donnay 1959). In the point groups with subgroup index, $s = 2$ (i.e. 1, 2, m, 222, $mm2$, 422, $4mm$, $\bar{4}2m$, 32, $3m$, 622, $6mm$, $\bar{6}2m$, 432, and $\bar{4}3m$) only hemihedry can occur. In the other point groups, the subgroup index is either 4 or 8 and the possible twin operations can be obtained, for example, by a coset decomposition (Flack 1987) [cf. chapter 4 in Giacovazzo (2002) and Koch (1999) for a listing]. Taking into account that 159 out of 230 space groups belong to merohedric point groups, the appearance of twinning cannot be considered a rare event.

5.1.1.2 *Twinning by metric merohedry*

The occurrence of crystal structures based on a **lattice** with a **metric symmetry**, (LMS) higher than the symmetry required by the point group of the structure is not rare, of which some examples shall be discussed below (Section 5.4.1.1). Nespolo and Ferraris (2000) have introduced the term **syngonic merohedry** for the classic merohedry described in Section 5.1.1.1 and defined **twinning by metric merohedry** to describe the case of twinning where the twin operation is a symmetry element belonging to the LMS but not to the syngony (crystal system) point group. A symmetry operation by metric merohedry brings to superposition non-equivalent lattice nodes, as does class II syngonic merohedry. The same authors split the class II of twins by merohedry into a **class IIA** (the previous II) and a **class IIB**. The twins by metric merohedry belong to the latter class.

Friedel (1926) suspected the possibility of twinning by metric merohedry, which he called **high-order merohedry**, in monoclinic crystals with vanishingly small obliquity (Section 5.1.1.3). However, he deemed that it was not possible to prove it within experimental error. Hurst *et al.* (1956) referred to Friedel (1926) to explain $\{001\}$ twinning in **staurolite** with an angle β close to 90°. Note, however, that according to Hawthorne *et al.* (1993), a monoclinic $\rightarrow$ orthorhombic phase transition may occur in staurolite.

5.1.1.3 *Twinning by pseudomerohedry*

If a lattice is pseudo-symmetric, that is, a lattice row or a lattice plane approximately corresponds to an extra element of symmetry, this element of pseudo-symmetry can act as twinning operation. Because of the centrosymmetry of the lattices, always a lattice symmetry plane (axis) is normal to a lattice symmetry axis (plane). The two **corresponding symmetry elements** generate equivalent twins when acting as twinning elements by merohedry. In case of pseudo-symmetry, the orthogonality becomes quasi-orthogonality and the two **corresponding pseudo-symmetry elements** generate non-equivalent twins that, after Friedel (1926), are known as **corresponding twins** [**conjugated twins** for some authors (e.g. Savitskii *et al.* 1996)]; the mineral **albite** offers a classical example. The oriented association generated by a pseudo-symmetry element is known as **twinning by pseudomerohedry** and shows the following characteristics:

- The lattice twin axis and the corresponding twin plane are only approximate symmetry elements of the crystal lattice. The twinning laws are expressed by the symbols of these axes and planes.
- The angle between the normal to the twin plane and the corresponding twin axis defines the **twin obliquity** ω. Le Page (2002) describes a method and programme to find the pseudo-symmetry elements of a lattice, if any, and calculate the twin obliquity.
- The twin operation does not systematically superimpose the lattice nodes belonging to the different individuals of the twin; the separation between nodes that would be superimposed for $\omega = 0^\circ$ is a function of *hkl*.
- The twin lattice corresponds to the lattice that is obtained by substituting a single node to the nodes that would be superimposed for zero obliquity.

The twinning by metric merohedry corresponds to the degeneration of the twinning by pseudomerohedry to zero obliquity.

5.1.2 *Twinning related to sublattice (pseudo)symmetry*

5.1.2.1 *Twinning by reticular merohedry*

In the presence of a sublattice displaying symmetry other than that of the crystal lattice, a symmetry element belonging to the sublattice point group, but not to the crystal point group, can act as twin element. If the twin symmetry element does not belong to the crystal lattice point group, the generated twin is known as **twin by reticular merohedry** (sometimes improperly quoted as non-merohedral twin), because the crystal symmetry is a merohedry of the sublattice symmetry. If the twin symmetry element does belong to the crystal lattice point group, a twin by merohedry is generated. If more than one twinning operation is active, both twinning by reticular merohedry and by merohedry may coexist.

The higher-symmetry sublattice represents the **twin lattice** defined in Section 5.1.1.1. It consists of a fraction $1/n$ of the crystal lattice nodes and n is known as **twin index** ($n \geq 2$). The nodes belonging to the twin lattice are overlapped by any allowed twin operation belonging to the superlattice symmetry. The Bravais cell of the twin lattice is the **twin cell**; it contains $m = gn$ nodes, where $g = 1, 2, 4$ according the twin Bravais lattice is P, C or I, F. With reference to a direct primitive crystal lattice, the twin lattice is characterized by a supercell that is m times multiple of the crystal cell. If the Bravais cell of the crystal lattice has multiplicity g', the Bravais twin cell is $m' = m/g' = gn/g'$ times multiple of the crystal cell. In the **reciprocal space**, the twin lattice is a **superlattice** based on a **subcell** that is m-times submultiple of the crystal cell. Note that in literature the use of sub- and super-lattice (cell) is reversed when the accent is put on the translations instead of on the set of nodes. Our use of sublattice is in agreement with the IUCr recommendation (Gruber 2002). A search for superlattices can be done by the Le Page (2002) method and related software developed by the same author.

Above the accent has been put on the common case of a sublattice with symmetry higher than that of the crystal lattice. However, cases of twinning due to an element of symmetry belonging to a sublattice with symmetry either equal (see Section 5.1.2.5) or lower than that of the crystal lattice are known, even if rare because of their high twin index. Friedel (1923, 1926) illustrated the latter case in twins of quartz with twin axis inclined on the three-fold axis. For a general discussion see Nespolo and Ferraris (2004).

To avoid confusion, reference to a crystal lattice primitive cell is suggested. That at least for two reasons: (i) the primitive cell of the crystal lattice could reveal a higher metric symmetry which might be hidden by a non-primitive cell, (ii) being, as said above, $m' = m/g'$ the sublattice might show a 'supercell' smaller than that a non-primitive cell of the crystal lattice. The following example can help to clarify the situation (cf. examples of twinning below). $Ba_2Na(Ni, Cu)_3O_6$ (Quarez *et al.* 2002) is F-centred orthorhombic with $a = 8.296$, $b = 14.369$, $c = 11.225$ Å, $V_o = 1338.1$ Å^3 ($g' = 4$). Due to $b = (3)^{1/2}a$, the crystal lattice shows a hexagonal P superlattice with $a = 8.296$, $c = 11.285$ Å, $V_h = 1/2V_o$ ($g = 1, n = 2$); thus

$m' = 1/2$. The perplexing situation of a sublattice with a 'supercell' smaller than that of the crystal cell, vanishes if the comparison is done with the primitive crystal cell which has a volume $\frac{1}{4} V_o$ and $\frac{1}{2} V_h$. The same orthorhombic crystal lattice considered as C-centred, would have a primitive hexagonal cell and twinning by metric merohedry might be expected.

The complexity of the indices (*hkl* for a plane, *uvw* for a direction) of a twin element increases with the twin index. As experimentally observed by Mallard (1885) (cf. Le Page 2002), values of n higher than 5–6 cannot be expected because the corresponding twin lattice would represent too small a fraction of lattice points and would not exert a suitable control on the crystallisation process. According to Friedel (1926), smaller is the twin index, higher is the probability of twin formation. Relative to the pseudo-cubic supercell of **staurolite**, Hurst *et al.* (1956) report a twinning with twin index 6. Beyond $n = 5$–6, the oriented associations of crystals, even if rare (Takeda 1975), can still be found and are classified in terms of coincidence site lattice theory (Section 5.1.3.1).

Note that in the twins by merohedry, the twin index is 1 and the twin lattice coincides with the crystal lattice. Twinning by metric merohedry (Section 5.1.1.2) can be considered the degeneration of twinning by reticular merohedry to index 1.

5.1.2.2 *Plurality of higher-symmetry sublattices*

In some cases, more than one sublattice allowing twinning can occur. For **staurolite** (Hurst *et al.* 1956), twinning relative to three different cells is known: (i) by pseudomerohedry, relative to a pseudo-orthorhombic basic cell ($a = 7.870$, $b = 6.623$, $c = 5.661$ Å, $\beta = 90.12°$); by reticular pseudomerohedry, (ii) relative to a treble pseudo-tetragonal subcell and, (iii) an 18-fold pseudo-cubic subcell. The latter two cells are obtained from the basic pseudo-orthorhombic cell by the 010/003/100 and 013/0$\bar{1}$3/300 transformations, respectively. The quoted sublattices explain the morphology of all twins observed in staurolite.

Takeda *et al.* (1967) investigated, by precession method, twinning by reticular pseudomerohedry in **djurleite**, CuS, as generated by the presence of a double pseudo-tetragonal subcell (obliquity 0.35°, twin index 2) and a quadruple pseudo-hexagonal subcell (obliquity 0.17°, twin index 4). These cells, in the order, are obtained from the basic orthorhombic cell ($a = 26.92$, $b = 15.71$, $c = 13.56$ Å) by 002/100/010 and 020/0$\bar{1}$2/100 transformations. The authors explain the non-space group absences (Section 5.2.3) observed for these twinnings and discuss the influence of the crystal structure on twinning. Friedel (1923) used various sublattices to explain the twins of quartz he studied morphologically.

5.1.2.3 *Twinning by reticular pseudomerohedry*

By analogy with twinning by pseudomerohedry, twinning by **reticular pseudomerohedry** may appear if a sublattice bearing pseudo-symmetry other than that of the crystal lattice does exist. A twin by reticular pseudomerohedry has a twin index higher than 1 and a non-zero obliquity. As for the twins by pseudomerohedry (Section 5.1.1.3), the twin lattice corresponds to the lattice that is obtained by substituting a single node to the nodes that would be superimposed for zero obliquity. Besides the upper limit of the **twin index** mentioned for the twinning by reticular merohedry, Mallard (1885) recognized also an upper limit of 5–6° for the **obliquity**. Due to the same mechanism discussed for pseudo-merohedry, **corresponding twins** can occur.

Donnay and Donnay's (1974) **twin-lattice symmetry (TLS)** and **twin-lattice quasi-symmetry (TLQS)** twins correspond to the twins by merohedry and pseudomerohedry, respectively. On the base of morphological observations only, the German school developed a twin nomenclature that is to some extent different and hardly useful in structural crystallography [cf. Tschermak (1921) and, for a modern English summary, Bloss (1971)].

Perovskite, $CaTiO_3$, is a good case to illustrate twinning by reticular pseudomerohedry. Ideally, the structure is cubic $Pm\bar{3}m$ (Section 4.2) with cell parameter a'_c, but natural perovskite, like many synthetic compounds showing perovskite-type structure, is usually orthorhombic with $a_o \sim b_o \sim 2^{1/2}a'_c$ and $c_o \sim 2a'_c$, that is, the cubic → orthorhombic cell transformation is $011/0\bar{1}1/200$. This cell transformation implies a space group *Pbnm* (No. 62) for the orthorhombic structure. Often however, orthorhombic perovskites are described with the orientation *Pnma* of the same No. 62 space group. That implies a cubic → orthorhombic transformation $0\bar{1}$ $1/200/011$ which leads to $a_o \sim c_o \sim 2^{1/2}a'_c$ and $b_o \sim 2a'_c$. In the following, we refer to the *Pbnm* orientation.

The orthorhombic (actually pseudo tetragonal) lattice admits a pseudocubic sublattice with unit cell parameter $a_c = 2a'_c$ (Fig. 5.1). Due to this pseudocubic

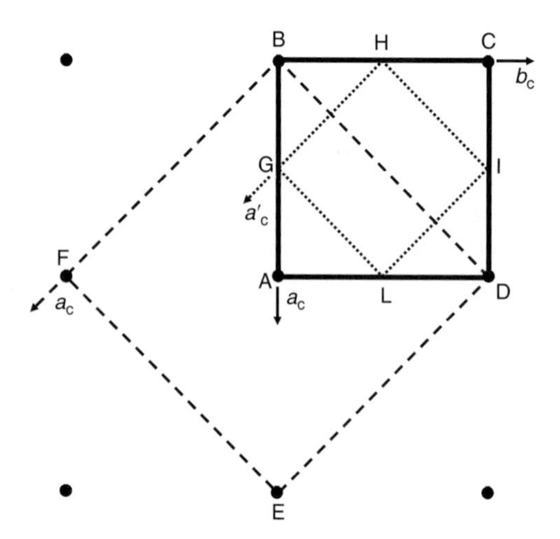

Fig. 5.1. GHIL (dotted lines) represents the bidimensional unit cell of an ideal cubic perovskite with edge a'_c. An orthorhombic perovskite has cell ABCD (full lines) with $a_o \sim b_o \sim 2^{1/2}a'_c$ and (not shown) $c_o \sim 2a'_c$. The filled circles represent the nodes of the orthorhombic lattice. FBDE (broken lines) is the base of the cubic supercell ($a_c = 2a'_c$) relative to the (pseudo)cubic sublattice embedded in the orthorhombic lattice.

sublattice, twinning by reticular pseudomerohedry must be expected. In fact, it is often observed particularly by transmission electron microscopy [e.g. Keller and Buseck (1994) and papers therein quoted].

Usually, the complex twinning of perovskite is explained with reference to the original ideal cubic cell, as presumably initiated by Kay and Bailey (1957), instead of with reference to the mentioned sublattice, as sometimes correctly done (e.g., Arakcheeva *et al.* 1997; Savitskii *et al.* 1996). The two (pseudo)cubic cells are parallel, thereby, the symbols expressing the twin laws are the same in both cases. However, the reference to a subcell (that does not exist in the orthorhombic lattice) instead of, to the correct supercell, can be misleading in properly understanding mechanism and consequences of twinning in this technologically important structure type.

The $\{100\}$ symmetry planes of the cubic sublattice are the most common twin elements acting by reticular pseudomerohedry in orthorhombic perovskites. The relative twin law (orthorhombic indexing) is either $\{110\}$ or $\{101\}$, according the structure is described as *Pbnm* or *Pnma*, as summarised by Arakcheeva *et al.* (1997) in their Table 4, together with other possible twins laws observed for perovskites.

5.1.2.4 *Twinning in rhombohedral and centred cubic lattices*

A non-primitive Bravais hexagonal sublattice of order 3 and symmetry $6/mmm$ is always embedded in an R rhombohedral Bravais lattice. The latter has symmetry $\bar{3}m$, a subgroup of $6/mmm$. Note that usually the structures belonging to the trigonal system (point groups 3, $\bar{3}$, 32, $3m$, and $\bar{3}m$) are referred to a hexagonal reference system corresponding to the edges of the hexagonal supercell. Consequent to the presence of a hexagonal sublattice, the crystal structures based on an R lattice are endemic candidates to twinning by reticular merohedry via the symmetry elements belonging to the $6/mmm$ point group of the sublattice, but not to the $\bar{3}m$ point group of the lattice.

Six differently oriented hexagonal settings can correspond to a rhombohedral lattice [cf. chapter 1 in Giacovazzo (2002)]; three of them are known as **obverse setting** and the other three as **reverse setting**. The twofold axis, which is parallel to the sixfold axis of the hexagonal sublattice and to the threefold axis of the R lattice, exchanges the obverse/reverse settings, this being the commonest twinning element of symmetry in rhombohedral structures. Unfortunately, this type of twin by reticular merohedry became known as **obverse/reverse twin** (e.g. Herbst-Irmer and Sheldrick 2002) and only seldom is properly recognized as a twin by reticular merohedry of order 3 (e.g. Araki 1991; Hodeau *et al.* 1992), with consequent misinterpretations (cf. examples below).

The I- and F-centred cubic cells are amenable to a primitive R cell. Thus, relatively to their relevant primitive cell, the cubic centred lattices can undergo twinning by reticular merohedry. **Spinels** are well known examples already described by Friedel (1926).

5.1.2.5 *Twinning by reticular polyholohedry*

Recently, Bindi *et al.* (2003) have reported a peculiar twinning in **melilite**. The diffraction data of a melilite $(Ca_{1.70}Sr_{0.07}Na_{0.20}K_{0.03})(Mg_{0.70}Fe^{2+}_{0.04}Al_{0.25}Fe^{3+}_{0.01})$ $(Si_{1.92}Al_{0.08})O_7$, collected by an area detector single-crystal diffractometer, could be indexed only by a tetragonal supercell ($a_s = 5^{1/2}a_b$, $c_s = c_b$), five-times larger than the basic cell ($a_b = 7.775$, $c_b = 5.032$ Å). This supercell is oriented such that its (100) symmetry plane corresponds to the ($1\bar{2}$ 0) plane of the basic cell and acts as twinning plane. The authors, on the basis of the non-space group absences, which are consistent with the eqn (5.1) given below, and the knowledge of the basic cell, were able to interpret the twin diffraction pattern and classified it as a 'non-merohedral twin' (twin index 5).

Actually, the oriented association reported by Bindi *et al.* (2003) represents a type of twining which does not correspond to the common reticular merohedry described in Section 5.1.2.1 because both the individual and the twin lattice have the same point group 4/*mmm* [a similar case occurs in Tamazyan *et al.* (2000)]. This peculiar kind of oriented association is called **twinning by reticular polyholohedry** by Ferraris and Nespolo (2003) to indicate that the twinning is originated by the presence of a sublattice showing the same, but differently oriented symmetry as the entire lattice. A symmetry element of the sublattice, which is not a symmetry element of the lattice, can act as twinning operator with results on the diffraction pattern of the same type observed for the common twinning by reticular merohedry.

We do not know other cases presenting the same type of twinning but presumably, the mentioned two cases are not unique. Note that the type of fivefold supercell described for melilite is independent of the *a* value; consequently, it occurs in all tetragonal crystals. Likely, this type of twinning has always a quite high twin index and, even if present, the crystal structure can be solved because the overlapped nodes represent a small fraction of the individual lattice (1/5 in the case of melilite).

5.1.3 *Beyond the classical twins*

Nespolo *et al.* (1999*a–c*; 2000*a*,*b*), besides the already mentioned twins by metric merohedry (Section 5.1.1.2) and reticular polyhedry (Section 5.1.2.5), have identified other peculiar types of twinning and defined **allotwinning**, **twinning by selective merohedry**, and **plesiotwinning**.

5.1.3.1 *Plesiotwinning*

As mentioned in Section 5.1.2.1, the symbol [*uvw*] of a direction or (*hkl*) of a plane acting as twinning element of symmetry is simple, when expressed in the reference system of the individual, and the twin index *n* is smaller than 5–6. However, even if rare, oriented associations of crystals with 'twin index' higher than 5–6 are known (cf. Sueno *et al.* 1971; Takéuchi *et al.* 1972*b*) and have been defined as **plesiotwins** by Nespolo *et al.* (1999*b*). The operation of plesiotwinning is expressed as a rotation around the normal to a lattice plane; this direction may be not rational, that is, not corresponding to a direct lattice row.

By analogy with twins, of which plesiotwins are substantially an extension to any value of the twin index, the **plesiotwin lattice** is defined as the lattice common to the plesiotwinned individuals. The **plesiotwin index** is the ratio of the number of crystal lattice nodes to the number of nodes superimposed by the **plesiotwin operation**. Plesiotwins can be geometrically explained by means of the **coincidence site lattice** (CSL) theory. A CSL is a lattice defined by the nodes that are shared by two individual lattices, one of which is rotated with respect to the other by an angle different from those belonging to crystallographic point groups (Ranganathan 1966).

5.1.3.2 *Allotwinning*

The oriented association of different compounds is defined as **epitaxy** in case of overgrowth (Royer 1928) and **syntaxy** in case of intergrowth (Ungemach 1935). The oriented association of two (or more) individuals differing only in their polytypic character, is a peculiar case of epitaxy/syntaxy. For this association, the term **allotwinning** has been introduced and the relevant laws of formation established by Nespolo *et al.* (1999*a*). The lattice common to the individuals is the **allotwin lattice**. It is defined by the two periodicities belonging to the layer stacked to form the polytypes and the shortest periodicity out of this layer that is in common to the individuals. The **allotwin symmetry element** relating the individuals is a symmetry element of the allotwin lattice, which may belong to the point group of one or more (but not all, obviously) individuals. In other words, it is sufficient that the allotwin symmetry element does not belong to at least one of the involved individuals.

5.1.3.3 *Selective merohedry*

In case of OD structures (cf. chapter 2), the family structure may correspond to a holohedry different from both those of the crystal lattice and the twin lattice. When the point group of the family structure is a subgroup of the point group of the crystal lattice and twinning is by class II merohedry (Section 5.1.1.1), one or more of the twin laws do not belong to the point group of the family structure. These twin operations produce a complex overlap of the family reciprocal sublattice and characteristic sequences of non-space group absences along family rows appear in the diffraction pattern. Nespolo *et al.* (1999*c*) introduced the definition of **selective merohedry** for this peculiar type of non-space group absences. These authors showed that, if recognized, selective merohedry can help to discriminate between the diffraction pattern belonging to the twin of an OD polytype and that belonging to an untwinned, different polytype [for details and examples see Ďurovič (1997*a*); Nespolo and Ďurovič (2002); Nespolo (1999)].

5.2 Consequences of twinning

At morphological level, twinning introduces re-entrant angles that, in principle, distinguish an individual (single crystal) from a twin. Except for special directions, non-scalar physical properties, being direction-dependent, show a discontinuous behaviour through the interface between two individuals of a twin (cf. Wadhawan 1997). Optical phenomena are the most exploited physical property used to detect twinning at morphological level. In particular, except for the twins by merohedry, the orientation of the optical indicatrix changes at the twinning interface and the extinction directions of the individuals are related by the twinning operation. That can be observed by a polarizing optical microscope, provided the incident beam of light is not normal to the twinning interface.

The most important consequences of twinning worth to be examined in this book are those related to the symmetry determination and crystal structure solution using diffraction data. As discussed above, a twin operation completely (merohedry) or partially (reticular merohedry), exactly (any merohedry) or approximately (pseudomerohedry) superimposes the diffracted intensities originated from different individuals. Often, instead of diffracted intensities it can be useful to refer to diffraction spots and peaks, or weighted nodes of the reciprocal lattice, that is, of the diffraction pattern.

The geometry of a twin diffraction pattern can reveal three distinctive features: (1) splitting of reflections; (2) inconsistencies in the symmetry; (3) non-space group systematic absences.

5.2.1 *Splitting of the diffraction pattern*

When the obliquity (Section 5.1.1.3) of a twin and the instrumental resolution are sufficiently large to show split reflections contributed by the different individuals, the presence of twinning by (reticular) pseudomerohedry is, in principle, easily recognized. The splitting, however, may appear for some classes of reflections only and at high angles only; it is thus advisable to explore the full diffraction pattern. The splitting appears at glance if the investigator records the diffraction pattern by a bidimensional detector. When a point-detector is used in the diffraction study, the detection of splitting can be more troublesome, but usually is manifested by an asymmetrical shape of the 'single' peaks and specifically, by the impossibility of indexing the whole diffraction pattern by using only one orientation matrix. In fact, the diffractions belong to at least two not exactly overlapped lattices.

A twinning that produces split diffractions is easier to recognize, but may be more difficult to treat because the peaks from different individuals are variably superposed. In particular, recovering individual diffracted intensities from partially superposed diffractions requires some **detwinning** procedure that is function of *hkl* and of the actual shape of the peak. Consequently, the procedure may be rather complex and approximate; thus, a poor structure refinement should be expected. On the other hand, the relative intensity of split equivalent reflections permits to obtain the volume ratio of the individuals.

5.2.2 *Symmetry of the diffraction pattern*

The symmetry of the twinning operation (Section 5.1) may appear either in the geometrical distribution of the spots only, or in the distribution of the diffracted intensities as well. A statistical analyses of the diffracted intensities (cf. Kahlenberg 1999), which is usually implemented in the crystallographic computing packages, in principle, can alert the presence of twinning. The following cases may occur (for non-space group systematic absences in general, see Section 5.2.3).

Class I twins by merohedry—In principle, the diffraction pattern of this type of twins is equivalent to that of an individual because only equivalent diffractions are overlapped. However, if the anomalous scattering is substantial (breakdown of the Friedel law) and the twinning is not recognized, the structure refinement is of low quality and the absolute structure cannot be established because the twin behaves like a racemate (cf. Flack and Bernardelli 1999).

Class IIA twins by merohedry—In this case, Laue non-equivalent diffractions are overlapped; the symmetry of the diffraction pattern complies with the Laue group of individuals unless the individuals of a complete twin have the same volume, in which case, (further complication) the higher symmetry of the lattice Laue group is simulated. In any case, an acceptable solution of the structure cannot be obtained without establishing the twinning law, with some exceptions if the twin element(s) of symmetry are closely valid also for the crystal structure, but the refinement would show anomalies. A reasonable solution and refinement of the crystal structure can be obtained when one individual accounts for most of the twin volume, an aspect holding for all type of twinning (cf. Section 5.4.2.3).

Class IIB twins by metric merohedry—Unless the individuals have the same volume, the Laue symmetry of the diffraction pattern is lower than the symmetry shown by the geometrical distribution of the nodes. This distribution complies with the symmetry of a lattice with higher metric symmetry (Section 5.4.1.1). This feature is a peculiar alert for twinning by metric merohedry, but the route for a structure solution is the same as for class IIA twins.

Twins by reticular merohedry—Only a fraction of the nodes are overlapped (Section 5.1.2.1) and, in general, the higher symmetry of the twin lattice is respected only by the weighted nodes belonging to this lattice, unless the individuals have the same volume. Except for the latter case, the following types of alert are expected from the diffraction pattern: (i) the geometrical distribution of the nodes has symmetry higher than the Laue symmetry of the diffraction pattern, (ii) a subset of nodes has a Laue symmetry higher than that of the full pattern, (iii) non-space group absences are systematically present.

5.2.3 *Non-space group systematic absences*

5.2.3.1 *The pattern of nodes in the direct space*

As a consequence of the twinning operations by reticular merohedry, the entire set of direct lattice nodes (UVW) of a twin by reticular merohedry with twin index n consists of two parts. (i) A twin Bravais lattice, which accounts for the subset of nodes shared by all j individuals forming the twin, (ii) nodes that belong (normally) to only one of the j individuals. The latter nodes are called **non-twin nodes**. In general, the entire set of nodes does not form a Bravais lattice; in fact, only the nodes belonging to the twin lattice necessarily satisfy the homogeneity condition. However, a Bravais lattice can always be obtained by placing nodes also in the appropriate void positions, such as to establish a lattice periodicity (see Figs 5.2 and 5.5).

As seen in Section 5.1.2.1, with reference to one individual, the twin Bravais cell contains $m = gn$ nodes, where g is the Bravais multiplicity of the cell. If the twin consists of j individuals, the twin Bravais cell contains g shared nodes and $jg(n-1)$ non-twin nodes. It turns out that necessarily the non-twin nodes 'centre' the twin Bravais (Figs 5.2, 5.3 and 5.5); consequently, its corresponding diffraction patterns must show some kind of systematic absences.

5.2.3.2 *The pattern of nodes in the reciprocal space*

In the coordinate system of the twin cell, the set of atom coordinates belonging to an individual j is that of the reference individual as transformed by the twin operations. In particular, the coordinates contain fractional translation components $(p_j/p'_j, q_j/q'_j, r_j/r'_j)_n$ (p_j, p'_j, q_j, q'_j, r_j, and r'_j are integers, including null values). The structure factor of the twin is expressed by

$$F(HKL) = \Sigma_j G(HKL)_j \Sigma_n \exp\{2\pi i[(p_j/p'_j)H + (q_j/q'_j)K + (r_j/r'_j)L]_n\}, \tag{5.1}$$

where, $G(HKL)_j$ is the contribution of the individual j that does not depend on the fractional components generated by the twinning operations; it accounts also for the fractional volume v_j of the jth individual and, if it is the case, for a non-primitive twin cell (thus the summation is over n and not m nodes in the eqn (5.1) and derived expressions). The following cases occur.

1. If $[(p_j/p'_j)H + (q_j/q'_j)K + (r_j/r'_j)L]_n$ is integer for any j and n, $F(HKL) = n\Sigma_j G(HKL)_j$ holds. Consequently, each HKL node that satisfies this condition receives a contribution $nG(HKL)_j$ from each j individual. These are the overlapped nodes belonging to the twin lattice.
2. If the values of $[(p_j/p'_j)H + (q_j/q'_j)K + (r_j/r'_j)L]_n$ are such that $\Sigma_n \exp\{2\pi i[(p_j/p'_j)H + (q_j/q'_j)K + (r_j/r'_j)L]_n\} = 0$ for any j, $F(HKL) = 0$, that is, the corresponding HKL node has zero weight (is absent).
3. If $\Sigma_n \exp\{2\pi i[(p_j/p'_j)H + (q_j/q'_j)K + (r_j/r'_j)L]_n\} = 0$ for all j but $j - j'$ ($j' < j$) individuals, $F(HKL) = [n - (j - j')]\Sigma_{j-j'} G(HKL)_{j-j'}$ holds.

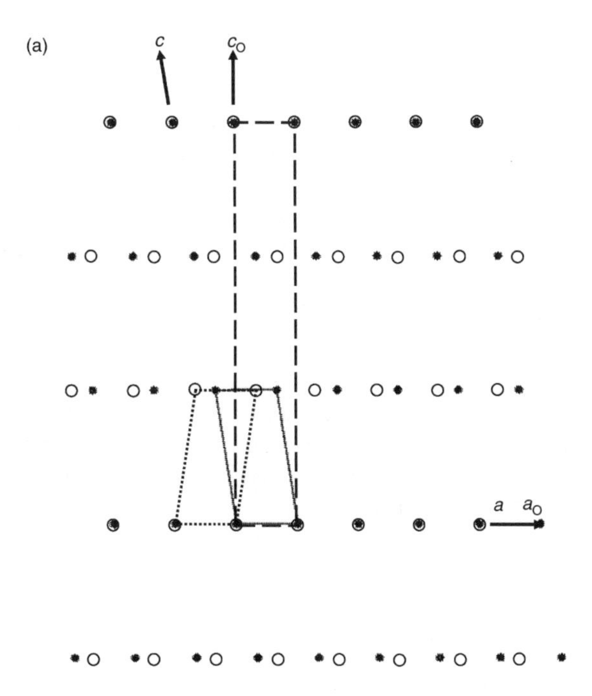

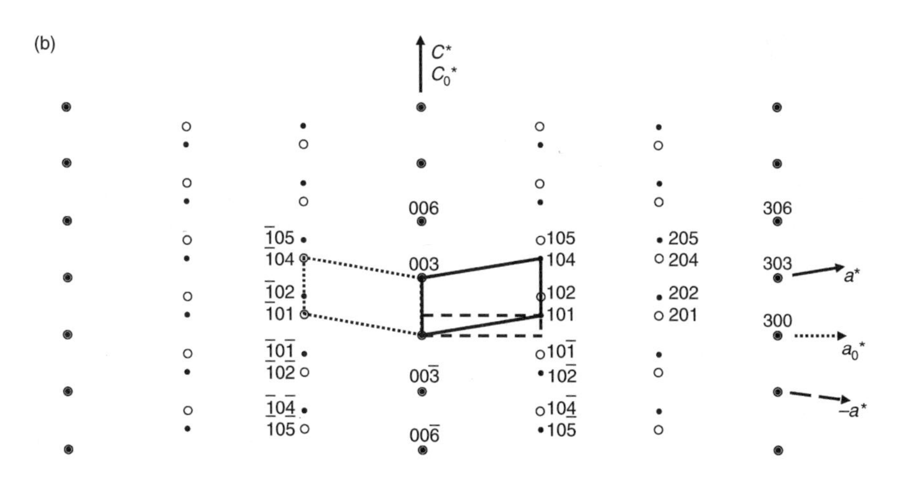

Fig. 5.2. Twinning by reticular merohedry in layer titanosilicates. The upper part (a) represents the direct lattice (010) plane. Filled and open circles represent the nodes of individual 1 and 2, respectively; the corresponding unit cell is full and dotted lines, in the order. The twin lattice contains the double nodes (filled and open circles superimposed) only; the twin cell is shown by broken lines. The lower part (b) represents the reciprocal (010)* plane and symbols are as in the upper part of the figure. Indexing is referred to the twin cell (broken lines).

Consequently, the corresponding HKL node is contributed only by $(j - j')$ individuals; usually $j - j' = 1$.

Note that, a part the twins by class I merohedry, the superimposed diffractions are usually non equivalent, except for those corresponding to nodes lying on a twinning element of symmetry or equivalent according to Friedel law. These reflections have been called **twin-proof reflections** (Le Page *et al.* 1984). In practical cases, often $p'_j = q'_j = r'_j = t$ for any j, thus the expressions given above can be greatly simplified. For example, in the case of a primitive twin lattice and non-twin nodes with translation component only along [001], the condition $r'_j = t$ for any j necessarily holds because the lattice rows are periodic with period c/t. Thus, the eqn (5.1) becomes

$$F(HKL) = \Sigma_j G(HKL)_j \Sigma_n \exp\{2\pi i[(r_j/t)L]_n\}.$$

In this case, the diffractions are overlapped, single, or absent if, in the order, L is a multiple of t for any j, only for one value of j, or $\Sigma_n \exp\{2\pi i[(r_j/t)L]_n = 0$ for any j.

In the twins by reticular merohedry, most of the non-space group systematic absences are related to the presence of non-twin nodes. However, also the translations due to non-primitive lattices, glide planes and screw axes occurring in the individual [they effects are embedded in $G(HKL)_j$] may concur to generate non-space group systematic absences. In particular, only this type of non-space group absences can occur in twins by class II merohedry (class I is not affected by non-space group absences). Tables reporting (some of) the possible non-space group absences in twins by merohedry can be found in Araki (1991), Giacovazzo (2002), Koch (1999), and Okamura (1971). The latter author uses expressions like eqn (5.1) to interpret the twin diffraction patterns, but only for particular cases.

5.2.3.3 *Occurrence of non-space group systematic absences*

Given that the twin cell is m times larger than the individual cell (primitive, for sake of simplicity), normalizing to 1 the number of nodes of the individual in the reciprocal space, the number of nodes referred to the twin cell is m. Because a non-primitive cell is used, the diffraction pattern of an individual shows $m - 1$ nodes with zero weight, that is, $m - 1$ systematic absences. At least a part of these absences must correspond to non-space group systematic absences, because the twin cell represents a sublattice.

The twin diffraction pattern consists of the overlapped diffraction patterns of j individuals. For each individual, a fraction $1/n$ of diffractions is superimposed to the same fraction of diffractions contributed by each of the other $j-1$ individuals. Instead, a fraction $(n - 1)/n$ of diffractions does not superimpose; thus, for j individuals, a total fraction $j[(n - 1)/n]$ of non-superimposed diffractions is contributed to the twin diffraction pattern. To recover the $m - 1$ absences occurring in the twin cell reference, it must be at least

$$j[(n - 1)/n] = m - 1. \tag{5.2}$$

Taking into account that $j \leq s \geq n$ (s is the twin order; $j = s$ in a complete twin; cf. Section 5.1.1.1), and $m \geq n$, the following cases are possible. (i) For $j < n$, eqn (5.2) cannot hold and non-space group systematic absences must occur, (ii) for $j \geq n$ the eqn (5.2) may hold. In the latter case, one might conclude that no systematic absences occur in the twin diffraction pattern. However, this conclusion can be valid only if space group absences do not occur in the individual diffraction pattern, that is, the crystal structure belongs to a primitive symmorphic space group. In the other cases, as mentioned above (Section 5.2.3.2), the pattern of the individual space group systematic absences is transformed into a non-space group pattern of systematic absences by the twinning operations (cf. gjerdingenite-Fe, Section 5.4.2.2). In conclusion, a twin diffraction pattern not affected by non-space group systematic absences cannot be excluded *a priori*, but most likely it is very rare.

5.2.4 *Further twinning alerts*

Besides the alerts for the presence of twinning mentioned above, including the statistical analyses of the diffracted intensities, the following further aspects can help in detecting a twinning through its diffraction features; spectroscopic methods can help too.

5.2.4.1 *Looking for smaller individuals*

Except for polysynthetic twins with submicrometric domains, a careful search may lead to find an individual suitable for single-crystal X-ray diffraction, particularly using modern in-house diffractometers and, even better, synchrotron radiation. Electron diffraction allows investigation of even smaller domains, but it must be kept in mind that a twin diffraction pattern appears as soon as, the electron beam has a diameter larger than the single domains, or is not parallel to the interface between twins.

5.2.4.2 *Powder diffraction data*

Twinning concerns single crystals and does not affect powder diffraction patterns. A reduced cell obtained from a single-crystal that is a multiple of the reduced cell obtained from powder diffraction is indicative of twinning by lattice merohedry. This type of alert is valid any time a cell multiple of an expected one is found.

5.2.4.3 *Unusual structure features*

In general, modular structures are at risk of structural disorder because of possible stacking faults in the sequence of building modules. Note that, as shown by Nespolo and Ferraris (2001) for the polytypes of layer silicates, in some cases, residual peaks in a difference electron density map may be due to a different precision in measuring streaked and unstreaked diffractions (**Ďurovič effect**; cf. an example in Ferraris *et al.* (2001*d*). In particular, structural disorder shown by a (moderately) refined structure, unreasonable residual peaks in a difference electron density map, and unusual bonds lengths may be due to a minor presence of twinning accompanying a dominant individual whose substantial contribution to the diffraction pattern allows the solution, but not a good refinement (cf. Flippen-Anderson *et al.* 2001). As consequence of using the twin supercell, an unusually large number of unit formulae per cell can sound as an alert for twinning (cf. Dunitz 1964). Systematic, in terms of hkl indices, appearance of observed intensities much higher than the calculated ones ($I_{obs} \gg I_{calc}$) in hard-to-refine structure may be indicative of a sublattice of overlapped nodes, that is, of a twin reciprocal lattice.

5.2.4.4 *Patterson function*

Nowadays calculation of the Patterson function is no longer a routine procedure because the structure solution is usually achieved through direct methods (Giacovazzo 2002). Anyway, if twinning affects the diffracted intensities, as far as the symmetry added by twinning is only approximate for the crystal structure, two effects are expected in a Patterson function calculated with non-detwinned intensities:

- presence of short interatomic vectors consequent to the overlapped individual structures;
- absence (or non-canonical appearance) of the Harker sections corresponding to the twinning elements of symmetry.

5.2.4.5 *Metric symmetry; sub- and supercells*

After a cell has been found, it should be a normal practice to compare its metric symmetry with the (expected) crystal symmetry and look for sub- and supercells by a suitable procedure (e.g. Le Page 2002). If the symmetry of the conventional Bravais cell or that of a derived primitive cell is higher than the holosymmetry pertaining to (or expected for) the crystal, twinning by metric symmetry must be suspected. The presence of supercells with a symmetry higher than that of the crystal lattice favours twinnning by lattice merohedry.

Note that in presence of twinning by reticular merohedry, one could by chance (or by choice) collect the diffraction data based on the individual cell, which is always a subcell of the twin cell, even if it cannot index the entire diffraction pattern (Section 5.2.1). If the fraction of the overlapped diffractions is small (high twin index) and/or one individual dominates the diffraction pattern, a structure solution may be obtained but refinement problems shall be met (cf. Section 5.4.2.3).

5.3 Detwinning procedure

The effects consequent to the superimposition of lattices must be carefully considered, particularly when non-equivalent diffracted intensities are overlapped. In practice, preliminary to attempting the structure solution, an existing twin must detected, its twinning law determined and then the superimposed diffracted intensities detwinned (deconvoluted). Only after these preliminary and crucial stages, the sets of diffracted intensities belonging to different individuals can be managed by ad hoc crystallographic software, like that described by Herbst-Irmer and Sheldrick (1998).

The twin laws are normally expressed by matrices (Section 5.1; cf. chapter 4 in Giacovazzo 2002); these matrices, together with their inverse ones, relate the cell parameters and Miller indices of the twin to those of the individual and those of the individuals between them (cf. examples below). After these relationships are established, one can obtain the indices $(hkl)_j$ relative to an intensity $I(hkl)_j$ that has been diffracted by the individual j and has contributed, together with those of the other $j-1$ individuals, to a given intensity $I(KHL)$ of the twin diffraction pattern (hkl and HKL indices are referred to the individual and twin cell, respectively). Note that the number of individuals constituting a complete twin and the number of intensities $I(KHL)$ which are equivalent, according to the twin law, is the same.

As example useful to understand the problem, let us consider twinning by hemihedry, of any type. One can write the following equations (Grainger 1969), where v_1 is the fractional volume of the individual 1,

$$\begin{aligned} I(K'H'L') &= v_1 I(hkl)_1 + (1-v_1)I(hkl)_2 \\ I(K''H''L'') &= (1-v_1)I(hkl)_1 + v_1 I(hkl)_2. \end{aligned} \tag{5.3}$$

Except when the two individuals have the same volume (i.e. $v_1 = 0.5$), the eqn (5.3) can be solved for the intensities diffracted by each individual as follows:

$$\begin{aligned} I(hkl)_1 &= I(K'H'L') + [v_1/(1-2v_1)][I(K'H'L') - I(K''H''L'')] \\ I(hkl)_2 &= I(K''H''L'') - [v_1/(1-2v_1)][I(K'H'L') - I(K''H''L'')]. \end{aligned} \tag{5.4}$$

In class I twins by merohedry, as mentioned above (Section 5.1.1.1), $I(H'K'L')$ and $I(H''K''L'')$ result from the overlapping of crystallographically equivalent diffractions; thus, apart from a scale factor, they do not differ from those of the individual. In this case, provided the correct crystal point group is used, even if the twinning is not recognized, a solution of the crystal structure can be obtained. The goodness of the refinement depends on how much the anomalous scattering breaks the Laue law and makes $I(hkl)_1 \neq I(hkl)_2$, as discussed by Catti and Ferraris (1976).

Equation (5.4) can be generalized to twins composed of more than two individuals. However, in cases of twinning by merohedry other than hemihedry, some of the overlapping intensities may be equivalent thus requiring a specific analysis of the equations corresponding to eqn (5.4) [cf. Catti and Ferraris (1976); chapter 4 in Giacovazzo (2002) and references therein].

5.3.1 *Finding the twin laws*

5.3.1.1 *(Pseudo)merohedry*

In twins by any type of (pseudo)merohedry, the cell is the same for the individual and twin. The twin operator is an element of (pseudo)symmetry belonging to the lattice but not to the crystal. Usually, the possible twin laws are among few possible ones because the lattice group point is known and the possible crystal point groups are the subgroups of the lattice point group belonging to the same crystal system (merohedry) plus those belonging to lower symmetry crystal systems in case of twinning by metric merohedry.

5.3.1.2 *Reticular (pseudo)merohedry*

In twins by reticular (pseudo)merohedry, the determination of the twin law(s) is strictly related to finding the cell of the individual. If this cell is already known for reasons mentioned above, the detwinning procedure is greatly simplified. Other information, like morphology and optical orientation of the individuals, can help, otherwise hunting the individual cell and its different orientations in the diffraction pattern is based on a trial and error procedure or the ability to decompose the twin cell into subcells with the guide of the systematic non-space group absences (Section 5.2.3). Trial and error may be more advantageous in programming automatic routines, particularly in presence of split reflections. It may be useful to note the following:

(1) Only the twin cell can index the whole diffraction pattern.
(2) The reciprocal cell of the individual is m-times larger than that of the twin (cf. Section 5.2.3.1).
(3) The reciprocal lattice of the individual should not contain systematic non-space group absences.
(4) The twin operations must be found among those operations that are symmetry operations for the twin lattice but not for the crystal structure.
(5) A twin operation belonging also to the individual lattice (but not to the crystal structure) generates twinning by merohedry that, in case, coexists with twinning by reticular merohedry.

Summarizing, in front of a diffraction pattern obtained from a twin consisting of j individuals, the problem consists in splitting of the pattern into j subsets of nodes. These subsets, often only geometrically, are related by one or more (pseudo)symmetry elements (twin operations). Usually, the key for the solution is the finding of a subcell of the twin cell which, when oriented in j different ways, defines j lattices each one belonging to one individual. The (pseudo)symmetry elements that bring the j subcells to a same orientation are the twinning elements. The matrices expressing the orienting operations represent the twin laws.

5.3.2 *Partial structure solution*

As mentioned in Section 5.2.3.2, in a twin by reticular merohedry the diffractions corresponding to nodes lying on a twinning element of symmetry or equivalent by

Friedel law are not affected by twinning in the sense that, apart from a scale factor, their intensity is the same as in the individual. This limited number of **twin-proof diffractions** (Araki 1991; Le Page *et al.* 1984) can be used to calculate projections of the Patterson function and obtain some information on correct interatomic vectors. This information can be a starting point towards further efforts in solving the crystal structure.

5.4 Examples

In this section, some worked examples of detwinning are reported. In some case, the published interpretation of the twin was not orthodox in terms of the nomenclature discussed in this chapter. Thus, the examples serve also to show the advantages of adopting an appropriate procedure and nomenclature.

5.4.1 *Twinning by merohedry*

5.4.1.1 *Twinning by metric merohedry in kristiansenite*

Kristiansenite, $Ca_2ScSn(Si_2O_7)(Si_2O_6OH)$, is one of few known scandium mineral. Its diffraction pattern showed $2/m$ Laue symmetry according to a monoclinic cell with $a = 10.028, b = 8.408, c = 13.339$ Å, $\beta = 109.10°$ and the intensity statistics indicated $C2$ as space-group (Ferraris *et al.* 2001*c*). However, the structure refinement was not satisfactory ($R = 0.092$), the polarity could not be determined, and a hypothesis of twinning by merohedry was not confirmed. Instead, the hypothesis of triclinic symmetry for the structure and occurrence of $\{010\}$ **twinning by metric merohedry** was successful; the crystal structure refined to $R = 0.0242$. The almost identical volume of the two individuals forming the twin, explains the observed $2/m$ Laue group. Following the definition given by Nespolo and Ferraris (2000), kristiansenite has been the first case where twinning by metric merohedry has been recognized. The reasonable solution of the structure in $C2$ is due to its strong pseudo-monoclinicity, the largest deviation from the triclinic atom positions in the monoclinic model being 0.1 Å.

5.4.1.2 *Undetected twinning by metric merohedry*

In literature, several crystal structures that have been solved using data collected from non-recognized twins by metric merohedry.

$C_{19}H_{33}LiN_3Tl$ Herbest-Irmer and Sheldrick (1998) interpret as 'twin by pseudomerohedry' (actually by reticular merohedry) the twin found in monoclinic $P2_1/c$ $\mathbf{C_{19}H_{33}LiN_3Tl}$ ($a = 13.390, b = 25.604, c = 13.390$ Å, $\beta = 112.30°$). These authors, followed by Giacovazzo (2002), noted that the monoclinic lattice admits an orthorhombic C-centred cell ($a = 14.900, b = 22.252, c = 25.604$ Å) and

interpreted the twin as originated by {010} twinning (orthorhombic reference). Actually, the structure is monoclinic, but its lattice is metrically orthorhombic as shown by $a = c$ in the monoclinic setting. The C-centred cell is in fact nothing else than the conventional cell of the orthorhombic lattice, of which the monoclinic cell is the primitive. In conclusion, the apparently monoclinic lattice is metrically orthorhombic and its $(10\bar{1})$ [(010) in the orthorhombic reference] plane can act as twinning element by metric merohedry. The same case occurs in the opioid peptide reported by Flippen *et al.* (2001).

$S_3(Ru_{0.336}Pt_{0.664})CuO_6$ and Sr_3CuPtO_6 The multiple twinning observed in the **hexagonal-related perovskite**, **$S_3(Ru_{0.336}Pt_{0.664})CuO_6$** (Friese *et al.* 2003) has been explained by a combination of rhombohedral **obverse/reverse twinning** plus a threefold axis, being the lattice pseudo-trigonal according to the basic monoclinic cell adopted for the structure refinement ($R12/c$, $a = b = 9.595$, $c = 11.193$ Å, $\gamma = 120°$). Actually the interpretation of twinning would be orthodox and easier by recognizing that the primitive of the conventional C-centred monoclinic cell ($a = 9.294$, $b = 9.595$, $c = 6.679$ Å, $\beta = 92.63°$), given by the authors in appendix, is metrically rhombohedral with $a = 6.679$ Å and $\alpha = 91.83°$ (transformation $110/\bar{1}10/001$). Thus, this is a case of twinning by metric merohedry and possible twinning elements of symmetry can be anyone belonging to the lattice point group $\bar{3}m$ and not to the crystal point group $2/m$. Because of the endemic presence of a hexagonal sublattice within a rombohedral lattice (Section 5.1.2.4), twins by reticular merohedry can be expected as well. The title compound compound is strictly related to **Sr_3CuPtO_6** (Hodeau *et al.* 1992) ($a = 9.317$, $b = 9.720$, $c = 6.685$ Å, $\beta = 91.95$, $C2/c$) where the primitive rhombohedral cell was recognized even if the observed twinning was not strictly discussed in terms of metric merohedry.

Fluorocyclohexane/thiourea The low-temperature phase of the **fluorocyclohexane/thiourea** inclusion compound (Yeo *et al.* 2001) is monoclinic with orthorhombic metric symmetry ($a = 27.52$, $b = 15.718$, $c = 12.33$ Å, $\beta = 90°$, $P2_1/n$) and the high-temperature phase is rhombohedral $R\bar{3}c$ ($a = 15.971$, $c = 12.495$ Å). Both phases show twinning which, for the rhombohedral phase, has been interpreted by using the obeverse/reverse formalism. The twinning of the low-temperature phase is typically by metric merohedry, but the correct [100] twofold twinning axis has been introduced only via analogy with the high-temperature phase.

η^8-cyclooctatetraenyl[hydrotris(pyrazol-1-yl)borato]titan(III) The twinning in crystals of the title compound presented by Herbst-Irmer and Sheldrick (1998) as example 2 of twinning by pseudomerohedry, is actually by metric merohedry; in fact, the C-centred monoclinic cell ($a = 10.220$, $b = 11.083$, $c = 7.538$ Å, $\beta = 96.85°$, Cm) has a metrically rhombohedral primitive cell with $a = 7.538$ Å and $\alpha = 94.64°$.

5.4.2 *Twinning by reticular (pseudo)merohedry*

5.4.2.1 *Twinning in layer titanosilicates*

This example refers to the bafertisite-like **heterophyllosilicates** reported in Table 4.2 and has been developed by Ferraris and Németh (2003).

Twinning in the direct space The **layer titanosilicates** listed in Table 4.2 have in common a two-dimensional cell $ab \sim 5.4 \times 7.0$ Å (or multiples) corresponding to the periodicity in the $(HOH)_B$ bafertisite-like layer. Practically, all these compounds are either monoclinic or triclinic; the latter, however, show always an angle γ very close to 90°. For most of them, the value of the third periodicity and of the β angle is such that $c \sin(\beta - 90) \sim a/n$ $(n = 3, 4, 6)$ holds. This relation implies that a direction $[uvw]$ with periodicity $c_o = nc \sin\beta$ and normal to the ab plane does exist. The supercell with parameters a, b and c_o is (pseudo)orthorhombic if $\alpha = 90°$ (monoclinic members) and (pseudo)monoclinic (angle $\alpha_m \neq 90°$) in the triclinic members with $\gamma \sim 90°$.

As observed experimentally, the occurrence of the supercell favours twinning by reticular (pseudo)merohedry with $\{100\}$ twinning law in both monoclinic and triclinic members; in monoclinic members only, the supercell favours also $\{010\}$, or $\{001\}$, twinning. Actually, a more or less large (pseudo)orthorhombic cell occurs often also in the triclinic members because $c_o \sin(\alpha_m - 90°) \sim b/n$ $(2 < n < 5)$ holds. In case of not null obliquity, the twinning by pseudomerohedry may appear as pairs of **corresponding twins** (Section 5.1.1.3) because also the pseudo-symmetry axis that is nearly normal to the mentioned pseudo-mirror planes can act as twinning operator.

In order to realize the conditions for reticular merohedry, we consider the following slightly idealized cell: $a = 5.383$, $b = 7.08$, $c = 11.74$ Å, $\alpha = 93.84$, $\beta = 98.79$, $\gamma = 90°$ where $c \sin(\beta - 90) = a/3 = 1.794$ Å holds; consequently, a non-primitive rectangular supercell with parameters $a_o = a = 5.383$ Å, $c_o = 3c \sin\beta = 34.806$ Å does exist. For the following discussion, the third axis can be either orthogonal to the ac plane (an orthorhombic supercell exists) or inclined towards c_o and normal to a (a monoclinic supercell exists). In both cases, the three-dimensional cell based on the $a_o c_o$ rectangular cell has centring nodes (filled circles) at $\pm(2/3, 0, 1/3)$ (Fig. 5.2).

Due to the higher supercell symmetry, the (100) plane can act as twinning element by reticular merohedry. The original lattice with nodes represented by filled circles in Fig. 5.2 (individual 1) is duplicated by the (100) twin mirror plane into the lattice with nodes represented by open circles (individual 2). The nodes (UVW) with integer W (capital symbols refer to the supercell) are double, those at $W = 1/3$ and 2/3 are single. On the whole, 1/3 of the nodes belonging to one individual are superimposed, thus the twin index is 3. The entire set of nodes describing the twin in the direct space can be split into two subsets, each one described by a multiple cell: one cell (filled circles) is centred at $\pm(2/3, 0, 1/3)$ and the other (open circles) is centred at $\pm(1/3, 0, 1/3)$.

Twinning in the reciprocal space For an orthorhombic cell, the values of the reciprocal axes in the a^*c^* plane are: $a^* = 0.1880$ Å^{-1}, $c^* = 0.0862$ Å^{-1}, $\beta^* = 81.21°$; $a_o^* = a^* \sin\beta^* = 0.1858$ Å^{-1}, $c_o^* = c^*/3 = 0.0287$ Å^{-1}. In a monoclinic cell, c^* and c_o^* would slightly change their values but maintaining the ratio 1/3. In the reciprocal space, the (100) twinning mirror plane corresponds to the plane (100)*.

In the weighted reciprocal lattice (i.e. the lattice with nodes weighted by their diffracted intensity), the non-zero nodes are obtained by duplicating, via the twinning mirror plane, the nodes of one individual (e.g. that represented by the full-line cell in Fig. 5.2). Taking into account also the zero-weight nodes (i.e. the systematic non-space group absences), the twin lattice is based on a rectangular twin cell with parameters a_o^* and c_o^*. *HKL* indices relative to the twin lattice are shown in Fig. 5.2. It can be noted that:

- the nodes contributed by each individual have both H and K multiple of 3;
- all other nodes are contributed by one individual only;
- in the twin lattice, along [001]* the periodicity becomes three times smaller than in the individual, i.e. $c_o^*/c^* = 1/3$;
- along [001]* at least one every three nodes is systematically absent.

The systematic absences correspond to the *HKL* nodes for which, at the same time, $2H + L \neq 0$ (mod3) and $H + L \neq 0$ (mod3) holds, that is, $L = 0$ (mod3) if $H \neq 0$ (mod3) and $L \neq 0$ (mod3) if $H = 0$ (mod3). The same systematic absences occur for any value of K. The matrices that transform the indices of one individual into those of the other (twinning laws) and the *HKL* indices referred to the twin cell into those of an individual are:

- $(hkl)_2 = (\bar{1}00/010/\frac{2}{3}01)(hkl)_1$
- $(hkl)_1 = (100/010/\frac{1}{3}, 0, \frac{1}{3})(HKL)$
- $(hkl)_2 = (\bar{1}00/010/\frac{1}{3}, 0, \frac{1}{3})(HKL)$.

The same matrices transform the cell parameters in the direct space.

The systematic non-space group absences Equation (5.1) in this case becomes ($n = 3$, including the null translations; $j = 2$)

$$F(HKL) = G(HKL)_1\{1 + \exp[2\pi i(2H + L)/3] + \exp[-2\pi i(2H + L)/3]\} + G(HKL)_2 \times \{1 + \exp[2\pi i(H + L)/3] + \exp[-2\pi i(H + L)/3]\}.$$

The following conditions hold:

(1) If at the same time $2H + L = 0$ (mod3) and $H + L = 0$ (mod3), that is, both H and L are multiple of 3, the contributions of the two individuals are superimposed.

(2) If at the same time $2H + L \neq 0$ (mod3) and $H + L \neq 0$ (mod3), the HKL node has zero weight, that is, the corresponding diffraction is absent.

(3) If neither of the conditions 1 and 2 are valid, the HKL node is contributed by only one individual.

5.4.2.2 *Twinning in gjerdingenite-Fe*

Gjerdingenite-Fe, $K_2[(H_2O)_2(Fe,Mn)][(Nb,Ti)_4(Si_4O_{12})_2(OH,O)_4]\cdot 4H_2O$, is a member of the **labuntsovite group** (Chukanov *et al.* 2002). The mineral is monoclinic $C2/m$ ($a = 14.529$, $b = 13.943$, $c = 7.837$ Å, $\beta = 117.61°$); its crystal structure was solved and refined (Raade *et al.* 2002) using diffraction data collected form a {001} twin simulating an orthorhombic F-centred lattice. The same type of twinning occurs in other members of the labuntsovite group.

Twinning in the direct space Capital letters are used for the symbols when referred to the orthorhombic twin lattice. Figure 5.3 shows that the full set of nodes UVW of the monoclinic C-centred lattice of gjerdingenite-Fe (filled circles at $V = 0$ and squares at $V = 1/2$) contains a subset of nodes that can be matched by a quadruple orthorhombic F-centred supercell (broken lines in the figure). The remaining nodes of the entire set further centre the F-cell at (3/4, 0, 1/4) plus F-centring operations. Thus, the whole lattice can be split into two identical orthorhombic F-centred sublattices with a relative shift of (3/4, 0, 1/4); in Fig. 5.3, the cell of the second sublattices is shown by dotted-broken lines. The cell parameters of the monoclinic C-centred

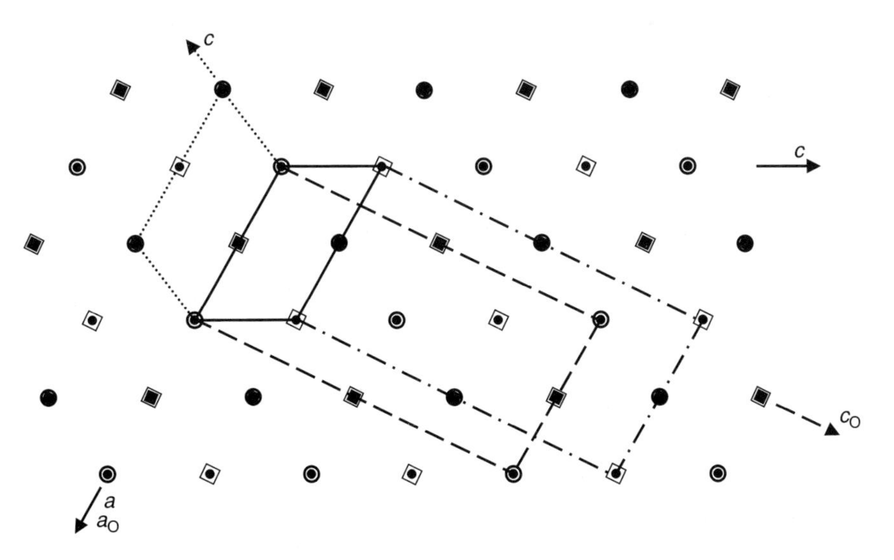

Fig. 5.3. Twinning by reticular merohedry in gjerdingenite-Fe (direct lattice). Filled and open symbols represent the individual 1 (full-line C cell) and 2 (dotted-line C cell) in the order. Two twin cells are represented by broken and dotted-broken lines, respectively. Circles and squares represent nodes a 0 and 1/2 level, respectively.

cell are obtained from those of the orthorhombic F-centred supercell ($a_O = 14.529$, $b_O = 13.943$, $c_O = 27.779$ Å) by the transformation (1, 0, 0/0, 1, 0/$-\frac{1}{4}$, 0, $\frac{1}{4}$) and has parameters $a = a_O = 14.529$, $b = b_O = 13.943$, $c = 7.837$ Å$\sim c_O/3.5$ and $\beta = 117.61°$. The monoclinic (001) and ($\bar{4}01$) planes correspond to the orthorhombic (001) and (100) symmetry planes, respectively. Each of these two planes can act as a twinning element by reticular merohedry (in fact obliquity is zero) but, because (010) is mirror plane of the individual, they generate the same twin by reticular merohedry.

The (001) twinning mirror plane brings to overlap the nodes belonging to the ($U0W$) rows with W even (filled and open symbols of the same type in Fig. 5.3). The rows with W odd do not overlap but are instead paralleled by a row at $V = 1/2$ (filled and open symbols of different type). Thus, 50 per cent of the nodes are superimposed and the resulting twin index is 2.

Twinning in the reciprocal space Figure 5.4 shows the ($H0L$) and ($H1L$) planes of the reciprocal lattice, twinned according to the (001) plane. Indices referred to the monoclinic cells of the two individuals (upper line) and to the cell of the orthorhombic

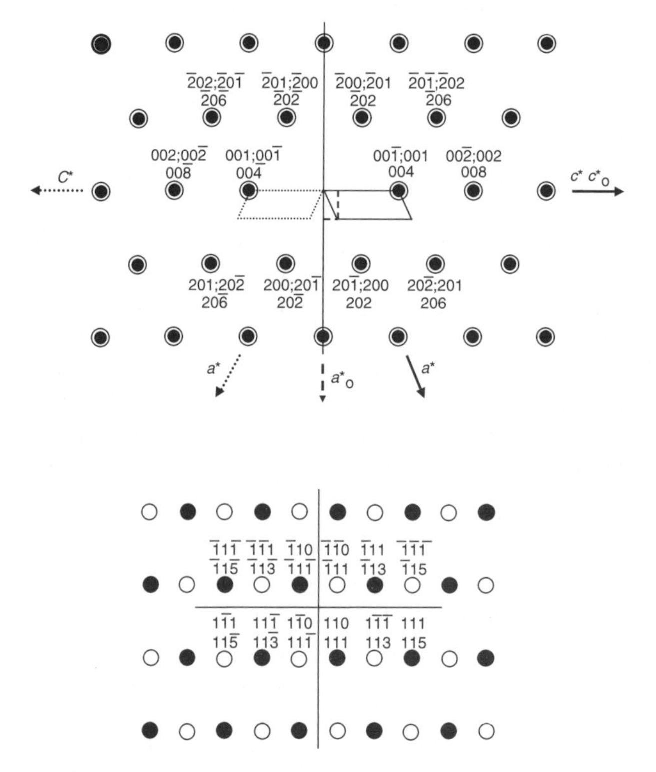

Fig. 5.4. Twinning by reticular merohedry in gjerdingenite-Fe (reciprocal lattice). Symbols as in Fig. 5.3. Indices referred to the monoclinic cell of each individual (upper line for each row) and to the twin cell (lower line) are shown for the plane with $K = 0$ (top) and $K = 1$ (bottom).

twin lattice (lower line) are given; the K index is the same for all types of indexing. For K even, each node of one individual (cell with full lines in Fig. 5.4) overlaps a non-equivalent node of the second individual (dotted line cell); for K = odd, the nodes of the two individuals intercalate such to halve the periodicity along c^*. According to the indexing referred to the twin cell (broken lines), besides the space-group systematic absences due to the F-centred lattice (H, K and L are either all even or all odd), non-space group systematic absences are observed for $3H + L \neq 0 \pmod 4$ and K even.

The systematic non-space group absences The non-space group systematic absences are related to the presence of the non-twin nodes occurring within the orthorhombic F-centred supercell at (3/4, 0, 1/4) (Fig. 5.3). The extra absences for the individual indexed according to the twin cell can be easily obtained by noting that, as mentioned above, the non-twin nodes belong to a second F-centred sublattice that is shifted by (3/4, 0, 1/4) relatively to the twin lattice. The two sublattices diffract in opposition of phase when $3H + L \neq 0 \pmod 4$.

Equation (5.1) in this case becomes ($j = 2, n = 2$)

$$F(HKL) = G(HKL)_1\{1 + \exp[2\pi i(3H + L)/4]\} + G(HKL)_2\{1 + \exp\{2\pi i[(3H + L)/4 + K/2]\}\}.$$

The occurrence of non-space group absences in the twin follows different patterns according K is even or odd.

- K even—The two individuals diffract in phase. Keeping in mind that, due to the F-centring, H, K and L must be either all odd or all even and thus $3H + L$ is always even, the contributions of both individuals is systematically absent if $3H + L \neq 0 \pmod 4$; consequently, the observed extra absences are proved.
- K odd—The two individuals diffract in opposition of phase and never both contributions can be systematically absent at the same time. In fact, the contribution from the first individual is absent when $3H + L \neq 0$ (mod4) and that from the second individual is absent when $3H + L \neq 0$ (mod4) $- 2K$. Consequently, in the reciprocal planes with K odd the contributions of the two individuals never overlap but intercalate along the [001]* direction; thus, only the space group absences due to the F-centring are observed.

5.4.2.3 *A case with a dominant individual*

If the volume of one individual is substantially dominant in a twin by reticular hemihedry, the diffractions contributed only by the smaller individual are weak or missing. This can create either further complications or, luckily, a simplification as happened for the structural characterization of **hydroxylclinohumite**, $Mg_9[SiO_4]_4(OH, F)_2$, a new monoclinic member of the **humite group** (Ferraris *et al.* 2002) with $a = 4.7480$, $b = 10.2730$, $c = 13.6894$ Å, $\alpha = 100.72°$, space group $P2_1/b$ (a unique axis). As recognized later, the diffraction data had been collected from a twin consisting of two individuals with 0.94:0.06 volume ratio and could be indexed only after doubling the c cell parameter of the expected cell given above. The structure was solved, but unsatisfactorily refined to $R = 0.12$.

Lacking of OH/F ordering, the only crystal-chemical feature which could justify a cell parameter $c' = 2c$, and the occurrence of non-space group systematic absences corresponding to the nodes with, at the same time, K even and L odd, suggested the presence of twinning by reticular merohedry. Inspection of the direct lattice (Fig. 5.5), revealed the presence of an A-centred orthorhombic supercell with $a_o = a$, $b_o = b$, $c_o = 4c \sin \alpha$, and further centring at $\pm(0, 2/3, 1/3)$. Consequently, the presence of $\{001\}$ twinning by reticular merohedry with twin index 2 and very small obliquity $0.09°$, being $\alpha_o = 90.09°$, was considered. This type of twinning overlaps 1/2 of the nodes belonging to the smaller individual (represented by filled circles in Fig. 5.5) with 1/2 of the nodes of the second individual (represented by open circles). *A posteriori*, it became clear that the double monoclinic cell, instead of the quadruple orthorhombic twin cell, had been detected in the diffraction pattern for the following reasons. First, the diffractions from the smaller individual (6% of the total volume) are very weak or absent. Second, the orthorhombic twin cell is affected by a high presence of absences originated, in the order, by the A-centring (space group absences) and the further centring at $\pm(0, 2/3, 1/3)$ (non-space group absences). Third, the chosen double monoclinic cell could have some crystal-chemical reason as mentioned above.

As shown by Fig. 5.5 in the direct space, for the reasons given above, the 'misinterpretation' of the diffraction pattern brought to consider the A-centred ACFH cell

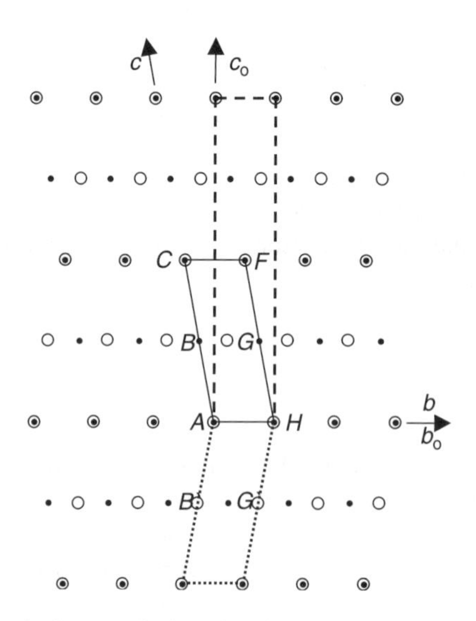

Fig. 5.5. Twinning by reticular merohedry of hydroxylclinohumite in direct space. The nodes of the two individuals forming the twin are represented by filled and open circles, respectively. ABGH and ACFH are the crystal unit cell and the supercell used to index the diffraction pattern of the twin, respectively. The real twin supercell is shown by broken lines. The dotted-line cell is related by the (001) twin plane to the full-line cell.

as twin supercell; this cell, however, does not account for very weak diffractions (see below). After detwinning, the structure was anisotropically refined to $R = 0.026$ and even the hydrogen atom, which is disordered on two positions, was detected.

The non-space group absences can be obtained by eqn (5.1). However, maybe more instructively, one can observe that, in the direct space, the full set of the twin nodes consists (Fig. 5.5) of an A-centred lattice (open circles only; larger individual 1) plus a (0, 0, 1/2)-centred lattice (filled circles only; smaller individual 2). The diffractions of the first lattice are systematically absent when $K + L$ is odd; those of the second lattice when L is odd. Consequently, the twin diffraction pattern shows: (i) systematic absences, when K is even and L is odd, (ii) overlapped diffractions, when both K and L are even, (iii) single diffractions due to the individual 1 only, when both K and L are odd, (iv) single diffractions due to the individual 2 only, when K is odd and L is even. Being the individual 2 the small one, only very weak or absent diffractions are observed when K is odd and L is even.

A case similar to that here described is reported by Schmalle *et al.* (1993) for $\mathbf{La_2Ti_2O_7}$, a perovskite-related **ferroelectric insulator**.

5.4.3 *Miscellaneous cases*

As already mentioned (Section 5.1.1.2), crystal structures showing twinning generated by different types of (pseudo)symmetries are not rare. $\mathbf{Cu_8GeS_6}$ (Onoda *et al.* 1999) is othorhombic pseudo-tetragonal ($a = 7.0445$, $b = 6.9661$, $c = 9.8699$ Å) and shows a pseudo-cubic superlattice (a close to 9.9 Å and angles close to 90°) that has been interpreted also as pseudo-rhombohedral by the authors. The crystal structure could be refined only after interpreting the multiple twinning generated both by pseudomerohedry (pseudo-tetragonal cell) and reticular pseudomerohedry (pseudo-cubic subcell).

Well presented cases of twinning by merohedry (crystal point group 6, lattice point group 6/*mmm*) are in Kahlenberg and Böhm (1998) and Makarova *et al.* 1993), similarly for reticular merohedry (crystal point group 2, sublattice point group *mmm*) in Keller *et al.* (2001).

5.5 Conclusions

The implementation in crystallographic software of routines dealing with twinning has increased the number of crystal structures solved by diffraction data collected from twinned crystals. In some cases, for example structures of superconductors and perovskite derivatives, this capability has brought benefits also for materials of high technological value.

However, the impact of twinning on the diffraction data is not straightforward and needs to be properly digested not only by people developing the relevant software, but also by the users. Unfortunately, the necessary crystallographic knowledge, on one side is spread in tens of original papers and, on the other side, only seldom is taught in basic crystallographic courses dedicated to structural crystallographers (Nespolo and Ferraris 2003).

This chapter has integrated basic knowledge and worked examples to provide the structural crystallographers and, in general, materials scientists, with a short, but quite complete, survey of the problems related to twinning, that is, an aspect of modularity at crystal scale. Actually, twinning is not the only modular aspect at crystal scale to hinder an accurate solution of the structural structures. In general, all kinds of domain, into which a 'single crystal' is normally fragmented, pose problems on the route of properly determining a crystal structure. However, domain structure is more a matter of electron microscopy except for twinning, which, instead, would completely prevent a classic crystal structure determination.

Note added in proofs In the forthcoming volume D of the International Tables (IT) for Crystallography, the chapter 3.3 "Twinning of crystals" (Hahn & Klapper 2003) presents the basic concepts and definitions of twinning, as well as the morphological (according to the German school mentioned in section 5.1.2.3), genetic and lattice classification of twins, but not the effect of twinning on diffraction and crystal-structure determinations. Thus, the present chapter well complements the IT chapter.

References

Abrahams, S.C. and Bernstein, J.L. (1971). Rutile: Normal probability plot analysis and accurate measurements of crystal structure. *Journal of Chemical Physics*, **55**, 3206–11.

Adiwidjaja, G., Friese, K., Klaska, K.-H. and Schlüter, J. (1997). The crystal structure of gordaite $NaZn_4(SO_4)(OH)_6Cl \cdot 6H_2O$. *Zeitschrift für Kristallographie*, **212**, 704–07.

Adolphe, C. and Laruelle, P. (1968). Structure cristalline de $FeHo_4S_7$ et de certains composés isotypes. *Bulletin de la Société Française de Minéralogie et Cristallographie*, **91**, 219–32.

Akimoto, S.-I. and Sato, Y. (1968). High-pressure transformation in Co_2SiO_4 olivine and some geophysical implications. *Physics of the Earth and Planetary Interiors*, **1**, 498–505.

Alberti, A. (1979). Possible 4-connected frameworks with 4-4-1 unit found in heulandite, stilbite, brewrsterite, and scapolite. *The American Mineralogist*, **64**, 1188–98.

Alberti, A. and Gottardi, G. (1975). Possible structures in fibrous zeolites. *Neues Jahrbuch für Mineralogie, Monatshefte*, 396–411.

Alberti, A., Cruciani, G., Galli, E., Merlino, S., Millini, R., Quartieri, Q., Vezzalini, G., and Zanardi, S. (2002). The crystal structure of tetragonal and monoclinic polytypes of tschernichite, the natural counterpart of synthetic zeolite beta. *Journal Physical Chemistry B*, **106**, 10277–84.

Aleksandrov, K.S. and Beznosikov, B.V. (1997). Architecture of perovskite-like crystals. *Crystallography Reports*, **42**, 556–66.

Alexander, E. and Hermann, K. (1929). Die 80 zweidimensionalen Raumgruppen. *Zeitschrift für Kristallogaphie*, **70**, 328–45.

Allmann, R. (1984). Die Struktur des Sursassits und ihre Beziehung zur Pumpellyit- und Ardennitstruktur. *Fortschritte der Mineralogie, Beiheft*, **62**, 3–4.

Allmann, R., Lohse, H.H., and Hellner, E. (1968). Die Kristallstruktur des Koenenits, eine Doppelschichtstruktur mit zwei inkommensurablen Teilgittern. *Zeitschrift für Kristallographie*, **126**, 7–22.

Andersson, S. (1983). Eine Beschreibung komplexer anorganischer Kristallstrukturen. *Angewande Chemie*, **95**, 67–80.

Andersson, S. and Hyde, B.G. (1974). Twinning on the unit cell level as a structure-building operation in the solid state. *Journal of Solid State Chemistry*, **9**, 92–101.

Angel, R.J. (1986). Polytypes and polytypism. *Zeitschrift für Kristallographie*, **176**, 193–204.

Angel, R.J. and Burnham, C.W. (1991). Pyroxene-pyroxenoid polysomatism revisited: A clarification. *The American Mineralogist*, **76**, 900–3.

Angel, R.J., Price, G.D., and Yeomans, J. (1985). The energetics of polytypic structures: further applications of the ANNNI model. *Acta Crystallographica*, **B41**, 310–19.

Antipov, E.V., Khasanova, N.R., Pshirkov, J.S., Putilin, S.N., Bougerol, C., Lebedev, O.I., Van Tendeloo, G., Baranov, A.N., and Park, Y.W. (2002). The superconducting bismuth-based mixed oxides. *Current Applied Physics*, **2**, 425–30.

Apostolov, A. and Bassi, G. (1971). Refinement of the structure of calcium ferrite. *Yearbook of the Faculty of Physical Sciences, University of Sofia*, **63**, 177–87.

Arakcheeva, A.V., Karpinskii, O.G. and Lyadova, V.Ya. (1991). Crystal structure of a $CaFe_3AlO_7$ aluminum–calcium ferrite of variable composition. *Soviet Physics Crystallography*, **36**, 332–6.

Arakcheeva, A.V., Pushcharovskii, D.Yu., Gekimyants, V.M., Popov, V.A., and Lubman, G.U. (1997). Crystal strcuture and microtwinning of natural orthorhombic perovskite CaTiO3. *Crystallography Reports*, **42**, 46–54.

Araki, T. (1991). Crystal structure determination for crystals with twinning by hemihedry or pseudohemihedry. *Zeitschrift für Mineralogie*, **194**, 161–81.

Armbruster, Th. (2002). Revised nomenclature of högbomite, nigerite, and taffeite minerals. *European Journal of Mineralogy*, **14**, 389–95.

Armbruster, Th. and Gunter, M.E. (2001). Crystal structures of natural zeolites. *Reviews in Mineralogy and Geochemistry*, **45**, 1–67.

Armbruster, Th. and Hummel, W. (1987). (Sb, Bi, Pb) ordering in sulfosalts: crystal-structure refinement of a Bi-rich izoklakeite. *The American Mineralogist*, **72**, 821–31.

Armbruster, Th., Makovicky, E., Berlepsch, P., and Sejkora, J. (2003). Crystal structure, cation ordering and polytypic character of diaphorite $Pb_2Ag_3Sb_3S_8$, a PbS based structure. *European Journal of Mineralogy*, **15**, 137–46.

Artioli, G. and Galli, E. (1994). The crystal structures of orthorhombic and monoclinic palygorskite. *Materials Science Forum*, **166–169**, 647–52.

Artioli, G., Smith, J.V., and Pluth, J.J. (1986). X-ray structure refinement of mesolite. *Acta Crystallographica*, **C42**, 937–42.

Aurivilius, B. (1983). The crystal structures of two forms of $BaBi_2S_4$. *Acta Chemica Scandinavica*, **A37**, 399–407.

Backhaus, K.O. (1979*a*). OD interpretation of the crystal structure of TeCl4. *Kristall und Technik*, **14**, 1157–62.

Backhaus, K.O. (1979*b*). OD interpretation of polytypism and polytypic twinning in WO_2Cl_2. *Kristall und Technik*, **14**, 1163–4.

Bailey, S.W. (1984). Classification and structures of the micas. *Reviews in Mineralogy*, **13**, 1–12.

Bakker, M. and Hyde, B.G. (1978). A preliminary electron microscope study of chemical twinning in the system $MnS + Y_2S_3$, an analogue of the mineral system $PbS + Bi_2S_3$ (galena + bismuthinite). *Philosophical Magazine*, **A38**, 615–28.

Balič Žunič, T. and Engel, P. (1983). Crystal structure of synthetic $PbTlAs_3S_6$. *Zeitschrift für Kristallographie*, **165**, 261–9.

Balič Žunič, T. and Makovicky, E. (1993). Contributions to the crystal chemistry of thallium sulphosalts. I. The O-D nature of imhofite. *Neues Jahrbuch für Mineralogie, Abhandlungen*, **165**, 317–30.

Balič Žunič, T. and Makovicky, E. (1996). Determination of the centroid or 'the best centre' of a coordination polyhedron. *Acta Crystallographica*, **B52**, 78–81.

Balič Žunič, T., Petersen, O.V., Bernhardt, H.-J., and Micheelsen, H.I. (2002*a*). The crystal structure and mineralogical description of a Na-dominant komarovite from the Ilimaussaq alkaline complex, South Greenland. *Neues Jahrbuch für Mineralogie, Monatshefte*, 497–514.

Balič Žunič, T., Topa, D., and Makovicky, E. (2002*b*). The crystal structure of emilite, $Cu_{10.7}Pb_{10.7}Bi_{21.3}S_{48}$, the second 45 Å derivative of the bismuthinite-aikinite solid-solution series. *The Canadian Mineralogist*, **40**, 239–45.

Bando, Y., Sekikawa, Y., Yamamura, H., and Matsui, Y. (1981). Crystal structure of $Ca_4YFe_5O_{13}$ by combining 1 MeV high resolution electron microscopy with convergent-beam electron diffraction. *Acta Crystallographica*, **A37**, 723–8.

Barbier, J. (1989). New spinelloid phases in $MgGa_2O_4$–Mg_2GeO_4 and $MgFe_2O_4$–Mg_2GeO_4 systems. *European Journal of Mineralogy*, **1**, 39–46.

Barbier, J. (1998). Crystal structures of sapphirine and surinamite analogues in the MgO–Ga_2O_3–GeO_2 system. *European Journal of Mineralogy*, **10**, 1283–93.

Barbier, J. and Hyde, B.G. (1988). Structure of sapphirine: its relations to the spinel, clinopyroxene and β-gallia structure. *Acta Crystallographica*, **B44**, 373–7.

Bärnighausen, H. (1976). *Proceedings 12th Rare Earth Research Conference*, **1**, 404.

Baronnet, A., Belluso, E., and Ferraris, G. (1994*a*). HRTEM study of carlosturanite: new polysomes and microstructural relationships with associated minerals. *Abstracts of the IMA 16th General Meeting (Pisa 1994)*, p. 30–31.

Baronnet, A., Mellini, M., and Devouard, B. (1994*b*). Sectors in polygonal serpentine. A model based on dislocations. *Physics and Chemistry of Minerals*, **21**, 330–343.

Battle, P.D., Graeme, R.B., Sloan, J., and Vente, J.F. (1998). Commensurate and incommensurate phases in the system $A_4A'Ir_2O_9$ (A = Sr, Ba; A' = Cu, Zn). *Journal of Solid State Chemistry*, **136**, 103–14.

Bayer, G. and Hoffmann, W.W. (1965). Über Verbindungen vom Na_xTiO_2-Typ. *Zeitschrift für Kristallographie*, **121**, 9–13.

Bednorz, J.G. and Müller, K.A. (1986). Possible high T_c superconductivity in the barium–lanthanum–copper–oxygen system. *Zeitschrift für Physik*, **B64**, 189–93.

Belli Dell'Amico, D., Calderazzo, F., Marchetti, F., and Merlino, S. (1998). Silver hydrogen sulfate, $Ag(O_3SOH)$: Preparation, and the OD character of the crystal structure. *Chemistry of Materials*, **10**, 524–30.

Belluso, E. and Ferraris, G. (1991). New data on balangeroite and carlosturanite from alpine serpentinites. *European Journal of Mineralogy*, **3**, 559–66.

Belokoneva, E.L. (2003). OD family of Ca, Mg-borates of the kurchatovite group. *Crystallography Reports*, **48**, 256–9.

Belov, N.V., Gavrilova, G.S., Solovieva, L.P., and Khalilov, A.D. (1978). The refined structure of lomonosovite. *Soviet Physics Doklady*, **22**, 422–4.

Bendersky, L.A., Chen, R., Fawcett, I.D. and Greenblatt, M. (2001). TEM studies of the electron-doped-layered $La_{2-2x}Ca_{1+2x}Mn_2O_7$: Orthorhombic phase in the $0.8 < x < 1.0$ composition range. *Journal of Solid State Chemistry*, **157**, 309–23.

Bente, K. and Kupčík, V. (1984). Redetermination and refinement of the structure of tetrabismuth tetracopper enneasulphide, $Cu_4Bi_4S_9$. *Acta Crystallographica*, **C40**, 1985–6.

Berger, R. (1987). A phase-analytical study of the Tl–Cu–Se system. *Journal of Solid State Chemistry*, **70**, 65–70.

Bergerhoff, G., Berndt, M., Brandenburg, K., and Degen, T. (1999). Concerning inorganic structure types. *Acta Crystallographica*, **B55**, 147–56.

Berlepsch, D. (1996). Crystal structure and crystal chemistry of the homeotypes edenharterite ($TlPbAs_3S_6$) and jentschite ($TlPbAs_2SbS_6$) from Lengenbach, Binntal (Switzerland). *Schweizerische Mineralogische und Petrographische Mitteilungen*, **76**, 147–57.

Berlepsch, P., Miletich, R., and Armbruster, T. (1999). The crystal structure of synthetic KSb_5S_8 and $(Tl_{0.598}K_{0.402})Sb_5S_8$ and their relation to parapierrotite ($TlSb_5S_8$). *Zeitschrift für Kristallographie*, **214**, 57–63.

Berlepsch, P., Makovicky, E., and Balič Žunič, T. (2000). Contribution to the crystal chemistry of Tl-sulfosalts. VI. Modular-level structure relationship between edenharterite $TlPbAs_3S_6$ and jentschite $TlPbAs_2SbS_6$. *Neues Jahrbuch für Mineralogie, Monatshefte*, 315–32.

Berlepsch, P., Makovicky, E., and Balič Žunič, T. (2001*a*). Crystal chemistry of meneghinite homologues and related sulfosalts. *Neues Jahrbuch für Mineralogie, Monatshefte*, 115–35.

Berlepsch, P., Makovicky, E., and Balič Žunič, T. (2001*b*). Crystal chemistry of sartorite homologues and related sulfosalts. *Neues Jahrbuch für Mineralogie, Abhandlungen*, **176**, 45–66.

Berlepsch, P., Armbruster, T., Makovicky, E., Hejny, C., Topa, D., and Graeser, S. (2001*c*). The crystal structure of (001) twinned xilingolite, $Pb_3Bi_2S_6$, from Mittal-Hohtenn, Valais, Switzerland. *The Canadian Mineralogist*, **39**, 1653–63.

Berlepsch, P., Miletich, R., Makovicky, E., Balič Žunič, T. and Topa, D. (2001*d*). The crystal structure of synthetic $Rb_2Sb_8S_{12}(S_2){\cdot}2H_2O$, a new member of the hutchinsonite family of merotypes. *Zeitschrift für Kristallographie*, **216**, 272–7.

Berlepsch, P., Armbruster, T., and Topa, D. (2003*a*). Structural and chemical variation in rathite, $Pb_8Pb_{4-x}(Tl_2As_2)_x(Ag_2As_2)As_{16}S_{40}$: modulations of a parent structure. *Zeitschrift für Kristallographie*, **217**, 581–90.

Berlepsch, P., Armbruster, T., Makovicky, E., and Topa, D. (2003*b*). Another step toward understanding the true nature of sartorite: Determination and refinement of a ninefold superstructure. *The American Mineralogist*, **88**, 450–61.

Bertaut, E.F. and Blum, P. (1956). Determination de la structure de Ti_2CaO_4 par la methode selfconsistante d'approche directe. *Acta Crystallographica*, **9**, 121–6.

Bertaut, E.F., Blum, P., and Sagnieres, A. (1959). Structure du ferrite bicalcique et de la brownmillerite. *Acta Crystallographica*, **12**, 149–59.

Betti, E. (1998). *Cristallochimica di silicati con formula generale* $X_{16}(Si_2O_7)_4$ $(O, OH, F)_8$. Thesis for the Master Degree at Università degli Studi di Pisa.

Bevan, D.J.M. and Mann, A.W. (1972). Intermediate fluorite-related phases in the Y_2O_3–YF_3 system—Examples of one-dimensional ordered intergrowth. *Journal of Solid State Chemistry*, **5**, 410–8.

Bevan, D.J.M. and Mann, A.W. (1975). The crystal structure of $Y_7O_6F_9$. *Acta Crystallographica*, **B31**, 1406–11.

Beznosikov, B.V. and Aleksandrov, K.S. (2000). Perovskite-like crystals of the Ruddlesden–Popper series. *Crystallographic reports*, **45**, 792–8.

Bhide, V. and Gasperin, M. (1979). A new GTB-type thallium niobate. *Acta Crystallographica*, **B35**, 1318–21.

Bindi, L., Rees, L.H., and Bonazzi, P. (2003). Twinning in natural melilite simulating a fivefold superstructure. *Acta Crystallographica*, **B59**, 156–8.

Birkedal, H., Burgi, H.-B., Komatsu, K., and Schwarzenbach, D. (2003). Polymorphism and stacking disorder in tris(bicyclo[2.1.1]hexeno)benzene. *Journal of Molecular Structure*, **647**, 233–42.

Bloss, F. D. (1971). *Crystallography and crystal chemistry – an introduction*. New York: Holt, Rinehart and Winston.

Boggs, R.C., Howard, D.G., Smith, J.V., and Klein, G.L. (1993). Tschernichite, a new zeolite from Goble, Columbia County, Oregon. *The American Mineralogist*, **78**, 822–6.

Bonaccorsi, E., Merlino, S., and Pasero, M. (1988). Trikalsilite: its structural relationships with nepheline and tetrakalsilite. *Neues Jahrbuch für Mineralogie, Monatshefte*, **1988**, 559–67.

Bonaccorsi, E., Merlino, M., and Pasero, M. (1990). Rhönite: structural and microstructural features, crystal chemistry and polysomatic relationships. *European Journal of Mineralogy*, **2**, 203–18.

Bonatti, S. and Gottardi, G. (1966). Un caso di polimorfismo a strati in sorosilicati: perrieirte e chevkinite. *Periodico di Mineralogia*, **35**, 69–91.

Bonazzi, P., Menchetti, S., and Sabelli, C. (1987). Structure refinement of kermesite: Symmetry, twinning, and comparison with stibnite. *Neues Jahrbuch für Mineralogie, Monatshefte*, 557–67.

Bonazzi, P., Borrini, D., Mazzi, F., and Olmi, F. (1995). Crystal structure and twinning of Sb_2AsS_2, the synthetic analogue of pääkkönenite. *The American Mineralogist*, **80**, 1054–8.

Bonazzi, P., Bindi, L., and Parodi, G. (2003). Gatelite-(Ce), a new REE-bearing mineral from Trimouns, French Pyrenees: Crystal structure and polysomatic relationships with epidote and törnebohmite-(Ce). *The American Mineralogist*, **88**, 223–8.

Bordet, P., Bougerot Chaillout, C., Grey, I.E., Hodeau, J.L., and Isnard, O. (2000). Structural characterization of the engineered scavenger compound, H-$Li_2Ti_3O_7$. *Journal of Solid State Chemistry*, **152**, 546–53.

Botkovitz, P., Brec, R., Deniard, P., Tournoux, M., and Burr, G. (1994). Electrochemical and neutron diffraction study of a prelithiated hollandite-type Li_xMnO_2 phase. *Molecular Crystals and Liquid Crystals*, **244**, 233–8.

Boullay, P., Hervieu, M., and Raveau, B. (1997). A new manganite with an original composite tunnels structure: $Ba_6Mn_{24}O_{48}$. *Journal of Solid State Chemistry*, **132**, 239–48.

Bovin, J.O. and Andersson, S. (1977). Swinging twinning on the unit cell level as a structure-building operation in the solid state. *Journal of Solid State Chemistry*, **20**, 127–33.

Bragg, W.L. (1930). The structure of silicates. *Zeitschrift für Kristallographie*, **74**, 237–305.

Braun, M., Tuffentsammer, W., Wachtel, H., and Wolf, H.C. (1999). Tailoring of energy levels in lead chloride based layered perovskites and energy transfer between the organic and inorganic planes. *Chemical Physics Letters*, **303**, 157–64.

Brauner, K. and Preisinger, A. (1956). Struktur und Entstehung des Sepioliths. *Tschermaks Mineralogische Petrographische Mitteilung*, **6**, 120–40.

Bravais, M. (1851). Etudes cristallographiques. Troisiemme partie. Des macle et des hemitropies. *Journal de l'Ecole Polytechnique*, **XX**, 248–76.

Bronger, W. and Böttcher, P. (1972). Über Thiomanganate und -kobaltate der schweren Alkalimetalle: $Rb_2Mn_3S_4$, $Cs_2Mn_3S_4$, $Rb_2Co_3S_4$, $Cs_2Co_3S_4$. *Zeitschrift für anorganische und allgemeine Chemie*, **390**, 1–96.

Brown, D.B., Zubieta, J., Vella, P.A., Wrobleski, J.T., Watt, T., Hatfield, W.E., and Day, P. (1980). Solid-state and electronic properties of a mixed-valence two-dimensional metal, KCu_4S_3. *Inorganic Chemistry*, **19**, 1945–50.

Buerger, M.J. (1945). The genesis of twin crystals. *The American Mineralogist*, **30**, 469–82.

Buerger, M.J. (1954). The diffraction symmetry of twins. *Anaias de Academia Brasilera de Ciencias*, **26**, 11–12.

Burns, P.C., Cooper, M.A., and Hawthorne, F.C. (1994). Jahn-Teller distorted $Mn^{3+}O_6$ octahedra in fredrikssonite, the fourth polymorph of $Mg_2Mn^{3+}(BO_3)O_2$. *The Canadian Mineralogist*, **32**, 397–403.

Burns, P.C., Grice, J.D., and Hawthorne, F.C. (1995). Borate minerals. I. Polyhedral clusters and fundamental building blocks. *The Canadian Mineralogist*, **33**, 1131–51.

Bursill, L. and Grzinic (1980). Incommensurate superlattice ordering in the hollandites $Ba_xTi_{8-x}Mg_xO_{16}$ and $Ba_xTi_{8-2x}Ga_{2x}O_{16}$. *Acta Crystallographica*, **B36**, 2902–13.

Bursill, L.A., Hyde, B.G., Terasaki, O., and Watanabe, D. (1969). On a new family of titanium oxides and the nature of slightly reduced rutile. *Philosophical Magazine*, **20**, 347–59.

Buseck, P.R. (ed.) (1992). Minerals and reactions at the atomic scale: transmission electron microscopy. *Reviews in Mineralogy vol. 27, Mineralogical Society of America*. pp. 508.

Buss, B. and Krebs, B. (1971). Crystal structure of tellurium tetrachloride. *Inorganic Chemistry*, **10**, 2795–2800.

Byström, A.M. (1949). The crystal structure of ramsdellite, an orthorhombic modification of MnO_2. *Acta Chemica Scandinavica*, **3**, 163–73.

Calabrese, J.C., Jones, N.L., Harlow, R.L., Herron, N., Thorn, D.L., and Wang, Y. (1991). Preparation and characterization of layered lead halide compounds. *Journal of the American Chemical Society*, **113**, 2328–30.

Callegari, A., Mazzi, F., and Tadini, C. (2003). Modular aspects of the crystal structures of kurchatovite and clinokurchatovite. *European Journal of Mineralogy*, **15**, 277–82.

Calvo, C. and Faggiani, R. (1974). A re-investigation of the crystal structure of chevkinite and perrierite. *The American Mineralogist*, **59**, 1277–85.

Cámara, F., Garvie, L.A.J., Devouard, B., Groy, Th.L. and Buseck, P.R. (2002). The structure of Mn-rich tuperssuatsiaite: A palygorskite-related mineral. *The American Mineralogist*, **87**, 1458–63.

Cameron and Papike (1981). Structural and chemical variations in pyroxenes. *The American Minerlogist*, **66**, 1–50.

Cannillo, E., Mazzi, F., and Rossi, G. (1972). Crystal structure of götzenite. *Soviet Physics Crystallography*, **16**, 1026–30.

Capitani, G. and Mellini, M. (2003). The modulated crystal structure of the "*odd*", $m = 17$, antigorite polysome. *The American Mineralogist*, in press.

Carré, D. and Laruelle, P. (1973). Structure cristalline du sulfure d'erbium et de lanthane, $Er_9La_{10}S_{27}$. *Acta Crystallographica*, **B29**, 70–73.

Carré, D. and Laruelle, P. (1974). Structure cristalline du sulfure néodyme et d'ytterbium, $NdYbS_3$. *Acta Crystallographica*, **B30**, 952–54.

Catti, M. and Ferraris, G. (1976). Twinning by merohedry and X-ray crystal structure determination. *Acta Crystallographica*, **A32**, 163–5.

Cellai, D., Carpenter, M.A., and Heaney, P.J. (1992). Phase transitions and microstructures in natural kaliophylite. *European Journal of Mineralogy*, **4**, 1209–20.

Ceolin, R., Toffoli, P., Khodadad, P., and Rodier, N. (1977). Structure cristalline du sulfure mixte de cerium et de bismuth $Ce_{1.25}Bi_{3.78}S_8$. *Acta Crystallographica*, **B33**, 2804–6.

Cervelle, B.D., Cesbron, F.P., and Sichère, M.C. (1979). La chalcostibite et la dadsonite de Saint-Pons, Alpes de Haute Provence, France. *The Canadian Mineralogist*, **17**, 601–5.

Chao, G.Y., Grice, J.D., and Gault, R.A. (1991). Silinaite, a new sodium lithium silicate hydrate mineral from Mont Saint-Hilaire, Quebec. *The Canadian Mineralogist*, **29**, 359–62.

Charkin, D.O., Dytyatiev, O.A., Rakounov, A.B., Dolgikh, V.A., and Lightfoot, P. (2003*a*). A new family of layered bismuth oxohalides. *Russian Journal of Inorganic Chemistry*, **48**, 149–56.

Charkin, D.O., Dytyatiev, O.A., Dolgikh, V.A., and Lightfoot, P. (2003*b*). A new family of layered bismuth compounds II: the crystal structures of $Pb_{0.6}Bi_{1.4}Rb_{0.6}O_2Z_2$, Z = Cl, Br, and I. *Journal of Solid State Chemistry*, **173**, 83–90.

Chen, X.-A., Wada, H., Sato, A., and Mieno, M. (1998). Synthesis, electrical conductivity, and crystal structure of $Cu_4Sn_7S_{16}$ and structure refinement of Cu_2SnS_3. *Journal of Solid State Chemistry*, **136**, 144–151.

Chen, X.-A., Wada, H., and Sato, A. (1999). Preparation, crystal structure and electrical properties of Cu_4SnS_6. *Materials Research Bulletin*, **34**, 239–47.

Chernov, A.N., Ilyukhin, V.V., Maksimov, B.A., and Belov, N.V. (1971). Crystal structure of innelite—$Na_2Ba_3(Ba, K, Mn)(Ca, Na)Ti(TiO_2)_2(Si_2O_7)_2(SO_4)_2$. *Soviet Physics Crystallography*, **16**, 65–69.

Chiari, G., Giustetto, R., and Ricchiardi, G. (2003). Crystal structure refinement of palygorskite and Maya Blue from molecular modelling and powder synchrotron diffraction. *European Journal of Mineralogy*, **15**, 21–33.

Choe, W., Lee, S., O'Connel, P., and Covey, A. (1997). Synthesis and structure of new Cd–Bi–S homologous series: a study in intergrowth and the control of twinning patterns. *Chemistry of Materials*, **9**, 2025–30.

Choi, K.-S. and Kanatzidis, M.G. (2000). Sulfosalts with alkaline earth metals. Centrosymmetric vs acentric interplay in $Ba_3Sb_{4.66}S_{10}$ and $Ba_{2.62}Pb_{1.38}Sb_4S_{10}$ based on the Ba/Pb/Sb ratio. Phases related to the arsenosulfide minerals of the rathite group and the novel polysulfide $Sr_6Sb_6S_{17}$. *Inorganic Chemistry*, **39**, 5655–62.

Chopin, C., Brunet, F., Gebert, W., Medenbach, O., and Tillmanns, E. (1993). Bearthite, $Ca_2Al[PO_4]_2(OH)$, a new mineral from high-pressure terranes of the western Alps. *Schweizerische Mineralogische und Petrographische Mitteilungen*, **73**: 1–9.

Christiansen, C.C. (2003). *A modular and crystal-chemical investigation of the rosenbuschite and låvenite–wöhlerite groups*. PhD. thesis, Geological Museum, Faculty of Science, University of Copenaghen.

Christiansen, C.C. and Rønsbo, J.G. (2000). On the structural relationship between götzenite and rinkite. *Neues Jahrbuch für Mineralogie, Monatshefte*, **2000**, 496–506.

Christiansen, C.C., Makovicky, E., and Johnsen, O.N. (1999). Homology and typism in heterophyllosilicates: an alternative approach. *Neues Jahrbuch für Mineralogie Abhandlungen*, **175**, 153–89.

Christiansen, C.C., Makovicky, E., and Johnsen, O. (2003). Modularity of the låvenite–wöhlerite and rosenbuschite groups: An OD polytypic study. *The Canadian Mineralogist*, in press.

Christy, A.G. and Putnis, A. (1988). Planar and line defects in sapphirine polytypes. *Physics and Chemistry of Minerals*, **15**, 548–58.

Chukanov, N.V., Pekov, I.V., and Khomyakov, A.P. (2002). Recommended nomenclature for labuntsovite-group minerals. *Euopean Journal of Mineralogy*, **14**, 165–73.

Chukanov, N.V., Moiseev, M.M., Pekov, I.V., Lazebnik, K.A., Rastsvetaeva, R.K., Zayakina, N.V., Ferraris, G., and Ivaldi, G. (2003). Nabalamprophyllite Ba(Na,Ba) $\{Na_3Ti[Ti_2O_2Si_4O_{14}](OH,F)_2\}$, a new layer titanosilicate of the lamprophyllite group from Inagli and Kovdor alkaline-ultramafic massifs, Russia. *Zapiski Vserossiyskogo Mineralogicheskogo Obschestva*, in press. (In Russian).

Chukhrov, F.V., Gorshkov, A.I. and Drits, V.A. (1989). *Hypergene oxides of manganese*. Nauka, Moscow. p. 206 (In Russian).

Cocco, G. and Mazzi, F. (1959). La struttura della brochantite. *Periodico di Mineralogia*, **28**, 121–49.

Cocco, G., Fanfani, L., and Zanazzi, P.F. (1967). The crystal structure of fornacite. *Zeitschrift für Kristallographie*, **124**, 385–97.

Collin, G., Boilot, J.P., Colomban, Ph., and Comes, R. (1986). Host lattices and superionic properties in *beta*- and *beta*" alumina. I. Structures and local correlations. *Physical Review*, **B34**, 5838–49.

Collomb, A., Wolfers, P., and Obradors, X. (1986). Neutron diffraction studies of some hexagonal ferrites: $BaFe_{12}O_{19}$, $BaMg_2$–W and $BaCo_2$–W. *Journal of Magnetism and Magnetic Materials*, **62**, 57–67.

Colville, A.A. and Geller, S. (1971). The crystal structure of brownmillerite, Ca_2FeAlO_5. *Acta Crystallographica*, **B27**, 2311–15.

Compagnoni, R., Ferraris, G., and Fiora, L. (1983). Balangeroite, a new fibrous silicate related to gageite from Balangero, Italy. *The American Mineralogist*, **68**, 214–19.

Cook, R. and Schäfer, H. (1983). Preparation and crystal structure of alkaline earth seleno-(telluro)-antimonites and bismuthides ABX_3 with A = Sr, Ba; B = Sb, Bi and X = Se, Te. In Metsetaard, R., Heijlingers, H.J., and Schoomen, J. (eds.). *Studies in Inorg. Chem. 3*. Elsevier, Amsterdam.

Cordier, G. and Schaefer, H. (1979). Zur Darstellung und Kristallstruktur von $BaSb_2Se_4$. *Z. Naturforschung, Teil B. Anorganische Chemie*, **34**, 1053–56.

Cordier, G., Schwidetzky, C., and Schaefer, H. (1984). New SbS_2 strings in the $BaSb_2Se_4$ structure. *Journal of Solid State Chemistry*, **54**, 84–88.

Cox, D.E., Torardi, C.C., Subramanian, M.A., Gopalakrishnan, J., and Sleight, A.W. (1988). Structure refinements of superconducting $Tl_2Ba_2CaCu_2O_8$ and $Tl_2Ba_2Ca_2Cu_3O_{10}$ from neutron diffraction data. *Physical Review*, **B38**, 6624–30.

Curien, H. and Donnay, J.D.H. (1959). The symmetry of the complete twin. *The American Mineralogist*, **44**, 1067–70.

Dai, Y. and Post, J.E. (1995). Crystal structure of hillebrandite: a natural analogue of calcium silicate hydrate (CSH) phases in Portland cement. *The American Mineralogist*, **80**, 841–4.

Dai, Y., Post, J.E. and Appleman, D.E. (1995). Crystal structure of minehillite: twinning and structural relationships to reyerite. *The American Mineralogist*, **80**, 173–8.

Darriet, J. and Subramanian, M.A. (1995). Structural relationships between compounds based on the stacking of mixed layers related to hexagonal perovskite-type structures. *Journal of materials Chemistry*, **5**, 543–52.

Darriet, B., Bovin, J.O. and Galy, J. (1976). Un nouveau compose de l'antimoine III; $VOSb_2O_4$. Influence stéréochimique de la paire non liée E, relations structurales, mécanismes de la réaction chimique. *Journal of Solid State Chemistry*, **19**, 205–12.

Darriet, J., Grasset, F. and Battle, P.D. (1997). Synthesis, crystal structure and magnetic properties of $A_3A'RuO_6$ (A = Ca, Sr; A' = Li, Na). *Materials Research Bulletin*, **32**, 139–50.

de Villiers, J.P. and Buseck, P.R. (1989). Stacking variations and nonstoichiometry in the bixbyite–braunite polysomatic mineral group. *The American Mineralogist*, **74**, 1325–36.

Dittmar, G. and Schäfer, M. (1977). Darstellung und Kristallstruktur von $(NH_4)_2Sb_4S_7$. *Zeitschrift für anorganische allgemeine Chemie*, **437**, 183–7.

Divjakovič, V. and Nowacki, W. (1976). Die Kristallstruktur von Imhofit, $Tl_{5.6}As_{15}S_{25.3}$. *Zeitschrift für Kristallographie*, **144**, 323–33.

Dódony, I. and Buseck, P.R. (2001). Polysomatism and modules of gehlenite composition and structure in the Hanic phase ($Ca_5Al_6MgSiO_{17}$). *Physics and Chemistry of Minerals*, **28**, 428–34.

Dódony, I., Pósfai, M., and Buseck, P.R. (1996). Structural variations in shattuckite, a copper silicate mineral. In Pósfai, M., Papp, G. and Weiszburg, T. (eds). *Mineralogy and Museums 3, International Conference, Abstracts / Acta Mineralogica Petrographica* (Szeged), 37, Suppl., 32.

Dódony, I., Pósfai, M., and Buseck, P.R. (2002). Revised structure models for antigorite: an HRTEM study. *The American Mineralogist*, **87**, 1443–57.

Dollase, W.A. (1968). Refinement and comparison of the structures of zoisite and clinozoisite. *The American Mineralogist*, **53**, 1882–98.

Donnay, G. and Donnay, J.D.H. (1953). The crystallography of bastnäsite, parisite, röntgenite and synchisite. *The American Mineralogist*, **38**, 932–63.

Donnay, G. and Donnay, J.D.H. (1974). Classification of triperiodic twins. *The Canadian Mineralogist*, **12**, 422–5.

Donohue, P.C., Katz, L., and Ward, R. (1965). The crystal structure of barium ruthenium oxides and related compounds. *Inorganic Chemistry*, **4**, 306–10.

Dornberger-Schiff, K. (1956). On the order–disorder structures (OD-structures). *Acta Crystallographica*, **9**, 593–601.

Dornberger-Schiff, K. (1959). On the nomenclature of the 80 plane groups in three dimensions. *Acta Crystallographica*, **12**, 173.

Dornberger-Schiff, K. (1964). Grundzüge einer Theorie der OD-Strukturen aus Schichten. *Abhandlungen der deutschen Akademie der Wissenschaften zu Berlin. Klasse für Chemie, Geologie und Biologie*, **3**, 1–107.

Dornberger-Schiff, K. (1966). *Lehrgang über OD-Strukturen*. Berlin: Akademie-Verlag, 135 p.

Dornberger-Schiff, K. (1979). OD structures—a game and a bit more. *Kristall und Technik*, **14**, 1027–45.

Dornberger-Schiff, K. (1982). Geometrical properties of MDO polytypes and procedures for their derivation. I. General concept and applications to polytype families consisting of OD layers all of the same kind. *Acta Crystallographica*, **A38**, 483–91.

Dornberger-Schiff, K. and Ďurovič, S. (1975*a*). OD interpretation of kaolinite-type structures—I: symmetry of kaolinite packets and their stacking possibilities. *Clays and Clay Minerals*, **23**, 219–29.

Dornberger-Schiff, K. and Ďurovič, S. (1975*b*). OD interpretation of kaolinite-type structures—II: the regular polytypes (MDO-polytypes) and their derivation. *Clays and Clay Minerals*, **23**, 231–46.

Dornberger-Schiff, K. and Fichtner, K. (1972). On the symmetry of OD structures consisting of equivalent layers. *Kristall und Technik*, **7**, 1035–56.

Dornberger-Schiff, K. and Grell-Niemann, H. (1961). On the theory of order–disorder (OD) structures. *Acta Crystallographica*, **14**, 167–77.

Dornberger Schiff, K. and Klevtsova, R.F. (1967). On the relation between the monoclinic and the orthorhombic form of yttrium hydroxylchloride, $[YCl(OH)_2]_n$. *Acta Crystallographica*, **22**, 435–6.

Dornberger-Schiff, K. and Merlino, S. (1974). Order-disorder in sapphirine, aenigmatite and aenigmatite-like minerals. *Acta Crystallographica*, **A30**, 168–73.

Dornberger-Schiff, K., Liebau, F., and Thilo, E. (1955). Zur Struktur des β-Wollastonits, des Maddrellschen Salzes und des Natriumpolyarsenats. *Acta Crystallographica*, **8**, 752–4.

Dornberger-Schiff, K., Backaus, K.-O., and Ďurovič, S. (1982). Polytypism of micas: OD-interpretation, stacking symbols, symmetry relations. *Clays and Clay Minerals*, **30**, 364–74.

Dornberger-Schiff, K., Drits, V.A., Ďurovič, S., and Zvyagin, V.V. (1985). Special form of polytypism potentially occurring in astrophyllite structures. *Soviet Physics Crystallography*, **30**, 292–4.

Drits, V.A. (1997). Mixed-layer minerals. *EMU Notes in Mineralogy*, **1**, 153–90.

Drits, V.A., Silvester, E., Gorshkov, A.I., and Manceau, A. (1997). Structure of synthetic monoclinic Na-birnessite and hexagonal birnessite: I. Results from X-ray diffraction and selected-area electron diffraction. *The American Mineralogist*, **82**, 946–61.

Dunitz, J.D. (1964). The interpretation of pseudo-orthorhombic patterns. *Acta Crystallographica*, **17**, 1299–1304.

Ďurovič, S. (1979). Desymmetrization of OD structures. *Kristall und Technik*, **14**, 1047–53.

Ďurovič, S. (1981). OD-Charakter, Polytypie und Identifikation von Schichtsilikaten. *Fortschritte der Mineralogie*, **59**, 191–226.

Ďurovič, S. (1992). Layer stacking in general polytypic structures. *International Tables for X-ray Crystallography*, Vol. C, chapter 9.2.2. Dordrecht: Kluwer Acad. Publ., pp. 667–680.

Ďurovič, S. (1997*a*). Cronstedtite-1*M* and co-existence of 1*M* and 3*T* polytypes. *Ceramics-Silikáty*, **41**, 98–104.

Ďurovič, S. (1997*b*): Fundamentals of the OD theory. *EMU Notes in Mineralogy*, **1**, 3–28.

Ďurovič, S. and Weiss, Z. (1983). Polytypism of pyrophyllite and talc. Part I. OD interpretation and MDO polytypes. *Silikáty*, **27**, 1–18.

Ďurovič, S. and Weiss, Z. (1986). OD structures and polytypes. *Bulletin de Mineralogie*, **109**, 15–29.

Effenberger, H. and Pertlik, F. (1981). Ein Beitrag zur Kristallstruktur von α-CuSe (Klockmannit). *Neues Jahrbuch für Mineralogie, Monatshefte*, 197–205.

Egorov-Tismenko, Yu.K. (1998). On the seidozerite–nacaphite polysomatic series of minerals: Titanium silicate analogues of mica. *Crystallography Reports*, **43**, 271–77.

Egorov-Tismenko, Yu.K. and Sokolova, E.V. (1990). Homologous series seidozerite–nacaphite. *Mineralogiskii Zhurnal*, **12**, 40–49. (In Russian).

Egorov-Tismenko, Yu.K., Sokolova, E.V., and Smirnova, N.L. (1984). Kindred relationships of structural types on the example of arctite-, alunite-, and sulphohalite- like structures. *Acta Crystallographica*, **40** Supplement, C245.

Egorov-Tismenko, Yu.K., Yamnova, N.A., and Khomyakov, A.P. (1996). A new representative of a series of chain–sheet silicates with inverted tetrahedral fragments. *Crystallography Reports*, **41**, 784–8.

Elcombe, M.M., Kisi, E.H., Hawkins, K.D., White, T.J., Goodman, P., and Matheson, S. (1991). Structure determination for $Ca_3Ti_2O_7$, $Ca_4Ti_3O_{10}$, $Ca_{3.6}Sr_{0.4}Ti_3O_{10}$ and a refinement of $Sr_3Ti_2O_7$. *Acta Crystallographica*, **B47**, 305–14.

Engel, P. (1980). Die Kristallstruktur von synthetischem Parapierrotit, $TlSb_5S_8$. *Zeitschrift für Kristallographie*, **151**, 203–16.

Engel, P. and Nowacki, W. (1969). Die Kristallstruktur von Baumhauerit. *Zeitschrift für Kristallographie*, **129**, 178–202.

Engel, P. and Nowacki, W. (1970). Die Kristallstruktur von Rathit-II $[As_{25}S_{56}/Pb^{(VII)}_{6.5}Pb^{(IX)}_{12}]$. *Zeitschrift für Kristallographie*, **131**, 365–75.

Engel, P., Gostojic, M. and Nowacki, W. (1983). The crystal structure of pierrotite, $Tl_2(Sb, As)_{10}S_{16}$. *Zeitschrift für Kristallographie*, **165**, 209–15.

Era, M., Maeda, K. and Tsutsui, T. (1997). PbBr-based layered perovskite containing chromophore-linked ammonium molecule as an organic layer. *Chemistry Letters*, pp. 1235–6.

Ercit, T.S., Cooper, M.A., and Hawthorne, F.C. (1998). The crystal structure of vuonnemite, $Na_{11}Nb_2(Si_2O_7)(PO_4)_2O_3(F, OH)$, a phosphate-bearing sorosilicate of the lomonosovite group. *The Canadian Mineralogist*, **36**, 1311–20.

Euler, R. and Hellner, E. (1960). Über komplex zusammengesetzte sulfidische Erze VI. Zur Kristallstruktur des Meneghinits, $CuPb_{13}Sb_7S_{24}$. *Zeitschrift für Kristallographie*, **113**, 345–72.

Evans, H.T. Jr. and Allmann, R. (1968). The crystal structure and crystal chemistry of valleriite. *Zeitschrift für Kristallographie*, **127**, 73–93.

Evans, H.T. Jr. and Konnert, J.A. (1976). Crystal structure refinement of covellite. *The American Mineralogist*, **61**, 996–1000.

Evans, H.T. Jr. and Mrose, M.E. (1977). The crystal chemistry of the hydrous copper silicates, shattuckite and planchéite. *The American Mineralogist*, **62**, 491–502.

Fahey, J.J., Ross, M., and Axelrod, J.M. (1960). Loughlinite, a new hydrous sodium magnesium silicate. *The American Mineralogist*, **45**, 270–81.

Farber, L., Levin, I., Borisevich, A., Grey, I.E., Roths, R.S., and Davies, P.K. (2002). Structural study of $Li_{1-x}Nb_{1-x-3y}Ti_{x+4y}O_3$ solid solutions. *Journal of Solid State Chemistry*, **166**, 81–90.

Ferraris, G. (1997). Polysomatism as a tool for correlating properties and structure. *EMU Notes in Mineralogy*, **1**, 275–95.

Ferraris, G. (2002). Mineral and inorganic crystals. In Giacovazzo, G. (ed.). *Fundamentals of Crystallography*. Oxford University Press, Oxford, pp. 503–84.

Ferraris, G. and Németh, P. (2003). Pseudo-symmetry, twinning and structural disorder in layer titanosilicates. *21st European Crystallographic Meeting, Abstracts* pp. 41–2.

Ferraris, G. and Nespolo, M. (2003). Applied geminography. *21st European Crystallographic Meeting, Abstracts* p 148.

Ferraris, G., Mellini, M., and Merlino, S. (1986). Polysomatism and the classification of minerals. *Rendiconti della Società Italiana di Mineralogia e Petrografia*, **41**, 181–92.

Ferraris, G., Mellini, M., and Merlino, S. (1987). Electron-diffraction and electron-microscopy study of balangeroite and gageite: crystal structures, polytypism, and fiber texture. *The American Mineralogist*, **72**, 382–91.

Ferraris, G., Pavese, A., and Soboleva, S.V. (1995). Tungusite: new data, relationship with gyrolite and structural model. *The Mineralogical Magazine*, **59**, 535–43.

Ferraris, G., Ivaldi, G., Khomyakov, A.P., Soboleva, S.V., Belluso, E., and Pavese, A. (1996). Nafertisite, a layer titanosilicate member of a polysomatic series including mica. *European Journal of Mineralogy*, **8**, 241–9.

Ferraris, G., Khomyakov, A.P., Belluso, E., and Soboleva, S.V. (1997). Polysomatic relationships in some titanosilicates occurring in the hyperagpaitic alkaline rocks of the Kola Peninsula, Russia. *Proceedings of the 30th International Geological Congress 'Mineralogy'*, **16**, 17–27.

Ferraris, G., Khomyakov, A.P., Belluso, E., and Soboleva, S.V. (1998). Kalifersite, a new alkaline silicate from Kola Peninsula (Russia) based on a palygorskite–sepiolite polysomatic series. *European Journal of Mineralogy*, **10**, 865–74.

Ferraris, G., Prencipe, M., Sokolova, E.V., Gekimyants, V.M., and Spiridonov, E.M. (2000). Hydroxylclinohumite, a new member of the humite group: Twinning, crystal structure and crystal chemistry of the clinohumite subgroup. *Zeitschrift für Kristallographie*, **215**, 169–73.

Ferraris, G., Ivaldi, G., Pushcharovsky, D.Yu., Zubkova, N., and Pekov, I. (2001*a*). The crystal structure of delindeite, $Ba_2\{(Na, K, \square)_3(Ti, Fe)[Ti_2(O, OH)_4Si_4O_{14}](H_2O, OH)_2\}$, a member of the mero-plesiotype bafertisite series. *The Canadian Mineralogist*, **39**, 1306–16.

Ferraris, G., Belluso, E., Gula, A., Soboleva, S.V., Ageeva, O.A., and Borutskii, B.E. (2001*b*). A structural model of the layer titanosilicate bornemanite based on seidozerite and lomonosovite modules. *The Canadian Mineralogist*, **39**, 1667–75.

Ferraris, G., Gula, A., Ivaldi, G., Nespolo, M., and Raade, R. (2001*c*). Crystal structure of kristiansenite: a case of class IIB twinning by metric merohedry. *Zeitschrift für Kristallographie*, **216**, 442–8.

Ferraris, G., Gula, A., Ivaldi, G., Nespolo, M., Sokolova, E., Uvarova, Y., and Khomyakov, A.P. (2001*d*): First structure determination of an MDO-2O mica polytype associated with a 1M polytype. *European Journal of Mineralogy*, **13**, 1013–23.

Ferraris, G., Belluso, E., Gula, A., Khomyakov, A.P., and Soboleva, S.V. (2003). The crystal structure of seidite-(Ce), $Na_4(Ce,Sr)_2Ti(OH)_2(Si_8O_{18})(O,OH,F)_4 \cdot 5H_2O$, a modular microporous titanosilicate of the rhodesite group. *The Canadian Mineralogist*, **41**, 1183–92.

Fichtner, K. (1977*a*). Zur Symmetriebeshreibung von OD-Kristallstrukturen durch Brandtsche und Ehresmannsche Gruppoide. *Beiträge zur Algebra und Geometrie*, **6**, 71–99.

Fichtner, K. (1977*b*) A new deduction of a complete list of OD groupoid families for OD structures consisting of equivalent layers. *Kristall und Technik*, **12**, 1263–7.

Fichtner, K. (1979). On the description of symmetry of OD structures. I. OD groupoid family, parameters, stacking. *Kristall und Technik*, **14**, 1073–78.

Finger, L.W. and Hazen, R.M. (1991*a*). Crystal structures of $Mg_{12}Si_4O_{19}(OH)_2$ (phase B) and $Mg_{14}Si_5O_{24}$ (phase Anh B). *The American Mineralogist*, **76**, 1–7.

Finger, L.W. and Hazen, R.M. (1991*b*). Crystal chemistry of six-coordinated silicon: a key to understanding the earth's deep interior. *Acta Crystallographica*, **B47**, 561–80.

Fischer, K. (1961). Note on the false symmetry effect ('Templeton effect'). *Zeitschrift für Kristallographie*, **115**, 310–3.

Fischer, P., Roessli, B., Merot, J., Allensbach, P., Staub, U., Kaldas, E., Bucher, B., Karpinski, J., Rusiecki, S., Jilek, E., and Hewat, A.W. (1992). Neutron diffraction investigation of structures 'RE124' (RE = Dy, Ho, Er) and - 'Nd 247' superconductors, 2D antiferromagnetism in 'Dy1'. *Physica C*, **180**, 414–16.

Fischer, R. and Pertlik, F. (1975). Verfeinerung der Kristallstruktur der Schafarzikite, $FeSb_2O_4$. *Tschermaks Mineralogische Petrographische Mitteilung*, **22**, 236–41.

Flack, H.D. (1987). The derivation of twin laws for (pseudo-)merohedry by coset decomposition. *Acta Crystallographica*, **A43**, 564–8.

Flack, H.D. and Bernardelli, G. (1999). Absolute structure and absolute configuration. *Acta Crystallographica*, **A55**, 908–15.

Fleck, M., Kolitsch, U., and Hertweck, B. (2002). Natural and synthetic compounds with kröhnkite-type chains: review and classification. *Zeitschrift für Kristallographie*, **217**, 435–43.

Fleet, S.G. and Megaw, H.D. (1962). The crystal structure of yoderite. *Acta Crystallographica*, **15**, 721–8.

Flippen-Anderson, J.L., Deschamps, J.R., Gilardi, R.D., and George, C. (2001). Twins, disorders and other demons. *Crystal Engineering*, **4**, 131–9.

Foit, F.F. Jr., Phillips, M.W., and Gibbs, G.V. (1973). A refinement of the crystal structure of datolite, $CaBSiO_4(OH)$. *The American Mineralogist*, **58**, 909–14.

Foit, F.F., Robinson, P.D. and Wilson, J.R. (1995). The crystal structure of gillulyite, $Tl_2\,(As, Sb)_8S_{13}$, from the Mercur gold deposit, Tooele County, Utah, U.S.A. *The American Mineralogist*, **80**, 394–99.

Foulon, S., Ferriol, M., Brenier, A., Boulon, S., and Lecocq, S. (1996). Obtention (sic) of good quality $Ba_2NaNb_5O_{15}$ crystals: growth characterization and structure of Nd^{3+}-doped single crystal fibres. *European Journal of Solid State and Inorganic Chemistry*, **33**, 673–86.

Freed, R.L., Rouse, R.C., and Peacor, D.R. (1993). Ribbeite, a second example of edge-sharing silicate tetrahedra in the leucophoenicite group. *The American Mineralogist*, **78**, 190–4.

Friedel, G. (1923). Sur les macles du quartz. *Bulletin de la Société Française de Minéralogie*, **46**, 79–95.

Friedel, G. (1926). *Leçons de Cristallographie*. Nancy, Berger – Levrault.

Friese, K., Kienle, L., Duppel, V., Luo, H. and Lin, C. (2003). Single-crystal X-ray diffraction and electron-microscpy study of multiple-twinned $S_3(Ru_{0.336}Pt_{0.664})CuO_6$. *Acta Crystallographica*, **B59**, 182–9.

Gaines, R.V., Skinner, H.C.W., Foord, E.E., Mason, B., and Rosenzweig, A. (1997). *Dana's New Mineralogy*, 8th edition, New York: John Wiley and Sons.

Galli, E. and Alberti, A. (1971). The crystal structure of rinkite. *Acta Crystallographica*, **B27**, 1277–84.

Galli, E., Quartieri, S., Vezzalini, G., and Alberti, A. (1995). Boggsite and tschernichite-type zeolites from Mt. Adamson, Northern Victoria land (Antartica). *European Journal of Mineralogy*, **7**, 1029–32.

Galy, J. and Roth, R.S. (1973). The crystal structure of $Nb_2Zr_6O_{17}$. *Journal of Solid State Chemistry*, **7**, 277–85.

Ganne, M., Dion, M., Verbaere, A., and Tournoux, M. (1979). Sur une nouvelle famille structurale $M_2(II)M(I)Ta_5O_{15}$; structure crystalline de $CaTlTa_5O_{15}$. *Journal of Solid State Chemistry*, **29**, 9–13.

Gard, J.A. and Taylor, H.F.W. (1960). The crystal structure of foshagite. *Acta Crystallographica*, **13**, 785–93.

Gatehouse, B.M. and Grey (1982). The crystal structure of högbomite-8*H*. *The American Mineralogist*, **67**, 373–80.

Gatehouse, B.M., Grey, I.E., and Nickel, E.H. (1983). The crystal structure of nolanite, $(V, Fe, Ti, Al)_{10}O_{14}(OH)_2$ from Kalgoorlie, Western Australia. *The American Mineralogist*, **68**, 833–39.

Giacovazzo, C. Ed. (2002). *Fundamentals of Crystallography*. Oxford: Oxford University Press, pp. 825.

Gibbs, G.V. and Ribbe, P.H., (1969). The crystal structures of the humite minerals. I. Norbergite. *The American Mineralogist*, **54**, 376–90.

Gibbs, G.V., Ribbe, P.H. and Andersson, C.W. (1970). The crystal structures of the humite minerals. II. Chondrodite. *The American Mineralogist*, **55**, 1182–94.

Ginderow, D. (1978). Structures cristallines de $Pb_4In_9S_{17}$ et $Pb_3In_{6.67}S_{13}$. *Acta Crystallographica*, **B34**, 1804–11.

Giovanolli, R., Buerki, P., Giuffredi, M., and Stumm, W. (1975). Layer-structured manganese oxides hydroxides. IV. The buserite group; structure stabilization by transition elements. *Chimia*, **29**, 517–20.

Gonzales-Calbet, J.M. and Vallet-Regí, M. (1987). A new perovskite-type compound: . $Ca_4Fe_2Ti_2O_{11}$. *Journal of Solid State Chemistry*, **68**, 266–72.

Gonzales-Calbet, J.M., Sayagués, J., and Vallet-Regí, M. (1989). An electron diffraction study of new phases in the LaNiO3-x system. *Solid State Ionics*, **32/33**, 721–26.

Goodenough, J.B., Ruiz-Diaz, J.E., and Zhen, Y.S. (1990). Oxide-ion conduction in $Ba_2In_2O_5$ and $Ba_3In_2MO_8$ (M = Ce, Hf, or Zr). *Solid State Ionics*, **44**, 21–31.

Gostojič, M., Nowacki, W., and Engel, P. (1982). The crystal structure of synthetic $TlSb_3S_5$, *Zeitschrift für Kristallographie*, **159**, 217–24.

Gottardi, G. (1966). X-ray crystallography of rinkite. *The American Mineralogist*, **51**, 1529–35.

Gottardi, G. and Galli, E. (1985). *Natural Zeolites*. Berlin: Springer-Verlag, 409 p.

Graham, A.R., Thompson, R.M., and Berry, L.G. (1953). Studies of mineral sulphosalts: XVII - Cannizzarite. *The American Mineralogist*, **38**, 536–44.

Grainger, C.T. (1969). Pseudo-merohedral twinning: The treatment of overlapped data. *Acta Crystallographica*, **A25**, 427–34.

Green, B.S. and Knossow, M. (1981). Lamellar twinning explains the nearly racemic composition of chiral, single crystals of hexahelicene. *Science*, **214**, 795–7.

Grell, H. (1984). How to choose OD layers. *Acta Crystallographica*, **A40**, 95–99.

Grell, H. and Dornberger-Schiff, K. (1982). Symbols for OD groupoid families referring to OD structures (polytypes) consisting of more than one kind of layers. *Acta Crystallographica*, **A38**, 49–54.

Grenier, J.C., Darriet, J., Pouchard, M., and Hagenmuller, P. (1976). Mise en evidence d'une nouvelle famille de phases de type perovskite lacunaire ordonnée de formule $A_3M_3O_8(AMO_{2.67})$ (1976). *Materials Research Bulletin*, **11**, 1219–26.

Grenier, J.C., Darriet, J., Pouchard, M., and Hagenmuller, P. (1977). Characterisation physico-chimique du ferrite du calcium et de lanthane $Ca_2LaFe_3O_8$. *Materials Research Bulletin*, **12**, 79–86.

Grey, I.E. and Gatehouse, B.M. (1979). The crystal structure of nigerite-24*R*. *The American Mineralogist*, **64**, 1255–64.

Grice, J.D. (1991). The crystal structure of silinaite, $NaLiSi_2O_5.2H_2O$: a monophyllosilicate. *The Canadian Mineralogist*, **29**, 363–7.

Grice, J.D. and Dunn, P.J. (1991). The crystal structure of cianciulliite, $Mn(Mg, Mn)_2Zn_2(OH)_{10}\cdot 2\text{–}4H_2O$. *The American Mineralogist*, **76**, 1711–14.

Grice, J.D. and Ferraris, G. (2002). New minerals approved in 2001 by the Commission on New Minerals and Mineral Names International Mineralogical Association. *European Journal of Mineralogy*, **14**, 993–9.

Grice, J.D. and Ferraris, G. (2003). New minerals approved in 2002 by the Commission on New Minerals and Mineral Names International Mineralogical Association. *The Canadian Mineralogist*, **41**, 795–802.

Grice, J.D. and Hawthorne, F.C. (1989). Refinement of the crystal structure of leucophanite. *The Canadian Mineralogist*, **27**, 193–7.

Grice, J.D. and Hawthorne, F.C. (2002). New data on meliphanite, $Ca_4(Na, Ca)_4Be_4AlSi_7O_{24}(F, O)_4$. *The Canadian Mineralogist*, **40**, 971–80.

Grice, J.D. and Roberts, A.C. (1993). Harrisonite, a well-ordered silico-phosphate with a layered crystal-structure. *The Canadian Mineralogist*, **31**, 781–5.

Grice, J.D., Burns, P.C., and Hawthorne, F.C. (1994). Determination of the megastructures of the borate polymorphs pringleite and ruitenbergite. *The Canadian Mineralogist*, **32**, 1–14.

Griffen, D.T. and Ribbe, P.H. (1973). The crystal chemistry of staurolite. *American Journal of Science*, **273A**, 479–95.

Grin', Yu.N. (1992). The intergrowth concept as a useful tool to interpret and understand complicated intermetallic structures. In Parthé, E. (ed.). *Modern Perspectives in Inorganic Crystal Chemistry*. NATO ASI Series C – vol. **382**, 77–96. Kluwer Acad. Publ.

Grin', Yu.N., Yarmolyuk, Ya., P., and Gladyshevskii, E.I. (1982). The crystal chemistry of series of inhomogeneous linear structures. I. Symmetry and numeric symbols of the structures composed of fragments of the structure types $BaAl_4$, CaF_2, AlB_2, $AuCu_3$, Cu, alpha-Fe and alpha-Po. *Soviet Physics Crystallography*, **27**, 413–7.

Grobéty, B. (2003). Polytypes and higher-order structures of antigorites: A TEM study. *The American Mineralogist*, **88**, 27–36.

Gruber, B. (2002). Further properties of lattices. In Th. Hahn (ed.). *International Tables for Crystallography*, Vol. A, 5th ed. Dordrecht: Kluwer, 756–9.

Guan, Ya.S., Simonov, V.I., and Belov, N.V. (1963). Crystal structure of bafertisite, $BaFe_2TiO[Si_2O_7](OH)_2$. *Doklady Akademii Nauk SSSR*, **149**, 1416–9. (In Russian).

Guggenheim, S. and Eggleton, R.A. (1988). Crystal chemistry, classification and identification of modulated layer silicates. *Reviews in Mineralogy*, **19**, 675–725. Reprinted 1991.

Guinier, A., Bokij, G.B., Boll-Dornberger, K., Cowley, J.M., Ďurovič, S., Jagodzinski, M., Krishna, P. *et al.* (1984). Nomenclature of polytype structures. Report of the IUCr *Ad-hoc* Committee on the Nomenclature of Disordered, Modulated and Polytype Structures. *Acta Crystallographica*, **A40**, 399–404.

Haeni, J.H., Theis, C.D., Schlon, D.G., Tian, W., Pan, X.Q., Chang, H., Takéuchi, I. and Xiang, X.D. (2001). Epitaxial growth of the first five members of the $Sr_{n+1}Ti_nO_{3n+1}$ Ruddlesden-Popper homologous series. *Applied Physics Letters*, **78**, 3292–4.

Hahn, T. and Buerger, M.J. (1955). The detailed structure of nepheline, $KNa_3Al_4Si_4O_{16}$. *Zeitschrift für Kristallographie*, **106**, 308–88.

Hahn, Th. and Klapper, H. (2003). Twinning in crystals. In Authier, A. (Ed.). *International Tables for Crystallography, Volume D: Physical Properties of Crystals*. Kluwer Academic Publishers, Dordrecht, pp. 391–446.

Hahn, Th., Janovec, V. and Klapper, H. (1999). Bicrystals, twins and domain structures—a comparison. *Ferroelectrics*, **222**, 11–21.

Hamilton, J.D.G., Hoskins, B.F., Mumme, W.G., Borbidge, W.E., and Montague, M.A. (1989). The crystal structure and crystal chemistry of $Ca_{2.3}Mg_{0.8}Al_{1.5}Fe_{9.3}O_{20}$ (SFCA): solid solution limits and selected phase relationships of SFCA in the SiO_2-Fe2O_3-CaO(-Al_2O_3) system. *Neues Jahrbuch für Mineralogie Abhandlungen*, **161**, 1–26.

Hammond, R. and Barbier, J. (1991). Spinelloids in the nickel gallosilicate system. *Physics and Chemistry of Minerals*, **18**, 184–90.

Hanic, F., Handlovič, M., and Kaprálik, I. (1980). The structure of a quaternary phase $Ca_{20}Al_{32-2v}Mg_{2v}Si_vO_{68}$. *Acta Crystallographica*, **B36**, 2863–9.

Haradem, P.S., Chamberland, B.L., and Katz, L. (1980). The structure of the 27-layer polytype of $BaCrO_3$. *Journal of Solid State Chemistry*, **34**, 59–64.

Harneit, O. and Mueller-Buschbaum, H. (1992). Ein Beitrag Veber $KCuTa_3O_9$. *Journal of Alloys and Compounds*, **184**, 221–5.

Harre, N., Mercurio, D., Trolliard, G., and Frit, B. (1998). Crystal structure of $Ba_2La_4Ti_5O_{18}$, member $n = 6$ of the homologous series $(Ba, La)_nTi_{n-1}O_{3n}$ of cation deficient perovskite-related compounds. *European Journal of Solid State Inorganic Chemistry*, **35**, 77–90.

Harris, D.C. and Chen, T.T. (1975). Gustavite – two Canadian occurrences. *The Canadian Mineralogist*, **13**, 411–4.

Harris, D.C., Jambor, J.L., Lachance, G.R., and Thorpe, R.I. (1968). Tintinaite, the antimony analogue of kobellite. *The Canadian Mineralogist*, **9**, 371–82.

Hawthorne, F.C. (1994). Structural aspects of oxide and oxysalt crystals. *Acta Crystallographica*, **B50**, 481–510.

Hawthorne, F.C. and Sokolova, E. (2002). Simonkolleite, $Zn_5(OH)_8Cl_2(H_2O)$, decorated interrupted-sheet structure. *The Canadian Mineralogist*, **40**, 939–46.

Hawthorne, F.C., Ungaretti, L., Oberti, R., Caucia, F., and Callegari, A. (1993). The crystal chemistry of staurolite. II. Order–disorder and the monoclinic → orthorhombic phase transistion. *The Canadian Mineralogist*, **31**, 583–95.

Heesch, H. (1930). Zur Strukturtheorie der ebenen Symmetriegruppen. *Zeitschrift für Kristallographie*, **73**, 95–102.

Heinemann, S., Wirth, R., and Dresen, G. (2003). TEM study of a special grain boundary in a synthetic K-feldspar bicrystal: Manebach Twin. *Physics and Chemistry of Minerals*, **30**, 125–30.

Hejny, C., Gnos, E., Grobety, B., and Armbruster Th. (2002). Crystal chemistry of the polysome ferrohögbomite-$2N2S$, a long-known but newly defined mineral species. *European Journal of Mineralogy*, **14**, 957–67.

Henriksen, R.B., Makovicky, E., Stipp, S.L.S., Nissen, C., and Eggleston, C.M. (2002). Atomic-scale observations of franckeite surface morphology. *The American Mineralogist*, **87**, 1273–8.

Herbert, H.K. and Mumme, W.G. (1981). Unsubstituted benjaminite from the A W Mine, NSW: a discussion of metal substitutions and stability. *Neues Jahrbuch für Mineralogie, Monatshefte*, **1981**, 69–80.

Herbst-Irmer, R. and Sheldrick, G.M. (1998). Refinement of twinned structures with SHELXL97. *Acta Crystallographica*, **B54**, 443–9.

Herbst-Irmer, R. and Sheldrick, G.M. (2002). Refinement of obverse/reverse twins. *Acta Crystallographica*, **B58**, 477–81.

Hesse, K.-F. and Stuempel, G. (1986). Crystal structure of harstigite, $MnCa_6Be_4(SiO_4)_2(Si_2O_7)_2(OH)_2$. *Zeitschrift für Kristallographie*, **177**, 143–8.

Hesse, K.F., Liebau, F., and Merlino, S. (1992). Crystal structure of rhodesite, $HK_{1-x}Na_{x+2y}Ca_{2-y}\{lB, 3, 2^2_\infty\}[Si_8O_{19}]$ (6-z)H_2O, from three localities and its relation to other silicates with dreier double layers. *Zeitschrift für Kristallographie*, **199**, 25–48.

Higgins, J.B. and Ribbe, P.H. (1979). Sapphirine II: A neutron and X-ray diffraction study of Mg–Al and Al–Si ordering in monoclinic sapphirine. *Contributions to Mineralogy and Petrology*, **68**, 357–68.

Higgins, J.B., Ribbe, P.H., and Nakajima, Y. (1982). An ordering model for the commensurate antiphase structure of yoderite. *The American Mineralogist*, **67**, 76–84.

Higgins, J.B., LaPierre, R.B., Schlenker, J.L., Rohrman, A.C., Wood, J.D., Kerr, G.T., and Rohrbaugh, W.J. (1988). The framework topology of zeolite beta. *Zeolites*, **8**, 446–52.

Hinks, D.G., Dabrowski, B., Jorgensen, J.D., Mitchell, A.W., Richards, D.R., Pei, S., and Shi, D. (1988). Synthesis, structure, and superconductivity in the $Ba_{1-x}K_xBiO_{3-y}$ system. *Nature*, **333**, 836–8.

Hodeau, J.L., Tu, H.Y., Bordet, P., Fournier, T., Strobel, P., and Marezio, M. (1992). Structure and twinning of Sr_3CuPtO_6. *Acta Crystallographica*, **B48**, 1–11.

Hoffmann, C. and Armbruster, T. (1997). Clinotobermorite, $Ca_5[Si_3O_8(OH)]_2\cdot 5H_2O$–$Ca_5[Si_6O_{17}]\cdot 5H_2O$, a natural C-S-H(I) type cement mineral: determination of the substructure. *Zeitschrift für Kristallographie*, **212**, 864–73.

Hoffmann, C., Armbruster, T., and Giester, G. (1997). Acentric structure (P3) of bechererite $Zn_7Cu(OH)_{13}$ $[SiO(OH)_3SO_4]$. *The American Mineralogist*, **82**, 1014–8.

Holser, W.T. (1958*a*). Relation of symmetry to structure in twinning. *Zeitschrift für Kristallographie*, **110**, 249–65.

Holser, W.T. (1958*b*). Point groups and plane groups in a two-sided plane and their subgroups. *Zeitschrift fürKristallographie*, **110**, 266–81.

Holtstam, D. (2003). Synthesis and nonstoichiometry of magnetoplumbite. *Neues Jahrbuch für Mineralogie Monatshefte*, **2003**, 55–73.

Holtstam, D. and Norrestam, R. (1993). Lindqvistite, $Pb_2MeFe_{16}O_{27}$, a novel hexagonal ferrite mineral from Jakobsberg, Filipstad, Sweden. *The American Mineralogist*, **78**, 1304–12.

Hong, S.T. and Sleight, A.W. (1997). Crystal structure of 4H $BaRuO_3$: high-pressure phase prepared at ambient pressure. *Journal of Solid State Chemistry*, **128**, 251–5.

Horioka, K., Takahashi, K.I., Morimoto, N, Horiuchi, H., Akaogi, M., and Akimoto, S.I. (1981*a*). Structure of nickel aluminosilicate (phase IV): a high pressure phase related to spinel. *Acta Crystallographica*, **B37**, 635–8.

Horioka, K., Nishiguchi, M., Morimoto, N., Horiuchi, H., Akaogi, M., and Akimoto, S.I. (1981*b*). Structure of nickel aluminosilicate (phase IV): a high-pressure phase related to spinel. *Acta Crystallographica*, **B37**, 638–41.

Horiuchi, H. and Wuensch, B.J. (1976). The ordering scheme for metal atoms in the crystal structure of hammarite, $Cu_2Pb_2Bi_4S_9$. *The Canadian Mineralogist*, **14**, 536–9.

Horiuchi, H. and Wuensch, B.J. (1977). Lindströmite, $Cu_3Pb_3Bi_7S_{15}$: Its space group and ordering scheme for metal atoms in the crystal structure. *The Canadian Mineralogist*, **15**, 527–35.

Horiuchi, H., Horioka, K., and Morimoto, N. (1980). Spinelloid: a systematics of spinel-related structures obtained under high conditions. *Journal of the Mineralogical Society of Japan*, **14**, special issue 2, 253–64 (in Japanese).

Hubbard, B., Kuang, W., Moser, A., Facey, G.A., and Detellier, C. (2003). Structural study of Maya Blue, thermal and solid-state multinuclear magnetic resonance characterization of the palygorskite-indigo and sepiolite-indigo adducts. *Clays and Clay Minerals*, **51**, 318–26.

Huminicki, D.M.C. and Hawthorne, F.C. (2002). Refinement of the crystal structure of aminoffite. *The Canadian Mineralogist*, **40**, 915–22.

Hurst, V.J., Donnay, J.D.H., and Donnay, G. (1956). Staurolite twinning. *The Mineralogical Magazine*, **31**, 145–63.

Huvé, M., Michel, C., Maignan, A., Hervieu, M., Martin, C., and Raveau, B. (1993). A 70-K superconductor – the oxycarbonate $Tl_{0.5}Pb_{0.5}Sr_4Cu_2(CO_3)O_7$. *Physica C*, **205**, 219–24.

Hyde, B.G. and O'Keefe, M.O. (1973). Relations between the $DO_9(ReO_3)$ structure type and some 'bronze' and 'tunnel' structures. *Acta Crystallographica*, **A29**, 243–8.

Hyde, B.G., Andersson, S., Bakker, M., Plug, C.M., and O'Keeffe, M. (1979). The (twin) composition plane as an extended defect and structure-building entity in crystals. *Progress in Solid State Chemistry*, **12**, 273–327.

Hyde, B.G., Bagshaw, A.N., Andersson, S., and O'Keeffe, M.O. (1974). Some defect structures in crystalline solids. *Annual Review of Material Science*, **4**, 43–92.

Hyde, B.G., White, T.J., O'Keeffe, M., and Johnson, A.W.S. (1982). Structures related to those of spinel and the β-phase, and a possible mechanism for the transformation olivine–spinel. *Zeitschrift für Kristallographie*, **160**, 53–62.

Iglesias, J.E., Zuñiga, F.I., and Nowacki, W. (1977). $NaAsS_2$, a synthetic sulfosalt related to the NaCl type. *Zeitschrift für Kristallographie*, **146**, 43–52.

Iitaka, Y. and Nowacki, W. (1961). Refinement of the pseudo crystal structure of scleroclase, $PbAs_2S_4$. *Acta Crystallographica*, **14**, 1291–2.

Iitaka, Y. and Nowacki, W. (1962). A redetermination of the crystal structure of galenobismutite, $PbBi_2S_4$. *Acta Crystallographica*, **15**, 691–8.

Ijima, S. (1975). High-resolution electron microscopy of crystallographic shear structures in tungsten oxides. *Journal of Solid State Chemistry*, **14**, 52–65

Ilinca, G. and Makovicky, E. (1999). X-ray powder diffraction properties of pavonite homoloues. *European Journal of Mineralogy*, **11**, 691–708.

International Tables for X-ray Crystallography, Vol. I (1952). Birmingham: Kynoch Press.

Iordanidis, L., Schindler, J.L., Kannewurf, C.R., and Kanatzidis, M.G. (1999). $ALn_{1\pm x}Bi_{4\pm x}S_8$ (A = K, Rb; *Ln* = La, Ce, Pr, Nd). New semiconducting quaternary bismuth sulfides. *Journal of Solid State Chemistry*, **143**, 151–62.

Ishizawa, N., Marumo, F., Iwai, S., Kimura, M., and Kawamura (1980). Compounds with perovskite-type slabs. III. The crystal structure of a monoclinic modification of $Ca_2Nb_2O_7$. *Acta Crystallographica*, **B36**, 763–6.

Ishizawa, N., Suda, K., Etschmann, B.E., Oya, T., and Kodama, N. (2001). Monoclinic superstructure of $SrMgF_4$ with perovskite-type slabs. *Acta Crystallographica*, **C57**, 784–6.

Ito, T.I. (1950). *X-ray studies on polymorphism*. Maruzen Co., Tokyo. pp. 231.

Ito, J. (1967). A study of chevkinite and perrierite. *The American Mineralogist*, **52**, 1094–104.

Ivanova, T.I. and Frank-Kamenetskaya, O.V. (2001). Using the statistical probability-based model of an irregular mixed-layer structure for describing the real structure of chemically non-homogeneous crystals. *Journal of Structural Chemistry*, **42**, 126–43.

Iwasaki, H. (1972). On the diffraction enhancement of symmetry. *Acta Crystallographica*, **A28**, 253–60.

Iyi, N. and Göbbels, M. (1996). Crystal structure of the new magnetoplumbite-related compound in the system SrO–Al_2O_3–MgO. *Journal of Solid State Chemistry*, **122**, 46–52.

Iyi, N., Göbbels, M., and Matsui, Y. (1995). The Al-rich part of the system CaO–Al_2O_3–MgO. Part II-Structure refinement of two new magnetoplumbite-related phases. *Journal of Solid State Chemistry*, **120**, 364–71.

Iyi, N., Göbbels, M., and Kimura, S. (1998). The aluminum-rich part of the system BaO–Al_2O_3–MgO – II: Crystal structure of the β-alumina-related compound, $Ba_2Mg_6Al_{28}O_{50}$. *Journal of Solid State Chemistry*, **136**, 258–62.

Izokh, P.E., Joswig, W., Fursenko, D.A., Leont'eva, A.V., Vosel, S.V., Thomas, V.G., and Fursenko, B.A. (1998). Hydrothermal synthesis, crystal structure, and properties of $CuAl_2(Si_2O_7)(F, OH)_2$. *Journal of Solid State Chemistry*, **141**, 527–36.

Jarchow, O., Schröder, F., and Schulz, H. (1968). Kristallstruktur und Polytypie von WO_2Cl_2. *Zeitschrift für anorganische und allgemeine Chemie*, **363**, 58–72.

Jefferson, D.A. and Bown, M.G. (1973). Polytypism and stacking disorder in wollastonite. *Nature Physical Science*, **245**, 43–44.

Jeffery, J.W. (1953). Unusual X-ray diffraction effects from a crystal of wollastonite. *Acta Crystallographica*, **6**, 821–25.

Jobic, S., Le Boterf, P., Brec, R., and Ouvrard, G. (1994). Structural determination and magnetic properties of a new mixed valence tin chromium selenide: $Cr_2Sn_3Se_7$. *Journal of Alloys and Compounds*, **205**, 139–45.

Jørgensen, J.D., Veal, B.W., Paulikas, A.P., Nowicki, L.J., Crabtree, G.W., Claus, H., and Kwok, W.K. (1990). Structural properties of oxygen-deficient $YBa_2Cu_3O_{7-\delta}$. *Physics Review*, **B41**, 1863–77.

Jumas, J.C., Olivier-Fourcade, J., Philippot, E., and Maurin, M. (1980). Sur le système $SnS–Sb_2S_3$: Étude structurale de $Sn_4Sb_6S_{13}$. *Acta Crystallographica*, **B36**, 2940–5.

Jung, W. and Juza, R. (1973). Darstellung und Kristallstruktur des Zirkonwitridfluorids. *Zeitschrift f. anorganische und allgemeine Chemie*, **399**, 129–47.

Jupe, A.C., Cockroft, J.K., Barnes, P., Colston, S.L., and Hall, C. (2001). The site occupancy of Mg in the brownmillerite structure and its effect on hydration properties: an X-ray/neutron diffraction and EXAFS study. *Journal of Applied Crystallography*, **34**, 55–61.

Kahlenberg, V. (1999). Application and comparison of different tests on twinning by merohedry. *Acta Crystallographica*, **B55**, 745–51.

Kahlenberg, V. and Böhm, H. (1998). Crystal structure of hexagonal trinepheline—a new synthetic $NaAlSiO_4$ modification. *The American Mineralogist*, **83**, 631–7.

Kaiman, S., Harris, D.C., and Dutrizac, J.E. (1980). Stibivanite, a new mineral from the Lake George antimony deposit, New Brunswick. *The Canadian Mineralogist*, **18**, 329–32.

Kampf, A.R., Merlino, S., and Pasero, M. (2003). Order-disoder approach to calcioar-avaipaite, $[PbCa_2Al(F, OH)_9]$: The crystal structure of the triclinic MDO polytype. *The American Mineralogist*, **88**, 430–5.

Kanatzidis, M.G., McCarthy, T.J., Tanzer, T.A., Chen, L.-H., Iordanidis, L., Hogan, T., Kannewurf, C.R., Uher, C., and Chen, B. (1996). Synthesis and thermoelectric properties of the new ternary bismuth sulfides $KBi_{6.33}S_{10}$ and $K_2Bi_8S_{13}$. *Chemistry of Materials*, **8**, 1465–74.

Karup-Møller, S. and Makovicky, E. (1979). On pavonite, cupropavonite, benjaminite and 'oversubstituted' gustavite. *Bulletin de Minéralogie*, **102**, 351–67.

Karup-Møller, S. and Makovicky, E. (1992). Mummeite—a new member of the pavonite homologous series from Alaska Mine, Colorado. *Neues Jahrbuch für Mineralogie, Monatshefte*, **1992**, 555–76.

Kasper, J.S., Lucht, C.M., and Harker, D. (1950). The crystal structure of decaborane, $B_{10}H_{14}$. *Acta Crystallographica*, **3**, 436–55.

Kato, K., Morimoto, N., and Narita, H. (1976). Crystallographic relationships of the pyroxenes and pyroxenoids. *Journal of the Japan Association of Mineralogists, Petrologists and Economic Geologists*, **71**, 248–54.

Kato, T. and Takéuchi, Y. (1983). The pyrosmalite group of minerals. I. Structure refinement of manganpyrosmalite. *The Canadian Mineralogist*, **21**, 1–6.

Kato, T., Ito, Y., and Hashimoto, N. (1989). The crystal structures of sonolite and jerrygibbsite. *Neues Jahrbuch für Mineralogie, Monatshefte*, **1989**, 410–30.

Kawada, I. and Hellner, E. (1971). Dei Kristallstruktur der Pseudozell (subcell) von Andorit VI (Ramdohrit). *Neues Jahrbuch für Mineralogie, Monatshefte*, **1971**, 551–60.

Kay, H. and Bailey, P.C. (1957). Structure and properties of $CaTiO_3$. *Acta Crystallographica*, **10**, 219–26.

Keller, L.P. and Buseck, P.R. (1994). Twinning in meteoritic and synthetic perovskite. *The American Mineralogist*, **79**, 73–79.

Keller, E., Ketterer, J., and Krämer, V. (2001). Crystal structure and twinning of $Bi_4O_5Br_2$. *Zeitschrift für Kristallographie*, **216**, 595–99.

Khomyakov, A.P. (1995). *Mineralogy of hyperagpaitic alkaline rocks*. Oxford: Clarendon Press. p. 223.

Khomyakov, A.P., Nechelyustov, G.N., Sokolova, E.V., and Dorokhova, G.I. (1992). Quadruphite $Na_{14}CaMgTi_4[Si_2O_7]_2[PO_4]_2O_4F_2$ and polyphite $Na_{17}Ca_3Mg(Ti, Mn)_4[Si_2O_7]_2[PO_4]_6O_2F_6$, two new minerals of the lomonosovite group. *Zapiski Vserossiyskogo Mineralogicheskogo Obschestva*, **121**(3), 105–12. (In Russian).

Khomyakov, A.P., Nechelyustov, G.N., Sokolova, E., Bonaccorsi, E., Merlino, S., and Pasero, M. (2002). Megakalsilite, a new polymorph of $KAlSiO_4$ from the Khikina alkaline massif, Kola Peninsula, Russia: mineral description and crystal structure. *The Canadian Mineralogist*, **10**, 961–70.

Khomyakov, A.P., Kulikova, I.A., Sokolova, E., Hawthorne, F.C., and Kartashov, P.M. (2003). Paravinogradovite, $\sim Na_2[Ti_3^{4+}Fe^{3+}\{Si_2O_6\}_2\{Si_3AlO_{10}\}(OH)_4 \cdot H_2O$, a new mineral from Khibina alkaline massif, Kola Peninsula: Description and crystal structure. *The Canadian Mineralogist*, **41**, 989–1002.

Kihlborg, L. (1963). Least squares refinement of the structure of $Mo_{17}O_{47}$. *Acta Chemica Scandinavica*, **17**, 1485–7.

Kihlborg, L. and Hussain, A. (1979). Alkali metal location and tungsten off-centre displacement in hexagonal potassium and cesium tungsten bronzes. *Materials Research Bulletin*, **14**, 667–74.

Kihlborg, L. and Klug, A. (1973). The alkali metal distribution in the tetragonal potassium tungsten bronze structure. *Chemica Scripta*, **3**, 207–11.

Kimata, M. and Ii, N. (1981). The crystal structure of synthetic åkermanite, $Ca_2MgSi_2O_7$. *Neues Jahrbuch für Mineralogie, Monatshefte*, **1981**, 1–10.

Koch, E. (1999). Twinning. In A.J.C. Wilson and E. Prince (eds). *International Tables for Crystallography*, Vol. C, 2nd ed. Dordrecht: Kluwer. 10–14.

Kocman, V. and Nuffield, E.W. (1973). The crystal structure of wittichenite, Cu_3BiS_3. *Acta Crystallographica*, **B29**, 2528–35.

Kodĕra, M., Kupčík, V., and Makovicky, E. (1970). Hodrushite – a new sulphosalt. *The Mineralogical Magazine*, **37**, 641–8.

Kohatsu, I. and Wuensch, B.J. (1971). The crystal structure of aikinite, $PbCuBiS_3$. *Acta Crystallographica*, **B27**, 1245–52.

Kohatsu, J.J. and Wuensch, B.J. (1974). Prediction of structures in the homologous series $Pb_{3+2n}Sb_8S_{15+2n}$ (the plagionite group). *Acta Crystallographica*, **B30**, 2935–7.

Kohatsu, I. and Wuensch, B.J. (1976). The crystal structure of gladite, $PbCuBi_5S_9$, a superstructure intermediate in the series Bi_2S_3–$PbCuBiS_3$ (bismuthinite–aikinite). *Acta Crystallographica*, **B32**, 2401–9.

Kolesnikov, N.N., Korotkov, V.E., Kulakov, M.P., Lagvenov, G.A., Molchanov, V.N., Muradyan, L.A., Simonov, V.I., Tamazan, R.A., Shibaeva, R.P., and Shchegolev, I.F. (1989). Structure of superconducting single crystals of $TlBa_2(Ca_{0.8}Tl_{0.13})Cu_2O_7$, Tc = 80 K. *Physica C*, **162**, 1663–4.

Kostov, V.V. and Macíček, J. (1995). Crystal structure of synthetic $Pb_{12.65}Sb_{11.35}S_{28.35}Cl_{2.65}$ – A new view of the crystal chemistry of chlorine-bearing lead-antimony sulphosalts. *European Journal of Mineralogy*, **7**, 1007–18.

Krämer, V. (1980). Structure of bismuth indium sulphide $Bi_3In_5S_{12}$. *Acta Crystallographica*, **B36**, 1922–3.

Krämer, V. (1983). Lead indium bismuth chalcogenides III. Structure of $Pb_4In_2Bi_4S_{13}$. *Acta Crystallographica*, **C42**, 1089–91.

Krämer, V. and Reis, I. (1986). Lead indium bismuth chalcogenides. II. Structure of $Pb_4In_3Bi_7S_{18}$. *Acta Crystallographica*, **C42**, 249–51.

Krause, W., Bernhardt, M.-J., Gebert, W., Graetsch, H., Belendorff, K. and Petittjean, K. (1996). Medenbachite, $Bi_2Fe(Cu, Fe)(O, OH)_2(OH)_2(AsO_4)_2$, a new mineral species: its description and crystal structure. *The American Mineralogist*, **81**, 505–12.

Krebs, B., Brendel, C. and Schäfer, H. (1988). Neue Untersuchungen an α-Platindichlorid Darstellung und Struktur. *Zeitschrift für anorganische und allgemeine Chemie*, **561**, 119–31.

Krivovichev, S.V., Armbruster, T., Yakovenchuk, V.N., Pakhomovsky, Y.A., and Men'shikov, Y.P. (2003). Crystal structures of lamprophyllite-2*M* and lamprophyllite-2*O* from the Lovozero alkaline massif, Kola peninsula, Russia. *European Journal of Mineralogy*, **15**, 711–18.

Krypiakevich, P.I. and Gladyshevskii, E.I. (1972). Homologous series including the new structure types of ternary silicides. *Acta Crystallographica*, **A28** Suppl., S97.

Kupčík, V. (1967). Die Kristallstruktur des Kermesits, Sb_2S_2O. *Naturwissenschaften*, **54**, 114–15.

Kupčík, V. (1984). Die Kristallstruktur des Minerals Eclarit (Cu,Fe) $Pb_9Bi_{12}S_{28}$. *Tschermaks Mineralogische Petrographische Mitteilungen*, **32**, 259–69.

Kupčík, V. and Makovicky, E. (1968). Die Kristallstruktur des Minerals (Pb, Ag, Bi) $Cu_3Bi_5S_{11}$. *Neues Jahrbuch für Mineralogie, Monatshefte*, **1968**, 236–7.

Kupčík, V. and Steins, M. (1991). Verfeinerung der Kristallstruktur von Gustavit $Pb_{1.5}Ag_{0.9}Bi_{2.5}Sb_{0.1}S_6$. *Berichte Deutschland Mineralogische Gesellschaft*, **1990/2**, 151.

Kuypers, S., van Landuyt, J. and Amelinckx, S. (1990). Inconmmensurate misfit layer compounds of the type MTS_3 (M = Sn, Pb, Bi, Rare-earth elements, T = Nb, Ta)—a study by means of electron microscopy. *Journal of Solid State Chemistry*, **86**, 212–32.

Lachowski, E.E., Murray, L.W., and Taylor, H.F.W. (1979). Truscottite: composition and ionic substitutions. *The Mineralogical Magazine*, **43**, 333–6.

Lamire, M., Labbe, P., Goreaud, M., and Raveau, B. (1987). Refinement et nouvelle analyse de la structure de $W_{18}O_{49}$. *Revue de Chémie Minérale*, **24**, 369–81.

Landa-Canovas, A.R. and Otero-Diaz, L.C. (1992). A transmission electron microscopy study of the MnS–Er_2S_3 system. *Australian Journal of Chemisty*, **45**, 1473–87.

Le Bihan, M.Th. (1962). Étude structurale de quelques sulfures de plomb et d'arsenic naturels du gisement de Binn. *Bulletin de la Société Française de Minéralogie et Cristallographie*, **85**, 15–47.

Le Page, Y. (2002). Mallard's law recast as a Diophantine system: fast and complete enumeration of possible twin laws by [reticular] [pseudo] merohedry. *Journal of Applied Crystallography*, **35**, 175–181.

Le Page, Y., Donnay, J.D.H., and Donnay, G. (1984). Printing sets of structure factors for coping with orientation ambiguities and possible twinning by merohedry. *Acta Crystallographica*, **A40**, 679–84.

Lemoine, P., Carré, D., and Guittard, M. (1981). Structure de sulfure d'europium et d'antimoine, $Eu_3Sb_4S_9$. *Acta Crystallographica*, **B37**, 1281–4.

Lemoine, P., Carré, D., and Guittard, M. (1986*a*). Structure de sulfure d'europium et de cuivre Eu_2CuS_3. *Acta Crystallographica*, **C42**, 390–1.

Lemoine, P., Carré, D., and Guittard, M. (1986*b*). Structure du sulfure d'europium et de bismuth $Eu_{1.1}Bi_2S_4$. *Acta Crystallographica*, **C42**, 259–61.

Léone, P., Le Leuch, L.-M., Palvadeau, P., Molinié, P., and Moëlo, Y. (2003). Single crystal structures and magnetic properties of two iron- or manganese–lead–antimony sulfides: $MPb_4Sb_6S_{14}$(M:Fe,Mn). *Solid State Sciences*, **5**, 771–6.

Leonyuk, L., Maltsev, V., Babonas, G.-J., Reza, A., and Szymczak (1998). Polysomatic series and superconductivity in complex cuprates with ladder-type structure. *International Journal of Modern Phusics B*, **12**, 3110–12.

Leonyuk, L., Babonas, G.-J., Maltsev, V., and Rybakov, V. (1999*a*). Polysomatic series in the structures of complex cuprates. *Acta Crystallographica*, **A55**, 628–34.

Leonyuk, L., Babonas, G.-J., Maltsev, V., Vetkin, A., Rybakov, V., and Reza, A. (1999*b*). Structural features and anomalies in the temperature dependence of resistance in superconducting Bi-2212 single crystals. *Journal of Crystal Growth*, **198--199**, 619–25.

Lewis, J. jr. and Kupčík, V. (1974). The crystal structure of $Bi_2Cu_3S_4Cl$. *Acta Crystallographica*, **B30**, 848–52.

Libowitzky, E. and Armbruster, T. (1996). Lawsonite-type phase transitions in hennomartinite, $SrMn_2[Si_2O_7](OH)_2 \cdot H_2O$. *The American Mineralogist*, **81**, 9–18.

Lichtenberg, F., Herrnberger, A., Wiedenmann, K., and Mannhart, J. (2001). Synthesis of perovskite-related layered $A_nB_nO_{3n+2}{=}ABO_x$ type niobates and titanates and study of their structural, electric and magnetic properties. *Progress in Solid State Chemistry*, **29**, 1–70.

Liebau, F. (1985). *Structural Chemistry of Silicates*. Berlin: Springer-Verlag. 347 pp.

Likforman, A.,Guittard, M.,and Jaulmes, S.(1987). Structure du tétratriacontasulfure d'octadécaindium et d'heptaétain. *Acta Crystallographica*, **C43**, 177–9.

Lima-de-Faria, J., Hellner, E., Liebau, F., Makovicky, E., and Parthé, E. (1990). Nomenclature of inorganic structure types. Report of the IUCr Commission on Crystallographic Nomenclature, Subcommittee on the Nomenclature of Inorganic Structure Types. *Acta Crystallographica*, **A46**, 1–11.

Lindemann, W., Wögerbauer, R., and Berger, P. (1979). Die Kristallstruktur von Karpholith $(Mn_{0.97}Mg_{0.08}Fe^{2+}_{0.07})(Al_{1.90}Fe^{3+}_{0.01})SiO_6(OH)_4$. *Neues Jahrbuch für Mineralogie, Monatshefte*, **1979**, 282–7.

Liu, G. and Greedan, J.E. (1994). The synthesis, structure and characterization of a novel 24-layer oxide—$Ba_8V_7O_{22}$ with V(III), V(IV), and V(V). *Journal of Solid State Chemistry*, **108**, 371–80.

Lopes-Vieira, A. and Zussman, J. (1969). Further details on the crystal structure of zussmanite. *The Mineralogical Magazine*, **37**, 49–60.

Louisnathan, S.J. (1970). The crystal structure of synthetic soda melilite, $CaNaAlSi_2O_7$. *Zeitschrift für Kristallographie*, **131**, 314–21.

Lüke, H. and Eick, H. (1982). Crystal structure of Yb_5ErCl_{13} and Yb_6Cl_{13}. *Inorganic Chemistry*, **21**, 965–8.

Ma, C.B. (1974). New orthorhombic phases on the join $NiAl_2O_4$–Ni_2SiO_4: stability and implications to mantle mineralogy. *Contributions to Mineralogy and Petrology*, **45**, 257–79.

Ma, C.B. and Sahl, K. (1975). Nickel aluminosilicate, phase III. *Acta Crystallographica*, **B31**, 2142–3.

Ma, C.B. and Tillmanns, E. (1975). Nickel aluminosilicate, phase II. *Acta Crystallographica*, **B31**, 2139–41.

Ma, C.B., Sahl, K., and Tillmanns, K. (1975). Nickel aluminosilicate, phase I. *Acta Crystallographica*, **B31**, 2137–9.

Magnéli, A. (1953). Structures of the ReO_3 type with recurrent dislocations of atoms: 'Homologous series' of molybdenum and tungsten oxides. *Acta Crystallographica*, **6**, 495–500.

Makarova, I.P., Verin, I.A., and Aleksandrov, K.S. (1993). Structure and twinning of $RbLiCrO_4$ crystals. *Acta Crystallographica*, **B49**, 19–28.

Makovicky, E. (1976). Crystallography of cylindrite. I. Crystal lattices of cylindrite and incaite. *Neues Jahrbuch für Mineralogie, Abhandlungen*, **126**, 304–26.

Makovicky, E. (1981). The building principles and classification of bismuth–lead sulphosalts and related compounds. *Fortschritte der Mineralogie*, **59**, 137–90.

Makovicky, E. (1985*a*). Cyclically twinned sulphosalt structures and their approximate analogues. *Zeitschrift für Kristallographie*, **173**, 1–23.

Makovicky, E. (1985*b*). The building principles and classification of sulphosalts based on the SnS archetype. *Fortschritte der Mineralogie*, **63**, 45–89.

Makovicky, E. (1988). Classification of homologous series. *Zeitschrift für Kristallographie*, **185**, 512.

Makovicky, E. (1989). Modular classification of sulphosalts—current status. Definition and application of homologous series. *Neues Jahrbuch für Mineralogie, Abhandlungen*, **160**, 269–97.

Makovicky, E. (1992*a*). Crystal chemistry of complex sulfides and its applications. In E. Parthé (ed.). *Modern Perspectives in Inorganic Crystal Chemistry*. Kluwer Acad. Publ. 131–61.

Makovicky, E. (1992*b*). Crystal structures of complex lanthanide sulfides with built-in non-commensurability. *Australian Journal of Chemistry*, **45**, 1451–72.

Makovicky, E. (1993). Rod-based sulphosalt structures. *European Journal of Mineralogy*, **5**, 545–91.

Makovicky, E. (1994). Polymorphism in Cu_3SbS_3 and Cu_3BiS_3: The ordering schemes for copper atoms and electron microscope observations. *Neues Jahrbuch für Mineralogie, Abhandlungen*, **168**, 185–212.

Makovicky, E. (1995). Structural parallels between the high-pressure B phases and the leucophoenicite series. *The American Mineralogist*, **80**, 676–9.

Makovicky, E. (1997*a*). Modularity—different approaches. *EMU Notes in Mineralogy*, **1**, 315–43.

Makovicky, E. (1997*b*). Modular crystal chemistry of sulphosalts and other complex sulphides. *EMU Notes on Mineralogy*, **1**, 237–71.

Makovicky, E. (1997*c*). Prediction of modular crystal structures. In H. Burzlaff (ed.). *Proceeding Symposium on Predictability of Crystal Structures of Inorganic Solids, Hünfeld 27.10-30.10.1997*. Erlangen: Universität Erlangen-Nürnberg, pp. 107–24.

Makovicky, E. and Balič Žunič, T. (1993). Contributions to the crystal chemistry of thallium sulphosalts. II. $TlSb_3S_5$—the missing link of the lillianite homologous series. *Neues Jahrbuch für Mineralogie, Abhandlungen*, **165**, 331–44.

Makovicky, E. and Balič Žunič, T. (1995). The crystal structure of skinnerite, $P2_1/c$-Cu_3SbS_3, from powder data. *The Canadian Mineralogist*, **33**, 655–63.

Makovicky, E. and Balič Žunič, T. (1998). New measure of distortion for coordination polyhedra. *Acta Crystallographica*, **B54**, 766–73.

Makovicky, E. and Balić-Žunić, T. (1999). Gillulyite $Tl_2(As, Sb)_8S_{13}$: reinterpretation of the crystal structure and order–disorder phenomena. *The American Mineralogist*, **84**, 400–6.

Makovicky, E. and Hyde, B.G. (1981). Non-commensurate (misfit) layer structures. *Structure and Bonding*, **46**, 101–70.

Makovicky, E. and Hyde, B.G. (1992). Incommensurate, two-layer structures with complex crystal chemistry: minerals and related synthetics. In '*Incommensurate misfit sandwiched layered compounds*', A. Meerschaut (ed.). *Materials Science Forum*, **100–101**, 1–100. Trans. Tech Publ. Ltd.

Makovicky, E. and Karup-Møller, S. (1977*a*). Chemistry and crystallography of the lillianite homologous series. I. General properties and definitions. *Neues Jahrbuch für Mineralogie, Abhandlungen*, **130**, 264–87.

Makovicky, E. and Karup-Møller, S. (1977*b*). Chemistry and crystallography of the lillianite homologous series. Part II: Definition of new minerals: eskimoite, vikingite, ourayite and treasurite. Redefintion of schirmerite and new data on the lillianite–gustavite solid solution series. *Neues Jahrbuch für Mineralogie, Abhandlungen*, **131**, 56–82.

Makovicky, E. and Karup-Møller, S. (1984). Ourayite from Ivigtut, Greenland. *The Canadian Mineralogist*, **22**, 565–75.

Makovicky, E. and Karup-Møller, S. (1986). New data on giessenite from the Bjøkåsen sulfide deposit at Otoften, northern Norway. *The Canadian Mineralogist*, **22**, 21–25.

Makovicky, E. and Mumme, W.G. (1979). The crystal structure of benjaminite $Cu_{0.50}Pb_{0.40}Ag_{2.30}Bi_{6.80}S_{12}$. *The Canadian Mineralogist*, **17**, 607–18.

Makovicky, E. and Mumme, W.G. (1983). The crystal structure of ramdohrite, $Pb_6Sb_{11}Ag_3S_{24}$ and its implications for the andorite group and zinckenite. *Neues Jahrbuch für Mineralogie, Abhandlungen*, **147**, 58–79.

Makovicky, E. and Mumme, W.G. (1986). The crystal structure of isoklakeite, $Pb_{51.3}Sb_{20.4}Bi_{19.5}Ag_{1.2}Cu_{2.9}Fe_{0.7}S_{11.4}$. The kobellite homologous series and its derivatives. *Neues Jahrbuch für Mineralogie, Abhandlungen*, **153**, 121–48.

Makovicky, E. and Norrestam, R. (1985). The crystal structure of jaskolskiite, Cu_xPb_{2+x} $(Sb, Bi)_{2-x}S_5$ ($x \approx 0.2$), a member of the meneghinite homologous series. *Zeitschrift für Kristallographie*, **171**, 179–94.

Makovicky, E. and Skinner, B.J. (1975). Studies of the sulfosalts of copper. IV. Structure and twinning of sinnerite, $Cu_6As_4S_9$. *The American Mineralogist*, **60**, 998–1012.

Makovicky, E., Mumme, W.G., and Watts, J.A. (1977). The crystal structure of synthetic pavonite, $AgBi_3S_5$ and the definition of the pavonite homologous series. *The Canadian Mineralogist*, **15**, 339–48.

Makovicky, E., Johan, Z., and Karup-Møller, S. (1980). New data on bukovite, thalcusite, chalcothallite and rohaite. *Neues Jahrbuch für Mineralogie, Abhandlungen*, **138**, 122–46.

Makovicky, E., Mumme, W.G., and Hoskins, B.F. (1991). The crystal structure of Ag–Bi bearing heyrovskyite. *The Canadian Mineralogist*, **29**, 553–8.

Makovicky, E., Mumme, W.G., and Madsen, I.C. (1992). The crystal structure of vikingite. *Neues Jahrbuch für Mineralogie, Monatshefte*, **1992**, 454–68.

Makovicky, E., Leonardsen, E., and Moëlo, Y. (1994). The crystallography of lengenbachite, a mineral with the non-commensurate layer structure. *Neues Jahrbuch für Mineralogie, Abhandlungen*, **166**, 169–91.

Makovicky, E., Balić-Žunić, T., and Olsen, P.N. (1998). OD phenomena and polytypy in sulfosalts. 17th *General Meeting of the International Mineralogical Association*, Abstracts, A54.

Makovicky, E., Topa, D. and Balič Žunič, T. (2000): New concepts of modular crystallography derived from the latest sulfosalts. *Abstracts of the 31st International Geological Congress*, August 6–17, 2000, Rio de Janeiro

Makovicky, E., Balič Žunič, T., and Topa, D. (2001*a*). The crystal structure of neyite, $Ag_2Cu_6Pb_{25}Bi_{26}S_{68}$. *The Canadian Mineralogist*, **39**, 1365–76.

Makovicky, E., Topa, D., and Balič Žunič, T. (2001*b*). The crystal structure of paarite, the newly discovered 56 Å derivative of the bismuthinite–aikinite solid–solution series. *The Canadian Mineralogist*, **39**, 1377–82.

Makovicky, E., Søtofte, I., and Karup-Møller, S. (2002). The crystal structure of $Cu_4Bi_4Se_9$. *Zeitschrift für Kristallographie*, **217**, 597–604.

Makovicky, E., Balič Žunič, T., Karanovič, Lj., Poleti, D. and Pršek, J. (2003). Structure refinement of natural robinsonite, $Pb_4Sb_6S_{13}$: Cation distribution and modular description. *Neues Jahrbuch für Mineralogie*, in press.

Mallard, E. (1885). Sur la théorie des macles. *Bulletin de la Société Française de Minéralogie*, **8**, 457–69.

Mariolacos, K. (1976). The crystal structure of $Bi(Bi_2S_3)_9Br_3$. *Acta Crystallographica*, **B32**, 1947–9.

Mariolacos, K., Kupčík, V., Ohmasa, M., and Miehe, G. (1975). The crystal structure of $Cu_4Bi_5S_{10}$ and its relation to the structures of hodrushite and cuprobismutite. *Acta Crystallographica*, **B31**, 703–8.

Martin-Vivaldi, J.L. and Linares-Gonzales, J. (1962). A random intergrowth of sepiolite and attapulgite. *Clay and Clays Minerals*, **9**, 592–602.

Marumo, F. and Nowacki, W. (1965). The crystal structure of rathite-I. *Zeitschrift für Kristallographie*, **122**, 433–56.

Marumo, F. and Nowacki, W. (1967). The crystal structure of dufrenoysite, $Pb_{16}As_{16}S_{40}$. *Zeitschrift für Kristallographie*, **124**, 409–19.

Marumo, F. and Saito, Y. (1972). A study on the diffraction enhancement of symmetry. *Acta Crystallographica*, **B28**, 867–70.

Massa, W., Yakubovich, O.V., Kireev, V.V., and Mel'nikov, O.K. (2000). Crystal structure of a new vanadate variety in the lomonosovite group: $Na_5Ti_2O_2[Si_2O_7](VO_4)$. *Solid State Sciences*, **2**, 615–23.

Matheis, D.P. and Snyder, R.L. (1990). The crystal structures and powder diffraction patterns of the bismuth and thallium Ruddlesden–Popper copper oxide superconductors. *Powder Diffraction*, **5**, 8–24.

Matsushita, Y. and Takéuchi, Y. (1994). Refinement of the crystal structure of hutchinsonite, $TlPbAs_5S_9$. *Zeitschrift für Kristallographie*, **209**, 475–8.

Matzat, E. (1972). Die Kristallstruktur des Wittichenits, Cu_3BiS_3. *Tschermaks Mineralogisch-Petrographische Mitteilungen*, **18**, 312–6.

Matzat, E. (1979). Cannizzarite. *Acta Crystallographica*, **B35**, 133–6.

Mazzi, F., Ungaretti, L., dal Negro, A., Petersen, O.V., and Rønsbo, J.G. (1979). The crystal structure of Semenovite. *The American Mineralogist*, **64**, 202–10.

Mazzi, F., Galli, E., and Gottardi, G. (1984). Crystal structure refinement of two tetragonal edingtonites. *Neues Jahrbuch für Mineralogie, Monatshefte*, **1984**, 372–82.

McConnell, J.D.C. (1954). The hydrated calcium silicates riverseideite, tobermorite, and plombierite. *The Mineralogical Magazine*, **30**, 293–305.

McKie, D. (1963). The högbomite polytypes. *The Mineralogical Magazine*, **33**, 563–80.

Meerschaut, A., Palvadeau, P., Moëlo, Y., and Orlandi, P. (2001). Lead–antimony sulfosalts from Tuscany (Italy). IV. Crystal structure of pillaite, $Pb_9Sb_{10}S_{23}ClO_{0.5}$, an expanded monoclinic derivative of hexagonal $Bi(Bi_2S_3)_9I_3$, from the zinkenite group. *European Journal of Mineralogy*, **13**, 779–90.

Megaw, H.D. (1973). *Crystal Structures: a working approach.* Philadelphia: W.B. Saunders Co. pp 564.

Megaw, H.D. and Kelsey, C.H. (1956). Crystal structure of tobermorite. *Nature*, **177**, 390–1.

Meier, W.M., Olson, D.H., and Baerlocher, C. (1996). Atlas of Zeolite Structure Types, 4th ed. *Zeolites*, **17**, 1–230.

Mellini, M. and Merlino, S. (1979). Versiliaite and apuanite: derivative structures related to schafarzikite. *The American Mineralogist*, **64**, 1235–42.

Mellini, M. and Viti, C. (1994). Crystal structure of lizardite-1T and lizardite-2H1 from Coli, Italy. *The American Mineralogist*, **72**, 943–8.

Mellini, M., Amouric, M., Baronnet, A., and Mercuriot, G. (1981). Microstructures and non stoichiometry in schafarzikite-like minerals. *The American Mineralogist*, **66**, 1073–9.

Mellini, M., Ferraris, G., and Compagnoni, R. (1985). Carlosturanite: HRTEM evidence of a polysomatic series including serpentine. *The American Mineralogist*, **70**, 773–81.

Mellini, M., Trommsdorff, V., and Compagnoni, R. (1987). Antigorite polysomatism: behaviour during progressive metamorphism. *Contribution to Mineralogy and Petrology*, **97**, 147–55.

Men'shikov, Yu.P., Bussen, I.V., Goiko, E.A., Zabavnikova, N.I., Mer'kov, A.N., and Khomyakov, A.P. (1975). Bornemanite—a new silicophosphate of sodium, titanium, niobium and barium. *Zapiski Vserossiyskogo Mineralogicheskogo Obschestva*, **104**, 322–6. (In Russian).

Men'shikov, Yu.P., Khomyakov, A.P., Ferraris, G., Belluso, E., Gula, A., and Kulchitskaya, E.A. (2003). Eveslogite, $(Ca,K,Na,Sr,Ba)_{24}[(Ti,Nb,Fe,Mn)_6(OH)_6Si_{24}O_{72}]$ $(F,OH,Cl)_7$, a new mineral from the Khibina alkaline massif, Kola Peninsula, Russia. *Zapiski Vserossiyskogo Mineralogicheskogo Obschestva*, **132(1)**, 59–67. (in Russian).

Meng, D., Wu, X., Han, Y., and Meng, X. (2002). Polytypism and microstructures of the mixed-layer member B_2S, $CaCe_3(CO_3)_4F_3$ in the bastnaesite–(Ce)–synchisite–(Ce) series. *Earth and Planetary Science Letters*, **203**, 817–28.

Meng, D., Wu, X., Mou, T. and Li, D. (2001*a*). Microstructural investigation of new polytypes of parisite-(Ce) by high-resolution transmission electron microscopy. *The Canadian Mineralogist*, **39**, 1713–24.

Meng, D., Wu, X., Mou, T., and Li, D. (2001*b*). Determination of six new polytypes in parisite–(Ce) by means of high resolution electron microscopy. *The Mineralogical Magazine*, **65**, 797–806.

Merlino, S. (1980). Crystal structure of sapphirine-1Tc. *Zeitschrift für Kristallographie*, **151**, 91–100.

Merlino, S. (1984). Feldspathoids: their average and real structures. In W.L. Brown (ed.). *Feldspars and Feldspathoids, NATO ASI Series C*, **137**, 435–70.

Merlino, S. (1988*a*). The structure of reyerite, $(Na, K)_2Ca_{14}Si_{22}Al_2O_{58}(OH)_8{\bullet}6H_2O$. *The Mineralogical Magazine*, **52**, 247–56.

Merlino, S. (1988*b*). Gyrolite: Its crystal structure and crystal chemistry. *The Mineralogical Magazine*, **52**, 377–87.

Merlino, S. (1990*a*). OD Structures in mineralogy. *Periodico di Mineralogia*, **59**, 69–92.

Merlino, S. (1990*b*). Lovdarite: structural features and OD character. *European Journal of Mineralogy*, **2**, 809–17.

Merlino, S. (1997*a*). OD approach in minerals: examples and applications. *EMU Notes in Mineralogy*, **1**, 29–54.

Merlino, S. (ed.) (1997*b*). *Modular aspects of minerals. EMU Notes in Mineralogy*, vol. 1. Budapest: Eötvös University press. 448 pp.

Merlino, S. and Pasero, M. (1997). Polysomatic approach in the crystal chemical study of minerals. *EMU Notes in Mineralogy*, **1**, 297–312.

Merlino, S. and Perchiazzi, N. (1988). Modular mineralogy in the cuspidine group. *The Canadian Mineralogist*, **26**, 933–43.

Merlino, S. and Zvyagin, B.B. (1998). Modular features of sapphirine-type structures. *Zeitschrift für Kristallographie*, **213**, 513–21.

Merlino, S., Franco, E., Mattia, C.A., Pasero, M., and Gennaro, M.D. (1985). The crystal structure of panunzite (natural tetrakalsilite). *Neues Jahrbuch für Mineralogie, Monatshefte*, **1985**, 322–8.

Merlino, S., Orlandi, P., Perchiazzi, N., Basso, R., and Palenzona, A. (1989). Polytypism in stibivanite. *The Canadian Mineralogist*, **27**, 625–32.

Merlino, S., Pasero, M., and Khomyakov, A.P. (1990*a*). The crystal structure of lintisite, $Na_3LiTi_2(Si_2O_6)_2O_2{\cdot}2H_2O$, a new titanosilicate from Lovozero (USSR). *Zeitschrift für Kristallographie*, **193**, 137–48.

Merlino, S., Perchiazzi, N., Khomyakov, A.P., Pushcharovsky, D.Y., Kulikova, I.M., and Kuzmin, V.I. (1990*b*). Burpalite, a new mineral from Burpalinskii massif, north Transbajkal, USSR: its crystal structure and OD character. *European Journal of Mineralogy*, **2**, 177–85.

Merlino, S., Pasero, M., and Perchiazzi, N. (1993). Crystal structure of paralaurionite and its OD relationships with laurionite. *The Mineralogical Magazine*, **57**, 323–8.

Merlino, S., Pasero, M., and Perchiazzi, N. (1994*a*). Fiedlerite: revised chemical formula $[Pb_3Cl_4F(OH){\cdot}H_2O]$, OD description and crystal structure refinement of the two MDO polytypes. *The Mineralogical Magazine*, **58**, 69–78.

Merlino, S., Pasero, M., Artioli, G., and Khomyakov, A.P. (1994*b*). Penkvilksite, a new kind of silicate structure: OD character, X-ray single crystal ($1M$), and powder Rietveld ($2O$) refinements of two MDO polytypes. *The American Mineralogist*, **79**, 1185–93.

Merlino, S., Bonaccorsi, E., and Armbruster, T. (1999). Tobermorites: their real structures and OD character. *The American Mineralogist*, **84**, 1613–21.

Merlino, S., Bonaccorsi, E., and Armbruster, T. (2000*a*). The real structures of clinotobermorite and tobermorite 9 Å: OD character, polytypes, and structural relationships. *European Journal of Mineralogy*, **12**, 411–29.

Merlino, S., Pasero, M., and O. Ferro (2000*b*). The crystal structure of kukisvumite, $Na_6ZnTi_4(Si_2O_6)_2O_4.4H_2O$. *Zeitschrift für Kristallographie*, **215**, 352–6.

Merlino, S., Bonaccorsi, E., and Armbruster, T. (2001). The real structure of tobermorite 11 Å: normal and anomalous forms, OD character and polytypic modifications. *European Journal of Mineralogy*, **13**, 577–90.

Merlino, S., Perchiazzi, N., and Franco, D. (2003). Brochantite, $Cu_4SO_4(OH)_6$: OD character, polytypism and crystal structures. *European Journal of Mineralogy*, **15**, 267–75.

Michel, C., Hervieu, M., Tilley, R.J.D., and Raveau, B. (1984). $Ba_{0.15}WO_3$ a bronze with an original pentagonal tunnel structure. *Journal of Solid State Chemistry*, **52**, 281–91.

Miehe, G. (1971). Crystal structure of kobellite. *Nature Physics Science*, **231**, 133–134.

Miehe, G. and Graetsch, H. (1992). Crystal structure of moganite: A new structure type for silica. *European Journal of Mineralogy*, **4**, 693–706.

Miehe, G. and Kupčík, V. (1971). Die Kristallstruktur des $Bi(Bi_2S_3)_9J_3$. *Naturwissenschaften*, **58**, 219–20.

Mitchell, R.H. (2002). *Perovskites: Modern and ancient*. Thunder Bay (Canada): Almaz Press. 318 pp.

Mitzi, D.B. (1999). Synthesis, structure, and properties of organic–inorganic perovskites and related materials. *Progress in Inorganic Chemistry*, **48**, 1–121.

Mitzi, D.B. (2001). Templating and structural engineering in organic–inorganic perovskites. *Journal of the Chemical Society, Dalton Transactions*, 1–12.

Mitzi, D.B., Feild, C.A., Harrison, W.T.A., and Guloy, A.M. (1994). Conducting tin halides with a layered organic based perovskite structure. *Nature*, **369**, 467–9.

Mitzi, D.B., Wang, S., Field, C.A., Chess, C.A., and Guloy, A.M. (1995). Conducting layered organic–inorganic halides containing $\langle 110 \rangle$-oriented perovskite sheets. *Science*, **267**, 1473–6.

Mitzi, D.B., Chondroudis, K., and Kagan, C.R. (2001*a*). Organic–inorganic electronics. *IBM Journal of Research & Development*, **45**, 29–45.

Mitzi, D.B., Dimitrakopoulos, C.D., and Kosbar, L.L. (2001*b*). Structurally tailored organic–inorganic perovskites: optical properties and solution-processed channel materials for thin-film transistors. *Chemistry of Materials*, **13**, 3728–40.

Miyajima, H., Matsubara, S., Miyawaki, R., Yokoyama, K., and Hirokawa, K. (2001). Rengeite, $Sr_4ZrTi_4Si_4O_{22}$, a new mineral Sr–Zr analogue of perrierite from the Itogawa-Ohmi district, Niigata Prefecture, central Japan. *The Mineralogical Magazine*, **65**, 111–20.

Miyajima, H., Miyawaki, R., and Ito, K. (2002). Matsubaraite, $Sr_4Ti_5(Si_2O_7)_2O_8$, a new mineral Sr–Ti analogue of perrierite in jadeite from the Itogawa-Ohmi district, Niigata Prefecture, Japan. *European Journal of Mineralogy*, **14**, 1119–28.

Mochiku, T., Mihara, Y., Hata, Y., Kamisawa, S., Furuyama, M., Suzuki, J., Kadowaki, K., Metoki, N., Fujii, H., and Hirata, K. (2002). Crystal structure of magnetic superconductor $FeSr_2YCu_2O_{6+\delta}$. *Journal of the Physical Society of Japan*, **71**, 790–6.

Moëlo, Y. (1982). Contribution à l'étude des conditions naturelles de formation des sulfures complexes d'antimoine et plomb. Signification metallogenique. *Thèse de doctorat d'état no. 82-01, Université Pierre et Marie Curie. March 1982.*

Moëlo, Y., Jambor, J.L., and Harris, D.C. (1984*a*). Tintinaite et sulfosels associés de Tintina (Yukon). la cristallochimie de la série de la kobellite. *The Canadian Mineralogist*, **22**, 219–26.

Moëlo, Y., Oudin, E., Picot, P., and Caye, R. (1984*b*). L'uchucchacuaite, $AgMnPb_3Sb_5S_{12}$, une nouvelle espèce minérale de la série de l'andorite. *Bulletin de Minéralogie*, **107**, 597–604.

Moëlo, Y., Makovicky, E., and Karup-Møller, S. (1988). Sulfures complexes plombo-argentifères: Minéralogie et cristallochimie de la série andorite-fizelyite $(Pb, \mathrm{Mn, Fe, Cd, Sn})_{3-2x}$ $(Ag, \mathrm{Cu})_x(Sb, \mathrm{Bi, As})_{2+x}(\mathrm{S, Se})_6$. *Documents BRGM (Orlèans)*, **167**, 107 pp.

Moëlo, Y., Makovicky, E., Karup-Møller, S., Corvelle, B., and Maurel, C. (1990). La lévyclaudite, $Pb_8Sn_7Cu_3(Bi, Sb)_3S_{28}$, une nouvelle espèce à structure incommensurable, de la série de la cylindrite. *European Journal of Mineralogy*, **2**, 711–23.

Moëlo, Y., Meerschaut, A., and Makovicky, E. (1997). Refinement of the crystal structure of nuffieldite, $Pb_2Cu_{1.4}(Pb_{0.4}Bi_{0.4}Sb_{0.2})Bi_2S_7$: structural relationships and genesis of complex lead sulfosalt structures. *The Canadian Mineralogist*, **35**, 1497–508.

Moëlo, Y., Meerschaut, A., Orlandi, P., and Palvadeau, P. (2000). Lead–antimony sulfosalts from Tuscany (Italy). II-crystal structure of scainiite, $Pb_{14}Sb_{30}S_{54}O_5$, an expanded monoclinic derivative of $Ba_{12}Bi_{24}S_{48}$ hexagonal sub-type (zinkenite group). *European Journal of Mineralogy*, **12**, 835–46.

Moore, P.B. (1969). A novel octahedral framework: Gageite. *The American Mineralogist*, **54**, 1005–17.

Moore, P.B. (1970*a*). Edge-sharing silicate tetrahedra in the crystal structure of leucophoencite. *The American Mineralogist*, **55**, 1146–66.

Moore, P.B. (1970*b*). Manganostibite: a novel cubic close-packed structure type. *The American Mineralogist*, **55**, 1489–99.

Moore, P.B. (1978). Manganhumite, a new species. *The Mineralogical Magazine*, **42**, 133–6.

Moore, P.B. and Araki, T. (1974). Pinakiolite, $Mg_2Mn^{3+}O_2(BO_3)$; warwickite, $Mg(Mg_{0.5}Ti_{0.5})O[BO_3]$; wightmanite, $Mg_3(O)(OH)_5[BO_3].nH_2O$: crystal chemistry of complex 3 Å wallpaper structures. *The American Mineralogist*, **59**, 985–1004.

Moore, P.B. and Araki, T. (1983). Surinamite, *ca.* $Mg_3Al_4Si_3BeO_{16}$: its crystal structure and relation to sapphirine, *ca.* $Mg_{2.8}Al_{7.2}Si_{1.2}BeO_{16}$. *The American Mineralogist*, **68**, 804–10.

Moore, P.B. and Smith, J.V. (1969). High pressure modification of Mg_2SiO_4: crystal structure and crystallochemical and geophysical implications. *Nature*, **221**, 653–5.

Moore, P.B. and Smith, J.V. (1970). Crystal structure of β-Mg_2SiO_4: crystal–chemical and geophysical implications. *Physics of the Earth and Planetary Interiors*, **3**, 166–77.

Moritomo, Y., Asamitsu, A., Kuwahara, H., and Tokura, Y. (1996). Giant magnetoresistence of manganese oxides with a layered perovskite structure. *Nature*, **380**, 141–4.

Morosin, B., Gingley, D.S., Schirber, J.E., and Venturini, E.L. (1988). Crystal structure of $TlCa_2Ba_2Cu_3O_9$. *Physica C*, **156**, 587–91.

Mössner, B. and Kemmler-Sack, S. (1985). $9R$-Stapelvarianten von Typ $Ba_3(B, B')_2O_{9-y}$ mit $B, B' =$ Mo, W, V, Ti. *Journal of the Less-Common Metals*, **114**, 333–41.

Müller, W.F. (1976). On the stacking disorder and polytypism in pectolite and serandite. *Zeitschrift für Kristallographie*, **144**, 401–8.

Mumme, W.G. (1975*a*). Junoite, $Cu_2Pb_3Bi_8(S, Se)_{16}$, a new sulfosalt from Tennant Creek, Australia: its crystal structure, and relationship with other bismuth sulfosalts. *The American Mineralogist*, **60**, 548–58.

Mumme, W.G. (1975*b*). The crystal structure of krupkaite, $CuPbBi_3S_6$, from the Juno Mine at Tennant Creek, Northern Territory, Australia. *The American Mineralogist*, **60**, 300–8.

Mumme, W.G. (1976). Proudite from Tennant Creek, Northern Territory, Australia: its crystal structure and relationship with weibullite and wittite. *The American Mineralogist*, **61**, 839–52.

Mumme, W.G. (1980*a*). Weibullite, $Ag_{0.32}Pb_{5.02}Bi_{8.55}Se_{6.08}S_{11.92}$ from Falun, Sweden. A higher homologue of galenobismutite. *The Canadian Mineralogist*, **18**, 1–18.

Mumme, W.G. (1980*b*). The crystal structure of nordströmite, $CuPb_3Bi_7(S, Se)_{14}$ from Falun, Sweden: a member of the junoite homologous series. *The Canadian Mineralogist*, **18**, 343–52.

Mumme, W.G. (1986). The crystal structure of paderaite, a mineral of the cuprobismutite series. *The Canadian Mineralogist*, **24**, 513–21.

Mumme, W.G. (1989). The crystal structure of $Pb_{5.05}(Sb_{3.75}Bi_{0.28})Se_{10.72}Se_{0.28}$: boulangerite of near ideal composition. *Neues Jahrbuch für Mineralogie, Monatshefte*, **1989**, 498–12.

Mumme, W.G. (1990). A note on the occurrence, composition and crystal structures of pavonite homologous series members ^4P, ^6P, and ^8P. *Neues Jahrbuch für Mineralogie, Monatshefte*, **1990**, 193–204.

Mumme, W.G. (2003). The crystal structure of SFCA-II, $Ca_{5.1}Al_{9.3}Fe^{3+}_{18.7}Fe^{2+}_{0.9}O_{48}$ a new homologue of the aenigmatite structure-type, and structure refinement of SFCA-type, $Ca_2Al_5Fe_7O_{20}$. Implications for the nature of the "ternary-phase solid-solution" previously reported in the CaO-Al_2O_3-iron oxide system. *Neues Jahrbuch für Mineralogie, Abhandlungen*, **178**, 307–35.

Mumme, W.G. and Watts, J.A. (1980). $HgBi_2S_4$: crystal structure and relationship with the pavonite homologous series. *Acta Crystallographica*, **B36**, 1300–4.

Mumme, W.G., Welin, E., and Wuensch, B.J. (1976). Crystal chemistry and proposed nomenclature for sulfosalts in the system bismuthinite-aikinite (Bi_2S_3–$CuPbBiS_3$). *The American Mineralogist*, **61**, 15–20.

Mumme, W.G., Niedermayr, G., Kelly, P.R., and Paar, W.H. (1983). Aschamalmite, $Pb_{5.92}Bi_{2.06}S_9$, from Untersulzbach Valley in Salzburg, Austria—'monoclinic heyrovskyite'. *Neues Jahrbuch für Mineralogie, Monatshefte*, **1983**, 433–44.

Mumme, W.G., Clout, J.M.F., and Gable, R.W. (1998). The crystal structure of SFCA-I, $Ca_{3.18}Fe^{3+}_{14.66}Al_{1.34}Fe^{2+}_{0.82}O_{28}$, a homologue of the aenigmatite structure type, and new crystal structure refinements of β–CFF, $Ca_{2.99}Fe^{3+}_{14.30}Fe^{2+}_{0.55}O_{25}$, and Mg-free SFCA, $Ca_{2.45}Fe^{3+}_{9.04}Al_{1.74}Fe^{2+}_{0.16}O_{20}$. *Neues Jahrbuch für Mineralogie, Abhandlungen*, **173**, 93–117.

Nakai, I. and Appleman, D.E. (1981). The crystal structure of gerstleyite $Na_2(Sb, As)_8S_{13}{\cdot}2H_2O$: The first sulfosalt mineral of sodium. *Chemistry Letters* (Japan), 1327–30.

Nakata, H., Akimitsu, J., Katano, S., Minami, T., Ogita, N., and Udagawa, M. (1995). Structural phase transition in $Sr_2CuO_2(CO_3)$. *Physica C*, **255**, 157–66.

Náray-Szabó, S. (1929). The structure of staurolite. *Zeitschrift für Kristallographie*, **71**, 103–116.

Náray-Szabó, I. and Sasvári, K. (1958). On the structure of staurolite, $HFe_2Al_9Si_4O_{24}$. *Acta Crystallographica*, **11**, 862–5.

Németh, P., Ferraris, G., Belluso, E., and Khomyakov, A.P. (2002). A modular structural model of a new layer titanosilicate, $BaNa(Na, Ti)_4[(Ti, Nb)_2(OH)_2Si_4O_{14}](OH, F)_2{\cdot}3H_2O$, from Lovozero. *4th EMU School and Symposium on Energy Modelling in Minerals, Budapest June 30–July 5. Abstracts.*

Németh, P., Ferraris, G., and Gula, A. (2003). Complex twinning in epistolite and murmanite – Structure model for epistolite from Lovozero massif. *5th EMU School and Symposium on Ultrahigh Pressure Metamorphism, Budapest 2003, Abstr.*, 24.

Nespolo, M. (1999). Analysis of family reflections of mica polytypes, and its application to twin identification. *Mineralogical Journal*, **21**, 53–85.

Nespolo, M. and Ďurovič, S. (2002). Crystallographic basis of polytypism and twinning in micas. *Reviews in Mineralogy and Geochemistry*, **46**, 155–279.

Nespolo, M. and Ferraris, G. (2000). Twinning by syngonic and metric merohedry. Analysis, classification and effects on the diffraction pattern. *Zeitschrift für Kristallographie*, **215**, 77–81.

Nespolo, M. and Ferraris, G. (2001). Effects of the stacking faults on the calculated electron density of mica polytypes—The Ďurovič effect. *European Journal of Mineralogy*, **13**, 1035–45.

Nespolo, M. and Ferraris, G. (2003). Geminography—The science of twinning applied to the early-stage derivation of non-merohedric twin laws. *Zeitschrift für Kristallographie*, **218**, 178–81.

Nespolo, M., and Ferraris, G. (2004). Applied geminography—symmetry analysis of twinned crystals and definition of twinning by reticular polyholohedry. *Acta Crystallographica*, **A60**, in press.

Nespolo, M., Takeda, H., Ferraris, G., and Kogure, T. (1997). Composite twins of 1*M* mica: Derivation and identification. *Mineralogical Journal*, **19**, 173–86.

Nespolo, M., Kogure, T. and Ferraris, G. (1999*a*). Allotwinning: oriented crystal association of polytypes—some warnings on consequences. *Zeitschrift für Kristallographie*, **214**, 5–8.

Nespolo, M., Ferraris, G., Takeda, H., and Takéuchi, Y. (1999*b*). Plesiotwinning: oriented associations based on a large coincidence-site lattice. *Zeitschrift für Kristallographie*, **214**, 378–82.

Nespolo, M., Ferraris, G., and Ďurovič, S. (1999*c*). OD character and twinning—Selective merohedry in class II merohedric twins of OD polytypes. *Zeitschrift für Kristallographie*, **214**, 776–9.

Nespolo, M., Ferraris, G., and Takeda, H. (2000*a*). Twins and allotwins of basic mica polytypes: theoretical derivation and identification in the reciprocal space. *Acta Crystallographica*, **A56**, 132–48.

Nespolo, M., Ferraris, G., and Takeda, H. (2000*b*). Identification of two allotwins of mica polytypes in reciprocal space through the minimal rhombus unit. *Acta Crystallographica*, **B56**, 639–47.

Newsam, J.M., Treacy, M.M.J, Koetsier, W.T., and De Gruyter, C.B. (1988). Structural characterization of zeolite beta. *Proceedings of the Royal Society, London*, **A20**, 375–405.

Nguyen, H.D. and Laruelle, P. (1977). Étude structurale des polytypes à deux anions *L*Se F (*L* = Y, Ho, Er ...). IV. Structure cristalline du polytype monoclinique à huit couches du fluorosélénìure d'yttrium 'Y Se Fe' 8 *O*. *Acta Crystallographica*, **B33**, 3360–3.

Nguyen, H.D. and Laruelle, P. (1980). Étude structurale des polytypes à deux anions *L*Se F (*L* = Y, Ho, Er ...). V. Structure du polytype orthorhombique à quatorze couches du fluorosélénìure d'yttrium 'Y Se F' 14 *O*. *Acta Crystallographica*, **B36**, 1048–51.

Ni, Y., Hughes, J.M., and Mariano, A.N. (1993). The atomic arrangement of bastnäsite–(Ce), $Ce(CO_3)F$, and structural elements of synchisite–(Ce), röntgenite–(Ce), and parisite–(Ce). *The American Mineralogist*, **78**, 415–8.

Ni, Y., Post, J.E. and Hughes, J.M. (2000). The crystal structure of parisite–(Ce), $Ce_2CaF_2(CO_3)_3$. *The American Mineralogist*, **85**, 251–8.

Niizeki, W. and Buerger, M.J. (1957). The crystal structure of jamesonite, $FePb_4Sb_6S_{14}$. *Zeitschrift für Kristallographie*, **109**, 161–83.

Noël, H. and Padiou, J. (1976). Structure cristalline de $FeUS_3$. *Acta Crystallographica*, **B32**, 1593–5.

Nuffield, E.W. (1952). Studies of mineral sulpho-salts: XVI-cuprobismutite. *The American Mineralogist*, **37**, 447–52.

Nuffield, E.W. (1980). Cupropavonite from Hall's Vallery, Park Country, Colorado. *The Canadian Mineralogist*, **18**, 181–4.

Nyfeler, D., Armbruster, T., Dixon, R., and Bermanec, V. (1995). Nchwaningite, a new pyroxene-related chain silicate from the N'chwaning mine, Kalahari manganese field, South Africa. *The American Mineralogist*, **80**, 377–86.

Ogborne, D.M. and Weller, M.T. (1992). The structure of $Tl_2Ba_2Ca_3Cu_4O_{12}$. *Physica C*, **201**, 53–57.

Ogborne, D.M. and Weller, M.T. (1994). The structure of $TlBa_2Ca_3Cu_4O_{11}$. *Physica C*, **230**, 153–58.

Ohashi, Y. (1984). Polysynthetically-twinned structures of enstatite and wollastonite. *Physics and Chemistry of Minerals*, **10**, 217–29.

Ohashi, Y. and Finger, L.W. (1978). The role of octahedral cations in pyroxenoid crystal chemistry. I. Bustamite, wollastonite, and the pectolite–schizolite–serandite series. *The American Mineralogist*, **63**, 274–88.

Ohmasa, M. (1973). The crystal structure of $Cu_{2+x}Bi_{6-x}S_9$ ($x = 1.21$). *Neues Jahrbuch für Mineralogie, Monatshefte*, **1973**, 227–33.

Ohmasa, M. and Nowacki, W. (1970). A redetermination of the crystal structure of aikinite $[BiS_2|S|Cu^{IV}Pb^{VIII}]$. *Zeitschrift für Kristallographie*, **137**, 422–32.

Ohmasa, M. and Nowacki (1973). The crystal structure of synthetic $CuBi_5S_8$. *Zeitschrift für Kristallographie*, **137**, 422–32.

Okamura, P. (1971). An algebraic analysis of twinned reciprocal space for structure determination. *Mineralogical Journal*, **6**, 405–21.

O'Keefe, M. and Hyde, B.G. (1996). *Crystal Structures. I. Patterns and Symmetry*. Mineralogical Society of America Publication.

Olivier-Fourcade, J., Maurin, M., and Philippot, E. (1983). Étude cristallochimique du système $Li_2S–Sb_2S_3$. *Revue de Chimie Minérale*, **20**, 196–217.

Onoda, M., Chen, X.-A., Kato, K., Sato, A., and Wada, H. (1999). Structure refinement of Cu_8GeS_6 using X-ray diffraction data from a multiple-twinned crystal. *Acta Crystallographica*, **B55**, 721–725.

Orlandi, P., Meerschaut, A., Palvadeau, P., and Merlino, S. (2002). Lead–antimony sulfosalts from Tuscany (Italy). V. Definition and crystal structure of moëloite, $Pb_6Sb_6S_{14}(S_3)$, a new mineral from the Ceragiola marble quarry. *European Journal of Mineralogy*, **14**, 599–606.

Otto, H.H. and Strunz, H. (1968). Zur Kristallchemie synthetischer Blei-Wismut-Spiessglanze. *Neues Jahrbuch für Mineralogie, Abhandlungen*, **108**, 1–19.

Ozawa, T. and Nowacki, W. (1975). The crystal structure of, and the bismuth–copper distribution in synthetic cuprobismutite. *Zeitschrift für Kristallographie*, **142**, 161–76.

Ozawa, T. and Tachikawa, O. (1996). A transmission electron microscope observation of 138 Å period in Pb–As–S sulfosalts. *Mineralogical Journal*, **18**, 97–101.

Ozawa, T. and Takéuchi, Y. (1993). X-ray and electron diffraction study of sartorite—a periodic antiphase boundary structure and polymorphism. *Mineralogical Journal*, **16**, 358–70.

Pabst, A. (1959). False symmetry, the Templeton effect, in lawsonite. *Zeitschrift für Kristallographie*, **112**, 53–59.

Pabst, A. (1961). Supplementary note on False symmetry, the Templeton effect, in lawsonite. *Zeitschrift für Kristallographie*, **115**, 307–9.

Pacalo, R.E.G. and Parise, J.B. (1992). Crystal structure of superhydrous B, a hydrous magnesium silicate synthesized at 1400°C and 20 GPa. *The American Mineralogist*, **77**, 681–4.

Pagnoux, C., Verbaere, A., Kanno, Y., Piffard, Y., and Tournoux, M. (1992). The synthesis and crystal structures of novel antimony compounds: $A_4Sb_4O_8$ (X_4O_{12}) *(A; K, Rb, Cs, Tl; X = Si, Ge)*. *Journal of Solid State Chemistry*, **99**, 173–81.

Pagnoux, C., Verbaere, A., Kanno, Y., Piffard, Y., and Tournoux, M. (1993). Crystal structure of $Cs_3Sb_3O_6(Ge_2O_7)$; its relationship with $K_3Sb_3O_6(PO_4)_2$. *European Journal of Solid State Inorganic Chemistry*, **30**, 111–23.

Palmer, D.C. (1994). Stuffed derivatives of the silica polymorphs. *Reviews in Mineralogy*, **29**, 83–122.

Parise, J.B. and Ko, Y. (1994). $[C_4H_8N_2]$ $[Sb_4S_7]$.

Parise, J.B., Smith, P.P.K., and Howard, C.J. (1984). Crystal structure refinement of $Sn_3Sb_2S_6$ by high-resolution neutron powder diffraction. *Materials Research Bulletin*, **19**, 503–8.

Parthé, E. (1976). The CrB and related structure types interpreted by periodic unit-cell twinning of close-packed structures. *Acta Crystallographica*, **B32**, 2813–8.

Parthé, E. (1990). *Elements of Inorganic Structural Chemistry, A Course on Selected Topics*. Publ. K. Sutter Parthé, Petit-Lancy, Switzerland.

Parthé, E., Chabot, B.A., and Cenzual, K. (1985). Complex structures of intermetallic compounds interpreted as intergrowths of segments of simple structures. *Chimia*, **39**, 164–74.

Pašava, J., Pertlik, F., Stumpfl, E.F., and Zemann, J. (1989). Bernardite, a new thallium arsenic sulphosalt from Allchar, Macedonia, with a determination of the crystal structure. *The Mineralogical Magazine*, **53**, 531–8.

Pasero, M. and Reinecke, T. (1991). Crystal chemistry, HRTEM analysis and polytypic behaviour of ardennite. *European Journal of Mineralogy*, **3**, 819–30.

Pasero, M., Perchiazzi, N., Bigi, S., Franzini, M., and Merlino, S. (1997). $Pb_2Fe^{3+}Cl_3(OH)_4{\cdot}H_2O$, a newly discovered natural phase from Tuscany, Italy: physico-chemical data, crystal structure and OD character. *European Journal of Mineralogy*, **9**, 43–51.

Pavlov, P.V. and Belov, N.V. (1959). The structures of Herderite, Datolite and Gadolinite determined by direct methods. *Soviet Physics Crystallography*, **4**, 300–314.

Peacor, D.R. (1967). New data on nigerite. *The American Mineralogist*, **52**, 864–6.

Pekov, I.V, Chukanov, N.V., Ferraris, G., Ivaldi, G., Pushcharovsky, D.Yu., and Zadov, A.E. (2003). Shirokshinite, $K(NaMg_2)Si_4O_{10}F_2$, a new mica with octahedral Na from Khibiny massif, Kola Peninsula: descriptive data and structural disorder. *The European Journal of Mineralogy*, **15**, 447–54.

Pelloquin, D., Maignan, A., Caldes, M.T., Hervieu, M., Michel, C., and Raveau, B. (1993). The bismuth oxycarbonate $Bi_2Sr_{6-y}Cu_3O_{10}(CO_3)_2$. A new 40 K superconductor, second member of the series $(Bi_2Sr_2CuO_6)_n(Sr_2CuO_2CO_3)_{n'}$. *Physica C*, **212**, 199–205.

Pen, Z.Z. and Shen, T.C. (1963). Crystal structure of bafertisite, a new mineral from China. *Scientia Sinica*, **12**, 278–80. (In Russian).

Perez-Mato, J.M. and Iglesias, J.E. (1977). On simple and double diffraction enhancement of symmetry. *Acta Crystallographica*, **A33**, 466–74.

Perez-Mato, J.M., Zakhour-Nakhl, M., Weill, F., and Darriet, J. (1999). Structure of composites $A_{1+x}(A'_xB_{1-x})O_3$ related to the $2H$ hexagonal perovskite: relation between compositions and modulation. *Journal of Materials Chemistry*, **9**, 2795–2808.

Perrault, G. and Szymanski, J.T. (1982). Steacyite, a new name, and a re-evaluation of the nomenclature of 'ekanite'–group minerals. *The Canadian Mineralogist*, **20**, 59–63.

Perrault, G., Harvey, Y., and Pertsowsky, R. (1975). La yofortierite, un nouveau silicate hydraté de manganèse de St-Hilaire, P.Q. *The Canadadian Mineralogist*, **13**, 68–74.

Perrotta, A.J. and Smith, J.V. (1965). The crystal structure of kalsilite $KAlSiO_4$. *The Mineralogical Magazine*, **35**, 588–95.

Petrova, I.V., Kuznetsov, A.I., Belokoneva, Ye.L., Simonov, M.A., Pobedimskaya, Ye.A., and Belov, N.V. (1978). On the crystal structure of boulangerite. *Doklady Akademii Nauk SSSR*, **242**, 337–40. (In Russian).

Petrova, I.V., Bortnikov, N.S., Pobedimskaya, Ye.A., and Belov, N.V. (1979). The crystal structure of a new synthetic Pb,Sb-sulphosalt. *Doklady Akademii Nauk SSSR*, **244**, 607–9. (In Russian).

Pfitzner, A. (1994). Cu_3SbS_3: Zur Kristallstruktur und Polymorphie. *Zeitschrift für anorganische und allgemeine Chemie*, **620**, 1992–7.

Pfitzner, A. (1995). Cu_3SbS_3: Synthese und Kristallstruktur. *Zeitschrift für anorganische und allgemeine Chemie*, **621**, 685–8.

Pfitzner, A. (1998). Disorder of Cu^+ in Cu_3SbS_3: Structural investigations of the high- and low-temperature modification. *Zeitschrift für Kristallographie*, **213**, 228–36.

Piilonen, P.C., McDonald, A.M., and Lalonde, A.E. (2001). Kupletskite polytypes from the Lovozero massif, Kola Peninsula, Russia: Kupletskite-1*A* and kupletskite-*Ma*2*b*2*c*. *European Journal of Mineralogy*, **13**, 973–84.

Piilonen, P.C., Lalonde, A.E., McDonald, A.M., Gault, R.A., and Larsen, A.O. (2003*a*). Insights into astrophyllite-group minerals. I. Nomenclature, composition and development of a standardized general formula. *The Canadian Mineralogist*, **41**, 1–26.

Piilonen, P.C., McDonald, A.M., and Lalonde, A.E. (2003*b*). Insights into astrophyllite-group minerals. II. Crystal chemistry. *The Canadian Mineralogist*, **41**, 27–54.

Pluth, J.J., Smith, J.V., Pushcharovsky, D.Yu., Bram, A., Reikel, Ch., Weber, H.-P., Broach, R.W., and Semenov, E.I. (1997). Third-generation synchrotron X-ray diffraction of six-micrometer crystal of raite, $Na_3Mn_3Ti_{0.25}Si_8O_{20}(OH)_2 \cdot 10H_2O$, opens up new chemistry and physics of low-temperature minerals. *Proceedings of the National Academy of Sciences USA*, **94**, 12263–7.

Popov, V.A., Pautov, L.A., Sokolova, E., Hawthorne, F.C., McCammon, C., and Bazhenova, L.F. (2001). Polyakovite–(Ce), $(REE, Ca)_4(Mg, Fe^{2+})(Cr^{3+}, Fe^{3+})_2(Ti, Nb)_2Si_4O_{22}$, a new metamict mineral species from the Ilmen Mountains, southern Urals, Russia: Mineral description and crystal chemistry. *The Canadian Mineralogist*, **39**, 1095–104.

Portheine, J.C. and Nowacki, W. (1975). Refinement of the crystal structure of zinckenite, $Pb_6Sb_{14}S_{27}$. *Zeitschrift für Kristallographie*, **141**, 79–96.

Post, J.E. and Appleman, D.E. (1988). Chalcophanite, $ZnMn_3O_7 \cdot 3H_2O$: New crystal structure determinations. *The American Mineralogist*, **73**, 1401–4.

Post, J.E. and Bish, D.L. (1988). Rietveld refinement of the todorokite structure. *The American Mineralogist*, **73**, 861–9.

Post, J.E. and Bish, D.L. (1989). Rietveld refinement of the coronadite structure. *The American Mineralogist*, **74**, 913–7.

Post, J.E. and Burnham, C.W. (1986). Modeling tunnel-cation displacements in hollandites using structure-energy calculations. *The American Mineralogist*, **71**, 1178–85.

Post, J.E. and Veblen, D.R. (1990). Crystal structure determinations on synthetic sodium, magnesium, and potassium birnessite using TEM and the Rietveld method. *The American Mineralogist*, **75**, 477–89.

Post, J.E., Van Dreele, R.B., and Buseck, P.R. (1982). Symmetry and cation displacements in hollandites: Structure refinements of hollandite, cryptomelane and priderite. *Acta Crystallographica*, **B38**, 1056–65.

Price, G.D., Parker, S.C., and Yeomans, J. (1985). The energetics of polytypic structures: a computer simulation of magnesium silicate spinelloids. *Acta Crystallographica*, **B41**, 231–9.

Pring, A. (1990). Disordered intergrowths in lead–arsenic sulfide minerals and the paragenesis of the sartorite-group minerals. *The American Mineralogist*, **75**, 289–94.

Pring, A. (2001). The crystal chemistry of the sartorite group minerals from Lengenbach, Binntal, Switzerland—a HRTEM study. *Schweizerische Mineralogisch-Petrographische Mitteilungen*, **81**, 69–87.

Pring, A. and Graeser, S. (1994) Polytypism in baumhauerite. *The American Mineralogist*, **79**, 302–7.

Pring, A., Williams, T., and Withers, R. (1993). Structural modulation in sartorite: an electron microscope study. *The American Mineralogist*, **78**, 619–26.

Pring, A., Jercher, M., and Makovicky, E. (1999). Disorder and compositional variation in the lillianite homologous series. *The Mineralogical Magazine*, **63**, 917–26.

Pushcharovsky, D.Yu., Pekov, I.V., Pluth, J., Smith, J., Ferraris, G., Vinogradova, S.A., Arakcheeva, A.V., Soboleva, S.V., and Semenov, E.I. (1999). Raite, manganonordite–(Ce) and ferronordite–(Ce) from Lovozero massif: crystal structures and mineral geochemistry. *Crystallography Reports*, **44**, 565–74.

Quarez, E., Huve, M., Roussel, P., and Mentré, O. (2002). Polysynthetic twinning characterization and crystallographic refinement in $NaBa_2M_2^{2+}M^{3+}O_6$ (M = Ni, Cu). *Journal of Solid State Chemistry*, **165**, 214–27.

Raade, G., Ferraris, G., Gula, A., and Ivaldi, G. (2002). Gjerdingenite–Fe from Norway, a new mineral in the labuntsovite group: description, crystal structure and twinning. *The Canadian Mineralogist*, **40**, 1629–39.

Ramsdell, L.S. and Kohn, J.A. (1951). Disagreement between crystal symmetry and X-ray diffraction data as shown by a new type of silicon carbide, $10H$. *Acta Crystallographica*, **9**, 199–200.

Ranganathan, S. (1966). On the geometry of coincidence-site lattices. *Acta Crystallographica*, **21**, 197–9.

Rao, C.N.R. and Raveau, B. (1995). *Transition Metal Oxides*. New York: VCH Publishers.

Rastsvetaeva, R.K. and N.V. Chukanov (1999). Crystal structure of a new high-barium analogue of lamprophyllite with a primitive unit cell. *Doklady Chemistry*, **368** (4–6), 228–31.

Rastsvetaeva, R.K., Simonov, V.I., and Belov, N.V. (1968). Crystal structure of vinogradovite, $Na_4LiTi_4[Si_2O_6]_2[Si_4O_{10}]O_4{\cdot}nH_2O$. *Soviet Physics Doklady*, **12**, 1990–2.

Rastsvetaeva, R.K., Tamazyan, R.A., Sokolova, E.V., and Belakovskii, D.I. (1991). Crystal structures of two modifications of natural Ba,Mn–titanosilicate. *Soviet Physics Crystallography*, **36**, 186–9.

Rastsvetaeva, R.K., Tamezyan, R.A., Puscharovsky, D.Yu., and Nadeshina, T.M. (1994). Crystal structure and micro-twinning of K-rich nenadkevichite. *European Journal of Mineralogy*, **6**, 503–9.

Raveau, B., Michel, C., Hervieu, M., and Groult, D. (1991). *Crystal chemistry of high TC superconducting copper oxides*. Springer-Verlag, Berlin. 331 pp.

Reid, A.F. and Ringwood, A.D. (1969). $Na_2Al_2Ge_6O_{16}$. *Journal of Solid State Chemistry*, **1**, 6–9.

Rentzeperis, P.J. (1970). The crystal structure of alleghanyite $Mn_5[(OH)_2(SiO_4)_2]$. *Zeitschrift für Kristallographie*, **132**, 1–18.

Ribbe, P.H. (1982). The humite series and Mn-analogs. *Reviews in Mineralogy*, **5**, 231–74.

Ribbe, P.H. and Gibbs, G.V. (1971). Crystal structures of the humite minerals. III. Mg/Fe ordering in humite and its relation to other ferromagnesian silicates. *The American Mineralogist*, **56**, 1155–73.

Richard, P. and Perrault, G. (1972). Structure cristalline de l'ékanite de St-Hilaire, P.Q. *Acta Crystallographica*, **B28**, 1994–9.

Ringwood, A.E. and Mayor, A. (1970). The system Mg_2SiO_4–Fe_2SiO_4 at high pressures and temperatures. *Physics of the Earth and Planetary Interiors*, **3**, 89–108.

Ringwood, A.E., Reid, A.F., and Wadsley, A.D. (1967). High pressure $KAlSi_3O_8$, an aluminosilicate with sixfold coordination. *Acta Crystallographica*, **23**, 1093–5.

Rives, V. (ed.) (2001). *Layered double hydroxides: Present and future*. New York: Nova Science. 439 pp.

Robinson, K., Gibbs, G.V., and Ribbe, P.M. (1973). The crystal structures of the humite minerals IV. Clinohumite and titanoclinohumite. *The American Mineralogist*, **58**, 43–49.

Rocha, J. and Anderson, M.W. (2000). Microporous titanosilicates and other novel mixed octahedral–tetrahedral framework oxides. *European Journal of Inorganic Chemistry*, 801–18.

Rodier, N. (1973). Structure cristalline du sulfure mixte de thalium et de cérium TmCe S_3. *Bulletin de la Société Française de Minéralogie et Cristallographie*, **96**, 350–5.

Rogl, P. (1992). Competition between trigonal prisms and other coordination polyhedra in borides, carbides, silicides and phosphides. In E. Parthé (ed.). *Modern Perspectives in Inorganic Crystal Chemistry*. NATO ASI Series **C-382**, Kluwer Acad. Publishers, 267–78.

Röhring, G., Gies, H., and Marler, B. (1994). Rietveld refinement of the crystal structure of the synthetic porous zincosilicate VPI-7. *Zeolites*, **14**, 498–503.

Ross, M., Takeda, H., and Wones, D.R. (1966). Mica polytypes systematic description and identification. *Science*, **151**, 191–3.

Ross II, C.R., Armbruster, T., and Canil, D. (1992). Crystal structure refinement of a spinelloid in the system Fe_3O_4–Fe_2SiO_4. *The American Mineralogist*, **77**, 507–11.

Roth, R.S. and Wadsley, A.D. (1965). Multiple phase formation in the binary system Nb_2O_5–WO_3 III. The structures of the tetragonal phases $W_3Nb_{14}O_{44}$ and $W_8Nb_{18}O_{69}$. *Acta Crystallographica*, **19**, 38–42.

Rouse, R.C., Peacor, D.R., Dunn, P.J., Su, S.-C., Chi, P.H., and Yeates, H. (1994). Samfowlerite, a new Ca Mn Zn beryllosilicate mineral from Franklin, New Jersey: its characterization and crystal structure. *The Canadian Mineralogist*, **32**, 43–53.

Royer, L. (1928). Recherches expérimentales sur l'épitaxie ou orientation mutuelle de cristaux d'espèces différentes. *Bulletin de la Société Française de Minéralogie et Cristallographie*, **51**, 7–159.

Rozenberg, K.A., Rastvetaeva, R.K., and Verin, I.A. (2003). Crystal structure of surkhobite: New mineral from the family of titanosilicate micas. *Crystallography Reports*, **48**, 384–9.

Ruddlesden, S.N. and Popper, P. (1957). New compounds of the K_2NIF_4 type. *Acta Crystallographica*, **10**, 538–9.

Ruddlesden, S.N. and Popper, P. (1958). The compound $Sr_3Ti_2O_7$ and its structure. *Acta Crystallographica*, **11**, 54–55.

Rytwo, G., Tropp, D., and Serban, C. (2002). Adsorption of diquat, paraquat and methyl green on sepiolite: experimental results and model calculations. *Applied Clay Science*, **20**, 273–82.

Rziha, T., Gies, H., and Rius, J. (1996). RUB-7, a new synthetic manganese oxide structure type with a 2×4 tunnel. *European Journal of Mineralogy*, **8**, 675–86.

Sabelli, C. (1980). The crystal structure of chalcophyllite. *Zeitschrift für Kristallographie*, **151**, 129–140.

Sadanaga, R. and Takeda, H. (1968). Monoclinic diffraction patterns produced by certain crystals and diffraction enhancement of symmetry. *Acta Crystallographica*, **B24**, 144–9.

Salje, E.K.H. (2002). Mesoscopic structures in ferroelastic and co-elastic materials. *Ferroelectrics*, **267**, 113–20.

Santoro, A. (1974). Characterization of twinning. *Acta Crystallographica*, **A30**, 224–31.

Savitskii, D.I., Ubizskii, S.B., Vasilechko, L.O., and Matkovskii, A.O. (1996). Twin models for orthorhombic $LaGaO_3$ crystals. *Crystallographic Reports*, **41**, 859–63.

Sawada, H., Kawada, I., Hellner, E. and Tokonami, M. (1987) The crystal structure of senandorite (andorite VI). $PbAgSb_3S_6$. *Zeitschrift für Kristallographie*, **180**, 141–50.

Schmalle, H.W., Williams, T., Reller, A., Linden, A., and Bednorz, J.G. (1993). The twin structure of $La_2Ti_2O_7$: X-ray and transmission electron microscopy studies. *Acta Crystallographica*, **B49**, 235–44.

Sedlacek, P., Zedler, A., and Reinecke, K. (1979). OD interpretation of pyroxenes. *Kristal und Technik*, **14**, 1055–62.

Seshadri, R., Martin, C., Hervieu, M., and Raveau, B. (1997). Structural evolution and electronic properties of $La_{1+x}Sr_{2-x}Mn_2O_7$. *Chemistry of Materials*, **9**, 270–7.

Seung, D.Y., Gravereau, P., Trut, L., and Levasseur, A. (1988). Li_3AsS_3. *Acta Crystallographica*, **C54**, 900–2.

Shaked, H., Keane, P.M., Rodriguez, J.C., Owen, F.F., Hitterman, R.L., and Jorgensen, J.D. (1994). *Crystal Structures of the High T_c Superconducting Copper oxides*. Amsterdam: Elsevier Science, B.V.

Shekhtman, V.S. ed. (1993). *The Real Structure of High-T_c Superconductors*. Berlin: Springer.

Sheldrick, W.S. and Kaub, J. (1985). Darstellung und Struktur von $Rb_2As_8S_{13}{\cdot}H_2O$ und $(NH_4)_2As_8S_{13}{\bullet}H_2O$. *Zeitschrift für Naturforschung*, **40B**, 1130–3.

Shen, J. and Moore, P.B. (1982). Törnebohmite, $RE_2Al(OH)[SiO_4]_2$: crystal structure and genealogy of RE(III)Si(IV) $\leftrightharpoons$ Ca(II)P(V) isomorphisms. *The American Mineralogist*, **67**, 1021–8.

Shi, N., Ma, Z., Li, G., Yamnova, N.A., and Pushcharovsky, D.Yu. (1998). Structure refinement of monoclinic astrophyllite. *Acta Crystallographica*, **B54**, 109–14.

Shimakawa, Y., Kubo, Y., Manako, T., Nakabayashi, Y., and Igarashi, H. (1988). Rietveld analysis of $Tl_2Ba2Ca_{n-1}Cu_nO_{4+2n}$ ($n = 1$, 2 and 3) by powder X-ray diffraction. *Physica C*, **156**, 97–102.

Shpanchenko, R.V., Nistor, L., Van Tendeloo, G., Van Landuyt, J., Amelinckx, S., Abakumov, A.M., Antipov, E.V., and Kovba, L.M. (1995). Structural studies of new ternary oxides $Ba_8Ta_4Ti_3O_{24}$ and $Ba_{10}Ta_{7.04}Ti_{1.2}O_{30}$. *Journal of Solid State Chemistry*, **114**, 560–74.

Silvester, E., Manceau, A., and Drits, V.A. (1997). Structure of synthetic monoclinic Na-rich birnessite and hexagonal birnessite: II. Results from chemical studies and EXAFS spectroscopy. *The American Mineralogist*, **82**, 962–78.

Skowron, A. (1991). PhD. Thesis, Mc Master University.

Skowron, A. and Brown, I.D. (1990*a*). Refinement of the structure of robinsonite, $Pb_4Sb_6S_{13}$. *Acta Crystallographica*, **C46**, 527–31.

Skowron, A. and Brown, I.D. (1990*b*). Refinement of the structure of boulangerite, $Pb_5Sb_4S_{11}$. *Acta Crystallographica*, **C46**, 531–4.

Skowron, A. and Brown, I.D. (1990*c*). Structure of antimony lead selenide, $Pb_4Sb_4Se_{10}$, a selenium analogue of cosalite. *Acta Crystallographica*, **C46**, 2287–91.

Skowron, A. and Tilley, R.J.D. (1990). Chemically twinned phases in the Ag_2S–PbS–Bi_2S_3 system. Part 1. Electron microscope study. *Journal of Solid State Chemistry*, **85**, 235–50.

Skowron, A., Boswell, F.W., Corbett, J.M., and Taylor, N.J. (1994). Structure determination of $PbSb_2Se_4$. *Journal of Solid State Chemistry*, **112**, 251–4.

Sloan, J., Battle, P.D., Green, M.A., Rosseinsky, M.J., and Vente, J.F. (1998). A HRTEM study of the Rudlesden-Popper compositions $Sr_2LnMn_2O_7$ (Ln = Y, La, Nd, Eu, Ho). *Journal of Solid State Chemistry*, **138**, 135–40.

Smith, J.V. (1977). Enumeration of 4-connected 3-dimensional nets and classification of framework silicates. I. Perpendicular linkage from simple hexagonal net. *The American Mineralogist*, **62**, 703–9.

Smith, J.V. (1988). Topochemistry of zeolites and related materials. 1. Topology and geometry. *Chemical Review*, **88**, 149–82.

Smith, P.P.K. (1984). Structure determination of diantimony tritin hexasulphide, $Sn_3Sb_2S_6$, by high-resolution transmission electron microscopy. *Acta Crystallographica*, **C40**, 581–4.

Smith, P.P.K. and Hyde, B.G. (1983). The homologous series Sb_2S_3–nPbS: Structures of diantimony dilead pentasulphide, $Pb_2Sb_2S_5$. *Acta Crystallographica*, **C39**, 1498–502.

Smith, P.P.K. and Parise, J.B. (1985). Structure determination of $SnSb_2S_4$ and $SnSb_2Se_4$ by high-resolution electron microscopy. *Acta Crystallographica*, **B41**, 84–87.

Sokolova, E.V., Yamnova, N.A., Egorov-Tismenko, Yu.K., and Khomyakov, A.P. (1984). Crystal structure of arctite, a new sodium calcium barium phosphate $(Na_5Ca)Ca_6Ba(PO_4)_6F_3$. *Soviet Physics Doklady*, **29**, 5–8.

Sokolova, E.V., Egorov-Tismenko, Yu.K., and Khomyakov, A.P. (1988). The crystal structure of sobolevite. *Soviet Physics Doklady*, **33**, 711–4.

Sokolova, E.V., Egorov-Tismenko, Yu.K., and Khomyakov, A.P. (1989). The crystal structure of nacaphite. *Soviet Physics Doklady*, **34**, 9–11.

Sokolova, E.V., Kabalov, Yu.K., Ferraris, G., Schneider, J., and Khomyakov, A.P. (1999). Modular approach in solving the crystal structure of a synthetic dimorph of nacaphite, $Na_2Ca[PO_4]F$, from powder diffraction. *The Canadian Mineralogist*, **37**, 83–90.

Sokolova, E., Hawthorne, F.C., and Khomyakov, A.P. (2002). The crystal chemistry of fersmanite, $Ca_4(Na, Ca)_4(Ti, Nb)_4(Si_2O_7)O_8F_3$. *The Canadian Mineralogist*, **40**, 1421–8.

Springer, G. (1976). Falcondoite, nickel analogue of sepiolite. *The Canadian Mineralogist*, **14**, 407–9.

Srikrishnan, T. and Nowacki, W. (1974). A redetermination of the crystal structure of cosalite, $Pb_2Bi_2S_5$. *Zeitschrift für Kristallographie*, **140**, 114–36.

Stephenson, N.C. (1968). A structural investigation of some stable phases in the region Nb_2O_5–WO_3. *Acta Crystallographica*, **24**, 637–53.

Strunz, H. (1993). Sulfide classification. *Mineralienmagazin Lapis*, Extra Edition, **12/93**, I–IV.

Strunz, H. and Nickel, E.H. (2001). *Strunz Mineralogical Tables*. Stuttgart: E. Schweizerbart'sche Verlagsbuchhandlung. 870 pp.

Strunz, H. and Tennyson, C. (1978). *Mineralogische Tabellen*, 7th ed. Leipzig: Akad. Verlagsgeselschaft.

Subramanian, M.A., Calabrese, J.C., Torardi, C.C., Gopalakrishnan, J., Askew, T.R., Flippen, R.B., Morrissey, K.J., Chowdry, U., and Sleight, A.W. (1988). Crystal structure of the high-temperature superconductor $Tl_2Ba_2CaCu_2O_8$. *Nature*, **332**, 420–2.

Sueno, S., Takeda, H., and Sadanaga, R. (1971). Two-dimensional regular aggregates of layered crystals. *Mineralogical Journal*, **6**, 172–85.

Sugaki, A., Shima, H., Kitakaze, A., and Mizota, T. (1981). Hydrothermal synthesis of nukundamite and its crystal structure. *The American Mineralogist*, **66**, 398–402.

Sugaki, A., Kitakaze, A., and Shima, H. (1987). Synthesis of cosalite and its phase relations in the Cu–Pb–Bi–S quaternary system. *Proceedings 13th General Meeting. International Mineralogical Association, Varna, Sept. 1982 vol 1* Stuttgart: E. Schweizerbart. Verlagsbuchhandlung.

Sundberg, M.R., Lehtinen, M., and Kivekäs, R. (1987). Refinement of the crystal structure of ramsayite (lorenzenite). *The American Mineralogist*, **72**, 173–7.

Szymanski, J.Y. (1980). A redetermination of the structure of Sb_2VO_5, stibivanite, a new mineral. *The Canadian Mineralogist*, **18**, 333–7.

Szymanski, J.T., Owens, D.R., Roberts, A.C., Ansell, H.G., and Chao, G.Y. (1982). A mineralogical study and crystal-structure determination of nonmetamict ekanite, $ThCa_2Si_8O_{20}$. *The Canadian Mineralogist*, **20**, 65–75.

Tagai, T. and Joswig, W. (1985). Untersuchungen der Kationenverteilung in Staurolith durch Neutronenbeugung bei 100 K. *Neues Jahrbuch für Mineralogie, Monatshefte*, **1985**, 97–107.

Takagi, J. and Takéuchi, Y. (1972). The crystal structure of lillianite. *Acta Crystallographica*, **B28**, 649–51.

Takeda, H. (1975). Prediction of twin formation. *Journal of the Mineralogical Society of Japan*, **12**, 89–102. (In Japanese).

Takeda, H. and Sadanaga, R. (1969). New unit layers for micas. *Mineralogical Journal (Japan)*, **5**, 434–49.

Takeda, H., Donnay, J.D.H., and Appleman, D.E. (1967). Djurleite twinning. *Zeitschrift für Kristallographie*, **125**, 414–22.

Takeda, Y., Kanamura, F., Shimada, M., and Koizumi, M. (1976). Crystal structure of $BaNiO_3$. *Acta Crystallographica*, **32**, 2464–6.

Takéuchi, Y. (1971). Polymorphic or polytypic changes in biotites, pyroxenes and wollastonite. *Journal Mineralogical Society of Japan*, **10**, Spec. Paper **2**, 87–99 (in Japanese).

Takéuchi, Y. (1978) Tropochemical twinning: a mechanism of building complex structures. *Recent Progress in Natural Sciences in Japan*, **3**, 153–181.

Takéuchi, Y. (1997). *Tropochemical cell-twinning*. Tokyo: Terra Scientific Publishing Co. 319 pp.

Takéuchi, Y. and Takagi, I. (1974). The crystal structure of heyrovskyite ($6PbS–Bi_2S_3$). *Procedings of Japan Academy*, **50**, 75–79.

Takéuchi, Y, Ghose, S., and Nowacki, W. (1965). The crystal structure of hutchinsonite, $(Tl, Pb)_2 As_5S_9$. *Zeitschrift für Kristallographie*, **121**, 321–48.

Takéuchi, Y., Aikawa, N., and Yamamoto, T. (1972*a*). The hydrogen locations and chemical composition of staurolite. *Zeitschrift für Kristallographie*, **136**, 1–22.

Takéuchi, Y., Sadanaga, R., and Aikwa, N. (1972*b*). Common lattices and image sets of hexagonal lattices, and their application to composite electron-diffraction patterns of biotite. *Zeitschrift für Kristallographie*, **136**, 207–25.

Takéuchi, Y., Ozawa, T., and Takagi, J. (1974). Structural characterization of the high-temperature phase V on the $PbS–Bi_2S_3$ join. *Zeitschrift für Kristallographie*, **140**, 249–72.

Takéuchi, Y., Ozawa, T., and Takagi, J. (1979). Tropochemical cell-twinning and the 60 Å structure of phase V in the $PbS–Bi_2S_3$ system. *Zeitschrift für Kristallographie*, **150**, 75–84.

Tamazyan, R., Arnold, H., Molchanov, V.N., Kuzmicheva, G.M., and Vasileva, I.G. (2000). Contribution to the crystal chemistry of rare earth chalcogenides. III. The crystal structure and twinning of $SmS_{1.9}$. *Zeitschrift für Kristallographie*, **215**, 346–351.

Tan, K., Ko, Y., and Parise, J.B., (1994). A novel antimony sulfide templated by ethylenediammonium. *Acta Crystallographica*, **C50**, 1439–42.

Tan, K., Parise, J.B., Ko, Y., Dorovsky, A., Norby, P., and Hanson, J.C. (1996). Applications of synchrotron imaging plate system to elucidate the structure and synthetic pathways to open framework antimony sulfides. *Abstracts XVII Congress Internat. Union of Crystallogr., Seattle, August 1996, abstr. PS 10.10.12*, C-402.

Tang, K., Qian, Y.T., Yang, M.L., Zhao, Y.D., and Zhang, Y.H. (1997). Crystal structure of a new series of 1212 type cuprate $RuSr_2Ln_2O_2$. *Physica C*, **282**, 947–8.

Taviot-Guéo, C., Chopinet, C., Palvadeau, P., Léone, P., Mozdzierz, N., and Rouxe, J. (1999). Synthesis and structural characterization of $La_4Mn_3Ge_{5.2}Si_{0.8}O_{22}$, a new compound with the perrierite structure. *Journal of Solid State Chemistry*, **147**, 247–50.

Taylor, H.F.W. (1959). The dehydration of tobermorite. *Proceedings of the 6th National Conference on Clays and Clay Minerals, Berkeley, 1957*, 101–109, Pergamon.

Templeton, D.H. (1956). Systematic absences corresponding to false symmetry. *Acta Crystallographica*, **9**, 199–200.

Teneze, N., Mercurio, D., Trolliard, G., and Frit, B. (2000). Cation-deficient perovskite-related compounds $(Ba, La)_nTi_{n-1}O_{3n}$ ($n = 4$, 5, and 6). A Rietveld refinement from neutron powder diffraction data. *Materials Research Bulletin*, **35**, 1603–14.

Thomas, J.M., Jefferson, D.A., Mallinson, L.G., Smith, D.G., and Sia Crawford, E. (1978). The elucidation of the ultrastructure of silicates minerals by high resolution electron microscopy and X-ray emission microanalysis. *Chemica Scripta*, **14**, 167–79.

Thompson, J.B., Jr. (1970). Geometrical possibilities for amphibole structures: model biopyriboles. *The American Mineralogist*, **55**, 292–3.

Thompson, J.B., Jr. (1978). Biopyriboles and polysomatic series. *The American Mineralogist*, **63**, 239–49.

Thompson, J.B., Jr. (1981). *Polytypism in complex crystals: contrasts between mica and classical polytypes*. In M. O'Keeffe and A. Navrotsky (eds). *Structure and Bonding in Crystals II*, 167–96. New York: Academic Press.

Thompson, J.G., Winthers, R.L., Sellar, J., Barlow, P.J., and Hyde, B.G. (1990). Incommensurate composite modulated $Nb_2Zr_{x-2}O_{2x+1}$: $x = 7.1$–10.3. *Journal of Solid State Chemistry*, **88**, 465–75.

Tillmanns, E. and Gebert, W. (1973). The crystal structure of tsumcorite, a new mineral from the Tsumeb mine, S.W. Africa. *Acta Crystallographica*, **B29**, 2789–94.

Tofield, B.C., Greaves, C., and Fender, B.E.F. (1975). The $SrFeO_{2.5}$–$SrFeO_{3.0}$ system. Evidence of a new phase $Sr_4Fe_4O_{11}$ ($SrFeO_{2.75}$). *Materials Research Bulletin*, **10**, 737–46.

Tomas, A. and Guittard, M. (1980). Cristallochimie des sulfures mixtes de chrome et d'erbium. *Materials Research Bulletin*, **15**, 1547–56.

Tomeoka, K. Ohmasa, M., and Sadanaga, R. (1980). Crystal chemical studies on some compounds in the copper–bismuth sulfide (Cu_2S–Bi_2S_3) system. *Mineraleralogical Journal*, **10**, 57–70.

Topa, D., Balič Žunič, T., and Makovicky, E. (2000*a*). The crystal structure of $Cu_{1.6}Pb_{1.6}Bi_{6.4}S_{12}$, a new 44.8 Å derivative of the bismuthinite–aikinite solid–solution series. *The Canadian Mineralogist*, **38**, 611–6.

Topa, D., Makovicky, E., Balič Žunič, T., and Berlepsch, P. (2000*b*). The crystal structure of $Cu_2Pb_6Bi_8S_{19}$. *European Journal of Mineralogy*, **12**, 825–33.

Topa, D., Makovicky, E., Criddle, A., Paar, W.H., and Balič Žunič, T. (2001). Felbertalite, $Cu_2Pb_6Bi_8S_{19}$, a new mineral species from Felbertal, Salzburg Province, Austria. *European Journal of Mineralogy*, **13**, 961–72.

Topa, D., Makovicky, E., and Paar, W.H. (2002*a*). Composition ranges and exsolution pairs for the members of the bismuthinite–aikinite series from Felbertal, Austria. *The Canadian Mineralogist*, **40**, 549–869.

Topa, D., Makovicky, E., and Balič Žunič, T. (2002*b*). The structural role of excess Cu and Pb in gladite and krupkaite based on new refinements of their structure. *The Canadian Mineralogist*, **40**, 1147–59.

Topa, D., Makovicky, E., and Paar, W.H. (2003). Kupčíkite, a new Cu–Bi sulfosalt from Felbertal, Austria and its crystal structure. *The Canadian Mineralogist*, **41**, 1155–66.

Topsöe, H. (1884). Krystallographisch-chemische Untersuchungen homologer Verbindungen. *Zeitschrift für Kristallographie*, **8**, 246–96.

Tournoux, M., Garne, M., and Piffard, Y. (1992). HTB-like six-membered rings of octahedra in some new oxides: structural aspects and related properties. *Journal of Solid State Chemistry*, **96**, 141–53.

Treiman, A.H. and Peacor, D.R. (1982). The crystal structure of lawsonbauerite, $(Mn,Mg)_9Zn_4(SO_4)_2(OH)_{22}.8H_2O$, and its relation to mooreite. *The American Mineralogist*, **67**, 1029–34.

Tschermak, G. and Becke, F. (ed.) (1921). *Lehrbuch der Mineralogie*. Wien und Leipzig: Hölder. 751 pp.

Turner, S. and Buseck, P.R. (1979). Manganese oxide tunnel structures and their intergrowths. *Science*, **203**, 456–8.

Turner, S. and Buseck, P.R. (1981). Todorokites: a new family of naturally occurring manganese oxides. *Science*, **212**, 1024–7.

Turner, S. and Post, J.E. (1988). Refinement of the substructure and superstructure of romanechite. *The American Mineralogist*, **73**, 1155–61.

Uehara, S. (1998). TEM and XRD study of antigorite superstructures. *The Canadian Mineralogist*, **36**, 1595–605.

Ungemach, H. (1935). Sur la syntaxie et la polytypie. *Zeitschrift für Kristallographie*, **91**, 1–22.

Vainshtein, B.K., Fridkin, V.M., and Indenbom, V.L. (1994). *Structure of Superconductors in: Structure of Crystals*, 2nd ed. Berlin: Springer-Verlag, 416–29.

Van Landuyt, J. and Amelinckx, S. (1975). Multiple beam direct lattice imaging of new mixed-layer compounds of the bastnäsite-synchesite series. *The American Mineralogist*, **60**, 351–8.

Vaugney, J.T. and Poeppelmeier, K.R. (1991). Structural diversity in oxygen-deficient perovskites. *National Institute of Standards Special Publication*, **804**, 419–25.

Veblen, D.R. (1991). Polysomatism and polysomatic series: a review and applications. *The American Mineralogist*, **76**, 801–26.

Veblen, D.R. and Burnham, C.W. (1977). New biopyriboles from Chester, Vermont: II The crystal chemistry of jimthompsonite, and chesterite, and the amphibole–mica reaction. *The American Mineralogist*, **63**, 1053–63.

Veblen, D.R. and Buseck, P.R. (1979). Chain-width order and disorder in biopyriboles. *The American Mineralogist*, **64**, 687–700.

Veblen, D.R., Buseck, P.R., and Burnham, C.W. (1977). Asbestiform chain silicates: new minerals and structural groups. *Science*, **198**, 359–65.

Vicat, J., Fanchon, E., Strobel, P., and Duc Tran Qui (1986). The structure of $K_{1.33}Mn_8O_{16}$ and cation ordering in hollandite-type structures. *Acta Crystallographica*, **B42**, 162–7.

Volk, K. and Schäfer, H. (1979). $Cs_2Sb_8S_{13}$, ein neuer Formel- und Strukturtyp bei Thioantimoniten. *Zeitschrift für Naturforschung*, **34B**, 1637–40.

Volk, K., Cordier, G., Cook, R., and Schäfer, H. (1980). $BaSbTe_3$ und $BaBiSe_3$. Verbindungen mit BiSe-bzw. SbTe-Schichtverbänden. *Zeitschrift für Naturforschung*, **35B**, 136–40.

von Knorring, O., Petersen, O.V., Karup-Moller, S., Leonardsen, E.S., and Condliffe, E. (1992) Tuperssuatsiaite, from Aris phonolite, Windhoek, Namibia. *Neues Jahrbuch für Mineralogie Monatshefte*, **1992**, 145–52.

Wada, T., Ichinose, A., Izumi, F., Nara, A., Yamauchi, H., Asano, H., and Tanaka, S. (1991). Neutron powder diffraction study of the Pb-based copper-oxide containing thick fluorite blocks—$(Pb, Cu)Sr_2(Ho, Ce)_3Cu_2O_{11+z}$. *Physica C*, **179**, 455–60.

Wada, T., Nara, A., Ichinose, A., Yamauchi, H., and Tanaka, S. (1992). Homologous compound series containing multiple-MO_2-unit fluorite block, $(Fe, Cu)Sr_2(Y, Ce)_nCu_2O_{4+2n+z}$ ($n = 1, 2, 3, \ldots$, and z approximately 1). *Physica C*, **192**, 181–90.

Wadhawan, V.K. (1997). A tensor classification of twinning in crystals. *Acta Crystallographica*, **A53**, 546–55.

Wadsley, A.D. (1963). Inorganic non-stoichiometric compounds. In L. Mandelcorn, ed. *Non-stoichiometric compounds*, 98–209. New York: Academic Press.

Wang, L., Ni, Y., Hughes, J.M., Bayliss, P. and Drexler, J.W. (1994). The atomic arrangement of synchisite–(Ce), $CeCaF(CO_3)_2$. *The Canadian Mineralogist*, **32**, 865–71.

Wang, N. and Eppelsheimer, D. (1976). The ternary phases in the system Sn–Sb–S. *Chemie der Erde*, **35**, 179–84.

Wang, S.M. and Hwu, S. (1995). $La_4Ti_5Si_{4-x}P_xO_{22}$ ($x = 0, 1$)—A new family of 2-dimensional solids—Synthesis and structure of the first member ($m = 1$) of the mixed-valence titanium(III/IV) oxosilicates series, $La_4Ti(Si_2O_7)_2(TiO_2)_{4m}$. *Inorganic Chemistry*, **34**, 166–71.

Wang, S.M., Hwu, S., Paradis, J.A., and Whangbo M.-H. (1995). α-and β-$La_4Ti_9Si_4O_{30}$: Synthesis and structure of the second member ($m = 2$) of novel layered oxosilicate containing (110) rutile sheets. Electrical property and band structure characterization of the mixed-valence titanium(III/IV) oxosilicates series, $La_4Ti(Si_2O_7)_2(TiO_2)_{4m}$ ($m = 1, 2$). *Journal of the American Chemical Society*, **117**, 5515–22.

Wang, X. (1989). *Transmission electron microscope study of the minerals of the franckeite family*. PhD Thesis (in Chinese). Chinese Geological University, Beijing.

Wang, X. (1995). Synthesis and structure of a new microporous thioantimonate (III) $[H_3N(CH_2)_3NH_3]Sb_{10}S_{16}$. *European Journal of Solid State Inorganic Chemistry*, **32**, 303–12.

Wang, X. and Liebau, F. (1994). Synthesis and structure of $[CH_3NH_3]_2Sb_8S_{13}$: A nanoporous thioantimonate (III) with a two-dimensional channel system. *Journal of Solid State Chemistry*, **111**, 385–9.

Weber, L. (1929). Die Symmetrie homogener ebener Punktsysteme. *Zeitschrift für Kristallographie*, **70**, 309–27.

Weller, M.T., Pack, M.J., Knee, C.S., Ogborne, D.M., and Gormezano, A. (1997). Multiple layer structures and superconductivity. *Physica C*, **282**, 849–50.

Wenk, H.R. (1969). Polymorphism of wollastonite. *Contributions to Mineralogy and Petrology*, **22**, 238–47.

Wenk, H.R., Müller, W.F., Liddell, N.A., and Phakey, P.P. (1976). Polytypism in wollastonite. In Wenk *et al.*, eds. *Electron Microscopy in Mineralogy*. Berlin, Heidelberg, New York: Springer, pp. 324–31.

White, T.J. and Hyde, B.G. (1983). An electron microscope study of leucophoenicite. *The American Mineralogist*, **68**, 1009–21.

White, T.J. and ZhiLi D. (2003). Structural derivation and crystal chemistry of apatites. *Acta Crystallographica*, **B59**, 1–16.

Whitfield, H.J. (1980). Polymorphism in skinnerite, Cu_3SbS_3. *Solid State Communications*, **33**, 747–8.

Wilkens, J. and Müller-Buschbaum, H. (1991). $Ba_{12}Ir_{12-x}Nb_xO_{36}$ ($x = 2, 4$)—A new compound with 12*R*-perovskite stacking prototype. *Journal of Alloys and Compounds*, **176**, 141–6.

Winter, J.K. and Ghose, S. (1979). Thermal expansion and high-temperature crystal chemistry of the Al_2SiO_5 polymorphs. *The American Mineralogist*, **64**, 573–86.

Wolf, M., Hunger, H.-J., and Bewilogua, K. (1981). Potosiit - ein neues Mineral der Kylindrit-Franckeit-Gruppe. *Freiberg Forschungs-Hefte*, **C364**, 113–33.

Wong-Ng, W. (1997). The ICDD/PDF superconductor subfile. *Powder Diffraction*, **12**, 13–15.

Wong-Ng, W., Kaduk, J.A., Young, R.A., Jiang, F., Swartzendruber, L.J., and Brown, H.J. (1999). Investigation of $(Sr_{4-\delta}Ca_{\delta})PtO_6$ using X-ray Rietveld refinement. *Powder Diffraction*, **14**, 181–9.

Woodrow, P.J. (1967). The crystal structure of astrophyllite. *Acta Crystallographica*, **22**, 673–8.

Wu, M.K., Ashburn, J.R., Torng, C.J., Hor, P.H., Meng, R.L., Gao, L., Huang, Z.J., Wang,. Y.Q., and Chu, C.W. (1987). Suoperconductivity at 93 K in a new mixed-phase Yb–Ba–Cu–O compound system at ambient pressure. *Physical Review Letters*, **58**, 908–10.

Wu, X., Meng, D., Pan, Z., and Yang, G. (1998). Transmission electron study of new, regular, mixed-layer structures in calcium-rare-earth fluorocarbonate minerals. *The Mineralogical Magazine*, **62**, 55–64.

Xu, H. and Buseck, P.R. (1996). TEM investigations of the domain structure and superstructure in hillebrandite, $Ca_2SiO_3(OH)_2$. *The American Mineralogist*, **81**, 1371–4.

Yakubovich, O.V., Simonov, M.A., Belokoneva, E.L., Egorov-Tismenko, Y.K., and Belov, N.V. (1976). Crystalline structure of Ca,Mg-diorthotriborate (pyroborate) kurchatovite $CaMg[B_2O_5]$. *Doklady Akademii Nauk SSSR*, **230**, 837–40.

Yakubovich, O.V., Dem'yanetz, L.N., and Massa, W. (2000). A new Cu,Al fluoride $CuAl_2F_2$ (Si_2O_7) and its relations to topaz. *Zeitschrift Allgemeine Chemie*, **626**, 1514–18.

Yamanaka, T., Hirano, M., and Takéuchi, Y. (1985). A high temperature transition in $MgGeO_3$ from clinopyroxene ($C2/c$) type to orthopyroxene (*Pbca*) type. *The American Mineralogist*, **70**, 365–74.

Yamnova, N.A., Sarp, Kh., Yegorov-Tismenko, Yu.K., and Pushcharovsky, D.Yu. (1993). The crystal structure of jaffeite. *Soviet Physics Crystallography*, **38**, 464–7.

Yamnova, N.A., Egorov-Tismenko, Yu.K., and Khomyakov, A.P. (1996). Crystal structure of a new natural (Na,Mn,Ti)-phyllosilicate. *Crystallography Reports*, **41**, 239–44.

Yamnova, N.A., Egorov-Tismenko, Yu.K., Zlykhenskaya, I.V., and Khomyakov, A.P. (2000). Refined crystal structure of iron-rich triclinic astrophyllite. *Crystallography Report*, **45**, 585–90.

Yang, G., Pan, Z., and Wu, X. (1994). Transmission electron microscope study of the new regular stacking structure in the calcium rare-earth fluorcarbonate mineral series from southwest China. *Scientia Geologica Sinica*, **29**(4), 393–8. (In Chinese).

Yang, H., Prewitt, C.T., and Frost, D.J. (1997). Crystal structure of the dense hydrous magnesium silicate, phase D. *The American Mineralogist*, **82**, 651–4.

Yang, H., Kowzett, J., and Prewitt, C.T. (2001). Crystal structure of phase X, a high pressure alkali-rich hydrous silicate and its anhydrous equivalent. *The American Mineralogist*, **86**, 1483–8.

Yang, Z., Tao, K., and Zhang, P. (1996). Polysomatic features of huanghoite-zhonghuacerite series minerals. *Neues Jahrbuch für Mineralogie, Monatshefte*, 264–70.

Yang, Z., Fleck, M., Pertlik, F., Tillmanns, E., and Tao, K. (2001). The crystal structure of natural gugiaite, $Ca_2BeSi_2O_7$. *Neues Jahrbuch für Mineralogie, Monatshefte*, 182–6.

Yang, Z., Fleck, M., Smith, M., Tao, K., Song, R., and Zhang, P. (2002). The crystal structure of natural Fe-rich chevkinite–(Ce). *European Journal of Mineralogy*, **14**, 969–75.

Yang, Z., Tao, K., and Zhang, P. (1998). The symmetry transformations of modules in bastnaesite-vaterite polysomatic series. *Neues Jahrbuch für Mineralogie, Monatshefte*, 1–12.

Yeo, L., Harris, K.D.M., and Kariuki, B.M. (2001). Temperature-dependent structural properties and crystal twinning in the fluorocyclohexane/thiourea inclusion compound. *Journal of Solid State Chemistry*, **156**, 16–25.

Yoshiasa, A. and Matsumoto, T. (1985). Crystal structure refinement and crystal chemistry of pumpellyite. *The American Mineralogist*, **70**, 1011–19.

Žák, L., Frýda, J., Mumme, W.G., and Paar, W.M. (1994). Makovickyite, $Ag_{1.5}Bi_{5.5}S_9$, from Baita Bihorului, Romania: The ^4P natural mineral member of the pavonite series. *Neues Jahrbuch für Mineralogie, Abhandlungen*, **168**, 147–69.

Zakrzewski, M.A. (1984). Jaskolskiite, a new Pb–Cu–Sb–Bi sulfosalt from Vena deposit, Sweden. *The Canadian Mineralogist*, **22**, 481–7.

Zakrzewski, M.A. and Makovicky, E. (1986). Izoklakeite from Vena, Sweden, and the kobellite homologous series. *The Canadian Mineralogist*, **24**, 7–18.

Zhu, N.J., Guo, F.L., Yan, S.S.X., Chen, L.R., and Li, A.C. (1992). Study on crystal structure of diaoyudaoite. *Acta Chimica Sinica*, **50**, 527–32. (In Chinese).

Zoltai, T. (1981). Amphibole asbestos mineralogy. *Reviews in Mineralogy*, **9A**, 237–78.

Zvyagin, B.B. (1985). Polytypism in contemporary crystallography. *Soviet Physics Crystallography*, **32**, 394–9.

Zvyagin, B.B. (1987). Polytypism in contemporary crystallography. *Soviet Physics Crystallography*, **32**, 394–9.

Zvyagin, B.B. (1993). Modular aspects of crystal structures. *Crystallography Reports*, **38**, 54–60.

Zvyagin, B.B. (1997). Modular analysis of crystal structures. *EMU Notes in Mineralogy*, **1**, 345–372.

Zvyagin, B.B. and Merlino, S. (2003). The pyroxene-spinel polysomatic system. *Zeitschrift für Kristallographie*, **218**, 210–20.

Zvyagin, B.B. and Romanov, E.G. (1990). Fragmentary character of crystal structures and its manifestation in high-T super-conductors. *Preprints of Lebeder's Physics Institute of the Academic Sciences* USSR, Moscow, 42, 1–43. (In Russian).

Zvyagin, B.B. and Romanov, E.G. (1991). Possibility of symbolic description and systematic derivation of HTSC-like structures. *Soviet Physics Crystallography*, **36**, 75–80.

Zvyagin, V.V. and Vrublevskaya, Z.V. (1976). Polytypic forms of astrophyllite. *Soviet Physics Crystallography*, **21**, 542–45.

Subject index